Clinical Anatomy and Physiology for Allied Health Sciences

PAUL D. ANDERSON, M.S.

Associate Professor of Anatomy and Physiology
Massachusetts Bay Community College
Watertown, Massachusetts

Illustrated by Gayanne DeVry

1976 W. B. SAUNDERS • Philadelphia • London • Toronto

W. B. Saunders Company: West Washington Square
Philadelphia, Pa. 19105

12 Dyott Street
London, WC1A 1DB

833 Oxford Street
Toronto, Ontario M8Z 5T9, Canada

Library of Congress Cataloging in Publication Data

Anderson, Paul D

Clinical anatomy and physiology for allied health sciences.

Includes index.

1. Human physiology. 2. Anatomy, Human. I. Title
[DNLM: 1. Anatomy. 2. Physiology. QS4 A548c]

QP34.5.A52 612 74-25472

ISBN 0-7216-1234-2

Cover illustration: Reading clockwise, (A) osteoarthritis with Heberden's nodes, (B) bone and joint changes resulting from rheumatoid arthritis, (C) normal spinal vertebra and disk, (D) ruptured intervertebral disk with lateral nerve involvement, (E) ruptured intervertebral disk with spinal cord involvement, (F) longitudinal fracture of right humerus, and (G) acute osteomyelitis of humeral shaft. (A, B, F and G from Gartland, J. J.: Fundamentals of Orthopaedics, 2nd ed., Philadelphia, W. B. Saunders Co., 1974.)

Clinical Anatomy and Physiology for Allied Health Sciences ISBN 0-7216-1234-2

Last digit is the print number: 9 8 7 6 5 4 3 2 1

Preface

Many college students take a course in anatomy and physiology as part of their education for one of the health professions, such as nursing, inhalation therapy, or x-ray technology. The material in their anatomy and physiology course should provide the groundwork on which their professional courses can be based. A knowledge of how the body functions in normal situations is later applied to understanding what happens to the body during disease and to understanding what kinds of procedures can be used in treating the disease. For example, a student of nursing should learn in her anatomy and physiology course the factors that determine pulmonary capillary pressure. In her nursing courses this information can then be used to explain why left heart failure can lead to pulmonary edema and how pulmonary edema can be treated. Unfortunately, many texts contain a great deal of anatomy that the student will never use again and often do not adequately explain the physiologic processes that will serve as the basis for the student's professional studies.

The aim of this textbook is to present the core material of anatomy and physiology in such a way that students can understand it and put it to use. The depth of coverage is based on the recommendations of representatives of the various health professions. The names of important anatomic structures are given but the emphasis is on physiologic processes. Microscopic anatomy is included when it is relevant to understanding physiologic processes. Every attempt has been made to emphasize material that the students will use in professional courses, professional licensing examinations, and in their careers.

Two major themes dominate this text: homeostasis and pathology. Throughout the book, the student is shown how normal structure and function are maintained by dynamic counterbalancing forces. Pathology is viewed as a disruption in homeostasis. Accordingly, I have presented a large number of clinical topics and have contrasted them with specific normal processes.

The subject matter of human anatomy and physiology is an exceedingly large and complex body of knowledge to present in an introductory course, so I have concentrated on providing an intelligible, lucid, and up-to-date account of the important principles. I have emphasized unifying concepts and data considered critical to a basic understanding and working knowledge of the human body. To facilitate comprehension, an exceptional array of original illustrations has been provided by Gayanne DeVry.

Pertinent questions follow each chapter. These are intended to test knowledge and comprehension and to provoke discussion. Detailed outlines are also included, to assist the student in reviewing important material.

Although the instructor and the books used are significant in the phenomenon of education, it is the learner who must ultimately assimilate, integrate, and synthesize the information. It is my hope that this text will make that process an enjoyable one.

Paul D. Anderson

CONTENTS

Chapter 1

The Human Organism

In order to study and understand our body and its activities, we must first know something of the structure and functions of the body's basic unit, the living cell, and of its composition and normal working patterns. The human body is made up of trillions of such minute cells, and the study of these structures is called *cytology*. Groups of cells that are alike form *tissues*; *histology* concerns itself with the study of these. Tissues that serve a common function form *organs*, most of which are large enough to be clearly visible to the naked eye. Their structure is described and analyzed in *gross anatomy*, whereas consideration of their functions is the major concern of *physiology*. There is enormous overlap among these fields of science, and research into them grows more complex every year. The latter two, gross anatomy and physiology, are the chief topic of this textbook.

THE CHEMISTRY OF LIFE

A working definition of living things is that they are chemical machines that grow and reproduce themselves. Chemistry is the study of *matter*—the analysis of its structure and of changes occurring in that structure. Matter is anything that has weight (mass) and occupies space. Chemistry also considers the amounts and kinds of energy needed to effect changes in matter, and the physical laws that govern such processes. Through a knowledge of chemistry, we are able to understand some of the normal and abnormal functions of the body. All body processes—for example, the digestion of food, the production of urine, and the regularity of breathing—are based on chemical principles. Drugs, antiseptics, and other agents used in treating disease are chemicals.

In order to emphasize the importance of chemistry in our daily lives, this section will briefly describe some fundamental chemical concepts—atoms, elements, compounds, and mixtures—and how they are interrelated.

The Atomic Theory

It is difficult for a beginning science student to realize that such things as atoms actually exist, especially since no one has ever seen an atom, even with the most powerful microscope. Scientists such as Rutherford (1910) and Bohr (1915) developed concepts of the atom and of matter in general without having seen them. Their scientific investigations led them to conclude that an atom has a *structure*. Recognition of this fact is fundamental to even an elementary understanding of chemistry.

Atoms can be thought of as the "building blocks" of matter—the smallest complete units of which all matter is made. An atom has a nucleus that contains positively charged electric particles called *protons* and noncharged (neutral) particles called *neutrons* (Fig. 1-1). The positively charged proton (p) has weight and is commonly called the "fingerprint" of the atom because the number of protons in an atom's nucleus provides a means of identifying it, the *atomic number*. No two elements can have atoms with the same number of protons.

The neutron (n) has weight but is not charged; therefore, it is said to be neutral. The number of neutrons in the nucleus is usually the same or nearly the same as the number of protons. In heavier atoms, there are more neutrons than protons. Protons and neutrons together determine the weight of an atom *(atomic weight)*.

The nucleus of an atom is surrounded by one or more electrons (e) circling in

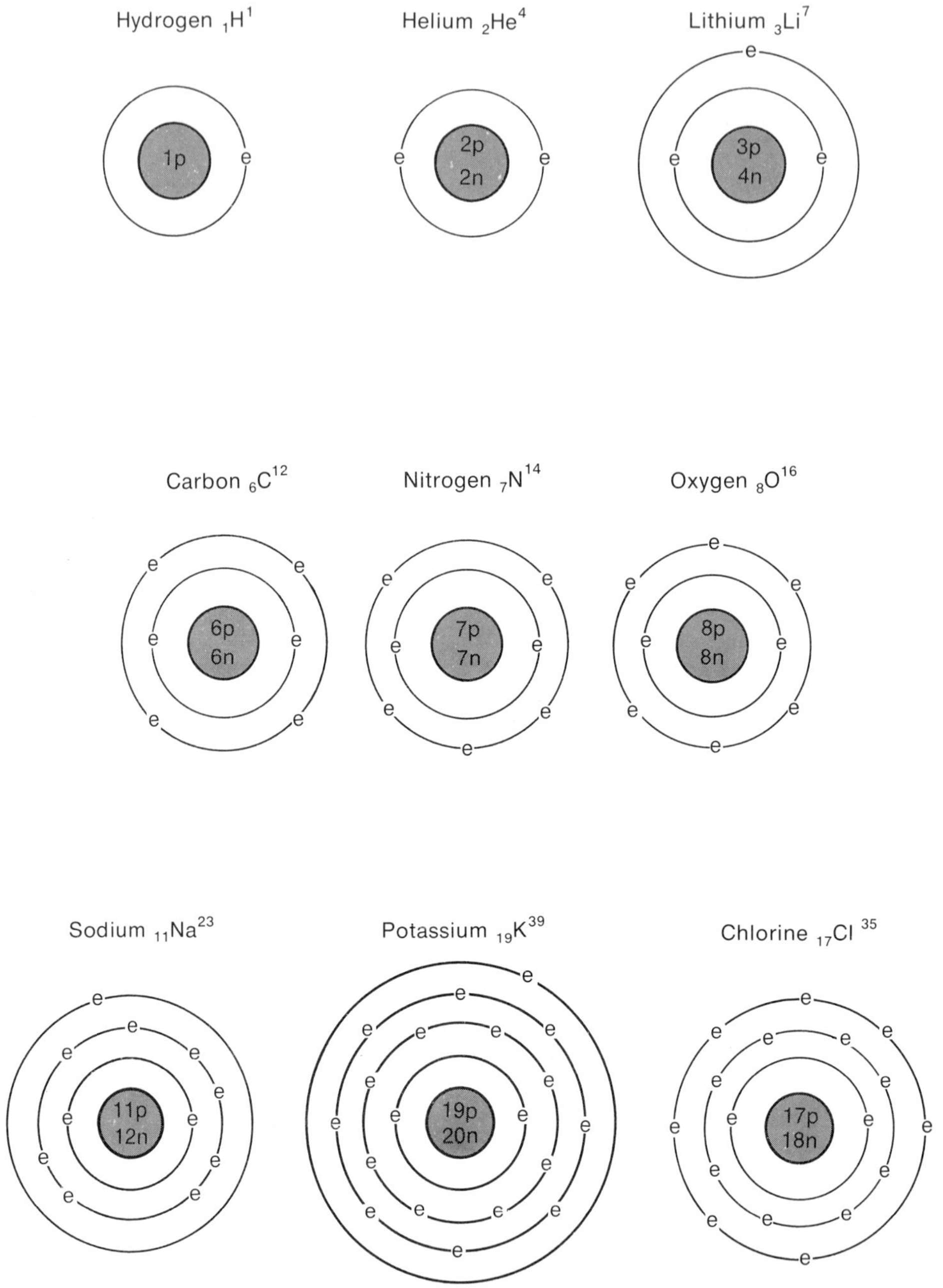

Figure 1–1 Diagrammatic representation of elements selected from the periodic table. Symbols: p, proton; n, neutron; e, electron.

well-defined orbits or "energy shells" (Figs. 1–1 and 1–3). Electrons are negatively charged particles, and there are as many present in an atom as there are protons in the nucleus. However, because the weight of the electrons is so minute, they are not considered in the total weight of the atom. The number of electrons in each energy shell is different in each chemical element, since the atoms of each element are of only one type. Each energy shell can contain only a certain number of electrons, and each is identified beginning with the shell closest to the nucleus and proceeding outward:

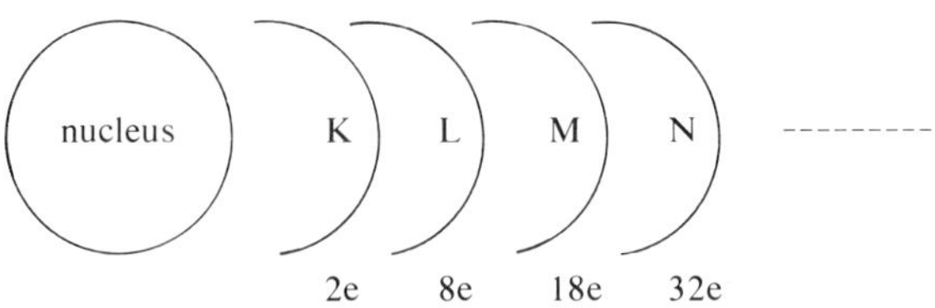

The first energy shell can contain no more than two electrons. The second energy shell can contain no more than eight electrons; the third no more than 18; the fourth no more than 32, and so on.

An atom of one substance or element differs from an atom of another in the pattern of arrangement of its particles. For example, an atom of hydrogen has one electron whirling around one proton in the nucleus, whereas an atom of oxygen has eight electrons whirling around eight protons in the nucleus (Fig. 1–1).

The *periodic table* lists the elements in order of the number of protons each carries in a single atom. A hydrogen atom has one proton and one electron, and thus its atomic number is 1. The helium atom has two protons; carbon has six; nitrogen seven; and chlorine 17 (Fig. 1–1). The atomic structures illustrated also show that the electrons are located in certain prescribed energy shells, indicated diagrammatically by circles. The number of neutrons present in an atom can be calculated by subtracting the atomic number from the atomic weight. Helium has an atomic weight of 4 and an atomic number of 2; it therefore has two protons in the nucleus and two electrons outside the nucleus. Thus two neutrons are present (4 – 2). The sodium atom has an atomic weight of 23 and an atomic number of 11; therefore, there are 12 (23 – 11) neutrons present in its nucleus.

Every phase of professional life requires a specific set of terms for everyday use. In chemistry, this special language is short, precise, and to the point. The chemical symbol is an abbreviation of the name of the element—the large "X" shown in the diagram below. The atomic number (Z) has the subscript position immediately preceding the symbol; the atomic weight (A) is placed in the superscript position following the symbol:

$$_{Z}X^{A}$$

In Figure 1–1, you can see the terminology in use:

$$_{6}C^{12}$$

This indicates that the element is carbon, that the atomic number is 6, and that the atomic weight is 12. Therefore, there are six protons in the nucleus, 6 electrons in the outer shells, and 6 (12 – 6) neutrons in the nucleus.

Some of the common chemical elements that will be referred to throughout this text are listed in Table 1–1.

Valence

An element's valence is its combining power in the formation of compounds. This power is entirely due to the number of electrons present in the atom's outermost shell. An atom is stable when its outermost shell is full; that is, when there are eight electrons in the outermost subshell and two electrons in the K shell (the shell

Table 1-1 COMMON CHEMICAL ELEMENTS

ELEMENT	SYMBOL	ATOMIC NUMBER	ATOMIC WEIGHT	VALENCE
Hydrogen	H	1	1	±1
Carbon	C	6	12	±4
Nitrogen	N	7	14	+3
Oxygen	O	8	16	−2
Sodium	Na	11	23	+1
Magnesium	Mg	12	24	+2
Phosphorus	P	15	31	−3
Chlorine	Cl	17	35	−1
Potassium	K	19	39	+1
Calcium	Ca	20	40	+2
Iron	Fe	26	56	+3
Cobalt	Co	27	59	+2
Copper	Cu	29	64	+1
Iodine	I	53	127	−1

closest to the nucleus). When all the shells of an atom contain their maximum number of electrons, the atom is stable or nonreactive. For example, He (helium) has two electrons in its K shell (its only shell) (Fig. 1-1); therefore, it will neither take on nor give off electrons—it will not react at all. Nitrogen, on the other hand, has two electrons in its K shell and five electrons in its outermost shell. To be stable, the outermost shell should contain eight electrons. Nitrogen can thus do one of two things: it can either give off its five outermost electrons and revert back to a stable K shell; or it can accept three electrons from a nearby atom to complete the outer shell. To simplify these concepts, an *octet theory* was formulated. It states that:

1. All outermost shells (except the K shell) will contain at most eight electrons; a full shell is stable and nonreactive.

2. If an atom gives off electrons from its outermost shell, it will acquire a positive (+) charge.

3. If an atom takes on electrons to complete its outermost shell, it will acquire a negative (−) charge.

4. In order to conserve as much energy as possible, the atom will give off or take on the least number of electrons that it can.

For example, nitrogen has five electrons in its outermost shell. It can either give off the five electrons and acquire a +5 charge or valence, or it can take on three electrons and acquire a −3 charge or valence. As the octet theory states, the least energy is lost when the three electrons are taken on; therefore, the common valence of N is −3.

There are always exceptions to rules, and hydrogen (±1) and carbon (±4) are prime examples. When the number of electrons in the outermost shells of reacting atoms is equal, an atom will either give off or take on electrons, depending on the atom with which it is reacting. For example, hydrogen, with only one electron in a K shell that needs two electrons, can either give off the electron or take on an electron:

1. H^+Cl^-—In this case, hydrogen gave off its electron and became +1.
2. $C^+H_4^-$—In this case, the hydrogen took on electrons and became −1.

The Chemical Bond

As we have seen, an atom is chemically stable when its outermost shell is completely filled. If this shell is not full, the atom will tend to react with any other element that will provide the shell with its full complement of electrons. An atom in this situation is called "charged" or an *ion.*

An *ionic* (or *electrovalent) bond* occurs when a chemical reaction involves a loss or gain of electrons in which electrons are transferred from one atom (+ charge) to another (− charge), thus forming a bond. The formation of sodium chloride (NaCl) typifies such an ionic bond. Sodium is a highly reactive metal with an atomic number of 11. Its outermost shell has one free electron that is easily removed in ionic bond formation (Fig. 1–2). Chlorine (Cl) is a gas of atomic number 17, with seven electrons in its outermost shell. When these two ions react with each other, the single electron in the outermost shell of the sodium atom is transferred to the outer shell of the chlorine atom, thus constituting a chemical reaction.

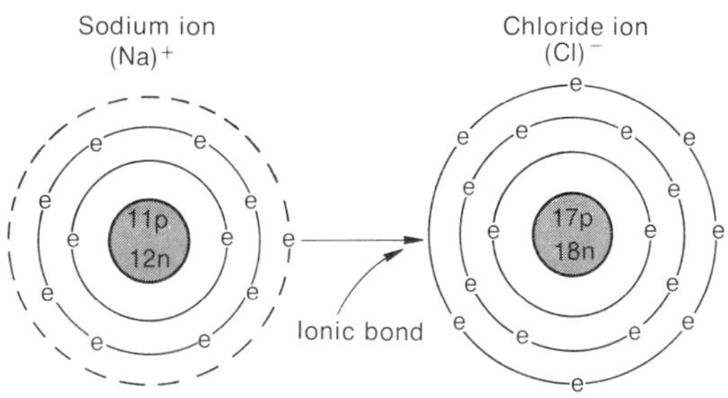

Figure 1–2 The union of oppositely charged ions to form a compound. Sodium loses an electron in its outer orbit to chlorine, making sodium positively charged. Chlorine accepts the electron in its previously incomplete outer orbit and becomes negatively charged. The combination of the two oppositely charged ions forms an ionic bond and a molecule of sodium chloride (table salt).

A *covalent bond* is a sharing of electrons between two atoms. Oxygen has six electrons in its outermost shell—two electrons short of a completed shell; therefore, oxygen has a combining power of −2 and tends to gain two electrons in a reaction. Figure 1–3 illustrates the formation of water (H_2O), which is an example of covalent bonding. One oxygen atom combines with two hydrogen atoms, each of which has one electron in its outer shell. In forming the compound H_2O, all three atoms are actually sharing electrons so that each atom has completed its outer electron shell to the maximum number—eight for oxygen and two for hydrogen. Most gases composed of only one element, such as hydrogen (H_2), are also covalently bonded.

Molecules, Compounds, and Mixtures

Up to this point we have discussed the atom and the element, and have mentioned the fact that two or more elements will react chemically to form a compound.

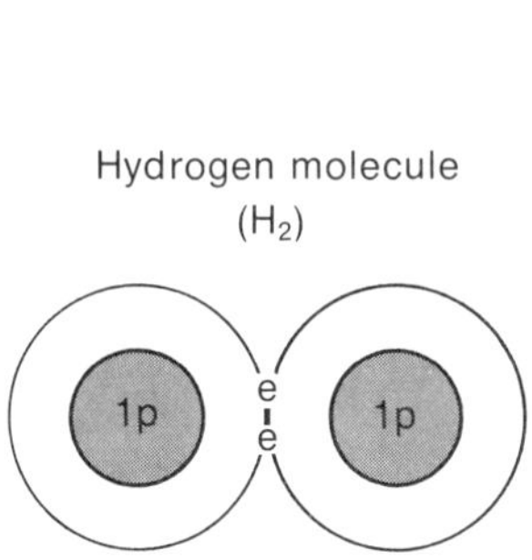

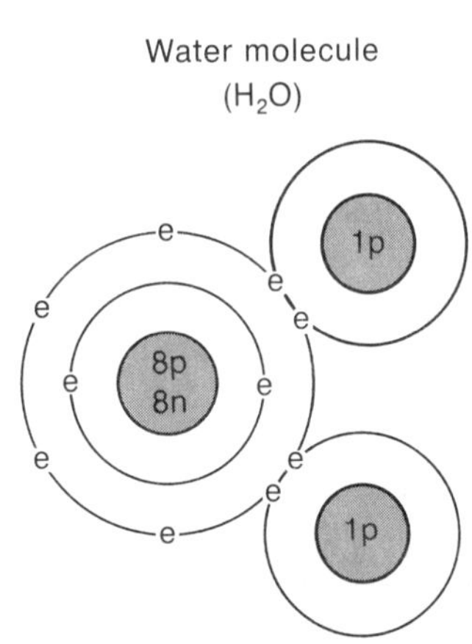

Figure 1–3 The formation of a molecule of water. The covalent bond between oxygen and hydrogen represents a set of electrons shared between the two. Also illustrated is covalent bonding forming a hydrogen molecule.

A *compound* is a union of atoms of two or more *different* elements. The term *molecule*, in contrast, is more general: a molecule is formed whenever two or more atoms join, whether these atoms are dissimilar or whether they are individual atoms of the same element. Nitric oxide (NO) is made up of one atom of nitrogen and one atom of oxygen. It is thus both a molecule and a compound. Hydrogen (H_2) and oxygen (O_2), on the other hand, both consist of two identical atoms. They are molecules but not compounds.

It is beneficial to the body to possess combinations of two or more dissimilar substances that do not unite chemically. Such a situation produces a *mixture*. Blood plasma, for example, is a mixture of many different substances that are essential in their separated form but not in one large chemical compound. The major difference between a mixture and a compound is that the components of a mixture can be separated by mechanical processes, as in salt water. Boiling the salt water will separate the salt from the water, a mixture, but will not break down or separate the sodium from the chlorine in salt (NaCl), a compound.

Radicals

Occasionally, two or more atoms may react as one. Oxygen has an atomic number of 8 and contains six electrons in its outermost shell; therefore, it has a valence of -2. Hydrogen has an atomic number of 1 and, consequently, a valence of ± 1. If these two ions are in close proximity, they may combine to form the hydroxyl radical, OH. If these two ions are combined in single quantity, there will be an imbalance of charges, resulting in one more negative charge than positive. Therefore, the proper way to symbolize the hydroxyl radical is OH^-.

There are other significant radicals, such as amine (NH_2^-) and carboxyl ($COOH^-$); these will mean much more later when we see that they provide the basic structure of foodstuffs.

Acids, Bases, and Salts

Most inorganic compounds (substances that do not contain carbon) can be classified into one of three groups:

1. Acid. An acid is any compound which, when it dissociates or breaks apart, will yield hydrogen ions (H^+) in solution. Examples are:

$$HNO_3 \rightarrow H^+ + NO_3^-$$

$$HCl \rightarrow H^+ + Cl^-$$

Some of the common acids are hydrochloric acid (HCl), sulfuric acid (H_2SO_4), nitric acid (HNO_3), carbonic acid (H_2CO_3), and phosphoric acid (H_3PO_4).

2. Base. A base is any compound which, when it dissociates, will yield hydroxyl ions (OH^-) in solution. Some examples are:

$$KOH \rightarrow K^+ + OH^-$$

$$NaOH \rightarrow Na^+ + OH^-$$

Bases are commonly called alkalis, and are represented by sodium hydroxide (NaOH), ammonium hydroxide (NH_4OH), potassium hydroxide (KOH), and calcium hydroxide ($Ca(OH)_2$).

3. Salt. A salt is any compound which, when it dissociates, does not yield a hydrogen ion or a hydroxyl ion. Examples:

$$KNO_3 \rightarrow K^+ + NO_3^-$$

$$NaCl \rightarrow Na^+ + Cl^-$$

When acids and bases are mixed, they neutralize each other, forming a salt and water:

$$\underset{\text{acid}}{HCl} + \underset{\text{base}}{NaOH} \rightarrow \underset{\text{salt}}{NaCl} + \underset{\text{water}}{H_2O}$$

H^+ and OH^- have a great affinity for each other and combine very rapidly to produce water. Some of the more important salts are the chlorides, phosphates, sulfates, and carbonates.

Ionization

We have already seen that when atoms gain or lose electrons, they become charged particles—ions—and the process involved is called *ionization*. Many inorganic compounds that dissolve in water tend to ionize; such compounds are called *electrolytes*. If the atom gains electrons, it has a negative electrical charge and is called an *anion*; if it loses electrons, its charge is positive and it is called a *cation*.

Sodium chloride (NaCl), for example, is ionized in water into sodium cations, Na^+, and chloride anions, Cl^-:

$$NaCl \rightleftharpoons Na^+ + Cl^-$$

Such an ion-containing solution is capable of conducting an electric current, and therefore is called an *electrolytic* solution. The properties and classes of electrolytes are determined by the kinds of ions they give off when they dissolve in water. Some familiar cations are calcium (Ca^{++}), magnesium (Mg^{++}), sodium (Na^+), and potassium (K^+). Some common anions are chlorine (Cl^-), sulfate (SO^{--}), and phosphate (PO_4^{--}).

Unlike inorganic substances, organic compounds (those containing carbon) show little tendency to dissociate into ions. Therefore, most organic compounds are considered nonelectrolytes.

Acidity-Alkalinity; Hydrogen Ion Concentration (pH)

Two very important ions to consider in studying the physiology of the human body are the hydrogen ion (H^+) and the hydroxyl ion (OH^-). When a solution contains more hydrogen than hydroxyl ions, it is called an *acid* solution; when it contains more hydroxyl than hydrogen ions, it is called an *alkaline* (or *basic*) solution. One way hydrogen ions and hydroxyl ions are produced is by the dissociation of water molecules:

$$H_2O \rightleftharpoons H^+ + OH^-$$

The concentration of hydrogen ions in a solution is represented by the symbol *pH*. A scale of pH values has been developed, ranging from 0.0 to 14.0. In pure water, the H^+ and OH^- ions are in equal concentration; thus, the pH of this neutral solution is at the center of the scale, or 7.0. (See bottom of page.)

It is essential to understand that a number on the pH scale is actually the result of dividing the numeral 1 by a mathematical value called a logarithm. The result is that the higher the concentration of H^+, and consequently the greater the acidity, the lower will be the pH value. Thus, pH 4 indicates a higher concentration of H^+, and a higher acidity, than does pH 5.

Most cells are extremely sensitive to

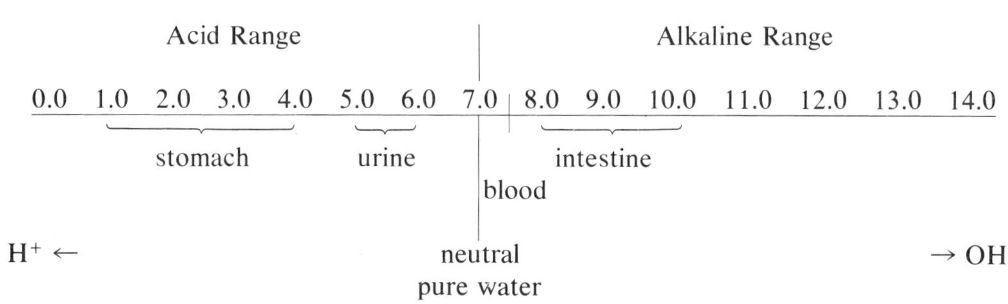

changes in the pH of their fluid. The pH of human blood plasma is usually maintained at a value between 7.34 and 7.44—that is, blood plasma is slightly alkaline. The normal burning of food by the cells releases CO_2, which forms carbonic acid when combined with water. The foods we commonly eat contain sodium, potassium, and calcium; and these substances form alkaline compounds within the body. When the normal limits of the blood plasma pH are greatly exceeded in either direction along the scale, *acidosis* (a pH below 6.8) or *alkalosis* (a pH above 7.8) can lead to serious illness and even death, unless a proper acid-base balance is restored.

Buffers

To alleviate these problems, blood plasma is well supplied with chemical compounds called *buffers*, which act to prevent any major changes in the body's pH. Buffers are, or can act as, weak acids and bases and the salts of these acids and bases. Sodium bicarbonate, carbonic acid, and proteins are active in major buffer systems of the body. Should the pH start to decrease (indicating an increase in H^+ concentration), proteins and sodium bicarbonate ($NaHCO_3$) combine with free H^+, thus holding the pH constant. On the other hand, an increase in pH is corrected by the action of other proteins and carbonic acid (H_2CO_3), which releases H^+ to the plasma, thus restoring the normal state. Plasma proteins, then, serve as a homeostatic mechanism for maintaining a constant pH.

Chemical Reactions

Chemical reactions can be classified into two types: (1) those in which there is electron transfer from one atom to another, known as oxidation-reduction reactions; and (2) those in which there is no electron transfer, called recombination reactions.

1. *Oxidation-reduction reactions* are concerned in the metabolism of foodstuffs and the production and utilization of energy. In the formation of salt, sodium provides the electrons (i.e., is oxidized), while chlorine accepts the electrons (i.e., is reduced). When any substance loses electrons, another substance must be available to accept them or the reaction will not proceed.

The simplest chemical reactions involve the combination of two atoms to form compounds with the release of energy:

$$2H_2 + O_2 \rightarrow 2H_2O + \text{energy}$$

Considerable energy is given off during the formation of a molecule of water, so the type of reaction is said to be an *exothermic synthesis*. The energy of the final compound is less than that of the separate elements. Conversely, the breakdown of water (hydrolysis) typifies an *endothermic analysis*. Energy is required to decompose the compound:

$$2H_2O + \text{energy} \rightarrow 2H_2 + O_2$$

2. Recombination reactions involve the exchange of atoms between molecules, resulting in neither a gain nor a loss of electrons. We mentioned before that when an acid and a base are mixed, they form a neutral salt and water:

$$HCl + NaOH \rightarrow NaCl + H_2O$$

Isotopes

No discussion of elements is complete without reference to the part they play in the diagnosis and treatment of disease. Isotopes are defined as forms of an element with the same number of protons but different numbers of neutrons, and thus different atomic weights. These different forms may be stable or unstable; however, since they are forms of the same element, they possess the same chemical properties. Unstable isotopes are said to be radioactive and are sometimes called radioisotopes.

A more detailed study of the structure and use of radioisotopes in medicine will be discussed in Chapter Eleven, Radiological Health.

MEASURING DEVICES

Not only does a scientist need a language through which he can communicate with his fellow workers, he also needs tools that will allow him to make accurate observations. He constantly needs to know "how much," "how far," "how high," and so forth. We are accustomed to the everyday English System that uses the pint, quart, ounce, pound, inch, foot, etc. Another, more universal, system is called the Metric System. The Metric System (Table 1–2) is based on the factor of 10. It may seem a little awkward at first, but it actually is an easier system to learn than the English System. Each unit is larger or smaller than the next units on the scale by a factor of 10.

The unit of length in the metric system is the *meter*. A meter (m) is a little longer than a yard. The prefix *milli-* indicates one one-thousandth of the quantity; a *millimeter* (mm) is one one-thousandth (0.001) of a meter. The prefix *centi-* indicates one one-hundredth of the quantity; a *centimeter* (cm) is one one-hundredth (0.01) of a meter. It would be advisable, at this point, just to sit down and memorize the basic system. It will be used throughout this text, and you are going to be confronted with the terminology no matter where you study.

The unit of weight (mass) in the metric system is the *gram*, abbreviated g or gm; it is equal to about 1/450th of a pound. The unit of volume is the *liter* (l); it is approximately one quart. You will constantly be measuring volumes in liters or milliliters (ml).

Table 1–2 THE METRIC SYSTEM

LENGTH

1 meter (m) = 39.37 inches (in) = 3.281 feet (ft)
1 centimeter (cm) = 0.01 m (1 in = 2.54 cm)
1 millimeter (mm) = 0.1 cm = 0.001 m
1 micron or micrometer (μ) = 0.001 mm (0.00004 in)
1 millimicron (nanometer–nm) = 0.001 μ
1 Angström unit (Å) = 0.0001 μ

WEIGHT

1 gram (g or gm) = 0.03527 ounce (oz) = 0.0022 pound (lb)
1 kilogram (kg) = 1000 g = 2.2 lb
1 decigram = 0.1 g
1 centigram (cg) = 0.01 g
1 milligram (mg) = 0.001 g
454 g = 1 lb
1 oz = 28.4 g

VOLUME

1 cubic centimeter (cm^3) = 1 milliliter (ml)
= 0.001 liter
1 liter (l) = 1000 cm^3 = 0.2643 U.S. gallons (gal) = 1 quart
1 liter = 0.220 Imperial gal

Temperature Measurement

The first temperature scale devised, and still the most common in some countries, is the Fahrenheit scale (Fig. 1–4). When we speak of the outside temperature as being 90° or 10° below zero, we are speaking in terms of degrees Fahrenheit (° F). On this scale, the freezing point of water is 32° and the boiling point of water is 212°. In science, the Celsius (Centigrade) temperature scale is the most commonly used. This scale fits in well with the metric system of weights and measurements.

Since the Fahrenheit scale is still more frequently used in the United States, it is necessary to determine equivalences between the two scales. Note in Figure 1–4 that most of the degree markings on the two scales are different for different temperatures. If Fahrenheit and Celsius thermometers are placed in ice water, the reading on the Fahrenheit scale will be 32° and the reading on the Celsius scale will be 0°. Conversely, if both temperatures are placed in boiling water, the Fahrenheit scale would read 212° and the Celsius scale 100°.

A formula combining both scales is as follows:

$$°F = (9/5\ °C) + 32 \text{ and } °C = 5/9\ (°F - 32)$$

The first formula is used to change Centigrade to Fahrenheit; the second formula achieves the reverse.

For example, to change 80° C to ° F:

$$F = (9/5\ C) + 32$$
$$F = 9/5\ (80) + 32$$
$$F = 144 + 32 = 176; \text{ therefore, } 80°\ C = 176°\ F$$

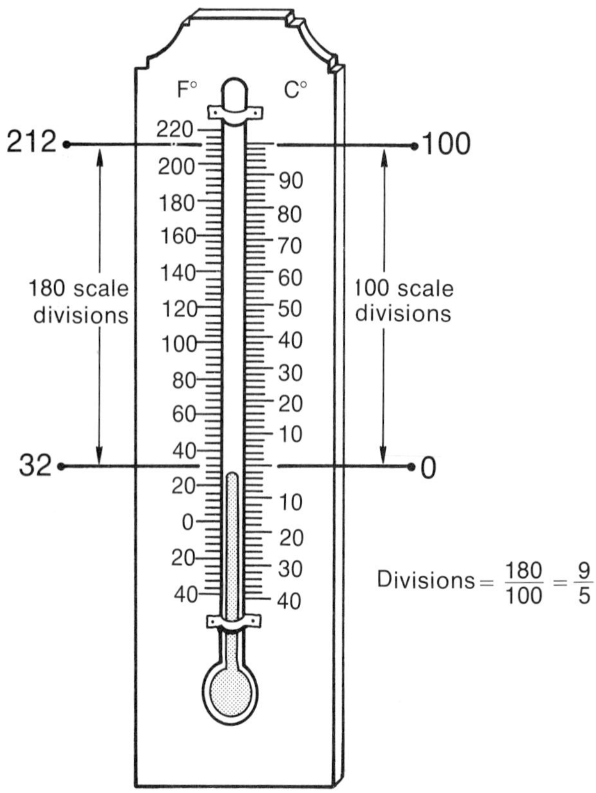

Figure 1–4 Comparison of Fahrenheit and Celsius temperature scales.

To change 50° F to °C:

$$C = 5/9\ (F - 32)$$
$$C = 5/9\ (50 - 32)$$
$$C = 5/9\ (18)$$
$$C = 10;\ \text{therefore,}\quad 50°\ F = 10°\ C$$

The only temperature on both scales that is the same is –40° (Fig. 1–4). Knowing this, a second formula can be used to convert between °F and °C by adding 40, multiplying by 5/9 or 9/5, and then subtracting 40:

For example, to change 80° C to °F:

$$80 + 40 = 120$$
$$120 \times 9/5 = 216$$
$$216 - 40 = 176$$

To change 50° F to °C:

$$50 + 40 = 90$$
$$90 \times 5/9 = 50$$
$$50 - 40 = 10$$

The results are the same no matter which system you use. Choose the one that is simplest for you.

ANATOMIC TERMINOLOGY: DIRECTIONS, PLANES, AND CAVITIES

In mammals and other animals there are large cavities throughout the body. Since it would be very awkward to discuss an organ as being situated in the "northeast" part of the abdomen, specific terms have been devised to indicate anatomic regions and directions.

A student just beginning the study of anatomy and physiology is like a person living in a foreign country and unable to speak the language. He or she should refer constantly to the Glossary and List of Prefixes at the end of this text to learn and understand the meaning and application of new terms. When reading a new assignment, it is a good idea to keep a record of the new terms used in that section. They invariably will be used again in later discussions.

Directions in the Body

For the purpose of these definitions, the anatomic position of a human is always regarded as standing upright, arms down, with the palms of the hands directed forward (Fig. 1–5):

1. *Superior*—toward the upper or head end of the body.
2. *Inferior*—away from the head, below, or lower; for example, heart is inferior to the head.
3. *Anterior*—toward the front; a *ventral* position.
4. *Posterior*—toward the back; a *dorsal* position; for example, the brain and spinal cord are in the back or dorsal part of the body.
5. *Medial*—toward the midline of the body.
6. *Lateral*—away from the midline of the body; for example, the arms are lateral to the main body trunk.
7. *Cranial*—near the head.
8. *Caudal*—near the tail region of the body.
9. *Proximal*—nearest the origin of a structure; toward the body trunk.
10. *Distal*—away from the body trunk; for example, the hand is distal to the forearm.

Terms of position and direction may be combined to indicate position in more than one plane: for example, *anterolateral* (front and side), *anteroposterior* (front and back).

The *aspect* of the body or of a body structure is that part facing in any of the preceding designated directions.

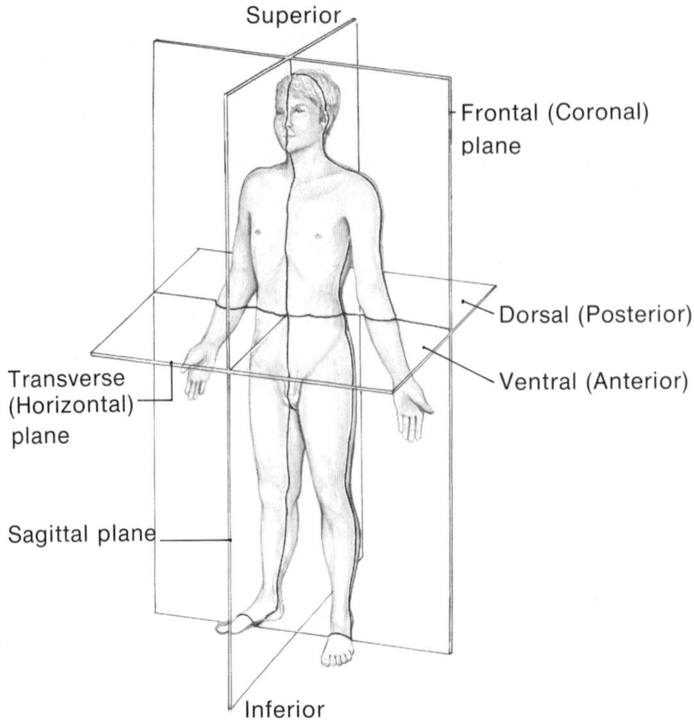

Figure 1–5 Body planes and directions.

Planes

For convenience of reference, the body has been divided into three imaginary planes (Fig. 1–5):

1. *Sagittal*—a lengthwise cut running from front to back, dividing the body into left and right sides. If the cut were directly in the middle, giving you exactly right and left halves, it would be a midsagittal or median section.

2. *Frontal* (coronal)—a lengthwise cut running from side to side, dividing the body into dorsal (posterior) and ventral (anterior) sections.

3. *Transverse* (horizontal)—a cross-cut through the body, dividing it into upper (superior) and lower (inferior) sections. There can be many such cross sections, each on a transverse plane.

Body Cavities

The body contains a few large internal spaces or cavities within which are various organs and systems (Fig. 1–6):

1. *Dorsal body cavity*. There is one major dorsal cavity that contains two smaller cavities: the cranial cavity, which contains the brain, and the spinal canal, which contains the spinal cord. Because these two subcavities are continuous and interconnected, and because both are encased entirely within a series of membranes called meninges, they are collectively regarded as one cavity.

2. *Ventral body cavity*. There are two large subdivisions of the ventral cavity, each much larger than the entire dorsal cavity.

a. *Thoracic cavity*. This is separated from the abdominal cavity by the broad, muscular diaphragm. It is lined with pleura (p. 349). The lungs and large vessels are located here, as is the heart, which actually is found in its own pericardial sac or cavity between the lungs (Fig. 1–6).

b. *Abdominal-pelvic cavity*. The upper region of this cavity contains the major organs of metabolism and also the kidneys; the lower pelvic region contains the urinary bladder, the rectum, and the internal parts of the reproductive system (Figs. 1–6 and 1–7). The abdominal-pelvic cavity and all organs within it are covered with peritoneum, a serous-type membrane. This membrane secretes a fluid that acts to protect and lubricate the organs and spaces contained within.

Abdominal Regions

The abdomen is so large that it is divided, for convenience of study, into nine regions (Fig. 1–7). A horizontal line at the level of the ninth rib creates the right hypochondriac, the epigastric, and the left hypochondriac regions. A horizontal line at the level of the iliac crest creates the right and left lumbar and the umbilical regions, and also the right and left inguinal and the hypogastric regions.

THE HUMAN BODY—AN ORGANISM

According to Webster, "... an organism is any living plant or animal made up of many complicated parts. . . ." The human organism fits this description to a "T." You have now finished the first chapter of this book, and have been presented with an assortment of facts that may seem completely unrelated. What does chemistry have to do with the planes and cavities of the body? I agree, of course, that these facts appear to be—and, in a sense, are—rather isolated. Still, we must start somewhere!

In the chapters that follow I shall build on the premise that a knowledge of the structure and function of a cell is of paramount importance in developing an understanding of the human organism, a mass of interrelated cells. Cellular mechanisms are 95 per cent chemical, and the introduction of chemical terminology in this chapter will be essential to your later understanding.

You will find that the cell is quite complicated—but also fascinating. This text will be clinically oriented, which means that you will learn a great deal about the abnormalities of the body *(pathology)* and about the many different

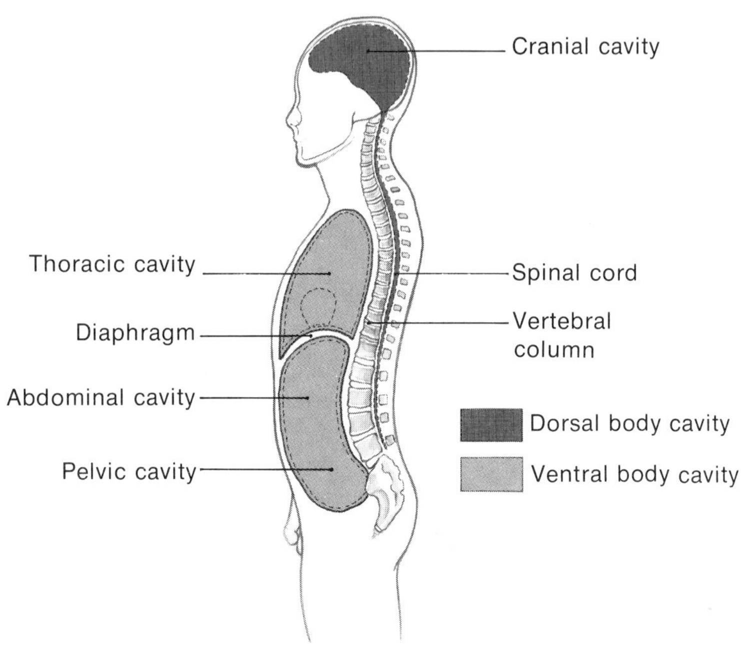

Figure 1–6 Sagittal section of the body, showing the dorsal and ventral body cavities.

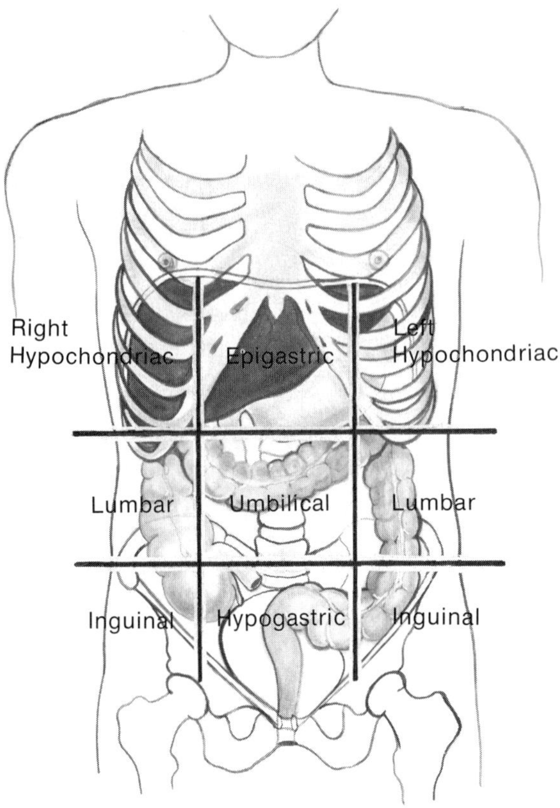

Figure 1–7 Abdominal regions.

methods of detecting disease. You cannot understand the abnormal, however, unless you first learn the normal. Therefore, we shall begin with the cell and develop the "human organism" from there. The human body is an organized assemblage of highly specialized tissues, which together perform the necessary functions that contribute toward the state of homeostasis—a constant internal environment.

OUTLINE

I. Introduction
  A. Related sciences
    1. Cytology
    2. Histology
    3. Anatomy
    4. Physiology
II. The Chemistry of Life
  A. Matter
    1. Occupies space and has weight
  B. Atomic theory
    1. Rutherford and Bohr
    2. Nuclear components
      a. Proton—positive charge, atomic number
      b. Neutron—neutral charge
      c. Atomic weight made up of addition of protons and neutrons
    3. Energy shells
      a. Electrons—negative charge
      b. Energy shell arrangement of electrons
    4. Periodic table
    5. Atomic structures
    6. Labeling configuration
  C. Valence
    1. Combining power of an atom
    2. Octet theory
      a. Eight electrons at most in all outer rings, except the first, which has two

- b. An atom that loses electrons acquires a positive charge
- c. An atom that gains electrons acquires a negative charge
- d. An atom will gain or lose electrons using the least energy possible

D. Chemical bond
1. Full outer shells imply stability
2. An ion is a charged atom
3. Ionic bonding
   - a. Electrical attraction of oppositely charged ions
4. Covalent bonding
   - a. Sharing of outer electrons

E. Molecules, compounds, and mixtures
1. Compound
   - a. Two or more dissimilar elements in a chemical reaction
2. Molecule
   - a. Two or more like or dissimilar elements
3. Mixture
   - a. Two or more dissimilar elements or substances that do not unite chemically
4. Radical
   - a. Two or more elements that act as one
   - b. Examples—OH^-, NH_2^-, $COOH^-$

F. Acids, bases, and salts
1. Acid—any compound giving off an H^+ ion at dissociation
2. Base—any compound giving off an OH^- radical at dissociation
3. Salt—any compound not giving off either H^+ or OH^- at dissociation

G. Ionization
1. Anion—negative ion
2. Cation—positive ion
3. Ion-containing solutions are called electrolytes

H. pH
1. The hydrogen ion concentration of a solution
2. Greater quantity of H^+ results in acidity from pH 1 to 7
3. Greater quantity of OH^- results in alkalinity from pH 7 to 14
4. Blood plasma pH is 7.3 to 7.44
   - a. Any decrease below 6.8 results in acidosis (and possibly death)
   - b. Any increase above 7.8 leads to alkalosis (and possibly death)

I. Buffers
1. Chemical compounds that prevent a major shift in pH
2. Weak acids or weak bases and their respective salts

J. Chemical reactions
1. Oxidation-reduction
   - a. One element gives off an electron—is oxidized
   - b. Another element takes on an electron—is reduced
   - c. Both independent particles must be available to complete the reaction
2. Recombination
   - a. Exchange of atoms between molecules—compounds split and recombine without a gain or loss of electrons

K. Isotopes
1. Forms of a chemical element with the same number of protons but different numbers of neutrons—thus different atomic weights
   - a. Increase in neutron number in the nucleus makes the atom unstable—radioactive

III. Measuring Devices

A. Scientific language
1. Metric vs. English System

B. Metric System
1. Length—from the angström (Å) to the meter (m)
2. Weight—from the milligram (mg) to the kilogram (kg)
3. Volume—from the milliliter (ml) to the liter (l)

C. Temperature measurement
   1. Fahrenheit scale—everyday use
   2. Celsius scale—universal use
   3. Conversion formula

$$F = (9/5\ C) + 32 \text{ and } C = 5/9\ (F - 32)$$

or

$$F = (40 + C)\ 9/5 - 40 \text{ and } C = (40 + F)\ 5/9 - 40$$

IV. Directions, Planes, and Cavities
   A. Directions in the body
      1. Superior–inferior
      2. Anterior–posterior
      3. Medial–lateral
      4. Cranial–caudal
      5. Proximal–distal
   B. Planes of the body
      1. Sagittal—left and right sections
      2. Frontal—dorsal and ventral sections
      3. Transverse—superior and inferior sections
   C. Body cavities
      1. Dorsal
         a. Combined cranial and spinal cavity covered with meninges
      2. Ventral
         a. Thoracic cavity lined with pleura—includes heart in its pericardial sac
         b. Abdominal-pelvic cavity lined with peritoneum
   D. Abdominal regions
      1. Abdominal area divided into nine basic anatomic regions
V. The Human Body—an Organism
   A. An organism is made up of many complicated parts
      1. Necessary items for a beginner
      2. Cellular mechanisms are primarily chemical
      3. Many cells make up a tissue, an organ, a system, an organism
      4. To learn the abnormal you must know the normal

STUDY QUESTIONS AND PROBLEMS

1. Of what is living matter composed?
2. Define tissue, organ, and system in reference to the entire organism.
3. Describe the different directions of the body and identify their opposites.
4. What are the three main body planes? Why is it important that we know them?
5. Correlate the lining of each body cavity with the cavity name.
6. Why is basic atomic chemistry important at this point in your education?
7. Describe an atom. What is in the nucleus and on the surrounding energy shells?
8. Are the electrons haphazardly grouped on the energy shells? If not, how are they arranged?
9. Define molecule, compound, and mixture, and show their interrelationships.
10. What is an ion? What are electrolytic solutions?
11. Define pH. Why is it important?
12. What is a buffer? Describe its action.
13. If normal body temperature is 98.6° F, what is it in °C?
14. What is an acid? If acid A has a pH of 3, and acid B has a pH of 5, which is the stronger acid? Why?
15. Most body reactions are of the oxidation-reduction type. What does this mean?

Chapter 2

The Living Cell

The term "cell" came into use through research undertaken by the English scientist, Robert Hooke. In 1665 he described how, while examining thin slivers of cork under his microscope, he had found them to be made up of many carefully arranged small chambers, which he called "cells." We now know that all Hooke actually saw were cellulose walls laid down by material that once had been alive. The modern meaning of "cell" is much more inclusive, as it refers to the contents of these chambers as well as to the walls that enclose them.

Only much later was the concept proposed that the cell is the fundamental unit of all living things—the *cell theory* as we call it today. This theory was developed through the simultaneous research of the German botanist, Schleiden, and the German zoologist, Schwann. The reports of their work were published in 1839. Since then, the study of the structure and function of the living cell has been a principal concern of biological research everywhere.

SIZES AND TYPES OF CELLS

Most cells are far too small to be seen without a microscope. Some, however, are large enough to be studied by the unaided eye. One example of a very large (*macroscopic*) cell is the unfertilized ostrich egg. Its yolk is three times the size of a hen's egg.

Some cells, especially nerve cells, are very long. The nerve cells that govern the movements of man's big toe, for example, are nearly three feet long. The nerve fibers in the neck of the giraffe are more than eight feet long.

The shape of a cell is usually determined by the function that cell must perform within the organism. That is why nerve cells, which must carry electrical impulses over long distances, usually have such long extensions. The cells of muscles must also be long so that muscular contractions can be carried out efficiently in a single direction. The tissues that provide the body's covering—the skin—and that line its cavities are made up of cells that tend to be relatively box-shaped and are arranged close together. The cells of connective tissues, on the other hand, are grouped loosely among the strong elastic fibers that support the body.

PROTOPLASM

The wall of every living cell—the *cell membrane*—encloses a clear, jellylike substance called *protoplasm*. Protoplasm is usually thought of as the living matter of the cell. It is 60 to 70 per cent water, 16 per cent protein, and 14 per cent fat, or *lipid*. The rest of the protoplasm is made up of dissolved gases, inorganic salts, metals, and a small amount of carbohydrate (sugar).

There are two principal kinds of protoplasm. These are the fluid substance of the cell's nucleus, the *nucleoplasm*, and the denser material outside the nucleus, the *cytoplasm* (Figs. 2-1 and 2-2). Within the cytoplasm there are substances or

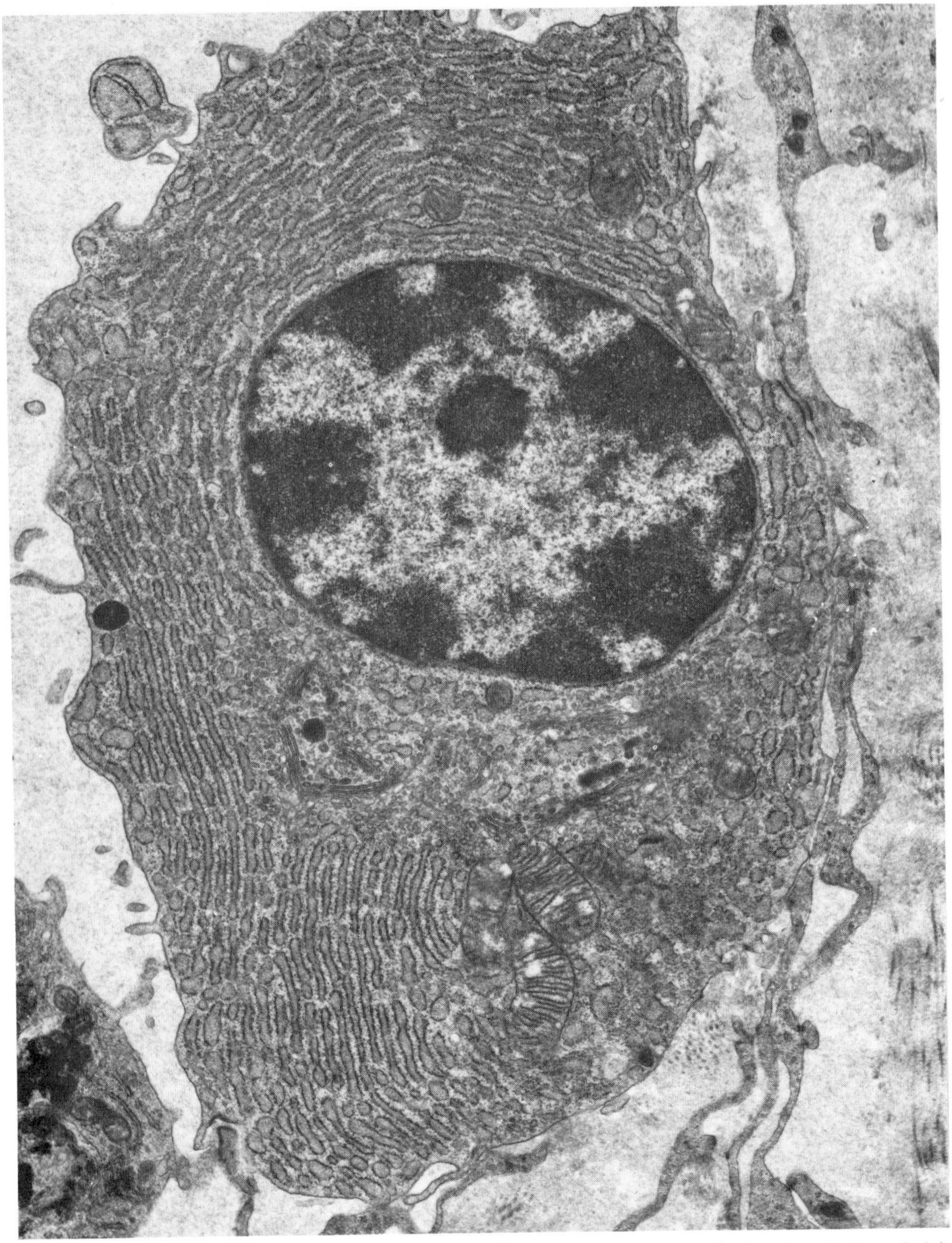

Figure 2–1 Section of a cell photographed through the electron microscope. Endoplasmic reticulum, Golgi apparatus, mitochondria, lysosomes, and a cell membrane are shown but not labeled. (From Warwick, R., and Williams, P. L.: Gray's Anatomy, 35th British ed. Philadelphia: W. B. Saunders Co., 1973.)

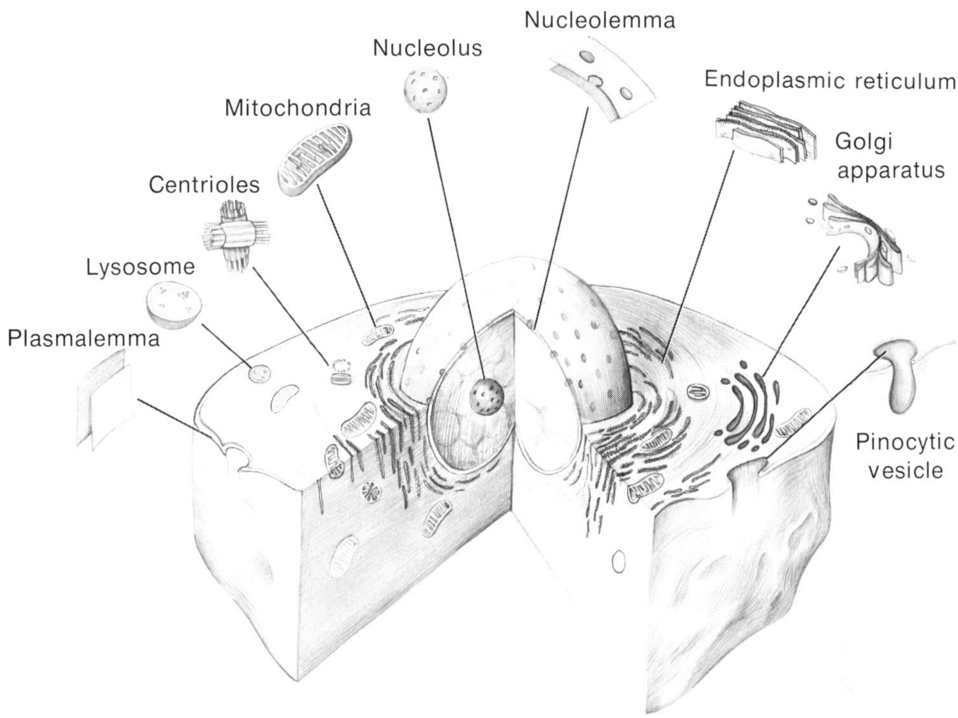

Figure 2–2 Parts of a cell as seen through the electron microscope. Enlargements of the various organelles are shown around the central figure.

structures sometimes called "cytoplasmic inclusions." Today, the preferred term for these minute bodies is *organelles* (Fig. 2–2). We will discuss these later on in this chapter.

PHYSIOLOGIC ASPECTS OF CELLULAR LIFE

Chemically, a cell is composed of such familiar materials as oxygen, hydrogen, carbon, nitrogen, and many other substances that may be found in nonliving structures. Within the cell, however, these materials combine to establish the unique distinction of being *alive*. To be alive, the cell must have all of the following functions or properties:

1. Respiration. We think of respiration as "breathing," but in our present discussion the term refers to the cell's ability to extract energy from food substances supplied to it. When a cell obtains energy without the aid of oxygen, its respiration is called *anaerobic;* when—as is the case with most cells—oxygen is required, *aerobic* respiration occurs.

2. Metabolism. This function of the living cell is actually rather nonspecific, for it refers to the life processes in general—all of the chemical processes that occur within the cell. *Anabolism* is the building of new cellular materials; *catabolism* is the breakdown of these materials into simpler substances. Metabolic functions include such processes as nutrition, digestion, and assimilation.

3. Growth. Normal cells continually increase in volume or mass until they divide, and this necessary growth is closely associated with the processes of metabolism. If the rate of anabolism is greater than that of catabolism, growth takes place.

4. Excretion. Like growth, this process is closely related to metabolism. The products of metabolism can be unneeded—even poisonous—and therefore must be disposed of, or excreted, from the cell.

5. Reproduction. Life cannot continue without reproduction, a physiologic function closely related to growth. Reproduction requires growth, since the embryo develops through the multiplication of cells—a form of growth. Growth, in turn, requires reproduction, because cells multiply by dividing—thus reproducing themselves. This ability of cells to multiply by means of mitotic division, as occurs in the growth of tissues, should not be confused with the ability of an organism to reproduce itself asexually, thus perpetuating a species. The physiology of sexual reproduction—the development of sperm and egg and the fertilization process that follows—will be discussed in Chapter 20.

6. Motility. In man, only two types of cells are capable of movement: the male sperm cell, which is propelled by its whiplike tail; and some kinds of white blood cells, which pass through capillary walls by a process called *ameboid movement.* Movement is common in unicellular ("one-celled") forms of life. For example, the ameba and the paramecium move, respectively, by means of a unique protoplasmic streaming and by the beating of hairlike appendages called *cilia.*

7. Irritability. This physiologic property involves the ability of the cell to respond to an external stimulus or to changes in the organism's internal environment. All living cells will respond to certain stimuli. Their ability to do so is regulated by the nucleus—the "control center" of every cell.

THE CELL MEMBRANE

At the outermost surface of the cell is a very thin, elastic membrane called the cell membrane, *plasma membrane,* or *plasmalemma.* This membrane is so thin that, although its existence had long been hypothesized, it was never actually seen until the invention of the electron microscope. The functions of the cell membrane will be discussed in more detail in Chapter 4.

CYTOPLASMIC ORGANELLES

As we already have seen (Fig. 2–2), the cytoplasm suspends or surrounds a number of organelles—minute organs, each enveloped in its own membrane. The first of these, the *mitochondria* (Figs. 2–2 and 2–3), are sometimes spherical, sometimes rod-shaped bodies ranging in size from 0.2 micrometer to 5 micrometers. Each mitochondrion is enclosed by a double membrane. The inner membrane is extended into many folds that project into the interior space, the *matrix,* of the mitochondrion. These shelflike folds are called *cristae.* Both membranes contain enzymes within their walls, as does the fluid matrix. These enzymes facilitate the breakdown of energy-rich molecules and thereby the conversion of the energy released into a biologically useful form (ATP). Recognizing these functions, it should be clear that mitochondria are at the center of the cell's respiratory function, since they oxidize the cell's foods to form carbon dioxide and water. The number of mitochondria in a given cell, and their location, provide an indication of the cell's metabolic activity. For example, one liver cell contains about 5000 mitochondria, whereas one brain cell has only 50. The mitochondria have been aptly called the "powerhouse" of the cell.

Another type of organelle, the *lysosome,* is an almost spherical body about the same size as the mitochondrion. Lysosomes are generally located toward the periphery of the cell. They contain a number of enzymes that digest the larger molecules of the cell's protoplasm—the carbohydrate, protein, and nucleic acid

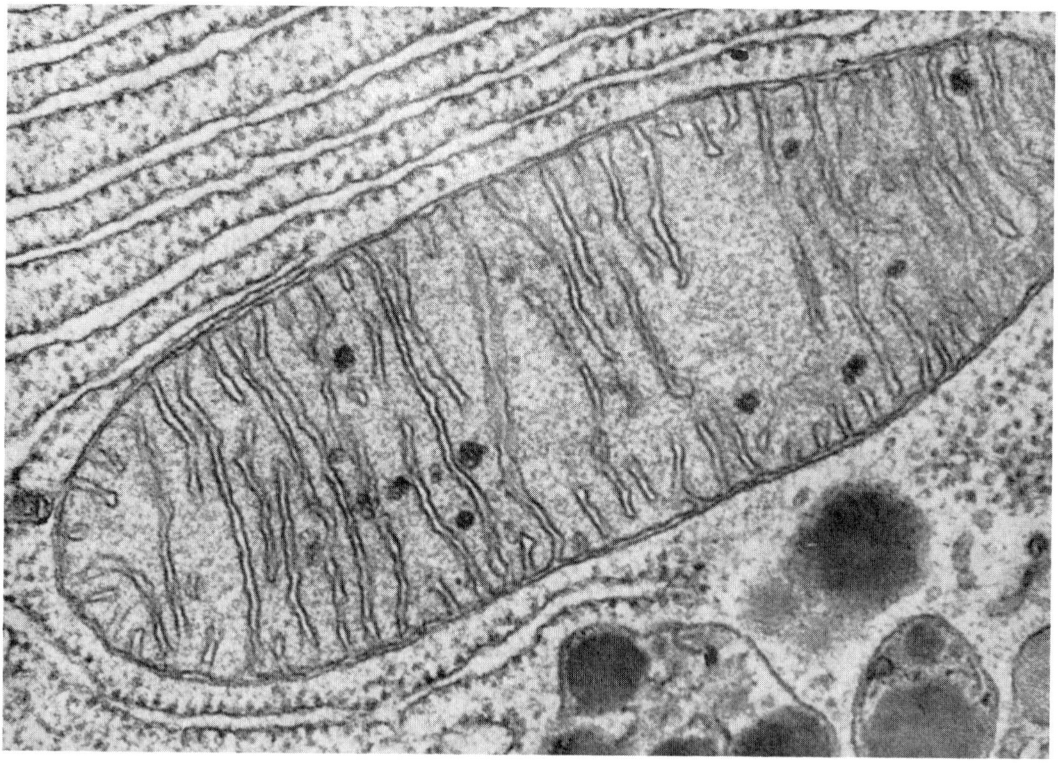

Figure 2–3 Photograph through the electron microscope of a typical mitochondrion from the pancreas of a bat, showing the cristae, matrix, and matrix granules. Endoplasmic reticulum is seen at the upper left and some lysosomes at the lower right. The mitochondrion is magnified 79,000 times. (From Bloom, W., and Fawcett, D. W.: A Textbook of Histology, 9th ed. Philadelphia: W. B. Saunders Co., 1970; courtesy of K. R. Porter.)

molecules. Because these enzymes are retained within the lysosome, they do not act to break down the cell itself. When materials within the cell are to be digested, they are first incorporated into a lysosome; then the initial breakdown of a complex food into a simpler form occurs. This simpler form is then further digested by the mitochondrion.

It is thought that when a cell dies the lysosomes break down, releasing the digestive enzymes that will then cause the cell to disintegrate. The theory has been advanced that this process is linked with aging.

The *Golgi apparatus,* another organelle, is made up of several parallel membranes arranged as flattened sacs. Its function is believed to be the "export" of substances produced by the cell, because it has been observed that the apparatus is more predominant and active in cells concerned principally with excretion. The precise function of the Golgi apparatus, however, is far from clear. A recent study, for example, indicates that this organelle is somehow involved with carbohydrate storage in cells concerned with metabolism.

The *endoplasmic reticulum,* often called "ER," consists of a network of irregular canals spread throughout the cytoplasm. Because the cavities of these canals are interconnected, it has been suggested that the ER may serve as an intracellular communication and transport system. Some cell physiologists have said that the endoplasmic reticulum is continuous from the inner cell membrane to the outer membrane of the nucleus, but this has not yet been established beyond question.

Two different kinds of ER have been observed—"rough" and "smooth." The rough type stores protein and transports it throughout the cell. Smooth ER is characterized by saclike structures that are generally more tubular than those seen in rough endoplasmic reticulum. The canals of smooth ER, according to some reports, store carbohydrates and steroids.

Closely associated with rough ER are some of the smallest (about 150 angstroms) structures of the cytoplasm, the *ribosomes* (Figs. 2–2 and 2–4), which manufacture protein. These organelles are found throughout the cell, and are composed of protein and nucleic acids (RNA).

In addition to the organelles just described, every animal cell contains two *centrioles,* which are cylindric arrangements of nine microtubules. The centrioles are positioned in the cytoplasm at right angles to each other, near the nucleus. Shortly before a cell divides, the centrioles duplicate. One pair migrates to the opposite side of the nucleus, and a "spindle" is formed between this pair. Electron microscopy has shown that there is a close similarity in structure between centrioles and cilia and whiplike structures called flagellae. This observation indicates that there may be an association in some cells between the centriole and cellular locomotion.

THE NUCLEUS

The nucleus, located near the center of the cell, plays a vital role in the cell's or-

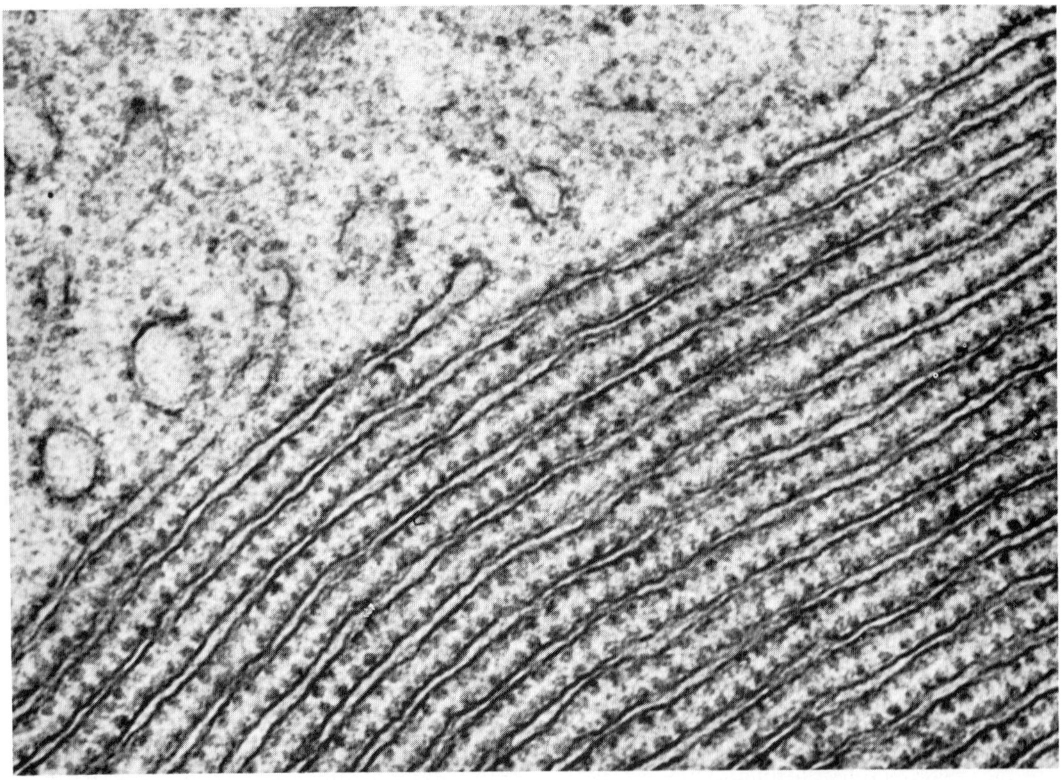

Figure 2–4 Photograph through the electron microscope of the rough endoplasmic reticulum of a pancreatic cell. Rough ER consists of parallel arrays of broad, flat sacs. Ribosomes are found on the outer surface of the limiting membranes of these sacs. Magnification: × 85,000. (From Bloom, W., and Fawcett, D. W.: A Textbook of Histology, 9th ed. Philadelphia: W. B. Saunders Co., 1970.)

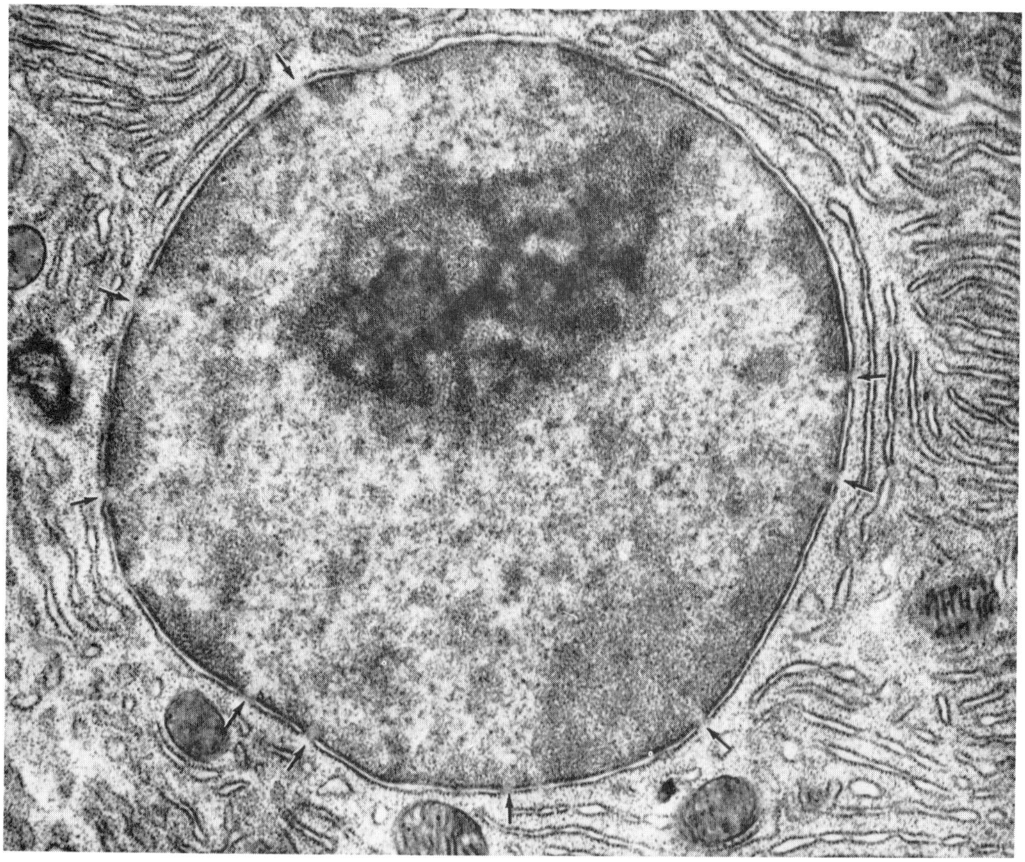

Figure 2–5 Photograph through the electron microscope of the nucleus of a pancreatic cell. The prominent nucleolus (dark area near top of nucleus) appears to be made up of fine granules closely packed in dense connecting strands that form loose, irregular network. At numerous points on the circumference of the nucleus (arrows), the nuclear membrane contains "pores" through which the nucleoplasm communicates with the cytoplasm. Magnification: × 16,000. (From Bloom, W., and Fawcett, D. W.: A Textbook of Histology, 9th ed. Philadelphia: W. B. Saunders Co., 1970.)

ganization and function. Just as the plasma membrane separates the cytoplasm from its external environment, so a double-layered, porous *nuclear membrane* (Fig. 2–5) confines the fluid *nucleoplasm* or nuclear sap.

Within the nuclear membrane are elongated structures called *chromosomes* (Fig. 2–6), which are made up of protein, called *histone,* and *deoxyribonucleic acid* (DNA). The chromosomes contain *genes*, which are the essential hereditary factors of the cell—and therefore of the entire organism. When a cell prepares to divide, the chromosomes change in appearance from long, thin strands to coiled, dense bodies. Every normal human cell contains 46 chromosomes.

A *nucleolus* (sometimes several of them) becomes prominent during the period preceding cell division. This spherical body is easily visible under the electron microscope (Fig. 2–5). Its function is the synthesis of several kinds of nucleic acids and protein molecules, both in and out of the nucleus.

MITOSIS

The commonest form of cell division is *mitosis*. In this process, each chromo-

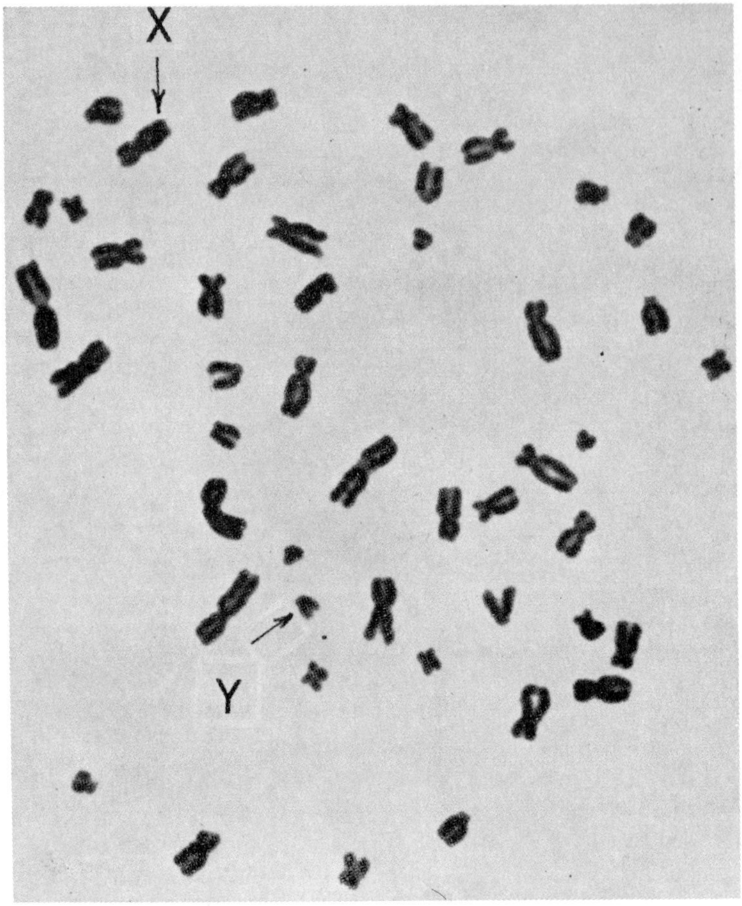

Figure 2–6 Photograph through the electron microscope of the 46 metaphase chromosomes from a cell of a human male. The cell was grown in tissue culture and treated by a technique that spreads out the 22 pairs of autosomes (nonsex chromosomes) and the 2 X and Y (sex) chromosomes so that their characteristic shapes and sizes can be studied. Magnification: × 1700. (From Bloom, W., and Fawcett, D. W.: A Textbook of Histology, 9th ed. Philadelphia: W. B. Saunders Co., 1970; courtesy of J. H. Tjio and T. T. Puck.)

some in the cell divides into identical halves, separates, and forms "daughter" nuclei. Thus each of the two daughter cells receives exactly the same number and kind of chromosomes that characterized the parent. The various stages of mitosis are illustrated in Figure 2–7; here is a brief account of what happens:

1. Interphase. When a "resting" nucleus is examined under a light microscope at its highest magnification, the nucleoplasm is seen as a dense, shapeless mass with little or no organization. The first indication that cell division is about to take place occurs when the centriole divides. At the same time, the substance of the chromosomes, called chromatin material, begins to form a long, spiral thread containing histone and DNA. These substances duplicate themselves to form dense threads.

2. Prophase. During early prophase, the centrioles migrate further toward opposite ends of the nucleus. Simultaneously, the dense *chromatids*—the two halves of each chromosome—become clearly distinct entities but are still held together by a connecting constriction or *centromere*. Later in prophase, the centrioles are located at the extreme opposite ends of the

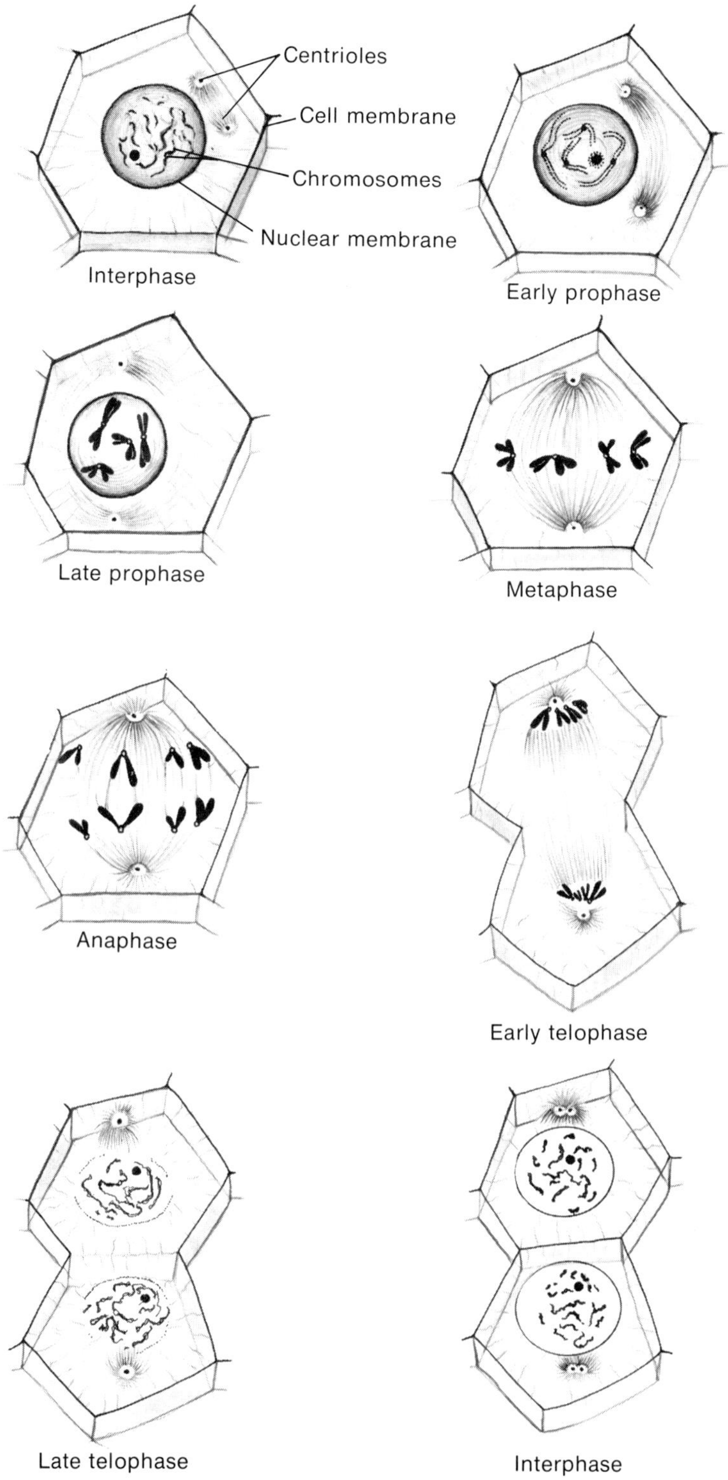

Figure 2–7 Schematic representation of mitosis of a cell containing four chromosomes. See text for an account of what happens during each of the various stages.

cell, where they send out spindle-shaped protein fibers that surround the nucleus, replacing the nuclear membrane that has begun to disintegrate. At this stage, if the centrioles and their fibers are stained carefully, a star-shaped apparatus called an "aster" can be seen at opposite ends of the cell.

3. Metaphase. By now, the nuclear membrane is gone, and the spindle fibers maintain the integrity of the nucleus. The chromosomes form a line across the middle of the cell and attach themselves to the spindle fibers. Microscopically, this gives the appearance of a solid, dark band across the cell called the "equatorial plate" (Figs. 2–7 and 2–8).

4. Anaphase. At this time, by a means not fully understood, the chromatids pull apart entirely and the now separate chromosomes begin their migration toward either end of the cell. According to one theory, the centrioles retract their fibers and the chromatids respond with them. Another theory proposes that the two chromatids of each chromosome are identical chemically—and therefore electrically. The law of electrical signs would then apply (opposites attract; like signs repel), and the identical chromatids would repel each other. Both theories have their adherents, but neither has been proved.

5. Telophase. Early in this final stage of mitosis, there is a "dimpling in" of the sides of the elongated cell, with an aggregation of chromosomes at each pole. As telophase progresses, new nuclear membranes form around the two sets of nuclear components, and the plasma membrane carries forward the division until cross-walls are laid down within the cell and between the two nuclei. Two cells are fully formed, and mitosis is complete. Each of the two new daughter cells now re-enters the "resting" stage of interphase—the period between cell divisions during which normal physiologic processes proceed. Since the chromosomes doubled during anaphase and subsequently divided into separate halves, the nuclei of the two new cells contain the same number of chromosomes as did the parent nucleus.

DNA

Each chromosome consists primarily of a long, spiral ladder or *double helix* of *deoxyribonucleic acid* (DNA). The sides of the ladder are formed by alternating units of deoxyribose (a five-carbon sugar) and phosphate; the rungs are pairs of nitrogenous bases. In most cases there are four kinds of bases: two *purines* (adenine and guanine) and two *pyrimidines* (thymine and cytosine). The purines and pyrimidines have double-ring and single-ring structures, respectively (Fig. 2–9). The four base substances can conveniently be referred to as A, G, T, and C.

The pairing (*coding*) of the bases is highly specific. A is constructed so that it will fit only with T; G will fit only with C. This specificity of pairing is important because the sequence of bases along one side of the ladder necessarily determines the sequence along the other side. Understanding this fact helps us understand how DNA duplicates itself during cell division.

The ingenious model that established the physical structure of the complex DNA molecule was developed by J. D. Watson and F. H. Crick. In their first paper, these Nobel prize winners proposed the double helix structure of DNA; they also postulated that the molecule was 1000 times longer than it was wide. Genes, they suggested, were made up of 500 to 2000 bases. The total number of bases per chain was estimated to be approximately 200,000 in the DNA of viruses.

Watson and Crick's discovery of the structure of DNA was particularly significant in that it showed that the coiled molecule could uncoil itself. The two chains of the molecule, their bases still intact, can, by uncoiling, make mirror images of themselves (Fig. 2–10). The eventual result is the formation of two new DNA molecules, each exactly like the "parent"—a process known as *replication*. This concept of the

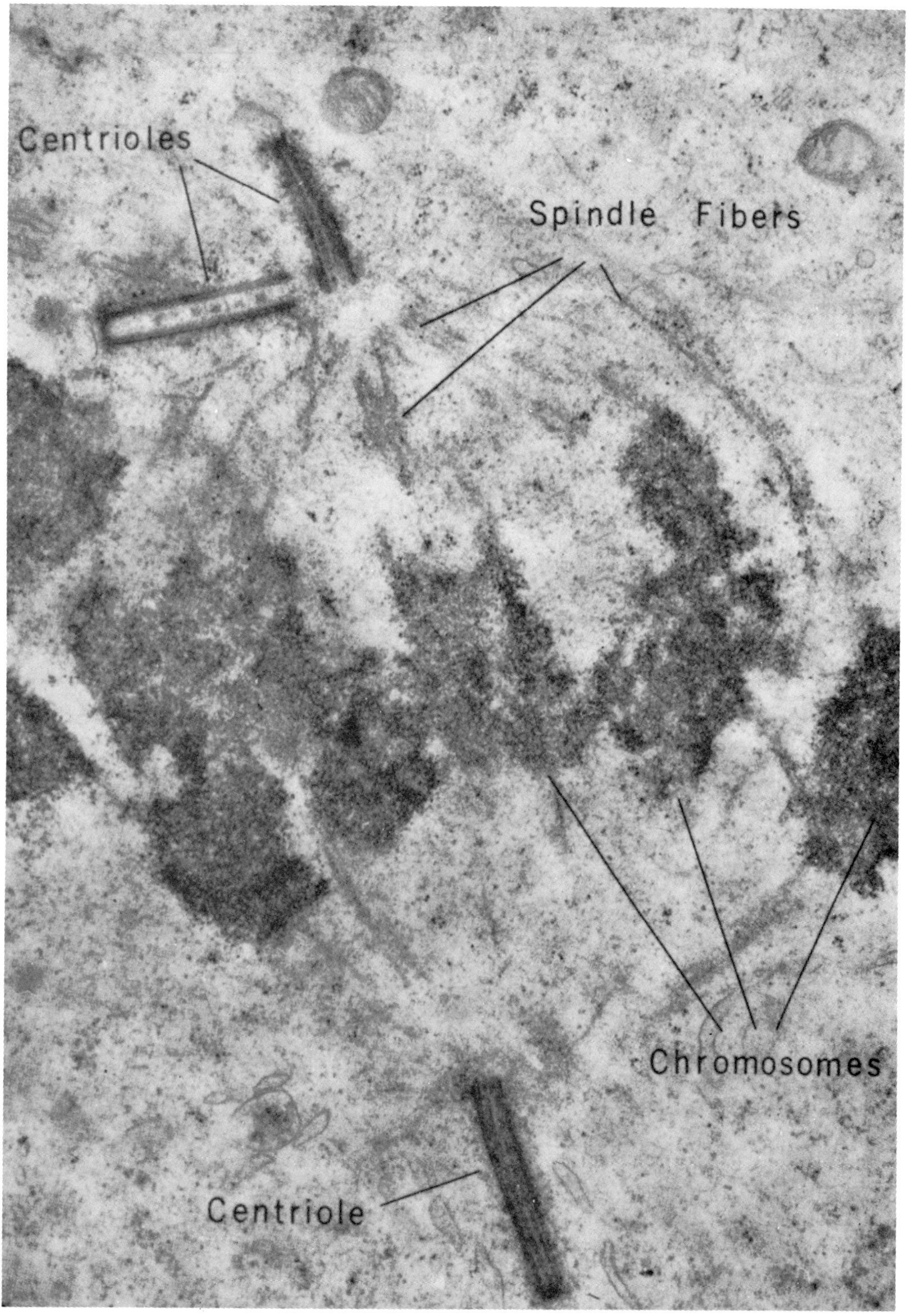

Figure 2–8 Metaphase in a spermatocyte division in the cock testis. The chromosomes are aligned across the middle of the figure. Above and below are three of the four centrioles. Magnification: × 26,000. (From Fawcett, D. W.: The Cell: Its Organelles and Inclusions. Philadelphia: W. B. Saunders Co., 1966; Courtesy of T. Nagano, M.D.)

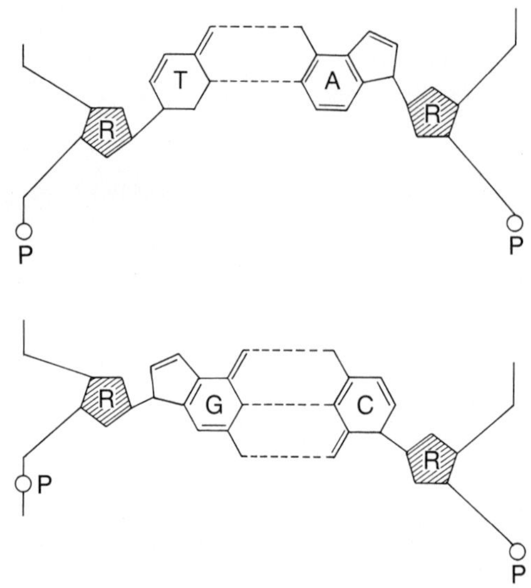

Figure 2–9 Diagram of nucleic acid (DNA) pairings. Phosphates connect ribose molecules. Abbreviations: A, adenine; T, thymine; G, guanine; C, cytosine; R, ribose sugar; P, phosphate radicals.

"unzippering" of corresponding nitrogenous bases and the matching of pairs provided the key to our present understanding of how chromosomes are duplicated.

The replication of DNA occurs at interphase, just before mitotic division. Thus the genetic code of each cell is duplicated every time the cell divides, the original DNA molecule acting as a template for the new one. The hereditary characteristics of the genes are therefore carried forward to each new generation of cells.

CLINICAL IMPLICATIONS

As we have seen, the cell's structure is infinitely complex and delicate. Therefore, it may be damaged by a wide variety of influences, including chemical poisons and bacterial toxins. When a cell becomes "sick," it does not function properly. Its cytoplasm changes, and the cell degenerates. Some kinds of cytoplasmic change and cell degeneration are characteristic of specific diseases, and thus their demonstration can be useful in the diagnosis of illness.

There are other diseases, however, that do not lead to observable cellular change, and these are more difficult to diagnose. A person who dies of tetanus (lockjaw) has nerve cells that *appear* to be normal—although we know that this is not the case. The same situation occurs in mental illness.

The process of cell death is called *necrosis*. Changes in the cell characteristic of necrosis occur after the cell has died but while it still remains in the body. These changes are caused by the action of enzymes.

OUTLINE

I. Introduction
  A. The cell
    1. Robert Hooke, 1665
      a. Cork
      b. Cellulose walls
    2. Schleiden and Schwann, 1839
      a. Cell theory
II. Sizes and Types of Cells
  A. Macroscopic
    1. Ostrich egg
  B. Cell shape correlates with function
    1. Nerve—long appendages
    2. Muscles—elongated
    3. Skin—box-shaped
    4. Connective tissue—fibrous
III. Protoplasm
  A. Living matter of cell
    1. 60 to 70 per cent water
    2. 16 per cent protein
    3. 14 per cent fat
    4. Trace elements
  B. Nucleoplasm
    1. Nuclear components
  C. Cytoplasm
    1. All components exclusive of nucleus
IV. Physiologic Aspects of Cellular Life
  A. Properties of life
    1. Respiration
    2. Metabolism
    3. Growth
    4. Excretion

5. Reproduction
6. Motility
7. Irritability

V. The Cell Membrane
   A. Not seen until after invention of electron microscope

VI. Cytoplasmic Organelles
   A. Mitochondria
      1. Matrix
      2. Cristae
      3. Enzymes in matrix responsible for cellular respiration
      4. "Powerhouse" of the cell
   B. Lysosome
      1. Contains enzymes
      2. Cellular digestion of complex compounds
      3. Aging
   C. Golgi apparatus
      1. Excretion
      2. Carbohydrate storage
   D. Endoplasmic reticulum
      1. Irregular canals that may serve for intracellular communication and transport, and may connect nucleus with the outside
      2. Rough and smooth
         a. Rough stores and transports protein
         b. Smooth stores carbohydrates
   E. Ribosomes
      1. Protein synthesis
   F. Centrioles
      1. Restricted to animal cells

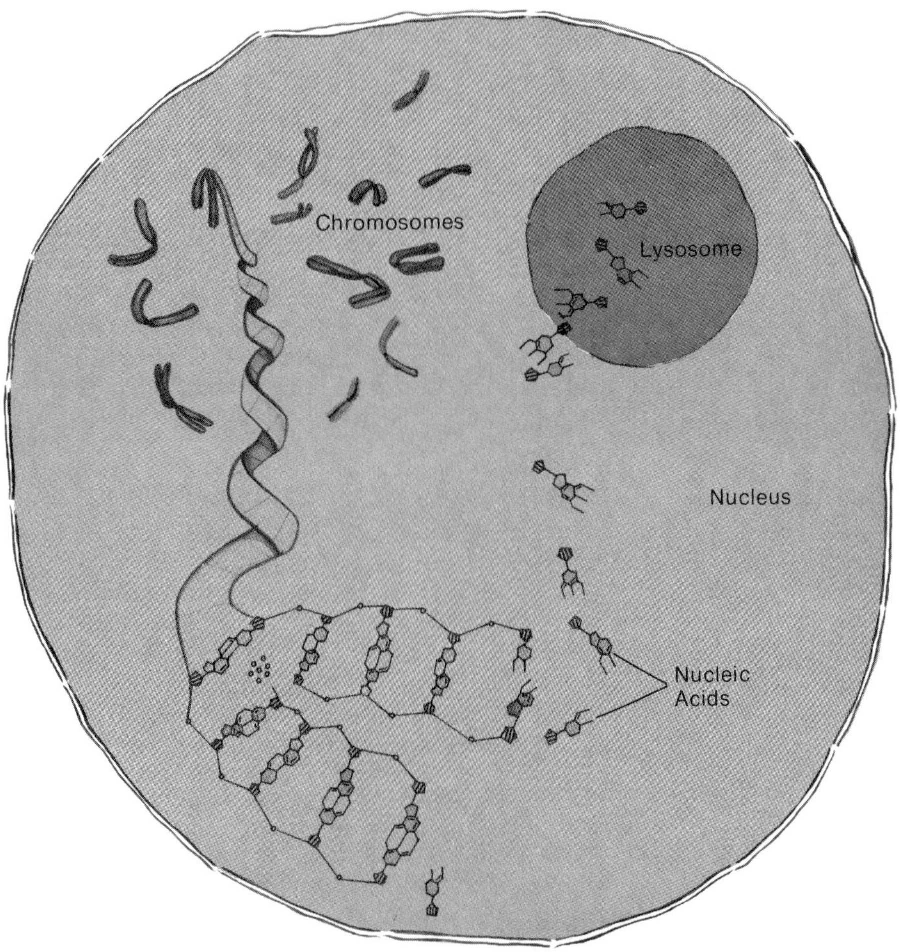

Figure 2–10 Diagram of DNA molecule, showing replication in the nucleus.

2. Nine microtubules
3. Two units positioned above the nucleus at right angles to each other
4. Cell division

VII. The Nucleus
- A. Nuclear membrane
  1. Double-layered, porous
- B. Chromosomes
  1. Histone and DNA
  2. Genes
  3. 46 in each human cell
- C. Nucleolus
  1. Prominent in period preceding cell division
  2. Nucleic acid and protein synthesis

VIII. Mitosis
- A. Interphase
  1. Visibly, an inactive phase
  2. Biochemically, the most active phase
- B. Prophase
  1. Centrioles migrate to positions at opposite ends of nucleus
  2. Chromosomes duplicate
  3. Nuclear membrane disintegrates
- C. Metaphase
  1. Chromosomes line up in midline of cell—"equatorial plate" phase
- D. Anaphase
  1. Chromatids pull apart
  2. Theories for chromatid separation
     - a. Fibers pull chromatids to each pole
     - b. Chromatids repel each other
- E. Telophase
  1. Cell wall "dimples in" to form two new cells
  2. Two new nuclear membranes form
  3. Daughter cells have same number of chromosomes as original parent cell

IX. DNA
- A. Double helix
  1. Deoxyribose (5-carbon sugar)
  2. Two purines
     - a. Adenine and guanine
  3. Two pyrimidines
     - a. Thymine and cytosine
- B. Coding
  1. Respective pairings
     - a. Adenine with thymine
     - b. Guanine with cytosine
- C. Watson and Crick
  1. Molecule 1000 times longer than wide
  2. Gene made up of 500 to 2000 bases
  3. Reproduction (replication) by uncoiling
  4. Replication occurs in interphase

X. Clinical Implications
- A. Chemical poisons and toxins
  1. Cytoplasm altered
  2. Cell degenerates
  3. Necrosis

STUDY QUESTIONS AND PROBLEMS

1. What are some of the physiologic activities performed by living cells?
2. What do the mitochondria have to do with metabolism of energy?
3. What kind of enzymes are found in lysosomes? How has the function of the lysosomes been linked with aging?
4. Describe the cellular membrane.
5. What do the ribosomes do? Where are they located?
6. Centrioles have what function in the cell? Discuss their role in cellular reproduction.
7. What is DNA? How is the DNA molecule constructed?
8. Describe the DNA coding mechanism.
9. Trace the chromosomes through the phases of mitosis.
10. What is the importance of the nucleus? What does the nucleolus do?
11. What is protoplasm?
12. Describe the properties of life and indicate which cellular organelles perform the cellular functions.

Chapter 3

The Tissues

The cell is the basic "building block" of all living things, and it is both useful and interesting to study its anatomy and physiology. Still, our real concern is the whole organism that results from the genetically determined arrangement of cells, and how that organism operates—or fails to operate—in health and disease. Therefore we should consider next the *tissues,* which are large masses or groups of cells similar in origin, structure, and function.

There are five basic classifications of tissue: *epithelium, connective tissue, muscle, nerve,* and *blood and lymph.* Although these five types are distinct and easily recognizable, they may be found in many different parts of the body and in many different combinations. Two tissues—epithelium and connective tissue—will be discussed in this chapter. The remaining types will be considered when we describe the systems that they compose.

EPITHELIUM

Epithelial tissue can be thought of as a complex protective membrane that covers the body and lines all the cavities and organs having a direct connection to the outside of the body. To put it simply, epithelium is a kind of sheet made up of a series of cells that abut each other. Between these cells there is a thin film of *intercellular fluid,* which provides a means for the exchange of nutritive and waste materials to and from the cells by diffusion to the capillaries, the smallest vessels of the circulatory and lymphatic systems. When examined under a microscope, epithelial cells can be seen to have a supporting *basement membrane,* which appears in prepared sections as a fine line.

Types of Epithelial Tissue

Four distinct types of epithelium occur in the human body. Their classification is based on the shapes and properties of the cells of which they are composed.

1. Squamous Epithelium. *Simple* squamous epithelium (Fig. 3–1, *A*) is made up of flat cells arranged in a single layer. The term "squamous" means scalelike, the cells of these tissues have irregular shapes. Each cell contains a large, prominent nucleus at its center. In its simplest form, squamous epithelial tissue lines the small saclike structures (*alveoli*) of the lung, which function in the exchange of oxygen and carbon dioxide during breathing. The lens of the eye is also made up of simple squamous epithelium.

A variation, *stratified* squamous epithelium, is arranged in layers. This tissue occurs, for example, in the skin and on the inside of the mouth. If you were to scrape the inside of your cheek and observe the material under a microscope, its stratified nature would be evident.

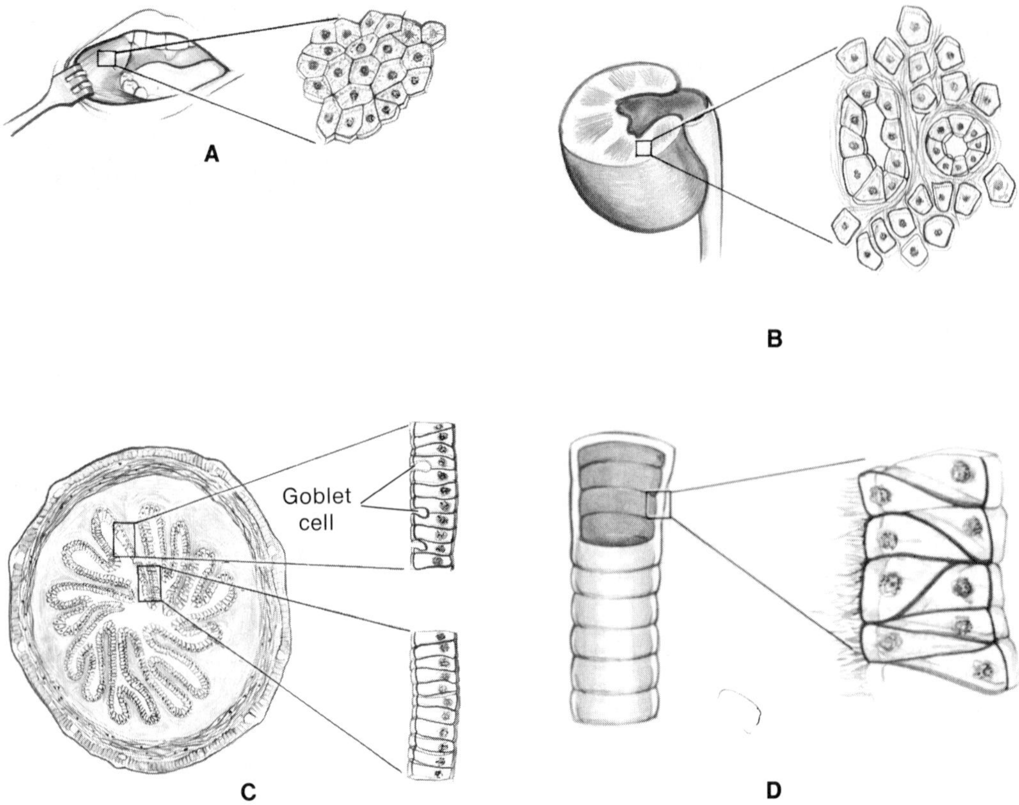

Figure 3–1 Three-dimensional diagrams illustrating the different types of epithelial tissue. *A*, Stratified squamous epithelium from the mouth. *B*, Cuboidal epithelium from the kidneys. *C*, Columnar epithelium (with goblet cells) from the intestine. *D*, Pseudostratified epithelium (with cilia) from the trachea.

Still another modification of squamous epithelium occurs in the mucous membrane that lines the urinary bladder, the ureters, and the urethra. This tissue, called *transitional* epithelium, has the special property of elasticity. It can be stretched, and, when relaxed, will resume its former shape. When transitional epithelium is relaxed, the cells appear microscopically as large clusters.

2. Cuboidal Epithelium. As its name suggests, cuboidal epithelium is made up of cube-shaped cells. This tissue lines the ducts of many glands and the kidney (Fig. 3–1, *B*). The nuclei of its cells are spherical, and usually are found in the center of the cell.

3. Columnar Epithelium. *Columnar* epithelial cells are compressed to form columnlike shapes. These tall, thin cells have a single nucleus, which usually can be found near the basement membrane. Epithelial tissues made up of these cells occur in the digestive tract, particularly in the intestine (Fig. 3–1, *C*). Columnar epithelium is concerned primarily with the secretion of digestive fluids and with the absorption of food materials.

4. Pseudostratified Epithelium. This fourth type of epithelium earned its name because on first glance it appears to be arranged in layers. This appearance is caused by variations in shape of each cell making up the tissue. Although some of the cells in contact with the basement membrane do not reach the surface of the tissue, most of the cells are tall and do reach the surface (Fig. 3–1, *D*). Pseudostratified epithelium is found most often in the respiratory tract, particularly in the trachea and in the bronchi of the lung.

Two modifications occurring in the cells of epithelial tissues are worth considering here—*goblet cells* and *cilia*. Goblet cells are flask-shaped and contain a mucoid secretion. Microscopically, they appear as "open" cells in their tissues (Fig. 3–1, *C*). Cilia, which were discussed briefly in Chapter 2, are hairlike appendages of the cell (Fig. 3–1, *D*). They wave continuously, producing currents in the fluids at the cell's surface. Both goblet cells and cilia occur in columnar and pseudostratified epithelium. Goblet cells aid digestion by secreting mucus for absorption; cilia aid in the passage of food materials down the intestinal tract. Goblet cells and cilia are also essential in the respiratory tract, where goblet cells add moisture to the air taken in and cilia clean the air of foreign particles that could otherwise clog the alveoli of the lungs.

Glands

Some epithelial tissues are made up of cells specifically organized to cause *secretion*. The mucus-secreting goblet cells just discussed cannot produce more than a small fraction of the many secretions needed if the body is to function properly. Therefore, in some epithelium secretory cells are so numerous that they form most of the membrane, combining both secretory and lining functions. Moreover, during development of the embryo, masses of secretory cells join together in distinct units and move deeper into the tissues, away from the surface membrane. Such groups of secretory cells are called *glands*. There are two basic types: *endocrine glands* and *exocrine glands*.

Endocrine Glands. These glands are sometimes called "glands of internal secretion" because they are situated far beneath the epithelial surface and have no ducts or passages by which their secretions can pass through the epithelium. Instead, they communicate directly with the circulatory system through the capillaries, which permits the distribution of their secretions throughout the body. The *thyroid* is an example of an endocrine gland.

Exocrine Glands. Like the endocrines, the exocrine glands also have broken away from the epithelial surface. However, the exocrines are equipped with ducts that carry their secretions to the tissue surface. The *salivary glands*, for example, are exocrines. The *pancreas* is a gland that is both endocrine and exocrine. It produces digestive enzymes that are passed to the intestine through the pancreatic duct (exocrine); but it also produces a hormone, *insulin*, which is transported through the body by the circulatory system (endocrine).

THE SKIN

Most of the body's contacts with its environment occur through the skin, or integument. As we have seen, living cells must be surrounded by fluid; therefore, the surface of the body in contact with air is made up of dead cells that form a protective covering for the inner, living cells.

Structure of Skin

The skin is made up of two distinct layers: the *epidermis* (cuticle) and the *dermis* (corium). These layers are illustrated in Figure 3–2.

Epidermis. The tissue of the epidermis is stratified squamous epithelium. On its external surface there is a network of ridges caused by the arrangement of the tissues that lie beneath. The ridges on the palmar surfaces of the fingers and hands and on the soles of the feet provide in-

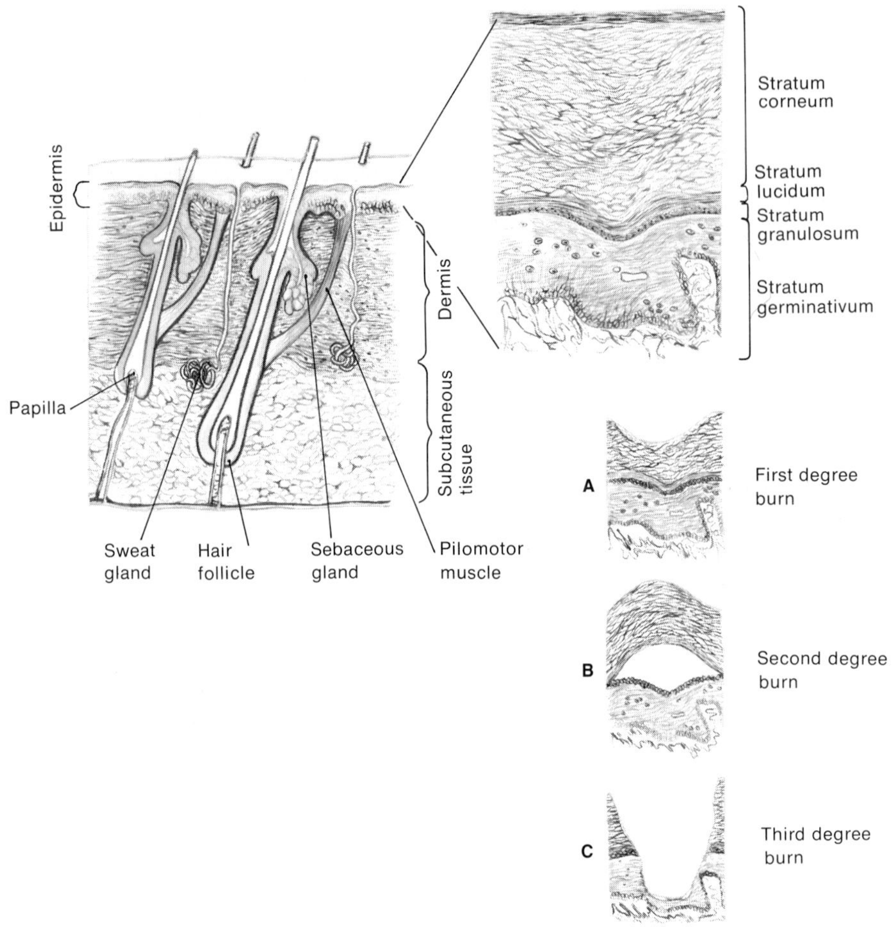

Figure 3–2 Diagram of a section of the skin to show its structure. The nucleated cell produced by the stratum germinativum dies (granulates) as it is forced outward to become the dead, scaly stratum corneum. The number of layers of the epidermis affected by the three types of skin burns is also shown. *A*, Only corneum cells are involved in first degree burns. *B*, Damage to the upper three layers occurs in second degree burns, forming a blister between layers 3 and 4. *C*, A third degree burn involves all epidermal layers and, therefore, usually requires a skin graft to replace the stratum germinativum.

creased resistance to objects, making it easier to walk, to grasp things, and to perform other motions involving contact with the environment. Each individual has his own distinct ridge patterns, such as fingerprints, and these can be used as a means of identification.

The epidermis is built up of a series of four layers, or *strata*:

1. Stratum Germinativum. This is the deepest of the four strata, and contains several layers of cells. One of its functions is to provide cells to the outer layers of the skin. As the cells of the stratum germinativum multiply and grow, cells previously formed are pushed upward toward the surface.

The stratum germinativum determines the thickness of the skin, since it is the only layer of the epidermis that contains living cells capable of division, leading to the development of the outer protective layers. This layer also contains the greatest number of pigment cells—*melanocytes*—in the skin.

Melanin, a pigment that ranges in

color from yellow to black, is found in all epidermal layers in the Negro race, and in the deeper layers in Caucasians. When the skin is exposed to ultraviolet radiation, the amount of melanin is increased and its color is darkened—a protective mechanism against radiation that causes tanning. Melanin is formed in the melanocytes by the action of an enzyme, tyrosinase, on an amino acid, tyrosine. When the enzyme is not present, the skin has no pigment (albinism). Skin blotches, such as freckles, are nothing more than irregular patches of melanin.

2. STRATUM GRANULOSUM. This is the layer of the epidermis directly above the stratum germinativum. Here the cells, as they progress upward, undergo a chemical degeneration of the nucleus. This leads to the death of the cell, which gives it microscopically a granular appearance. In this layer, the cells also change physically, becoming flattened or scalelike.

3. STRATUM LUCIDUM. In this next layer, the scalelike cells form a shield. They become hard, or *keratinized.* The process of keratinization of a cell involves the conversion of protoplasm into keratin, a hard, hornlike protein.

4. STRATUM CORNEUM. This outermost, horny layer of the epidermis consists of hard, scalelike cells that shield the body against the environment.

Dermis. The dermis, or corium (true skin), is the skin's inner layer. It is highly sensitive and contains numerous blood and lymph vessels, nerves, glands, and the hair follicles. Elevations on its surface, called *papillae,* create the ridges that protrude above the epidermis. The papillae also contain a network of tactile corpuscles, or nerve endings, that are responsible for the sensation of touch (Fig. 3–2).

The papillary and lower regions of the dermis also contain bundles of fibers that form an elastic connective tissue that can be stretched and extended. This property becomes evident when parts of the body become swollen, as in pregnancy. The skin stretches and becomes smooth and shiny. As an individual ages, his skin contains fewer elastic fibers, losing much of its extensibility.

Beneath the dermis is the *subcutaneous layer,* which contains many fat cells. These provide insulation against the loss of body heat and also store food. In the subcutaneous layer are also found blood and lymph vessels, nerves and nerve endings for the sensations of pressure and temperature, sweat glands, and hair follicles.

Blood Supply to the Skin

The arteries that supply blood to the skin form a network in the subcutaneous tissue and send branches to the papillae, the hair follicles, and the glands. The capillaries are so numerous that, when distended, they are capable of holding a large portion of all the blood contained in the body. When the skin is exposed to heat, or when the body temperature rises, the small arteries of the skin dilate, thus allowing more blood to flow through them *(vasodilation)* and permitting more heat to leave the body. On the other hand, when the skin is exposed to cold, the arteries constrict *(vasoconstriction)*; the blood flow through them slows down, thus preventing loss of body heat.

When blood vessels are damaged by disease or injury, there is a decrease or even a total loss of blood to an area of tissue, causing the death of the cells of that tissue. Such cell death can result in open sores or *ulcers.* Decubitus ulcers (bedsores) occur when patients spend prolonged periods in one position (Fig. 3–3). Patients requiring extended bed care may develop ulcers because of the pressure of their body weight on the skin around a bony prominence—for example, the elbow, the hip, or the spine. Good nursing care requires continual stimulation or manipulation of these areas to help return the blood

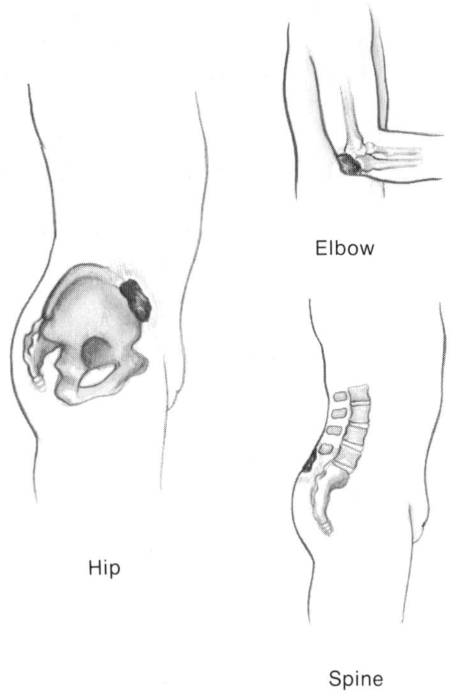

Figure 3–3 Areas of decubitus ulcer formation.

supply to normal and to prevent further complications.

Skin Appendages

Sweat Glands. The *sweat glands* are simple tubules. Their lower ends lie in the deeper part of the corium or in the subcutaneous tissue; their upper portions end in small openings in the skin's surface. The principal function of the sweat glands is to help regulate internal body temperature. Their pores (openings) are very numerous, and act as an emergency mechanism when it is necessary for the body to lose heat through evaporation or perspiration.

Sebaceous Glands. The sebaceous or oil glands are the most numerous of the skin appendages. These glands are found everywhere on the surface of the body except on the palm of the hand and the sole of the foot. They are generally associated with hair follicles, and their ducts usually open into the sides of the follicle. The glands themselves are located in the dermis. Their oily secretions, called *sebum,* lubricate the hair and the horny layer of the skin, keeping them soft and pliable. Persons with underactive sebaceous glands develop dry hair and skin; those with overactive glands have oily hair and skin. When the ducts of these glands become blocked, infection follows—a common cause of blackheads. Too much sebum on the skin leads to seborrhea; the excess sebum may collect as *dandruff. Acne* is a chronic inflammation of the sebaceous glands.

Ceruminous Glands. The ceruminous glands are modified sweat glands, and are found only in the skin of the passages leading into the ear (external canal). These glands secrete a waxy substance, *cerumen,* which helps to protect the eardrum and keeps the skin of the outer ear canal soft. Excessive amounts of this waxy cerumen can accumulate in the ear, impairing hearing or causing an irritating "cracking" sound when dry and loose.

Hair. Hair is another modification of the epidermis. Its follicles are formed by a downward growth of epidermal cells into the dermis, where germinal cells divide and are pushed upward until they eventually die and become keratinized. Each hair consists of a portion projecting above the skin's surface, the *shaft,* and a portion below the surface, the *root* (Fig. 3–2). The root is embedded in a deep indentation in the dermis—the *follicle,* which we have already discussed.

A hair usually has a central core, the *medulla,* which is surrounded by an outer layer, the *cortex.* Hair color depends on the amount of melanin in the cortex and medulla. White hair has no pigment at all. The hair follicle is supported and controlled by an arrector muscle which, when stimulated, either straightens the follicle, causing the shaft to project upward, or pinches the dermis, resulting in "goose flesh." Flattening and bending of the hair follicle are related to curly hair; cylindric, straight hair follicles are related to straight hair. Contrary to common belief, cutting or shaving hair does not stimulate its growth. Only stimulation of the epidermal cells in the hair follicle can cause hair to grow faster—or at all.

Nails. Nails are horny modifications of the corneal part of the epidermis. They protect the dorsal surfaces of the fingers and toes. The germinal layer beneath the nail is the nail bed. Nail growth occurs in the epithelium of the proximal portion of the nail bed. As a nail grows, it moves forward and becomes longer, but never loses its attachment to the outermost germinal layer. The crescent or "moon" at the base of the nail is called the *lunula.* It is generally overlapped by the soft *cuticle.*

Functions of the Skin

The many functions of the skin can be summarized in four broad categories:

1. Protection. The skin acts as a barrier to prevent bacteria and other foreign organisms from invading the body. Harmless bacteria reside on the skin, but if a person's habits of cleanliness are poor, pathogenic (disease-causing) bacteria may invade the tissues, leading to infection.

2. Fluid Retention. The skin also serves as a waterproof covering that prevents excessive water loss when the body is exposed to a dry environment. Perspiration and sebum form a film over the body, preventing dehydration and the absorption of chemical substances injurious to the tissues.

The importance of maintaining a normal water balance can be shown by a study of *burns.* Exposure to excessively high temperatures may produce local or general damage. Local damage can occur in one of three degrees: a burn may be *first degree* (an erythema or slight reddening of the skin); *second degree* (blistering); or *third degree* (destruction of the whole thickness of the skin). These degrees are illustrated in Figure 3–2. Burns result in a rapid loss of fluid from the underlying tissues and in a severe upset of the body's internal water balance. The resultant threat to life depends as much on the size of the area burned as on the burn's severity. The "rule of nines" gives the relative distribution of the total body surface area:

Head and neck	09%
Anterior trunk	18%
Posterior trunk	18%
Upper extremity	18%
Lower extremity	36%
Perineum	01%

When a patient dies within a day or two after being burned, his death is due to shock (systemic collapse). Death occurring later than that is due to absorption of toxins from dead tissues.

3. Sensation. The importance of the skin's role in providing sensory awareness of the environment and its changes cannot be minimized. The nerve endings of the skin, which are responsive to touch, pain, pressure, and temperature changes, supply the brain and the nervous system with information that enables the body to adjust to its constantly changing surroundings.

4. Absorption. At one time, the skin was thought to be impermeable to all substances. Now, however, we know that gases (oxygen, carbon dioxide, nitrogen, and ammonia) readily penetrate the skin and travel through the sweat and sebaceous glands. Lipid (fat) substances, such as vitamins A, D, and K, pass relatively easily through the skin. However, the skin is impermeable to electrolytes and water.

CONNECTIVE TISSUES

Connective tissues are found throughout the body. As their name implies, the principal function of these tissues is binding the body parts together. They form a framework for the internal organs; they also perform a variety of other functions;

ranging from protection against injury to storage of fat.

A fundamental difference between connective tissue and epithelial tissue can be seen in their cellular compositions. In epithelium, as we have seen, there are many cells directly adjacent to one another, separated only by a very small amount of intercellular substance (matrix). On the other hand, connective tissue contains few cells, and these are widely separated. The intercellular matrix is relatively abundant (Fig. 3–4) and usually determines the physical characteristics of a given connective tissue.

Connective Tissue Cells

In their embryonic stage, typical connective tissue cells are large and star-shaped, having many projections called *processes*. These cells are *fibroblasts*, and they arise from an early embryonic tissue, *mesenchyme*. This tissue develops into many different forms, including blood cells and muscle. As the connective tissues develop, the cells lose their star-shaped appearance and become widely separated by large amounts of intercellular materials. The adult connective tissue cell, which is responsible for the formation of fibers, is the *fibrocyte*. Other types of cells found in connective tissue are:

1. Histiocytes or Macrophages. These cells move about through the connective tissue, ingesting foreign materials, bacteria, and cellular debris—a process called *phagocytosis*.

2. Plasma Cells. These small, irregular cells are associated with the formation of *antibodies*, an important part of the body's defense against foreign substances.

3. Mast Cells. These are located near

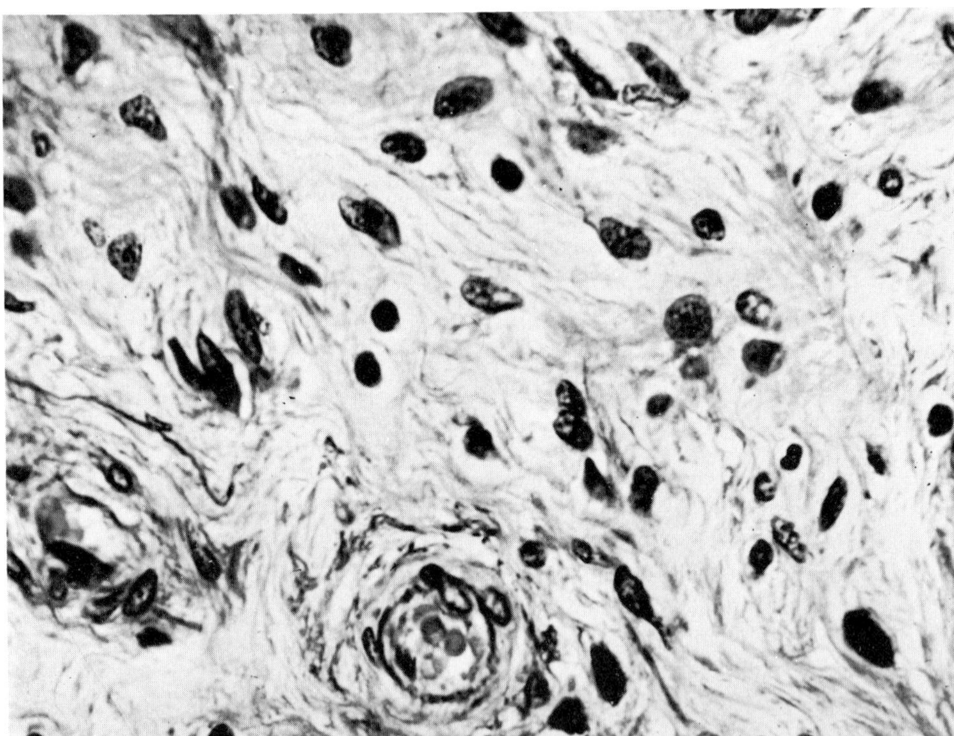

Figure 3–4 Loose connective tissue. (From Leeson, C. R., and Leeson, T. S.: Histology, 2nd ed. Philadelphia: W. B. Saunders Co., 1970.)

blood vessels and are involved in the production of *heparin,* an anticoagulant. They are also important in the production of *histamine* in allergic reactions.

4. Blood Cells. The white blood cells, such as lymphocytes, monocytes, and neutrophils, are often present in connective tissue, where their function is to destroy bacteria by phagocytosis (see Chapter 4).

5. Fat Cells. These specialized cells store fats and oils. Microscopically, they somewhat resemble a signet ring because the stored fat pushes the nucleus and the cytoplasm to one side of the cell (Fig. 3–6).

Connective Tissue Fibers

The fibers characteristic of connective tissue are found within the intercellular matrix. There are two general types:

1. Collagenous Fibers. These white fibers contain *collagen,* an albumin that is the principal structural protein of the body. This protein is synthesized by fibrocytes, and is the major constituent of skin, ligaments, tendons, cartilage, and bone. The fibers occur in bundles and are relatively nonelastic.

2. Elastic Fibers. Elastic fibers are yellow, and may occur singly or in bundles. They are highly elastic, and branch to unite with other fibers. They are both larger and straighter than collagenous fibers.

Types of Connective Tissue

1. Loose Connective (Areolar) Tissue. As its name suggests, this tissue is made up of cells separated by a semifluid matrix. Within this matrix is an irregular network of white collagenous and yellow elastic fibers (Fig. 3–5). The intercellular substance of areolar tissue allows the tissue to stretch without damage to the tissue structure. This tissue is the most abundant in the body. It is found beneath epithelial membranes and around blood vessels, nerves, and the ducts of glands. Areolar tissue is associated with maintaining the body's water balance, since it can absorb and hold a considerable amount of water. Because it contains many macrophages and white blood cells, this tissue also provides a second line of defense against bacteria.

2. Adipose Tissue. Areolar tissue beneath the epidermis is commonly filled with loosely packed fat cells held together in bundles by collagenous and elastic fibers (Fig. 3–6). Adipose (fat) tissue is found primarily as a kind of padding around the joints, as soft pads between the organs, around the kidney and the heart, and in the yellow marrow of long bones. Its fat cells provide a reserve food supply and also insulate the body against heat loss.

3. Dense Connective Tissue. Many parts of the body require a stronger type of connective tissue to enclose, restrain, or separate functioning structures. Such dense tissue can be identified by its characteristic cells, which are concerned primarily with producing collagenous and elastic fibers. In the strong, tough fibrous tissue, there are many more collagenous than elastic fibers. These are arranged as cords or sheets, resulting in a tissue that is pliant but yet has little elasticity. Since fibrous tissue contains few blood vessels, it may take a long time to heal when injured or strained.

There are four basic forms of dense connective tissue: *tendons,* which attach muscle to bone; *ligaments,* which connect the bones that form joints; *aponeuroses,* which are thin, tendinous sheets attached to flat muscles; and *fasciae,* the thin sheets of tissue that cover muscles and hold them in place.

4. Cartilage. Cartilage is a modified form of connective tissue in which cartilage cells (chondrocytes) lie in small capsules, the *lacunae,* which are surrounded by an irregular matrix made up of fibers and/or a gel. Cartilage develops from the

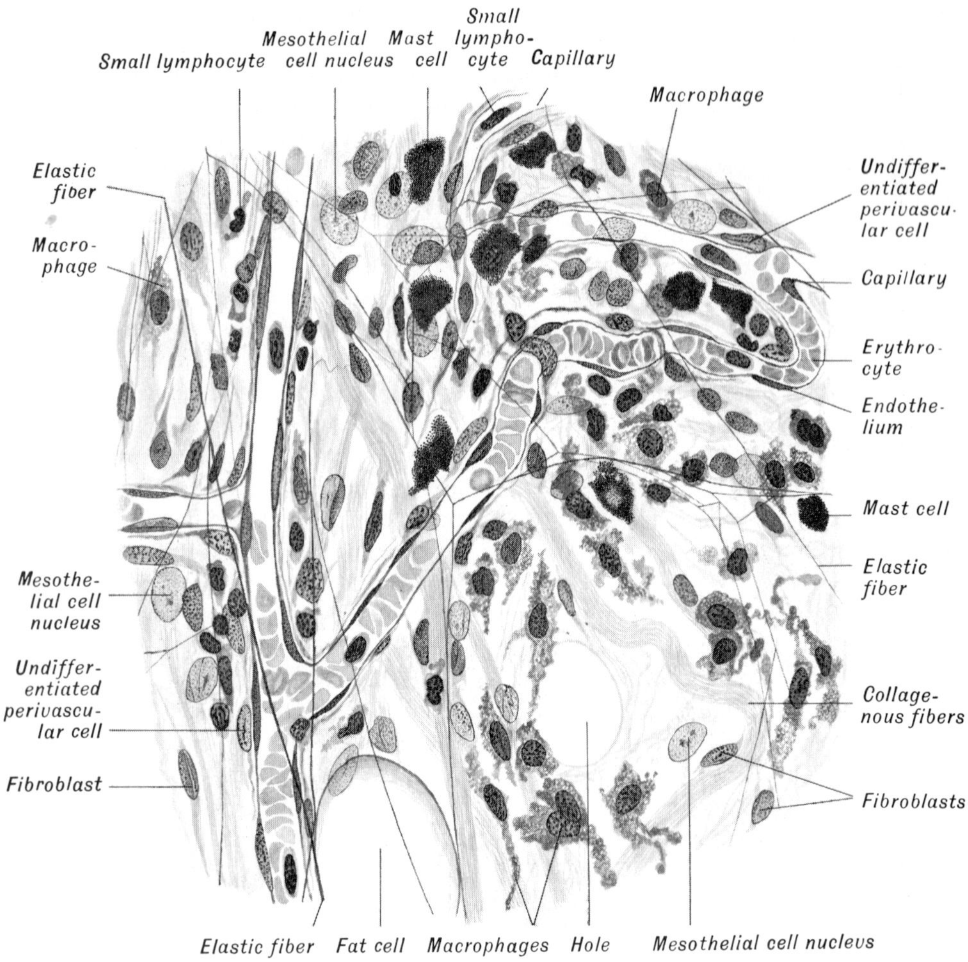

Figure 3–5 Fibrous connective tissue. (From Bloom, W., and Fawcett, D. W.: A Textbook of Histology, 9th ed. Philadelphia: W. B. Saunders Co., 1970.)

embryonic mesenchyme. Its adult form varies in the proportion and kinds of fibers, depending on the function it is to perform.

The most common form of cartilage, *hyaline* (Fig. 3–7, *C*), is a modified areolar tissue with a white or glossy appearance. It may contain any number of unevenly dispersed lacunae. Hyaline cartilage forms the skeleton in the embryo and covers the articulating surfaces of bones in the joint cavities. It also provides the structure for the nose and the connections of the ribs to the breast bone, and, in the respiratory tract, forms the ringlike trachea and bronchi.

Elastic cartilage (Fig. 3–7, *B*) contains many elastic fibers and is found wherever cartilage is required to move—for example, in the epiglottis and in the external ear.

Fibrous cartilage (Fig. 3–7, *A*) is less rigid than hyaline, but contains heavy bundles of collagenous connective tissue. It is found in the intervertebral disks that absorb shocks between the vertebrae of the backbone; it also reinforces the hyaline articular cartilages at the knee and hip.

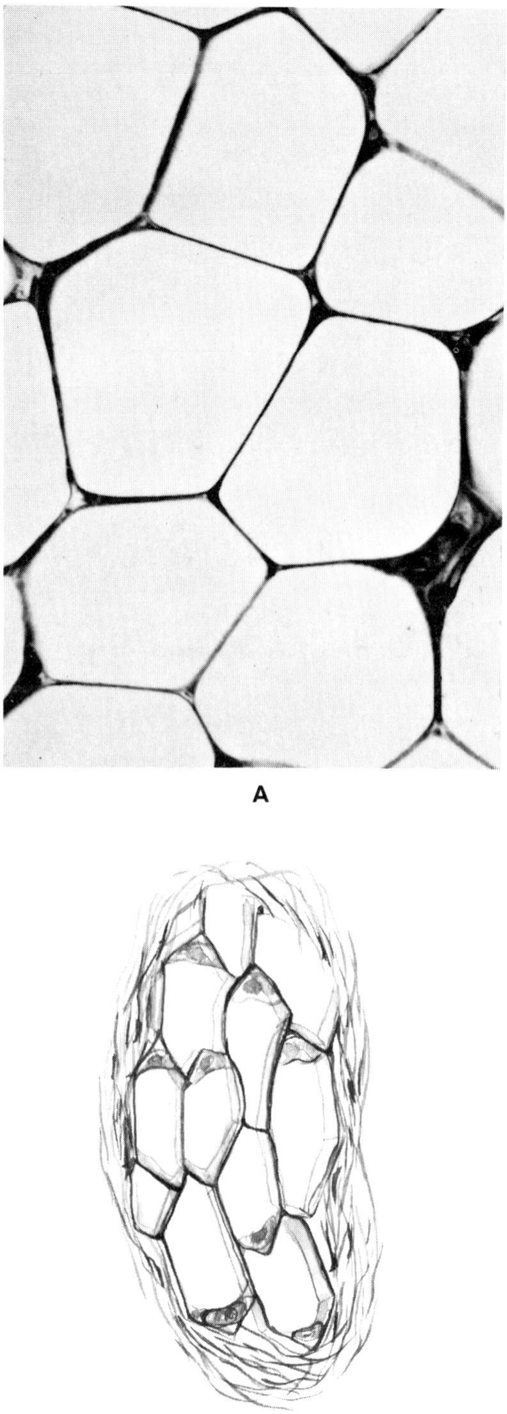

A

B

Figure 3–6 Adipose connective tissue. *A*, From a mouse. *B*, Human. (*A* from Bloom, W., and Fawcett, D. W.: A Textbook of Histology, 9th ed. Philadelphia: W. B. Saunders Co., 1970.)

Cartilage is not supplied with nerves, and rarely contains blood vessels. The *perichondrium*, a moderately vascular fibrous membrane, covers and nourishes cartilage, except in the articular regions.

CLINICAL IMPLICATIONS

Tissue transplantations involving skin, certain organs, and dismembered limbs have become increasingly successful in recent years owing to continued research in the fields of hematology (particularly in antigens) and plastics. Kidney, and blood vessel transplants are now relatively common—although not always successful—and severed arms and legs have been reconnected to the bodies of accident victims—again with limited success. Skin grafts have proved invaluable. Sometimes the skin is obtained from the individual who receives the graft (autotransplant); sometimes from another person (homotransplant).

A method routinely used to determine the success or failure of a skin transplant or of the reimplantation of a limb is the Lewis Triple Reaction:

1. Red line. When the skin is stroked with a dull instrument, the normal capillary response shows a red line.

2. Red flare. A more forceful stroke demands a deeper response involving some of the connective tissue. The normal reaction to such a stimulus is a widened red area or flare.

3. Red wheal. Continued forceful stroking of the skin area involved will induce a response of the vascular and nervous systems, causing a raised welt or red wheal to appear temporarily.

When there is a negative response to any of these tests, it is likely that the transplant did not "take" and that it will eventually die.

Normal cells of the organs and tissues of the body are controlled strictly by the laws of growth. The liver, for example, or a bone increases in size not through

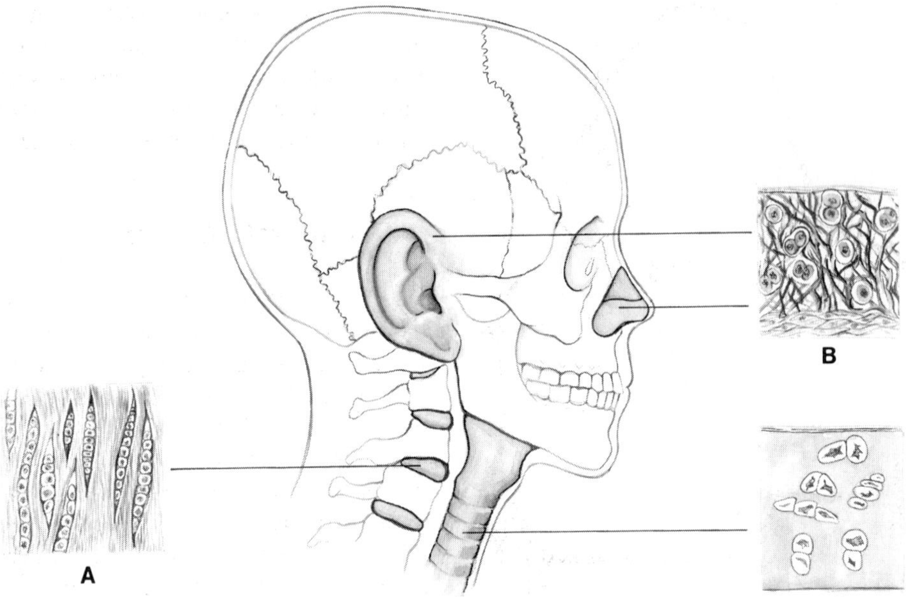

Figure 3–7 Schematic diagram, showing the three types of cartilage. *A,* Hyaline, found in the tracheal rings. *B,* Elastic, found in the ear and nose. *C,* Fibrous, found in the intervertebral disks.

growth of the individual cells of which it is composed but by a process of continual division and multiplication of these cells. Abnormal conditions may develop, however, under which masses of cells grow and proliferate without control, leading to the development of a *tumor* or *neoplasm.* This lack of control is particularly marked in malignant tumors (cancer). Thus cancer is essentially a disorder of cell growth, and the cells of the malignant tumor are merely modified normal cells. With these modifications come loss of the cell's specialized functions, increase in size, and uncontrolled growth into neighboring tissues (metastasis). We know that cellular growth is a chemical process, and it would appear that the abnormal growth we call cancer must reflect some basic change in this process. That is why most cancer research today is being done on a cellular-tissue level and why many cancer patients are being treated by chemotherapy.

Tumors may be subdivided into two broad categories: *benign,* nonpathogenic growths; and *malignant,* pathogenic tumors that, if untreated, eventually will kill the patient, regardless of their location.

Some of the benign tumors found in epithelial and connective tissues are:

1. PAPILLOMA. This is an epithelial tumor found in the skin, the mouth, and the urinary bladder.

2. ADENOMA. This epithelial tumor resembles a gland in its structure. It is found near the breast and near the thyroid gland.

3. FIBROMA. This is a tumor of fibrous tissue. It may be found anywhere in the body, but most often is directly under the skin.

4. LIPOMA. This soft, fatty tumor is found in the neck and on the shoulders, back, and buttocks, where fat deposits are abundant.

5. CHONDROMA. This tumor affects cartilage, and is found primarily at the ends of bones where some cartilage still persists.

Malignant tumors may be classified into two broad groups, *carcinoma* and *sarcoma.* Both are included under the common name cancer. Carcinoma is the most common, and its malignant cells are usually spread through the lymphatic system. These tumors may develop in the skin or in glandular organs—the breast or the liver, for example.

A sarcoma is a malignant tumor of the fibrous connective tissue, and can be found in any part of the body. The principal difference between sarcomas and carcinomas is that sarcomas tend to be localized and do not spread throughout the body. One common form of sarcoma is the *melanoma,* a malignant tumor originating in the melanocytes.

The final diagnosis in suspected cancer—the decision whether the neoplasm is benign or malignant—must be made through laboratory studies of the tumor tissues and cells. When a tumor is removed by a surgical procedure, its type is determined by examination in the pathology laboratory. In other instances, a small sample of the diseased tissue is removed and studied; this process is known as *biopsy.* Rational treatment may then be recommended by the physician or surgeon.

OUTLINE

I. Introduction
    A. The cell
        1. Building block of all tissues
    B. Tissue types
        1. Epithelium
        2. Connective tissue
        3. Muscle
        4. Nerve
        5. Blood and lymph
II. Epithelium
    A. Definition
        1. Covers the body and lines all cavities having a direct connection to the outside of the body
    B. Types
        1. Squamous
            a. Simple, stratified, or transitional
            b. Irregular shapes found in the mouth, lung, alveoli, lens of the eye, etc.
        2. Cuboidal
            a. Cube-shaped cells found in gland ducts and kidney tubules
        3. Columnar
            a. Elongated cells found in the intestines
        4. Pseudostratified
            a. False-layered elongated cells found in the respiratory tract
    C. Associated structures
        1. Goblet cells—flask-shaped cells containing a mucoid secretion
        2. Cilia—hairlike appendages of the cell concerned with filtration and motility
        3. Both goblet cells and cilia are found in columnar and pseudostratified epithelial tissues
    D. Glands
        1. Endocrine
            a. Internal secretion
            b. Ductless
            c. Example—thyroid
        2. Exocrine
            a. Ducts for external secretion
            b. Example—salivary glands
III. Skin
    A. Structure
        1. Epidermis—stratified squamous epithelium
            a. Stratum germinativum—deepest of four epidermal layers
                1) Produces the skin cells
                2) Contains melanocytes that produce melanin—skin pigment
            b. Stratum granulosum
                1) Cells undergo chemical degeneration, forming granules
                2) Cells appear flattened and are dead
            c. Stratum lucidum—second layer from the outside
                1) Made up of hard (keratinized) dead cells to

form a shield for protection
- d. Stratum corneum—outermost horny layer of dead cells
2. Dermis—true skin
   - a. Inner layer of tissues
   - b. Blood and lymph vessels
   - c. Nerve endings—sense of touch
   - d. Hair follicles
   - e. Elastic connective tissue
3. Subcutaneous layer
   - a. Fat cells
   - b. Deep nerve receptors for sensations of pressure and temperature
   - c. Sweat glands
   - d. Hair follicles

B. Blood supply
1. Network of capillaries
2. Vasodilation
   - a. Increased blood supply and heat loss
3. Vasoconstriction
   - a. Decreased blood supply and heat loss
4. Blood deficiency to cells—ulceration

C. Skin appendages
1. Sweat glands
   - a. Tubules
   - b. Regulate body temperature
2. Sebaceous glands
   - a. Oil—sebum
   - b. Lubricate skin ahd hair
   - c. Chronic inflammation—acne
   - d. Too much sebum—seborrhea
   - e. Blockage of sebaceous glands—blackheads
3. Ceruminous glands
   - a. Found only in ear canal
   - b. Secrete cerumen
   - c. Cerumen softens eardrum
4. Hair
   - a. Dermal outgrowth
   - b. Color depends on quantity of melanin in cortex and medulla of shaft
   - c. Supported by arrector muscles
5. Nails
   - a. Horny modifications of epidermis

D. Functions of skin
1. Protection
2. Fluid retention
   - a. Skin burns—first, second, third degree
   - b. "Rule of nines"
3. Sensation
4. Absorption

IV. Connective Tissues

A. Function
1. Support body parts and bind them together

B. Structure
1. Few cells, widely separated
2. Intercellular matrix
3. Fat cells—storage

C. Connective tissue cells
1. Fibroblasts—embryonic mesenchyme
2. Histiocytes—phagocytosis
3. Plasma cells—antibody production
4. Mast cells—heparin production
5. Blood cells—phagocytosis
6. Fat cells—storage of oils

D. Connective tissue fibers
1. Collagenous—white fibers, nonelastic
2. Elastic—yellow, highly elastic

E. Types of connective tissue
1. Loose (areolar)
   - a. Most abundant in body
   - b. Supports and maintains body's water balance
2. Adipose—padding around joints
   - a. Fat storage bundles
   - b. Reserve food supply
3. Dense connective tissue
   - a. More collagenous fibers than elastic fibers
   - b. Examples—tendons, ligaments, aponeuroses, fasciae
4. Cartilage
   - a. Hyaline

1) Modified areolar tissue with glossy appearance
2) Forms skeleton of embryo, covers bones in joint cavities, forms the nose and rings in respiratory tract

b. Elastic
1) Many elastic fibers
2) Forms epiglottis and external ear

c. Fibrous
1) Many collagenous fibers
2) Forms intervertebral disks

V. Clinical Implications
A. Tissue transplants
1. Skin graft
a. Lewis Triple Reaction—red line, red flare, red wheal

B. Tumors
1. Benign
a. Nonpathogenic growth
b. Examples—papilloma, adenoma, fibroma, lipoma, and chondroma

2. Malignant
a. Pathogenic
b. Carcinoma—cells spread through lymphatic system
c. Sarcoma—fibrous connective tissue growth; localized

3. Biopsy—diagnostic tool

STUDY QUESTIONS AND PROBLEMS

1. Define a tissue. What are the five basic types of tissues?
2. What is epithelium? How is it different from other tissues?
3. What is the principal function of connective tissue?
4. List at least five different types of cells found in connective tissue and describe their functions.
5. Name the two general categories of tumors.
6. Name three examples of benign tumors and tell where each is found.
7. What is the major difference between a malignant and a benign tumor?
8. How does carcinoma generally spread?
9. What is collagen? What is its function?
10. Discuss the structure and functions of skin.
11. Differentiate among the three degrees of skin burn.
12. What three tests compose the Lewis Triple Reaction? For what is the reaction used?
13. How does one form of cartilage differ from another?
14. Locate various types of cartilage in the body.
15. What is adipose tissue and what is its function?
16. If columnar epithelium is found in the intestine, what is its function?
17. Why do we have cilia and goblet cells in the respiratory tract?
18. Differentiate, structurally, between an endocrine and an exocrine gland. Give examples.
19. Why does a skin graft have to include at least part of the stratum germinativum?
20. Where does hair grow from? Why do some people have white hair?

Chapter 4

The Cell Membrane; Passive and Active Transport

In the last two chapters, we have discussed the structure of the cell, the several types of cells that make up the complex human organism, and how these cells are arranged into tissues that give the body its form and protect it from the outside environment. Now, with this background, we can discuss the life-supporting physiologic processes that take place within cells.

For the process we call *metabolism* to occur, food must enter the cell, and waste products must be discharged from it. (See Chapter 18 for more information on metabolism.) The cell, to perform these functions, must be a highly organized unit engaged in ceaseless chemical activities. These activities depend on a continuous supply of substances from the tissue fluids of the body's internal environment—substances that are provided by the "supply organs" and obtained from the outside environment. At the same time, the extracellular fluids continuously remove waste materials to the "eliminating organs" for disposal. A cell must maintain a constant internal environment for optimum growth and health; the term *homeostasis* has been coined to describe this state.

The maintenance of homeostasis depends on the movement of substances through the *membranes* that separate the cells. A cell membrane is much more than a passive covering that gives strength and shape to the cell. Rather, it is an active component of the living cell in that it regulates the traffic of substances being received and discharged. The exact nature of the physical processes governing the movement of materials across the cell membrane is not completely understood at the present time. However, the basic modes of transport are known, and they will be discussed in this chapter.

PASSIVE MODES OF TRANSPORT

Diffusion

A solid, a liquid, or a gas may be dissolved in a liquid. The material dissolved is called the *solute;* the liquid in which it is dissolved is the *solvent.* Molecules of the solute are dispersed throughout the solvent.

The term *diffusion* refers to the spreading or scattering of molecules of gases or liquids. If two gases are mixed, their molecules will move continuously and will soon become uniformly distributed. This movement of molecules is always directed toward an area where fewer molecules exist—in other words, from an area of greater concentration (parts per volume) to an area of lesser concentration of solute molecules. When we dissolve a cube of sugar in a beaker of water, the concentration of sugar in the area immediately adjacent to the dissolving cube remains high at first, but within a few hours the sugar will have scattered throughout the water until an even distribution of sugar per volume of water exists (Fig. 4–1, *A*). Thus we can define diffusion

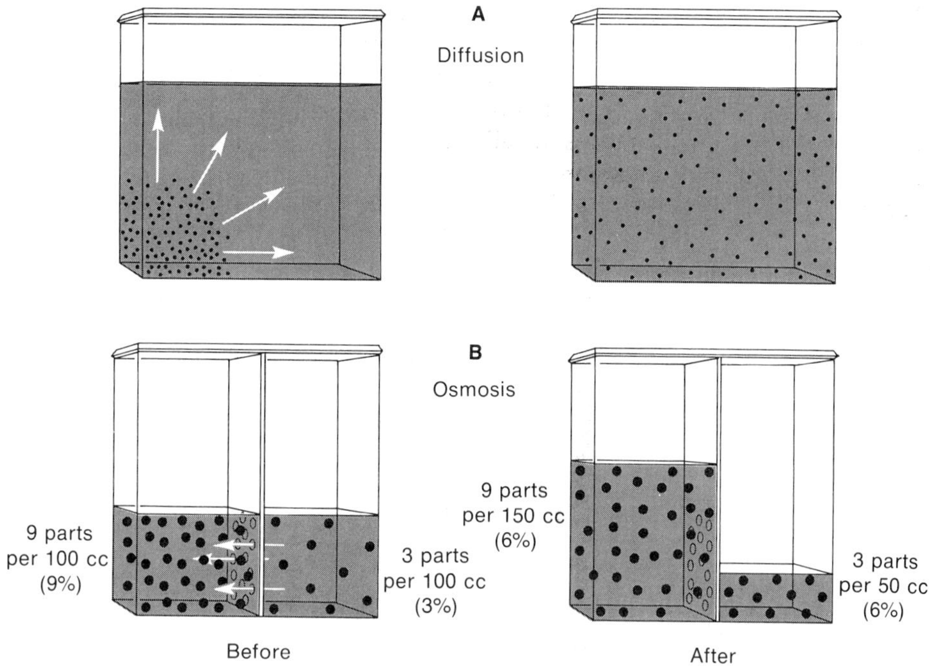

Figure 4–1 *A*, Diffusion. Diagram illustrating the diffusion or scattering of particles from an area of greater particle concentration to an area of lesser particle concentration, until equal parts per volume are achieved. *B*, Osmosis. Arrows indicate the passage of solvent (H_2O) from the less concentrated side to the more concentrated side to achieve equal parts per volume. In this case the particles could not penetrate the semipermeable membrane.

as the passage of solute particles from an area of higher concentration to an area of lesser concentration.

Diffusion occurs in all the body's life-maintaining processes. Oxygen in the blood moves from the blood plasma into the extracellular fluid and then into the cell itself as a result of diffusion. The exchange in the lungs of the gases oxygen and carbon dioxide during inspiration and expiration is another example of diffusion.

The Pore Theory

Nutrients and oxygen, available to the living cell in dissolved form, are able to cross the cell membrane and enter the cell by diffusion. The properties of the cell membrane, however, are extremely complex; other substances whose molecules are too large to diffuse across the membrane also find their way into the cell, and still others are not permitted to pass through. The cell membrane, therefore, has been called "selectively permeable" or *semipermeable*. Put simply, this means that it allows the passage of molecules up to a certain size or diameter, but restricts others. Sugar and water molecules are small enough to pass through the membrane as if it were not there; larger molecules are partially or completely prevented from passing. To explain how molecules too large to enter the cell by diffusion are permitted to cross the membrane, several theories have been advanced. One of these proposes that the membrane contains special holes or channels through which molecules up to a certain size might pass. This is known as the *pore theory*.

The discovery of the pore theory is a fascinating part of scientific history. Cell physiologists had long known of the cell membrane's permeability—that lipids (fats) and other substances soluble in lipids could move easily between the cell and its surrounding medium. Thus it was concluded that the outer boundary of the cell must contain lipids. But this was hardly enough to explain how substances of all kinds passed through the membrane, and thus the pore theory was formulated. The pores, it was proposed, were nonlipid patches in a lipid-containing membrane. The physical properties of the boundary made it necessary also to hypothesize the presence of a layer of protein in the membrane.

These theories stemmed from studies of cellular permeability conducted by Danielli at King's College in London in 1935. His experiments led him to propose the existence of inner and outer layers of protein separated by two lipid layers—the "double-layer membrane theory" (Fig. 4–2). Danielli further proposed that each of the four layers was approximately 20 angstroms thick (1 angstrom, you will remember, is one 100-millionth of a centimeter); therefore, the total thickness of the plasma membrane is approximately 80 angstroms.

The development of the electron microscope has made it possible to prove that Danielli's assumptions were essentially correct. Photographs taken through the electron microscope have shown that each cell membrane consists of two dense, dark layers through which electrons do not penetrate, separated by a lighter, less dense area (Fig. 4–3). As measured by these studies, the total thickness of the cell membrane is 75 to 100 angstroms. The inner and outer layers have been found to be composed of protein, as Danielli originally believed.

Osmosis

In diffusion, as we have seen, the molecules or particles move across the cell membrane according to a *concentration gradient* established by the difference in quantities of particles on the two sides of the membrane. In another process, *osmosis,* the same kind of concentration gradient exists, but the particles in the intracellular fluid are too large to pass through the membrane. The principle of homeostasis must be maintained, however, and therefore the water (solvent) penetrates the cellular membrane to equalize the concentrations on both sides of the membrane (Fig. 4–1).

The force that causes this movement of water is called *osmotic pressure,* and varies directly with the number of solute particles in the solution. For example, a thistle tube containing 20 per cent sugar (20 parts sugar per 100 parts solution) can be covered by a semipermeable membrane, inverted into a beaker containing 10 per cent sugar (10 parts sugar per 100 parts solution), and allowed to stand for a few hours (Fig. 4–4). The membrane is impermeable to sugar in this instance, and will

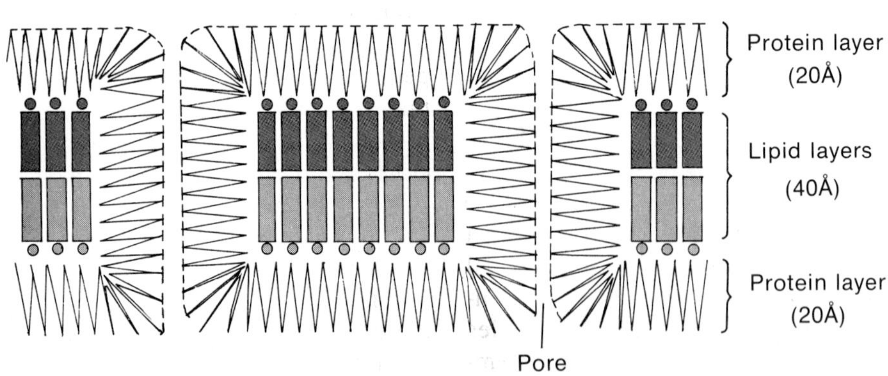

Figure 4–2 Diagram illustrating the layers of a cell membrane.

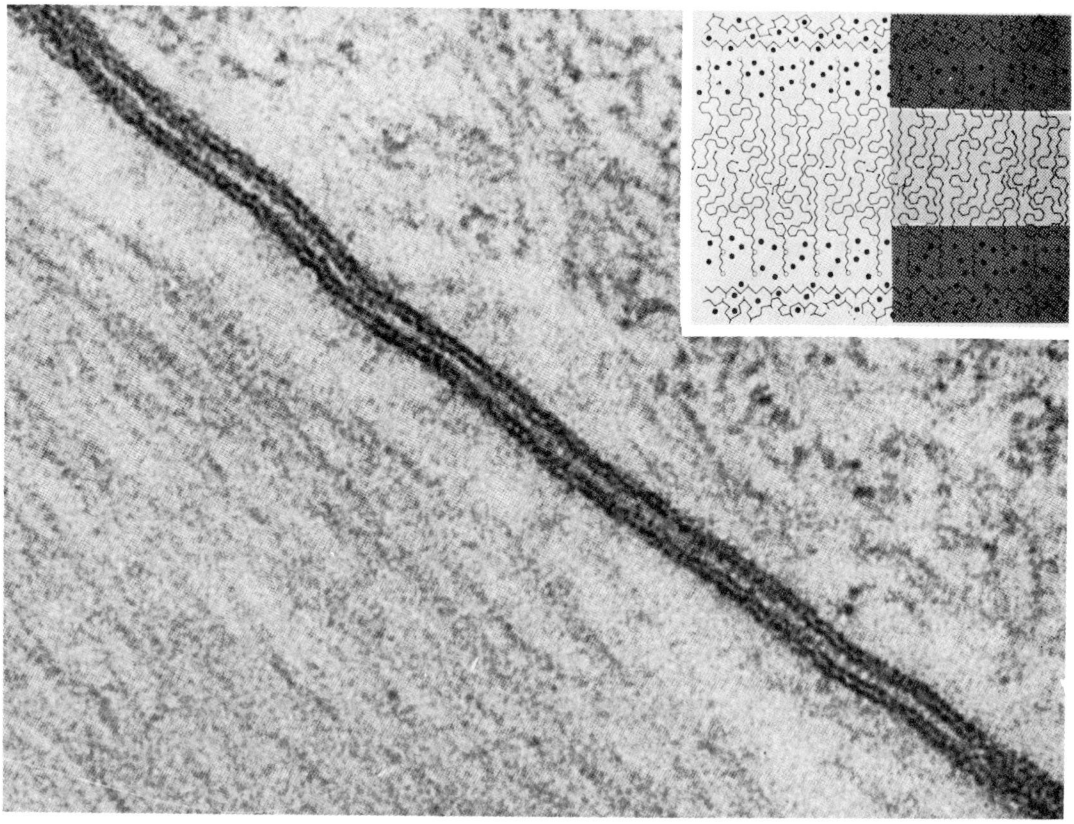

Figure 4–3 Photograph of adjoining cell membrane layers, taken with an electron microscope. (From Bloom, W., and Fawcett, D. W.: A Textbook of Histology, 9th ed. Philadelphia: W. B. Saunders Co., 1970.)

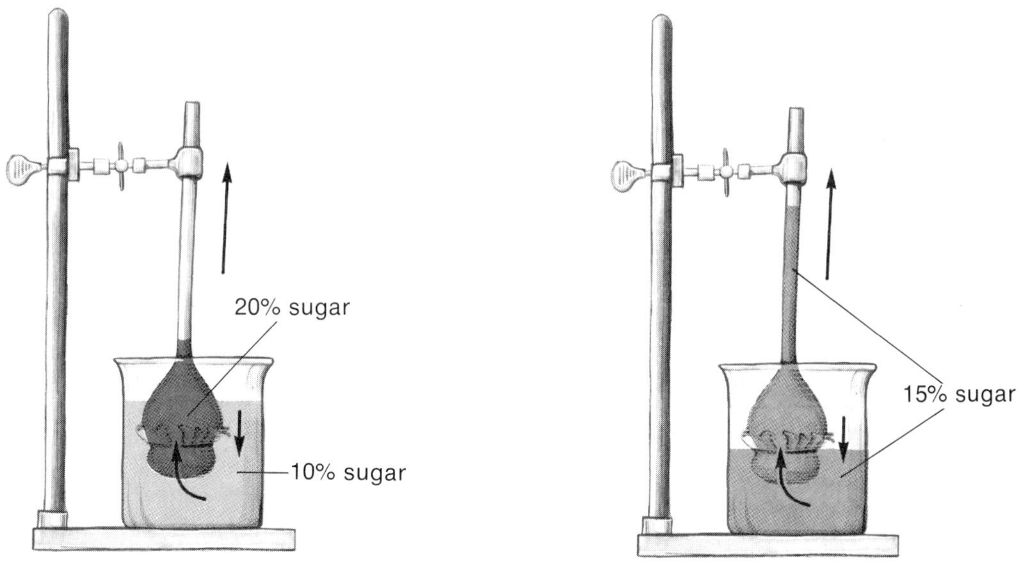

Figure 4–4 Simple osmometer demonstrating the passage of a solvent across a semipermeable membrane. Note the rise of the water level in the thistle tube coming from the beaker.

not allow diffusion to occur. However, the concentration of sugar is clearly greater in the thistle tube than in the beaker. Therefore, water will pass from the beaker into the tube, causing the level of fluid in the tube to rise.

Osmosis can be thought of as the passage of solvent from an area of lesser concentration of solute to one of greater concentration of solute. It represents an attempt to dilute the more concentrated solution in order to achieve equilibrium between the concentrations on both sides of the membrane.

The principle of osmosis can be demonstrated clearly in the laboratory by using red blood cells suspended in salt solutions of various concentrations (Fig. 4–5). That homeostasis is maintained in the bloodstream becomes evident when we observe that the concentration of dissolved substances, such as salts, in the blood plasma (the liquid portion of the blood) is the same as the concentration of those substances in red blood cells (the solid components of blood). When two solutions demonstrate the same concentrations of dissolved substances, they are said to be *isotonic* (*iso* = "same" or "equal").

In order to show the effect of osmosis, red blood cells may be placed in two separate solutions: one having a lower concentration of salts than the red cells, the other having a greater concentration. The solution having the greater concentration is called *hypertonic* (*hyper* = "greater than") to the blood cell, and osmotic pressure forces water to leave the cell and to pass into the solution, causing a shrinking or *crenation* of the cell (Fig. 4–5). On the other hand, a solution of less concentration than the blood cell is *hypotonic* (*hypo* = "less than") to the cell. In this situation, water enters the cell until the cell finally bursts—a condition called *hemolysis*.

The exchange of water and dissolved nutrients in the capillary network of the circulation is a prime example of osmotic pressure in action.

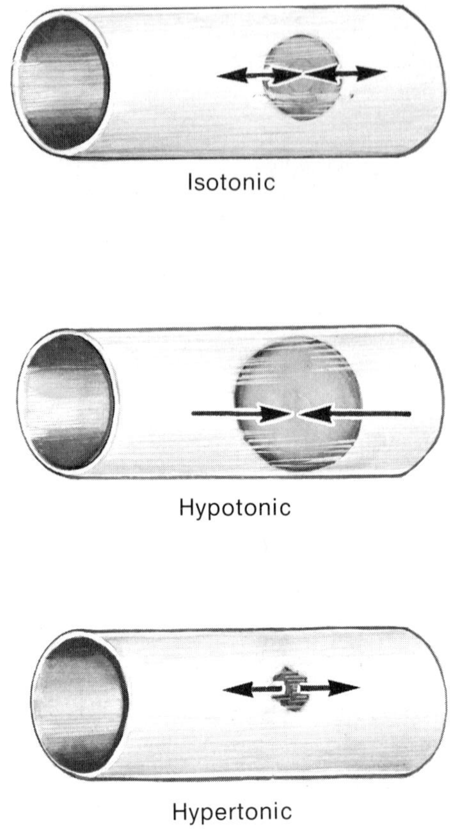

Figure 4–5 The effect of salt solutions of different concentrations on red blood cells. Isotonic: the cell is surrounded by a liquid of concentration equal to that of the cell. Hypotonic: the cell is surrounded by a liquid having a lower salt concentration than itself; liquid enters the cell and causes it to swell and eventually burst (hemolysis). Hypertonic: the cell is surrounded by a liquid having a greater salt concentration than itself; liquid from the cell passes out into the surrounding medium, causing the cell to shrink (crenation).

Filtration

Until now, we have considered only those physical processes involving the passage of particles or water through a membrane according to a concentration gradient. Another mode of passive transport, *filtration*, occurs when water and particles are literally forced through a permeable membrane from an area of higher pressure to an area of lower pressure of fluid. This force is *hydrostatic pressure*.

In the laboratory, the principle of hydrostatic pressure can be used to advantage when a fluid containing dissolved particles is poured through a filter paper in a funnel. The weight of the fluid in the funnel forces the fluid from an area of greater hydrostatic pressure in the funnel into an area of lower hydrostatic pressure in the beaker below. The filter paper is not permeable to some of the particles in the fluid, and these remain in the funnel (Fig. 4–6).

In the human body, this principle can be demonstrated by the action of the kidney capillaries. The hydrostatic pressure of blood is greater than that of the surrounding tissue fluids; therefore, water and substances dissolved in the blood filter through the capillary membrane into the tissue fluid. Some particles, such as blood cells and proteins, are too large to pass through the membrane, and they remain in the bloodstream. A small quantity of albumin may be forced through, but small amounts of protein in the urine (proteinuria) or of blood in the urine (hematuria) are insufficient evidence on which to base a diagnosis of kidney disease. Examples of diffusion, osmosis, and filtration in the body are shown in Figure 4–7.

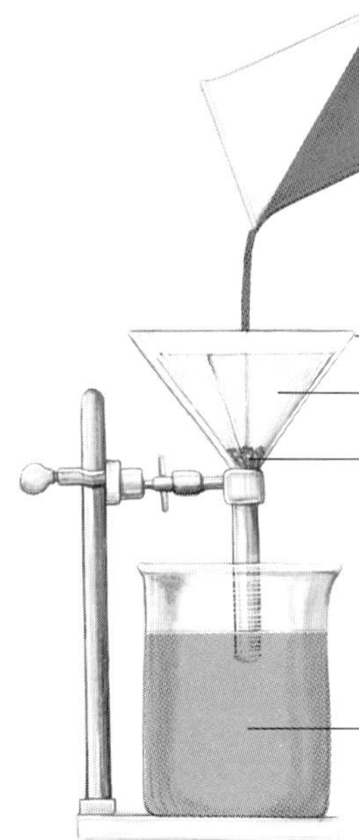

Figure 4–6 Filtration. The solution to be filtered is poured into the funnel, and the solid particles are separated from the clear solution (filtrate) by the filter paper.

ACTIVE TRANSPORT

The living cell membrane, as we have seen, exercises a selective control over which materials enter the cell and whether these substances are retained. Two of the three modes of *passive transport* across the membrane—diffusion and osmosis—operate in the direction of a concentration gradient of substances within and outside the cell. Another mechanism exists, however—one that operates *against* the concentration gradient. This is *active transport.*

When materials are actively transported across the cell membrane in a direction opposite to that which would be expected from the principles of diffusion and osmosis, energy must be utilized by the cell. The energy required is supplied by the cell's mitochondria in the form of adenosine triphosphate (ATP).

The basic mechanism of active transport is thought to be a carrier system (Fig. 4–8) that involves a number of enzymes called permeases. Sodium and potassium are exchanged across the cell wall by active transport in all cells of the body. Other substances, such as glucose and other sugars, are actively transported in only a few areas—particularly in the intestine and the kidney—in order to prevent loss of energy.

The mechanism that transports sodium ions (Na^+) and potassium ions (K^+) across cell membranes is essential in maintaining a "physiologic balance" between the concentrations of each of these ions within the cell and in the surrounding tissue fluids. It is important to remember that "physiologic balance" does not necessarily

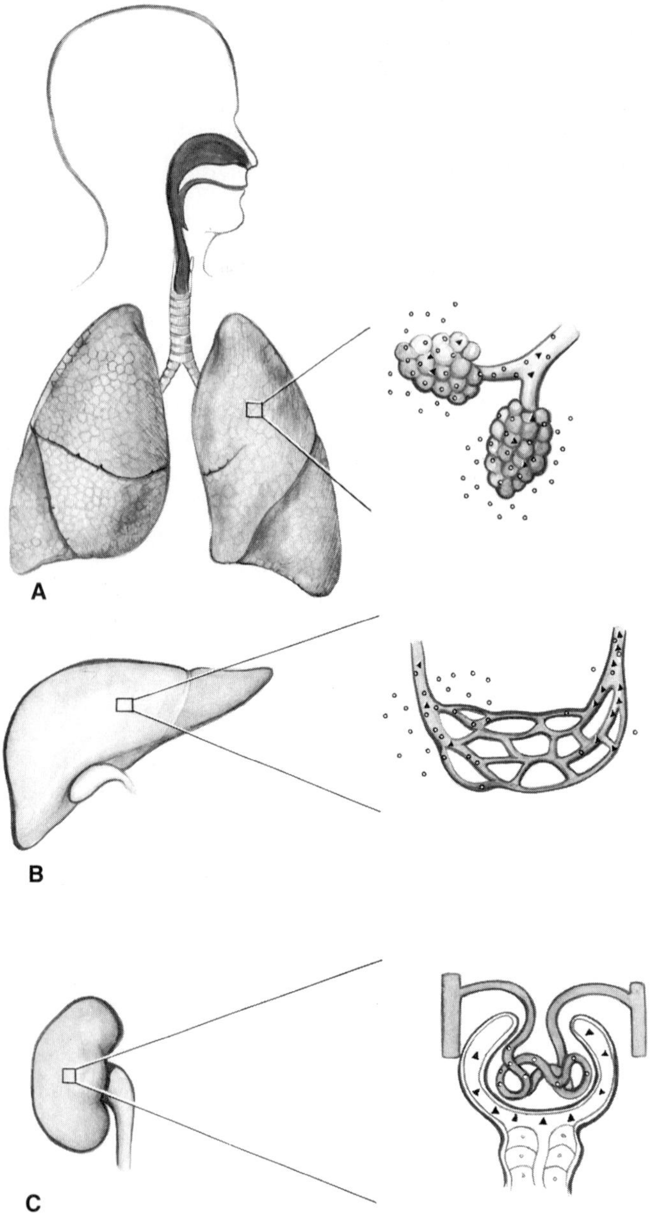

Figure 4–7 Types of permeability. *A,* Diffusion of gases in the lung. *B,* Osmosis: passage of nutrients and waste materials in and out of the capillaries by osmotic pressure. *C,* Filtration: passage of materials from the bloodstream into the kidney tubules by a filtration force.

mean "equal concentrations." On the contrary, in this situation proper cell function exists only if the concentration of Na^+ outside the cell is greater than the concentration within the cell, and, conversely, if there is a greater concentration of K^+ inside the cell than outside.

Under normal conditions, Na^+ tends to pass into the cell, where it is less concentrated, and K^+ tends to leak out into the extracellular fluid where it, likewise, is less concentrated. Proper cell function, however, requires that these ions be returned to their original locations. Homeostasis is maintained, according to a widely accepted theory, by the action of a so-

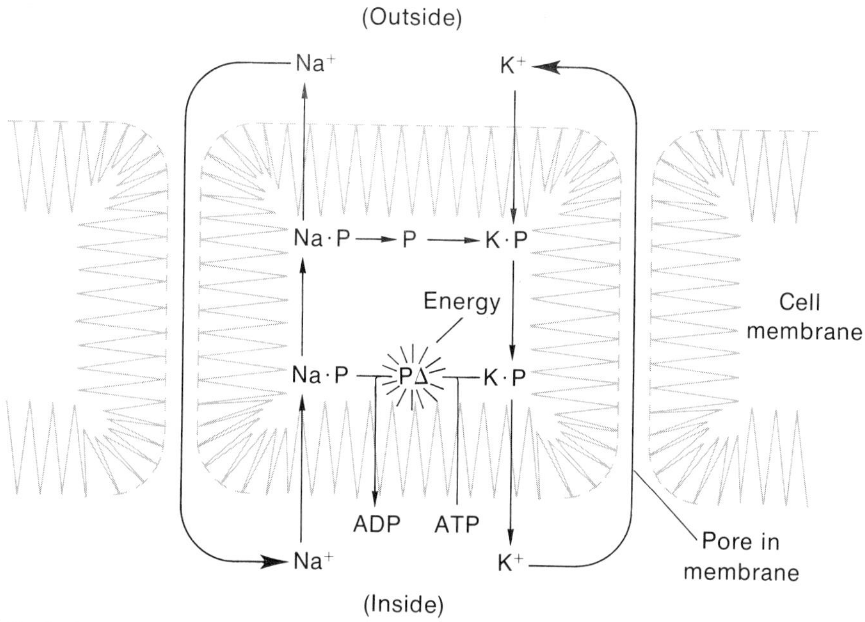

Figure 4–8 Active transport mechanism of sodium and potassium through the cell membrane, showing the permease carrier system. Sodium (Na^+) diffuses into the cell through a pore in the cell membrane, and is actively pumped out of the cell by a permease (P) carrier system with the aid of energy from ATP.

dium-potassium "pump" that shuttles these ions back and forth across the cell membrane by special "carriers." These "carriers" transport Na^+ out, and return unloaded for another sodium ion. According to this theory, each carrier is specific for a particular ion, and energy is expended as the carrier releases the ion at its appropriate destination. This mechanism of ion exchange is particularly important in the transmission of nerve impulses, and will be discussed again when we consider the physiology of the nervous system.

PHAGOCYTOSIS AND PINOCYTOSIS

The process of *phagocytosis* is also relevant to our present discussion of cell physiology. The transport of materials from the extracellular fluid into the cell's interior can be accomplished by engulfing them as a portion of the cell membrane folds inward (Fig. 4–9). The "pouch" that results from this process then breaks loose from the outer portion of the cell membrane and forms a vacuole within the cytoplasm. By this process of phagocytosis, solid particles such as bacteria and foreign bodies are kept separate from the cell's normal constituents.

Through phagocytosis, molecules present in the extracellular fluid that normally could not penetrate the cell membrane may gain entry along with ingested food particles. The principal cells of the body capable of performing a phagocytic action are the white blood cells and the macrophages (histiocytes).

Pinocytosis ("cell drinking") involves essentially the same mechanism as phagocytosis, the only difference being that a liquid rather than solid particles is ingested by the cell. A pinocytic vacuole contains small molecules, usually in a dissolved state. It is thought that the lysosomes may play a part in the breakdown of these vacuoles and in the digestion of their contents.

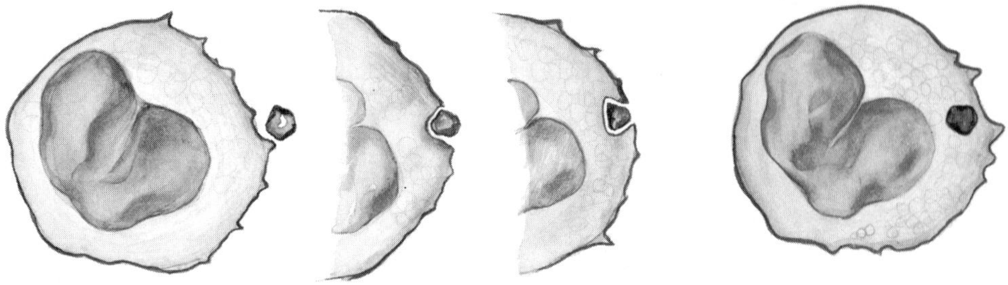

Figure 4–9 Phagocytosis. Portions of cytoplasm stream out, surround the particulate matter, and draw this material into the interior of the cell, where it is digested by cellular enzymes. Here the cell is a white blood cell (neutrophil) that is ingesting a bacterium.

CLINICAL IMPLICATIONS

The supreme importance of water, which makes up about 70 per cent of the body weight, has already been mentioned. Water is present in the blood vessels (in the form of plasma), in the extracellular fluids, and in the cells themselves. An exchange of fluid among these three major compartments goes on continually. Water lost from one compartment can be restored from another compartment.

For the most part, water balance is regulated by water loss, which averages about 2500 ml ($2\frac{1}{2}$ qts) per day in an adult. This loss occurs through the intestines, the lungs (in expired air), the skin (as perspiration), and the kidneys. In a healthy person, the main responsibility for removing excess fluid belongs to the kidneys. However, large amounts of water may be lost in severe and continued diarrhea or in profuse perspiration.

Electrolyte Balance

The electrolytes of the body exert an important influence on the amount of water contained in the various compartments because of the osmotic pressure they produce. The laws of physics dictate that whenever water is shifted within the body, electrolytes are also shifted. Thus an abnormal water balance leads to a depletion of electrolytes in one area and to an excessive accumulation in another area. Both changes are potentially hazardous to health.

To illustrate this principle, we can consider potassium, which is the major intracellular cation. The concentration of potassium within the cells is roughly 35 times greater than the concentration in the tissue fluids. Potassium deficiency is more common than potassium excess. Such deficiencies can result from vomiting, diarrhea, and excessive glandular activity, leading to severe acidosis that results in progressive mental disorientation, muscular weakness, and eventual paralysis. Excessive potassium (intoxication) is a different matter. Malfunction of the kidneys can elevate potassium levels; an increase in potassium blood levels from the normal average of 4.5 mEq per liter to just 8 mEq per liter is sufficient to cause heart stoppage (cardiac arrest). (mEq is the standard abbreviation for "milliequivalent," which is a measure of the chemical reactivity, or combining power, of ions.)

A slight increase in the concentration of sodium increases muscle excitability, resulting in convulsive contractions called *spasms*. Too low a concentration of calcium in the blood can lead to convulsions of the entire body *(tetany)*; too high a calcium level can cause the formation of stones *(calculi)* in the gallbladder and kidney.

Obviously, potassium, sodium, and calcium are cations of major clinical importance. Many anions can also cause serious illness if they are present in abnormal concentrations. These facts clearly illustrate the significance of electrolyte imbalance.

Dehydration

Normally the body takes in more water than is needed for nutrition and cellular metabolism, and the excess is excreted in the urine by the kidneys. However, if water intake is insufficient, the loss due to evaporation through the skin and other causes results in a negative water balance and the state known as *dehydration.* Generally, when the body's water loss reaches 15 per cent of body weight, death follows.

A decrease in body water level from any cause is seen first in the circulatory system, where there is a reduction in blood volume, carrying with it the possibility of *shock* (systemic collapse). Then the regulating mechanism of the hormonal system comes into play, withdrawing fluid from the extracellular tissues and finally from within the cell itself. It is this final depletion of cellular fluid that really defines cellular dehydration, creating a thirst that can be unendurable and fatal. These disturbances of water and salt balance may be caused by prolonged exposure, kidney failure, severe burns, or acute intestinal obstruction.

Edema

Edema may be defined as an abnormal collection of fluids in the extracellular spaces. Such fluids can move from one place to another, so that when the swollen (edematous) part is pressed, a "pit" is left. Accumulations of fluid are most marked and most readily recognized in the subcutaneous tissues. Edematous tissues have a pale, watery appearance, and, if the cause is not corrected, may come to resemble a jelly.

There are four principal causes of edema:

1. An increased permeability of the capillary walls, usually due to an inflammation.

2. A decrease in the osmotic pressure of the plasma caused by a loss or deficiency of the protein albumin, which acts to establish or maintain osmotic pressure.

3. Increased pressure inside the capillary, as occurs in chronic heart failure, in which blood accumulates in the vessels leading back to the heart.

4. Lymphatic obstruction, which may be caused by a tumor, an inflammation, or a parasitic infestation.

Although we usually think only of the unfavorable effects of edema, it can produce benefits as well. The shift of large amounts of fluid into the extracellular spaces may dilute bacterial poisons (as in infection); such a shift may also reduce a state of hypertonicity, thus protecting the cells from electrolyte imbalance.

OUTLINE

- I. Introduction
  - A. Metabolism—the sum of all the physical and chemical processes by which a living substance is produced and maintained, and also the transformation by which energy is made available for the uses of the organism
    1. Food enters the cell
    2. Waste leaves the cell
    3. This process depends on ceaseless chemical activities
  - B. "Physiologic balance"
    1. Stable external environment
    2. Homeostasis—constant internal environment
  - C. Cell membrane
    1. More than a passive covering
    2. Regulates movement of materials in and out of the cell
    3. Semipermeable
- II. Passive Modes of Transport
  - A. Diffusion
    1. Movement of solute from area of greater concentration of solute to area of lesser concentration of solute
       - a. Solute—substance dissolved

b. Solvent—liquid in which solute is dissolved
c. Concentration—parts per volume
d. Example—oxygen and carbon dioxide exchange in blood cells and lungs

B. Pore theory
1. Molecules up to a certain size pass through pores; larger ones are not permitted to pass
2. Danielli's "double-layer membrane theory"—membrane consists of inner and outer protein layers separated by two lipid layers

C. Osmosis
1. Passage of solvent through membrane from area of lesser concentration of solute to area of greater concentration of solute
a. Isotonic—equal concentrations of dissolved substances
b. Hypotonic–hypertonic—unequal concentrations
c. Hypotonic—hemolysis
d. Hypertonic—crenation

D. Filtration
1. Hydrostatic pressure—water and particles forced through permeable membrane from area of higher fluid pressure to area of lower fluid pressure
2. Examples
a. Filter paper–funnel
b. Kidney filtration–excretion

III. Active Transport
A. Carrier system
1. Sodium normally higher in concentration outside the cell
2. Potassium normally higher in concentration inside the cell
3. Carriers called permeases
4. Sodium and potassium exchange against a concentration gradient occurs in the presence of energy (ATP)

IV. Phagocytosis and pinocytosis
A. Phagocytosis
1. Engulfing solid particles through a folded portion of cell membrane
2. Formation of vacuoles inside the cell
3. Performed by white blood cells and macrophages

B. Pinocytosis
1. Engulfing liquid
2. Lysosome activity

V. Clinical Implications
A. Water content
1. 70 per cent of body weight
2. Water loss about 2500 ml per day
3. Water loss balanced by fluid intake

B. Electrolyte balance
1. If water is shifted, electrolytes in tissue fluids are shifted
a. Excessive potassium can be lethal
b. Increase in sodium increases muscle excitability

C. Dehydration
1. Decrease in body water level may promote systemic shock

D. Edema
1. Abnormal collection of fluids in extracellular spaces
2. Caused by:
a. Increased permeability of capillary walls
b. Decreased osmotic pressure
c. Increased capillary pressure
d. Lymphatic obstruction

STUDY QUESTIONS AND PROBLEMS

1. What does permeable mean? Semipermeable?
2. Define the three basic means of passive transport through a membrane and give an example of where each occurs.

3. What is meant by osmotic pressure?
4. How does filtration differ from osmosis? Where does it occur?
5. What does active transport mean? Describe the permease carrier system.
6. Why may crenation occur in the presence of a hypertonic solution? Are blood cells crenated in the urine in the case of kidney retentions?
7. How is energy used in the active transport system? In what form?
8. What does a cell membrane look like?
9. List at least two functions of cell membranes.
10. How does a cell phagocytize?
11. What happens to the food particles after phagocytosis?
12. Explain the importance of maintaining the body's electrolyte balance.
13. What is edema? How does it occur?
14. What is the pore theory? How was it discovered?
15. How is the body's water content normally maintained?

Chapter 5

The Skeletal System

The *skeleton,* the bony supporting structure of the body, is made up of more than 200 bones of many different shapes and sizes (Fig. 5–1). In man and other vertebrates, the skeleton is covered by soft tissues, such as muscle and fat, and therefore is called an *endoskeleton.* Many invertebrate animals, for example, lobsters, crayfish, and clams, are protected by another type of skeleton—an *exoskeleton,* which, as the name suggests, is on the outside of the soft tissues.

An endoskeleton has five principal functions:

1. Protection. The skull, for example, protects the brain; the vertebral column protects the spinal cord. The rib cage and the bones of the pelvis protect the organs contained in the thorax and the lower part of the abdomen.

2. Support. Most of the body's structures are soft, pliable, and delicate. The skeleton's rigid framework provides a means of suspension for body systems.

3. Muscle Attachment. Muscles are attached to bone by tendons. The rigidity of bone, combined with the elasticity and contractility of muscle, makes the arms and legs strong levers, tools, or weapons that can be used for work, locomotion, or defense.

4. Reservoir for Minerals. The *osseous* (bony) tissues of all parts of the skeletal system provide the body with a vast depot of such mineral salts as calcium and phosphates.

5. Hematopoiesis. During late fetal life, and continuing in some bones through adult life, the red marrow of long bones and of the ribs, the sternum, and the pelvic bones produces blood cells—a process called hematopoiesis.

It is important to remember that the skeleton is composed of *living* tissue. Its materials are constantly being dissolved and passed into the bloodstream. They are, in turn, constantly replenished from other tissues and from food.

FORMATION OF BONE

Osseous tissue develops from *mesenchyme,* embryonic connective tissue that is characterized by branching cells in a fluid matrix. The development of bone involves the transformation of mesenchyme into a fibrous tissue containing *fibroblasts* (embryonic fibrils) and collagenous and elastic fibers. This tissue may develop into any one of the several types of mature connective tissue. However, when blood vessels are introduced, their rich contents stimulate mitosis (cell division); the cells increase in size, change in shape from irregular to cuboid, and then may be classified as bone-forming cells or *osteoblasts* (Fig. 5–2). The osteoblast deposits a mucoprotein substance that incorporates the fibers, giving the resultant *osteoid tissue* a rubbery, elastic consistency.

Any cell that has the property of being able to lay down a matrix which calcifies without dying in the process is an osteoblast—a "bone builder." The precise origin of bone-forming cells is still controversial; their function is much better understood.

Bone formation begins early in embryonic development. It occurs by two methods: intramembranous ossification and cartilaginous (or endochondral) ossification.

Intramembranous Ossification

The bones of the skull are formed by this method. In the embryo, one of the layers covering the developing brain is fibrous in nature, and is composed of fibroblasts and collagenous fibers. In the areas where bone is to be formed, the fibroblasts differentiate into osteoblasts, which then lay down an intercellular substance for the deposition of calcium. From this matrix, ossification progresses in ra-

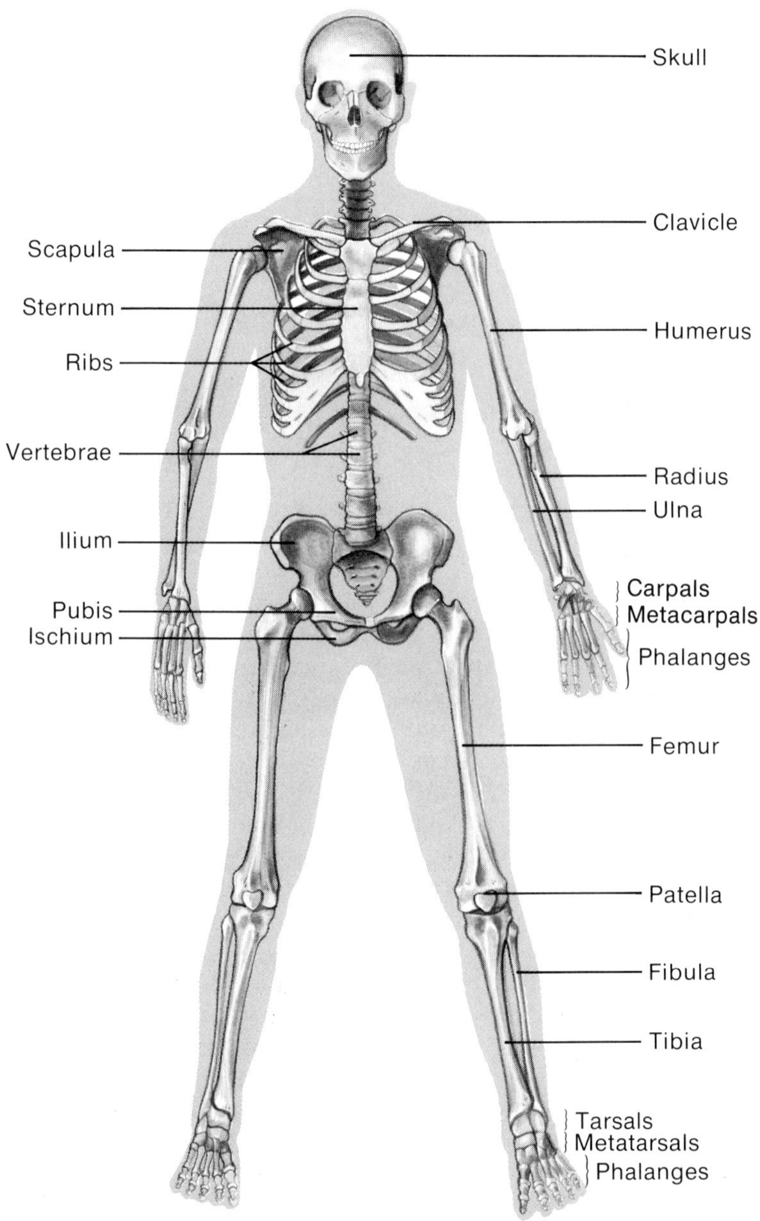

Figure 5–1 The human skeleton, anterior aspect.

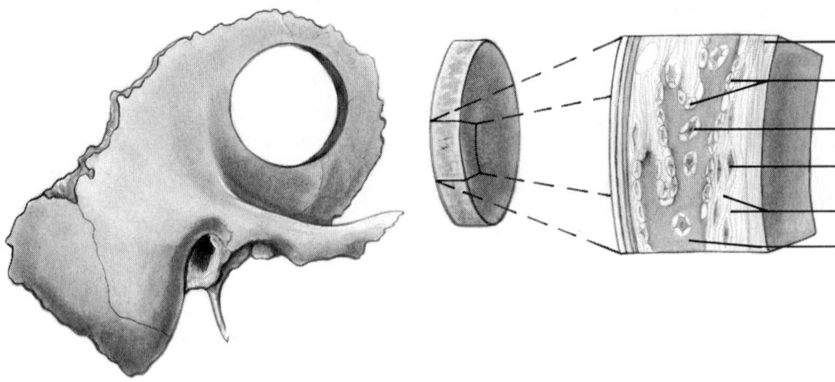

Figure 5–2 Developing intermembranous bone in the skull of the fetal cat.

diating columns of calcified fibers, between which the osteoblasts construct bone (Fig. 5–2).

The ossification of the skull is not complete at birth, but continues through early childhood until the structure has developed into a strong bony case to protect the brain.

Cartilaginous (Endochondral) Ossification

Most of the skeleton is formed by this method, which involves a gradual replacement of cartilage by osteoblasts. Cartilage is a specialized type of fibrous connective tissue that forms most of the temporary skeleton of the embryo. A skeleton of hyaline cartilage, a flexible, semitransparent substance, is fully formed by the second month of embryonic life. It is enveloped in a layer of dense fibrous connective tissue called the *perichondrium.* However, soon after this period, bone begins to form directly on the shaft of the cartilaginous "model." The previous perichondrial layer becomes the covering of the bone, the *periosteum* (Figs. 5–2, 5–4, and 5–8).

As the process of calcification proceeds, the cartilage cells swell and degenerate, ultimately leaving a linear arrangement of empty capsules *(lacunae)* (Fig. 5–7). Blood vessels then penetrate the periosteum, bringing osteoblasts to lay down a matrix of collagen and mucoprotein, two types of protein that serve as supporting substances. Through this process, the osteoblasts mature into *osteocytes* (Fig. 5–3).

The growth of a bone continues as it becomes thicker at the outer periosteum and longer as more cartilage is replaced by osteocytes. These changes make it necessary that the cavity within the bone increase in size, and bone material must be removed for this to occur. The *osteoclasts,* large cells that can dissolve bone, are responsible for this process, enlarging the bone's internal cavity while the osteoblasts continue to deposit new bone material outside. Thus the depth of the bone remains constant throughout most of a person's normal life while the bone is growing. Continued osteoclast activity with simultaneous gradual inhibition of osteoblasts occurs in some diseases and in old age. This leads to a reduction in the depth of the bone, resulting in brittleness and fractures.

Within the spaces of bony structures are areas where blood cells reproduce, mature, and enter the circulation. These areas are commonly called the bone marrow or *medullary* cavities (p. 64; Figs. 5–3 and 5–8, *A*).

In long bones—for example, those of the arms and legs—ossification proceeds in two directions. First, it begins at the shaft

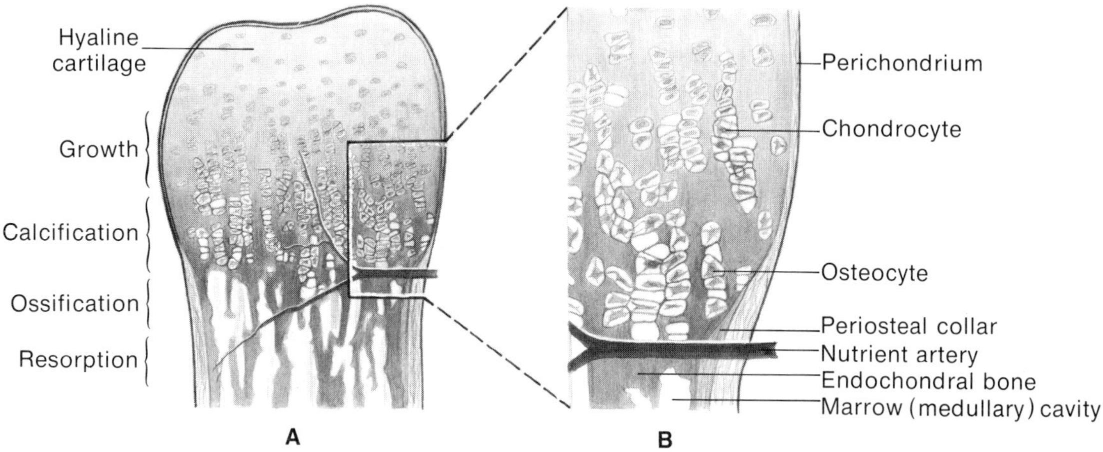

Figure 5–3 Endochondral ossification in longitudinal sections through the zone of epiphyseal growth of a fetal bone. *A*, Diagram of the zones of development, from unmodified cartilage to resorption by osteoclasts. *B*, High-power section of the zone of calcification, illustrating the conversion of a soft hyaline cartilage into hard bone through the replacement of chondrocytes by osteocytes and solidification of the extracellular matrix.

of the bone *(diaphysis)* and moves toward the head *(epiphysis)*; at the same time, ossification occurs in the opposite direction, from the epiphysis toward the diaphysis (Fig. 5–4). The epiphyseal cartilage persists until early adulthood, and continued growth of this layer by cell division provides for the lengthwise growth of the bone. Although skeletal growth begins early in fetal life, it is not complete until the person's twentieth to twenty-fourth year. When growth is complete, the epiphyseal cartilage cells cease to multiply (Fig. 5–5).

Chemical Influences on Bone Formation

The final stage in the formation of bone occurs when the osteoblasts draw upon mineral ions that diffuse from the

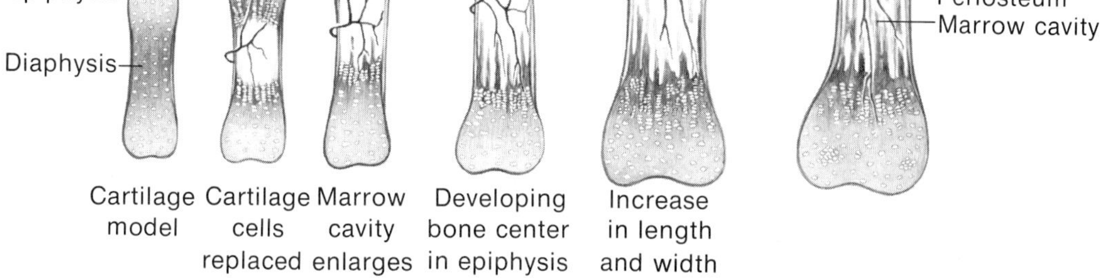

Figure 5–4 Diagrammatic representation of the development of a long bone.

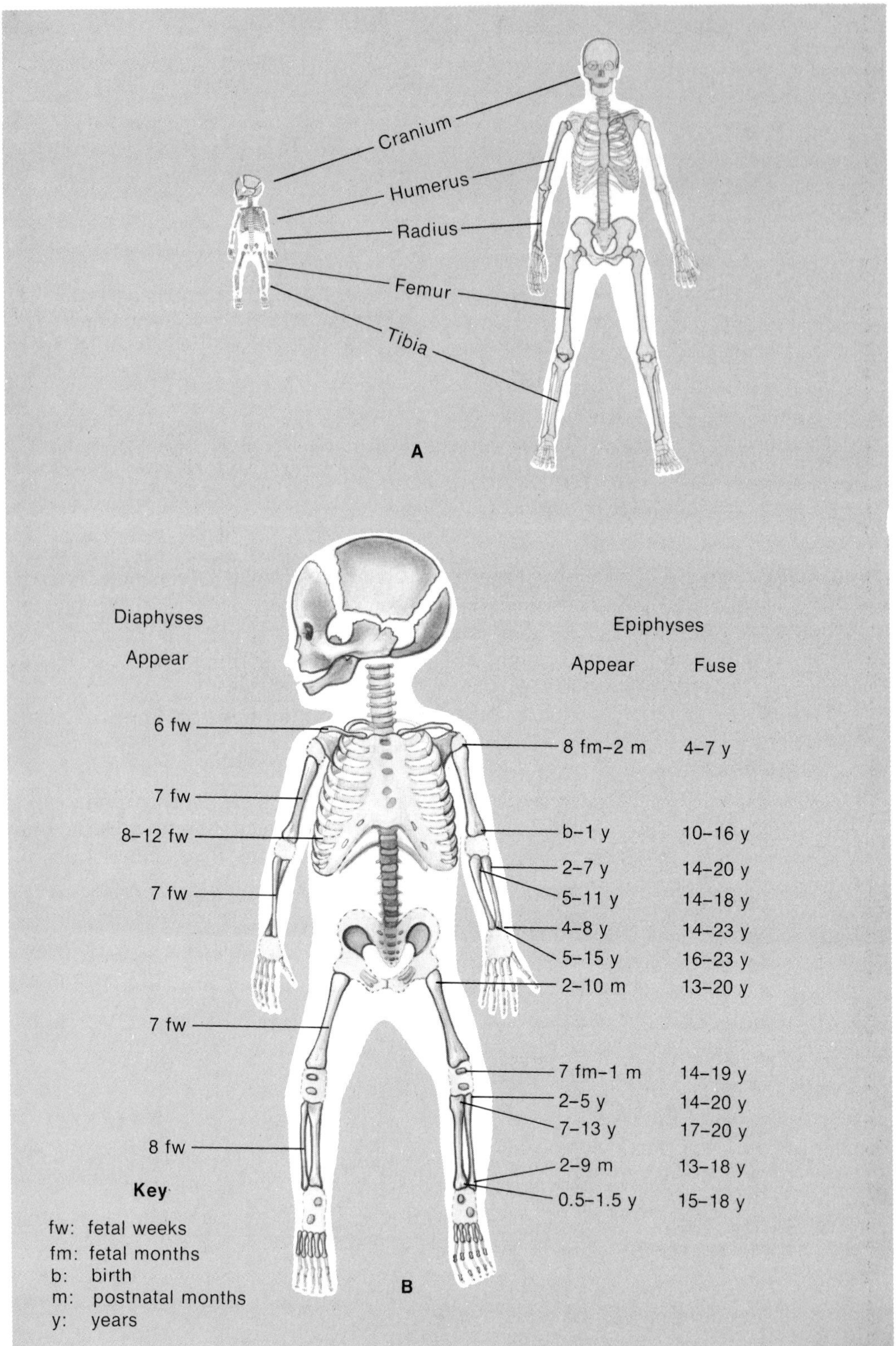

Figure 5–5 Comparative sketches of baby and adult skeletons. *A*, The relation of size and shape of individual bones to total body size is comparable in the baby and the adult. *B*, The initial time of development and final growth period of some of the bones of the human skeleton.

blood vessels into the *osteoid,* the intercellular matrix of young bone.

At this point, we should consider other processes necessary to the development of bone. *Vitamin D* and *alkaline phosphatase* are two important compounds necessary for the absorption and availability of the mineral substances essential for the deposition of bone. Vitamin D is thought to be the chief agent in the absorption of minerals through the intestine. Insufficient quantities of the vitamin may lead to loss of calcium in the feces. Alkaline phosphatase, an enzyme found in many tissues, breaks down calcium salts and frees the all-important calcium and phosphate ions necessary for the hardening of bone.

The third important regulatory substance needed for normal bone deposition and growth is *parathormone,* a hormone produced by the parathyroid glands. Parathormone is responsible for regulating the calcium and phosphorus levels of the blood, thereby maintaining homeostasis for bone growth.

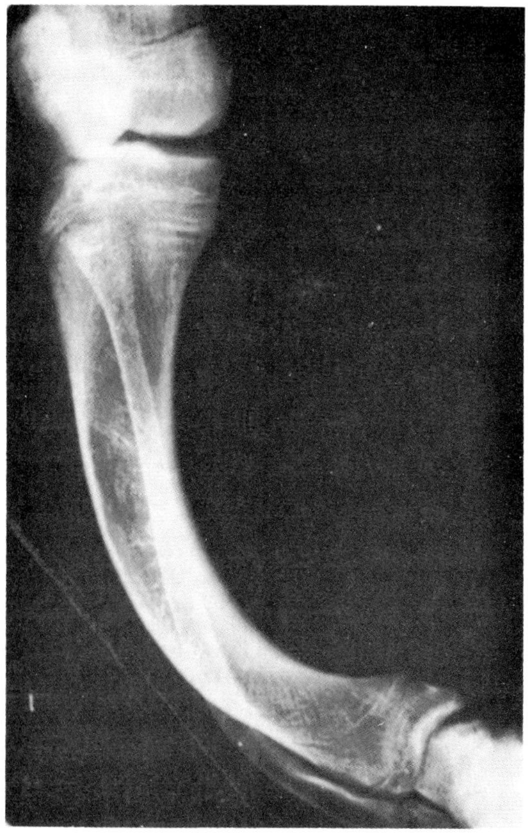

Figure 5–6 Deformity of a long bone produced by severe untreated rickets. (From Trueta, J.: Studies of the Development and Decay of the Human Frame. London: William Heinemann Medical Books, Ltd., 1968.)

Disorders of Bone Development

An insufficiency of any or all of the three factors just described will generally result in a weakening or softening of bone. When the deficiencies occur during the period of ossification, they lead to the disease state called *rickets* (Fig. 5–6). A similar deficiency state developing later in life is known as *osteomalacia.* Other metabolic disorders also can lead to skeletal deformities. For example, hypofunction of the anterior lobe of the pituitary gland in its production of growth hormone may cause *dwarfism.* Hyperfunction of the gland before maturity may cause *gigantism;* hyperfunction after maturity may cause *acromegaly,* a condition characterized by overgrowth of the bones of the hands, feet, and face.

THE STRUCTURE OF BONE

Microscopic Structure

Collagen and minerals together form a homogeneous material that, microscopically, can be seen as an open meshwork of plates and bars (spongy or *cancellous* bone), or as several relatively uniform concentric layers around each of the smallest blood vessels in *compact* bone.

The basic unit of structure of compact bone is the *osteon.* It consists of a *haversian canal* and from four to twenty *lamellae,* which are bony plates arranged concentrically around the canal (Fig. 5–7). The canals serve as passageways for nutrient materials to reach the living osteocytes, which are located in lacunae, spaces within the lamellae. In addition to the haversian canals, compact bone also contains *Volkmann's canals,* through which course blood vessels (Figs. 5–7 and 5–8, *C*). Volkmann's canals cross compact bone longitudinally, maintaining a blood supply between the periosteum and the haversian canals. Delicate channels called *canaliculi* radiate from the haversian canals, connecting the separate lacunae—thus forming a complete circulatory system from the marrow cavity to the outer periosteum (Fig. 5–7).

Microscopically, the epiphyses of long bones appear as a latticework of spongy cancellous bone (Fig. 5–8, *B*). At first sight, the meshes of this network seem to be arranged in no apparent order; more careful examination, however, has shown that the structural pattern of this bony tissue follows the lines of stress and strain to which the bone is exposed.

Gross Structure

As we have seen, a typical long bone consists of a shaft (the diaphysis) and two heads (epiphyses). The diaphysis is made up principally of compact bone (lamellae arranged to form a hard, solid mass). In the core of the diaphysis is a large space, the *medullary cavity* (Figs. 5–3 and 5–8, *A*), filled with yellow bone marrow, a hematopoietic tissue.

In adult bone there are two types of marrow: red and yellow. Red marrow is found in the vertebrae, the sternum, the ribs, and in the proximal epiphyses of long bones. As discussed earlier, its major function is the production of blood cells and platelets. Yellow marrow, found in the cavities of long bones, is made up mostly of adipose tissue (fat cells). Under normal conditions yellow marrow takes no part in red blood cell production in adulthood, but in certain abnormal circumstances it may replace red marrow when the supply of the latter becomes depleted.

Outside, the bone is covered with periosteum, a fibrous connective tissue membrane. Nutrients and osteoid components are supplied to the bone by the large *nutrient artery* (Fig. 5–3), as well as by many small blood vessels in the periosteum. These vessels permeate the compact bone and reach the medullary cavity through the haversian and Volkmann's canals.

The marrow cavities of bone are lined with a membrane, the *endosteum* (Fig. 5–

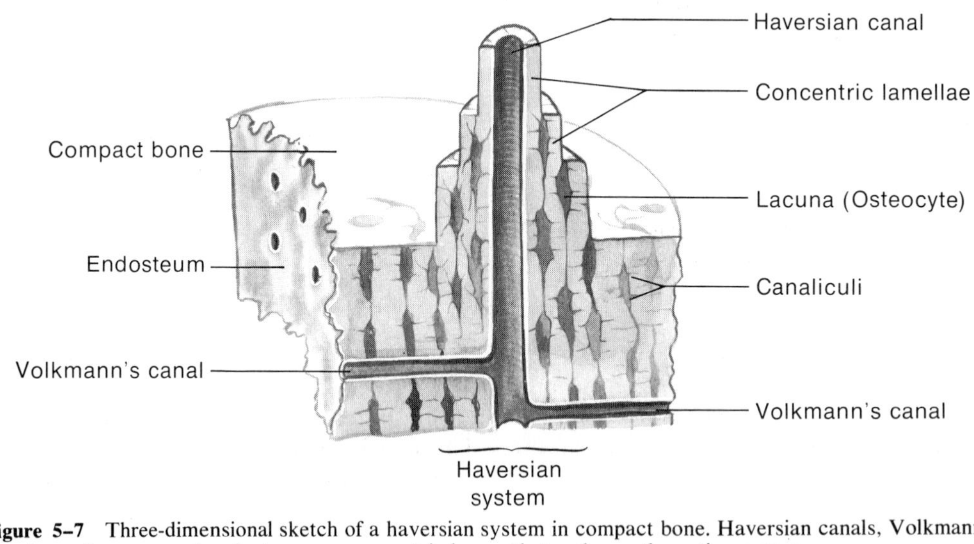

Figure 5–7 Three-dimensional sketch of a haversian system in compact bone. Haversian canals, Volkmann's canals, periosteum, and endosteum are concentric layers that make up the entire system.

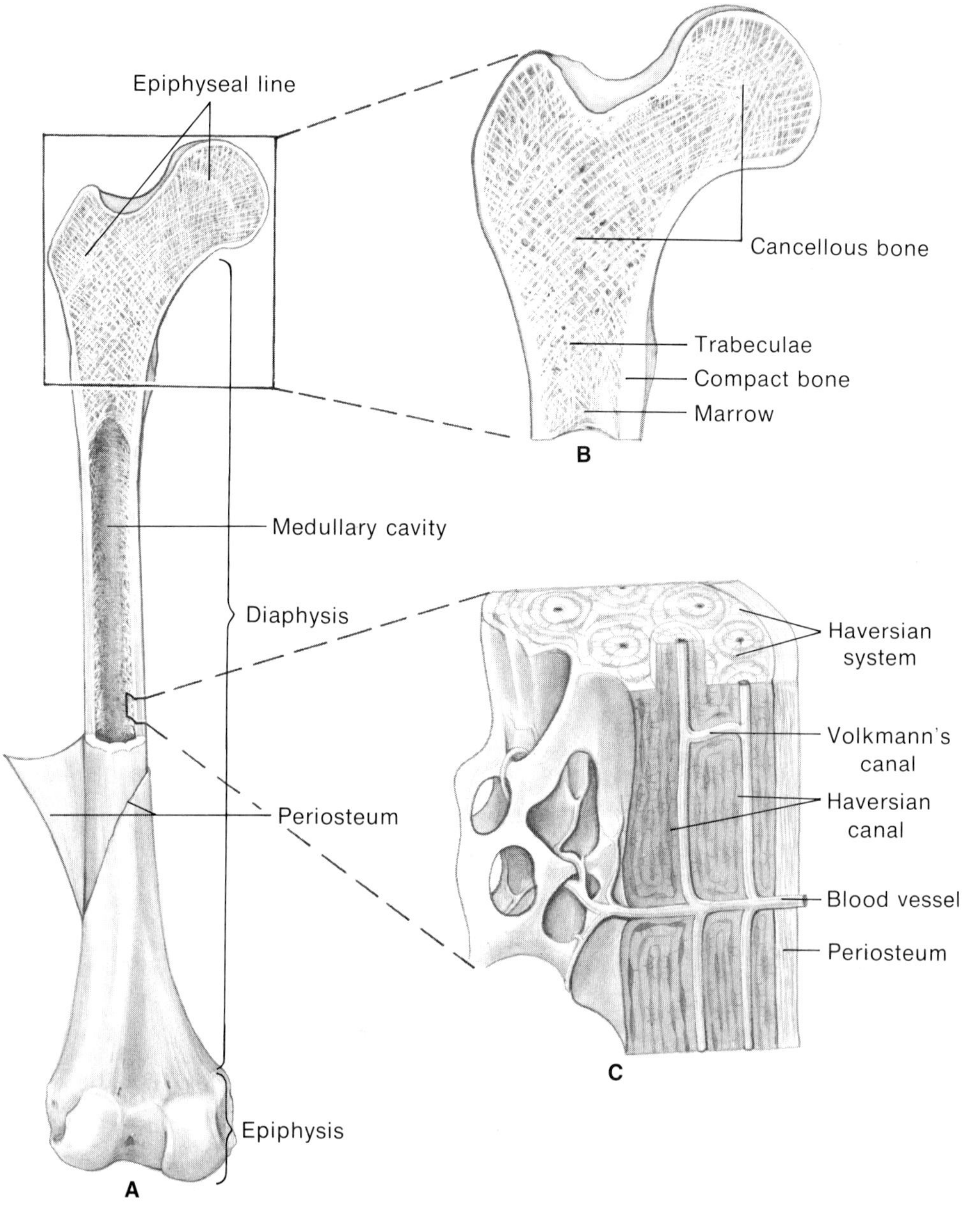

Figure 5–8 Schematic diagram of the structure of a typical long bone. *A*, Longitudinal section. *B*, Epiphyseal section. *C*, Diaphyseal section.

7), which is basically a single layer of cells—either osteoblasts or osteogenic cells.

TYPES OF BONES (Fig. 5–1)

Each bone is a distinct structural unit. However, all osteoid tissue has the same basic composition. Its pattern varies according to four basic skeletal functions:

1. Long Bones. Long bones are the largest in the body. Included in this group are the prominent bones of the arms and legs. Each has a shaft and two ends, as previously described. The general function of long bones is to act as levers. Examples are the *femur* of the leg and the *humerus* of the upper arm.

2. Short Bones. These differ from long bones in that they are smaller and have less prominent ends. They are basically cuboid in shape, and are composed of cancellous bone enclosed within a thin shell of compact bone. Their principal function is to provide strength. Examples are the *carpals* of the wrist and the *tarsals* of the ankle.

3. Flat Bones. These are platelike and usually have two broad surfaces (tables) and narrow edges by which one flat bone articulates with another. Each flat bone is thin, and is made up of two plates of compact bone that enclose between them a layer of cancellous bone. The general function of these bones is protection. Some examples: the *ribs,* the *scapula* at the back of the shoulder, and the *cranium.*

4. Irregular Bones. These are so named because of their blocklike, irregular shapes. Several surfaces may articulate with several other bones. Some of the bones of the skull are irregular, as are the vertebrae and the hip bones. The thinner parts of irregular bones consist of two thin plates of compact bone with cancellous bone between them. The general function of irregular bone is articulation.

A separate category of bone is the *sesamoid* (seedlike) bone. These bones develop in the capsules of joints or in tendons. They are small and round and are covered with cartilage or ligaments. Their function is to eliminate friction. An example is the *patella*, or knee cap.

THE SKELETON

Although the skeleton is a single functional entity, it may be divided for purposes of morphologic description into an *axial* skeleton and an *appendicular* skeleton (Fig. 5–9).

The Axial Skeleton

The divisions of the axial skeleton are the bones of the vertebral column, the thorax, and the skull. We will discuss each of these in some detail.

The Vertebral Column

The vertebral column or "backbone" (Fig. 5–10) is a long, flexible pillar that supports the trunk and neck in an upright position but also allows the body to make bending and twisting movements. In addition to these functions, the column protects the spinal cord and the roots of the spinal nerves (Fig. 5–11, *B*).

When viewed from the lateral aspect, a normal vertebral column exhibits four curves (Fig. 5–10, *C*). The *thoracic* and *sacral* curves are present at birth; the *cer-*

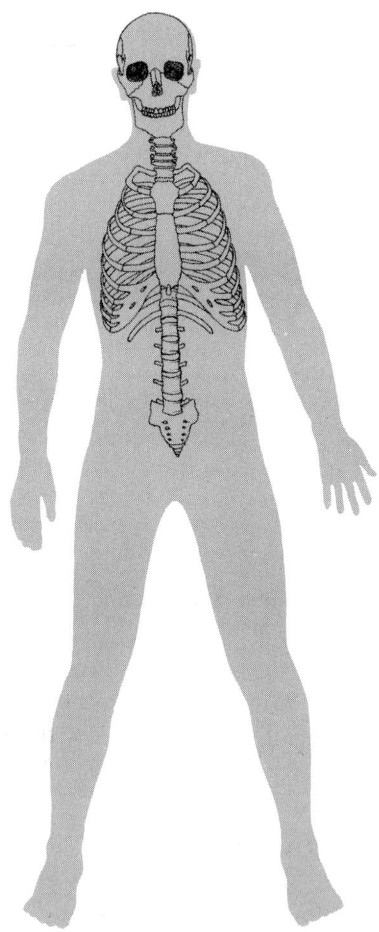

Figure 5–9 The axial skeleton. The bones of the head and trunk form the axial skeleton.

vical curve appears when the infant begins to hold up his head. This usually occurs at about three months, and the curve becomes more pronounced when the child begins to sit up—at about nine months. The last curve, the *lumbar* curve, appears when the child begins to walk. These curves provide the vertebral column with the resilience and spring necessary for such actions as walking and jumping.

The column is formed of a series of 26 *vertebrae* (see below), which are separated and cushioned by *intervertebral disks* or cartilages (Fig. 5–10, *B*). An inner canal is formed by a succession of *foramina* (passages) in the individual vertebrae and by the ligaments and disks connecting them.

The Vertebrae

All of the vertebrae are constructed on the same basic plan, although in different regions of the column there are characteristic specializations. A typical

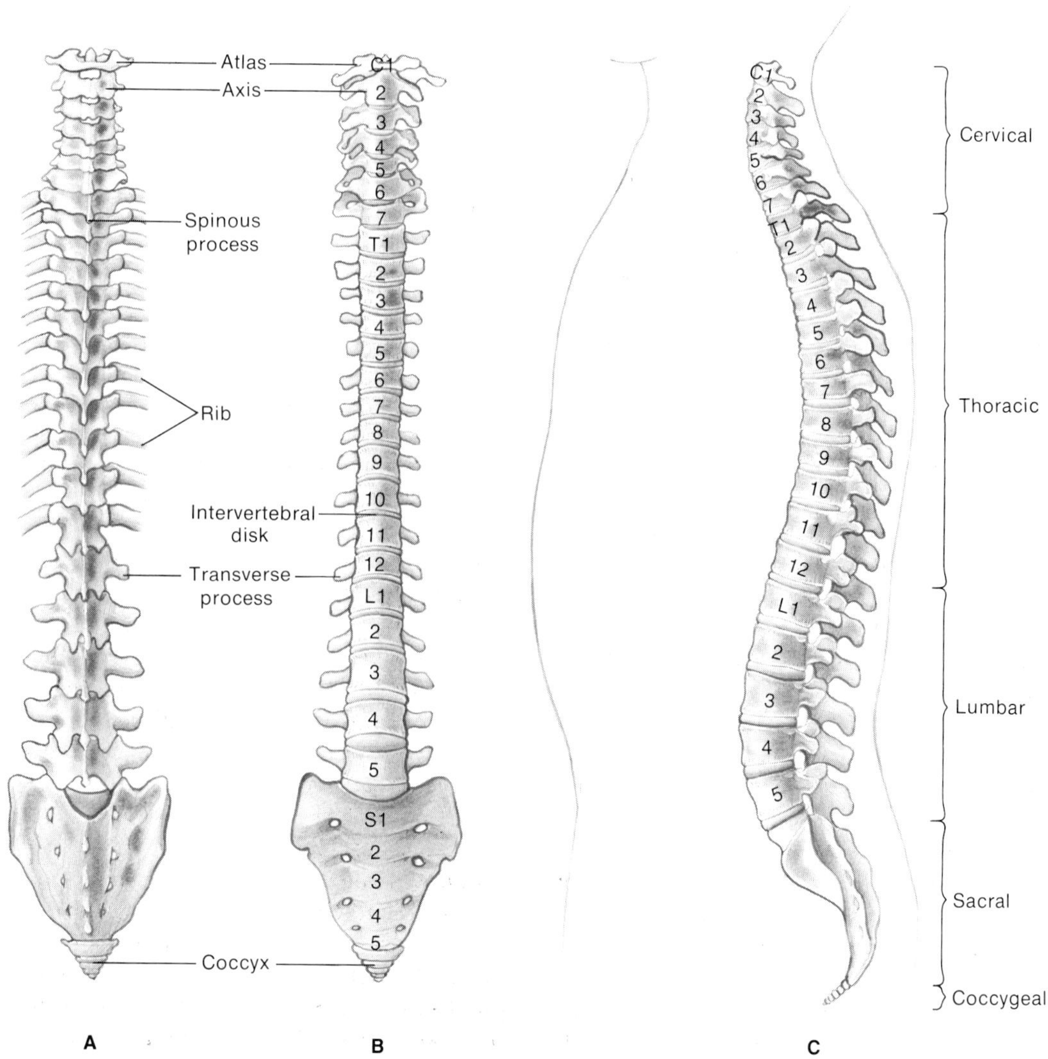

Figure 5–10 The vertebral column. *A*, Posterior aspect. *B*, Anterior aspect. *C*, Lateral aspect, showing curvatures.

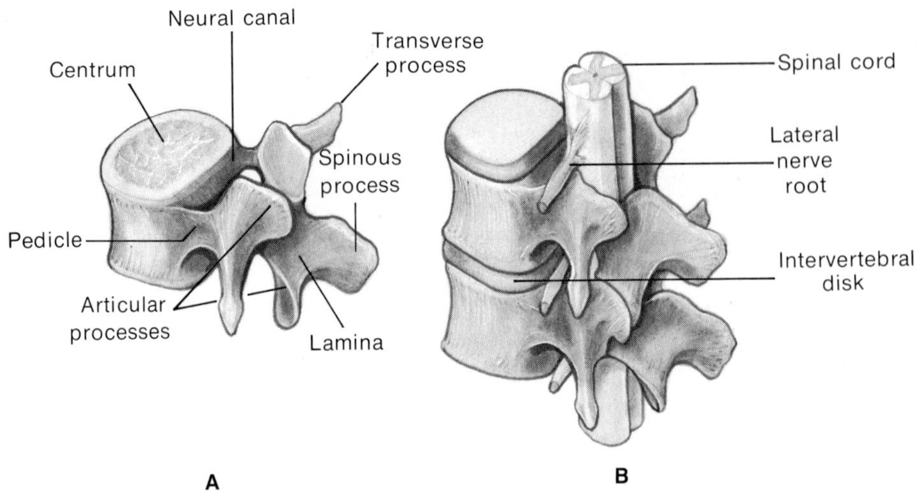

Figure 5–11 *A*, Diagonal view from the top of a typical vertebra. *B*, Two vertebrae held in position by an intervertebral disk to show position of the spinal cord and its peripheral nerves.

vertebra (Fig. 5–11, *A*) consists of five principal parts:

1. The *body* (or centrum) is the thick, disk-shaped anterior portion. Its upper and lower surfaces are roughened to allow for the attachment of intervening disks of fibrocartilage. Its anterior edge is pierced by small holes through which blood vessels enter to nourish the bone.

2. The *arch* projects dorsally from the body, enclosing a space (the *neural canal*) for the passage of the spinal cord. The arch bears three processes for muscle attachment: a *spinous process,* directed dorsally, and two *transverse processes,* one on either side of the vertebral canal or foramen.

3. Each vertebra has four *articular processes:* two superior and two inferior. These have smooth, curved surfaces for articulation with the vertebrae immediately above and below.

4. Two *pedicles* originate from the vertebral body. They are notched to allow passage of nerves to and from the spinal cord.

5. The *lamina* forms the roof (posterior wall) of the vertebral canal. It is the weakest point of the vertebra, and frequently a single lamina or several laminae must be removed surgically (laminectomy) following a traumatic injury to the column. This operation relieves pressure on the spinal cord, and may aid in the fusion of two or more adjacent vertebrae.

The column, which extends the full length of the back, is made up of three groups of movable vertebrae and two groups of fixed vertebrae (Fig. 5–10, *B*). The movable vertebrae include the seven *cervical* (in the neck), the 12 *thoracic* (in the thorax), and the five *lumbar* (in the loin). These remain separate throughout life. The fixed vertebrae are the five *sacral* and the four *coccygeal.* In the adult, the sacral vertebrae have fused to form the *sacrum,* and the coccygeal vertebrae have united to form the *coccyx* (tailbone). That is why we speak of man as having 26 vertebrae, not 33.

Cervical Vertebrae. These are the smallest vertebrae and are characterized by oblong bodies and very wide transverse processes. Each transverse process is pierced by a foramen through which the vertebral artery passes (Fig. 5–14). The spinous processes of the third, fourth, fifth, and sixth cervical vertebrae are forked (bifid) to cradle the strong ligaments of the head.

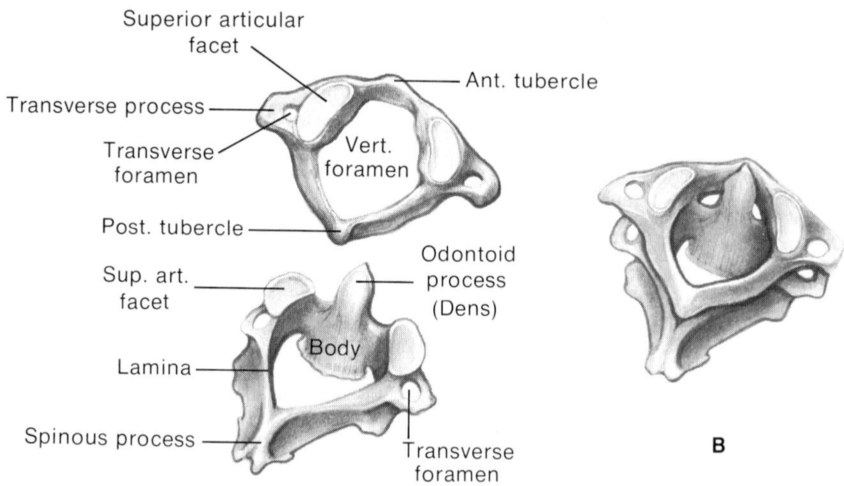

Figure 5–12 *A*, The atlas and axis, the first and second cervical vertebrae, seen diagonally from behind. *B*, Diagrammatic positioning of the dens of the axis up into the large vertebral formation of the atlas. This permits rotation of the atlas on the axis.

The first two cervical vertebrae are unique. The second, the *axis* (Fig. 5–13), has a small body and a projection of the centrum called the *dens* or *odontoid process*. This process passes through a canal in the first cervical vertebra to form the axis of rotation of the skull and the first cervical vertebra.

The first cervical, called the *atlas* (Fig. 5–12), supports the head by articulating with the condyles of the occipital bone (p. 77). This vertebra is a bony ring with no body and short, winglike transverse processes. Because of its articulation with the condyle, it is possible for a person to rock his head back and forth on the atlas. As he turns his head from side to side, the atlas moves with the cranium, and the two turn on the axis, using the odontoid process as a center of rotation.

The seventh cervical vertebra has a very prominent spinous process with a tubercle at its tip called the *vertebra prominens*, which can be seen and felt at the base of the neck.

Thoracic Vertebrae. The twelve vertebrae of this group increase progressively in size from the base of the neck to the lumbar region. Thoracic vertebrae have two distinguishing features: a long, spinous

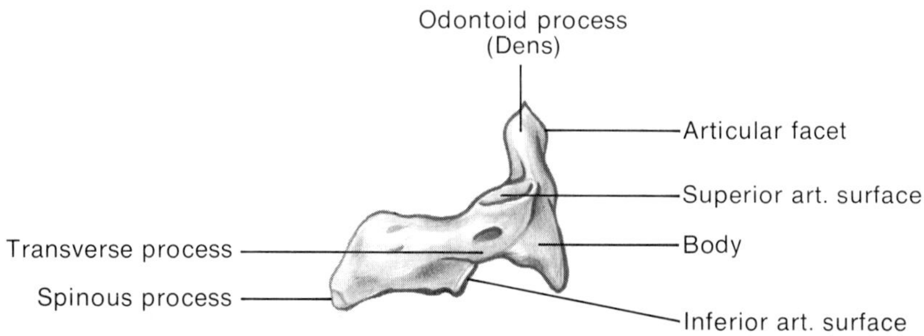

Figure 5–13 The axis, seen from the right side.

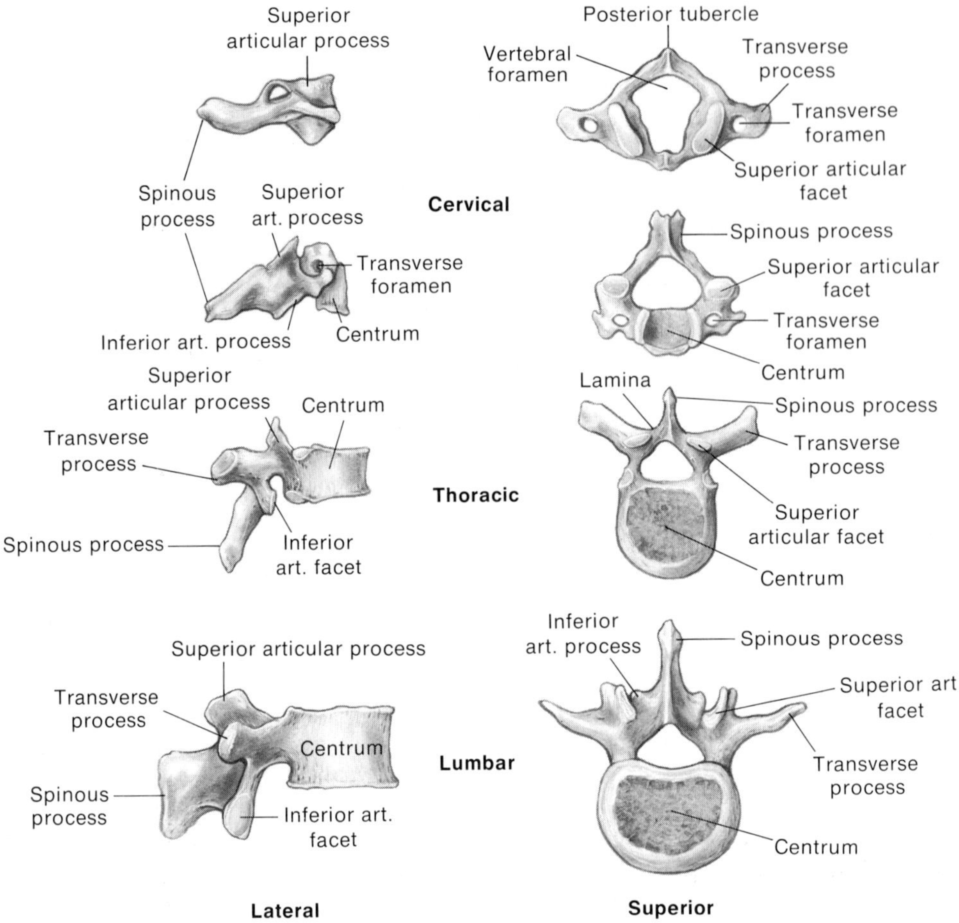

Figure 5–14 The atlas, lateral and superior aspects; a cervical, a thoracic, and a lumbar vertebra. Note the distinction between the processes and the facets: a process is a prominence or projection of bone; a facet is a small plane surface on a bone at the site where it articulates (joins) with another structure. A facet may or may not be situated on a process.

process, pointed and directed downward; and six articular facets—three on each side—which provide attachments for the ribs (Fig. 5–14).

Lumbar Vertebrae. These are the largest and strongest of the vertebrae. They lack spinous processes, but have short, thick projections for attachment of the powerful muscles of the back.

Sacrum. This triangular, curved bone (Fig. 5–15) forms the dorsal aspect of the pelvis and is wedged between the two hip bones.

Coccyx. The coccyx articulates with the tip of the sacrum and is slightly movable. This property permits an increase in size of the birth canal during delivery. In some newborn infants, the coccyx may curve outward, but it can be removed surgically without major discomfort at this time.

Injuries and Diseases of the Vertebral Column

The normal curves of the spine (Fig. 5–16, *A*) can become exaggerated as a result of injury, poor posture, or disease. When the posterior curvature is exagger-

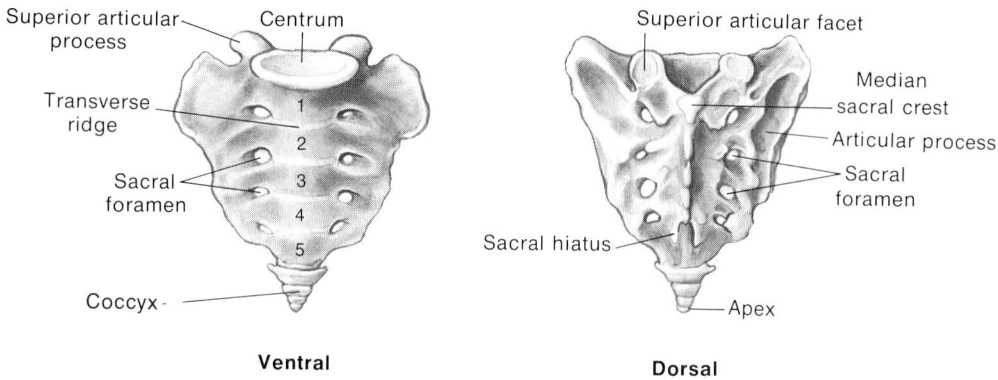

Figure 5-15 The sacrum and coccyx, dorsal and ventral aspects.

ated in the thoracic region, the condition is called *kyphosis* (hunchback). An exaggerated anterior curvature in the lumbar region causes *lordosis* (swayback). In *scoliosis* (curvature of the spine), there is a lateral curvature associated with such other abnormalities as herniation, osteoid tumors, and slipped intervertebral disks (Fig. 5-16, *B*).

The cartilaginous intervertebral disks between each pair of vertebrae (Fig. 5-17, *A*) act as cushions and shock absorbers. The vertebrae are bound together by strong ligaments, thick fibrous bands that strengthen joints. These are somewhat extensible, and the disks are compressible; therefore, movements of flexion, extension, abduction, and rotation are possible (p. 103). When the spinal column is subjected to violent trauma, fractures and dislocations can result. The most common fracture is to the lamina of the vertebral body itself. In addition to such immediate shocks, the complication to be dreaded most is injury to the spinal cord, which sometimes may be completely severed.

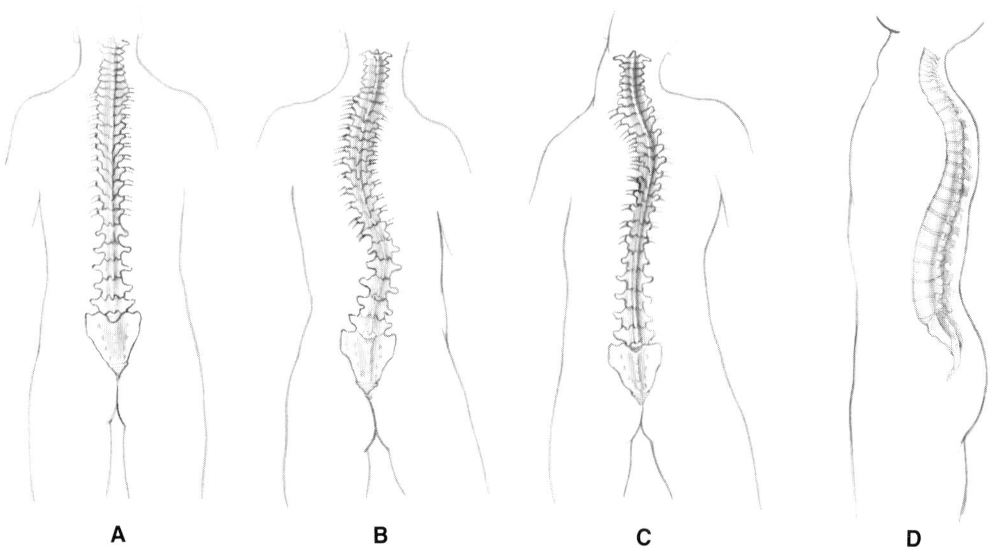

Figure 5-16 Dorsal and lateral positioning of the vertebral column. *A*, Dorsal view of a normal vertebral column. *B*, Scoliosis. *C*, Lordosis (swayback). *D*, Lateral view of an exaggerated posterior thoracic curvature—kyphosis (hunchback).

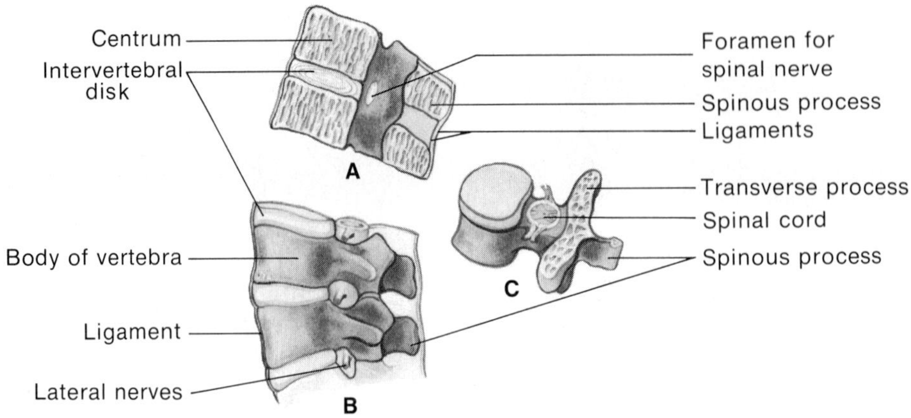

Figure 5–17 The vertebral column. *A*, Sagittal section through two vertebrae, showing the position of ligaments, intervertebral disks, and intervertebral foramina. *B*, Lateral aspect. *C*, Posterosuperior aspect, showing the position of the spinal cord and lateral nerves.

Such damage to the cord causes paralysis and loss of sensation below the point of injury. Thus, if the injury is high in the cervical region, respiration will no longer be possible, and death is inevitable.

The intervertebral disk is made up of a tough, fibrocartilaginous outer layer *(annulus fibrosus)* and a soft, putty-like interior material *(nucleus pulposus)*. If the annulus becomes injured or punctured, the pressure within the disk is decreased, and the nucleus may protrude (Fig. 5–18, *B*). Such a herniation can cause pressure on the spinal nerve associated with the affected vertebra, resulting in nerve root irritation (shooting leg pains) and possible compression fractures of the articular surfaces on the vertebrae above or below the injured disk.

Tuberculosis not infrequently attacks the bodies of the vertebrae, practically eating them away. If the disease is untreated, the bones eventually become so weak that they are crushed by the weight of the body parts above them, and a hunchback (Fig. 5–16, *D*) is produced. Such conditions

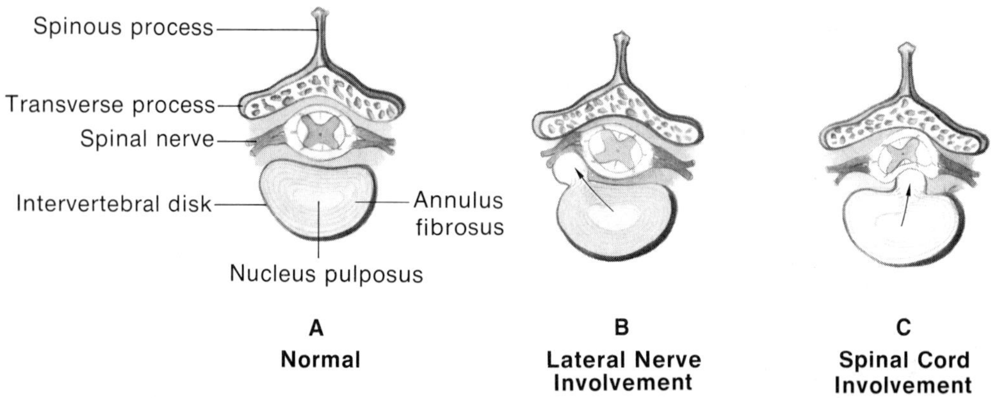

Figure 5–18 Spinal vertebrae, superior aspect. *A*, Normal positioning of the spinal cord within the vertebral canal. *B* and *C*, Diagrams of ruptured intervertebral disks exerting pressure on the lateral nerve and spinal cord, respectively. Arrows indicate protrusion of the nucleus pulposus after injury to the annular rings surrounding it.

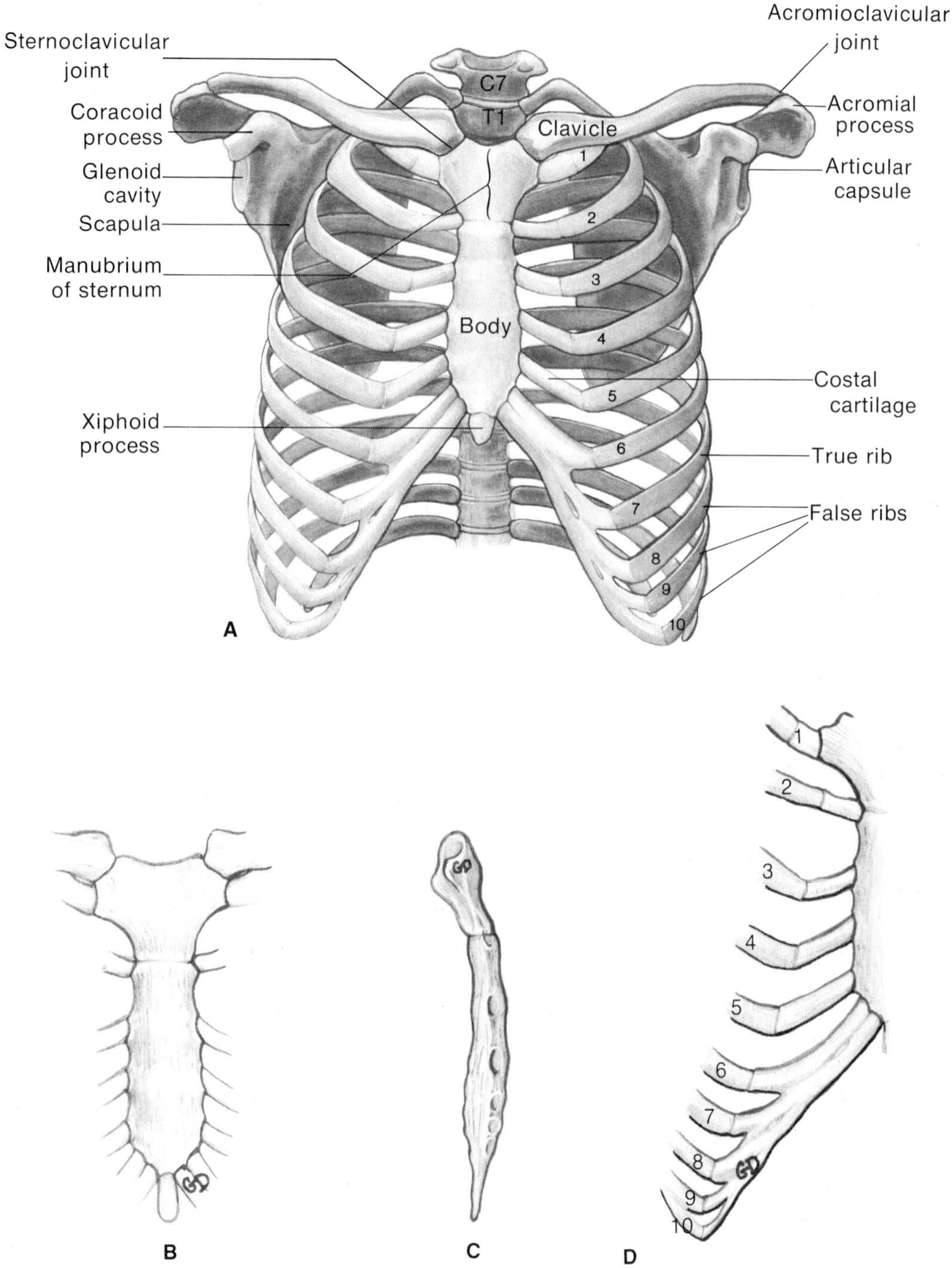

Figure 5–19 *A*, The rib cage, anterior aspect. *B*, Anterior surface of the sternum. *C*, Lateral view of the sternum. *D*, Costal cartilage, anterior aspect.

have become relatively rare owing to the more effective treatment of tuberculosis in children.

The Bones of the Thorax

The bony structure of the thorax, or chest, is made up of four principal groups: the sternum, the costal cartilages, the ribs, and the bodies of the thoracic vertebrae. These bones and cartilages are covered with muscles and skin; the floor of the thorax is formed by the diaphragm. As a functional unit, the thorax protects and supports the heart and lungs, plays a leading role in respiration, and helps support the bones of the pectoral (shoulder) girdle. As mentioned earlier, red blood corpuscles are formed in the red marrow of the ribs and sternum. The thorax in the adult is cone-shaped and has a broad muscular base.

Sternum (Fig. 5–19, *B* and *C*). This bone, sometimes called the breast bone, is centered at the front of the rib cage at the midline of the thorax. It is shaped like a sword with a short handle: the "handle" at the top is the *manubrium;* lying below is a long "blade," the *body* of the sternum (or *gladiolus*), then the tip, the *xiphoid process.* No ribs are attached to the xiphoid process, the manubrium and the gladiolus, however, are notched on either side to provide attachment for the first seven costal cartilages. The upper end of the sternum articulates with the clavicle and with the first rib on each side. The diaphragm, linea alba, and rectus abdominus muscles are attached to the xiphoid.

Costal Cartilages (Fig. 5–19, *D*). At the front of the thorax, each of the fixed ribs is attached in a different manner. However, each ends some distance short of the sternum and is continued by a piece of hyaline cartilage, the costal cartilage. The first through the seventh pairs of costal cartilages attach to the sternum and contain small *synovial cavities* (not shown in Figure 5–19). These cavities disappear in old age. The eighth, ninth, and tenth costal cartilages do not reach the sternum; instead, each attaches to the cartilage above.

Ribs (Figs. 5–19, *A* and 5–20). There are 12 pairs of ribs, forming the bony portions of the thoracic walls. These are long, slender, curved bones. Posteriorly, they are attached to the thoracic vertebrae; anteriorly, to the costal cartilages. The first seven pairs, as just described, are attached to the sternum by separate costal cartilages, and are called "true" ribs. The remaining five pairs are "false" ribs. Of these, the eighth, ninth, and tenth pairs are attached by their costal cartilages to the costal cartilage immediately above; the eleventh and twelfth have no cartilage attachments to the sternum and are called "floating" ribs.

A typical rib (Fig. 5–21, *A* and *B*) has a head, the larger facet of which articulates with the body of the corresponding thoracic vertebra; a short neck; a protuberance from the neck, or *tuberosity,* which articulates with the transverse process of the vertebra; and a shaft. The upper ribs are shorter than the lower ones, except for the floating ribs. From their posterior vertebral attachments, the curved ribs slope at a downward angle as well as outward, thus increasing the size of the thoracic cavity. The conical thorax is wider below than it is above, and is wider laterally than anteroposteriorly (front to back).

Clinical Considerations. Fracture of one or more ribs is relatively common (Fig. 5–21, *C*). Depending on the site and nature of the trauma, the rib or ribs will break—at the point of impact when hit by a small instrument with considerable force, or at the angle of the rib when a blow is applied over a large area on the front or back of the chest. In either case, the tips of the ribs will be driven inward, causing a decrease in thoracic space and an increase in pressure within the cavity. A free fractured rib tip may cause multiple abrasions

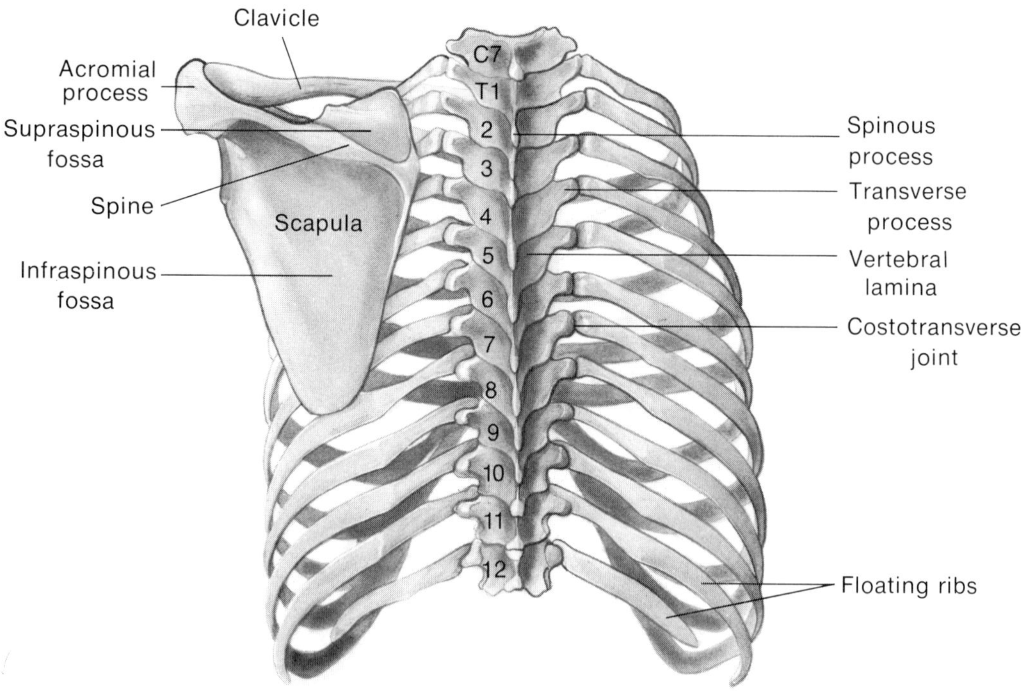

Figure 5–20 Rib cage and scapula, posterior aspect.

and possible puncture of a lung. Immediate attention is required to prevent a possible collapsed lung.

The Bones of the Skull

The skull provides the skeletal structure of the face and head. It is composed of 21 separate bones, which are joined together and therefore move as a unit, plus one freely movable bone, the *mandible* (lower jaw). Eight of the skull's bones form the *cranium;* the remaining 14 are *facial bones*. Except for the mandible and the ossicles of the ears, the bones of the adult human skull are immovably united by *sutures*. The bones of the cranium enclose and protect the brain and its associated structures, the special sense organs.

Cranial Sinuses. Within the bony structure of the cranium are cavities called air sinuses (Fig. 5–22), which communicate with the nasal passageways. Four of these sinuses, the *paranasal sinuses,* communicate with each nasal cavity: the *frontal* and *ethmoidal* open into the nasal cavity; the *sphenoidal* opens into the nasopharynx; and the *maxillary* opens on the lateral walls of each nasal passage. These sinuses are lined with a mucous membrane that serves also as periosteum and is therefore called *mucoperiosteum*. This membrane is continuous with the mucous membrane that lines the air passages. Inflammation of the mucoperiosteum may extend into any one of the sinuses, causing *sinusitis*. The sinuses are important as resonance chambers in the production of voice; furthermore, being hollow, they decrease the weight of the skull.

Cranial Sutures. As we have noted, the individual bones of the cranium are

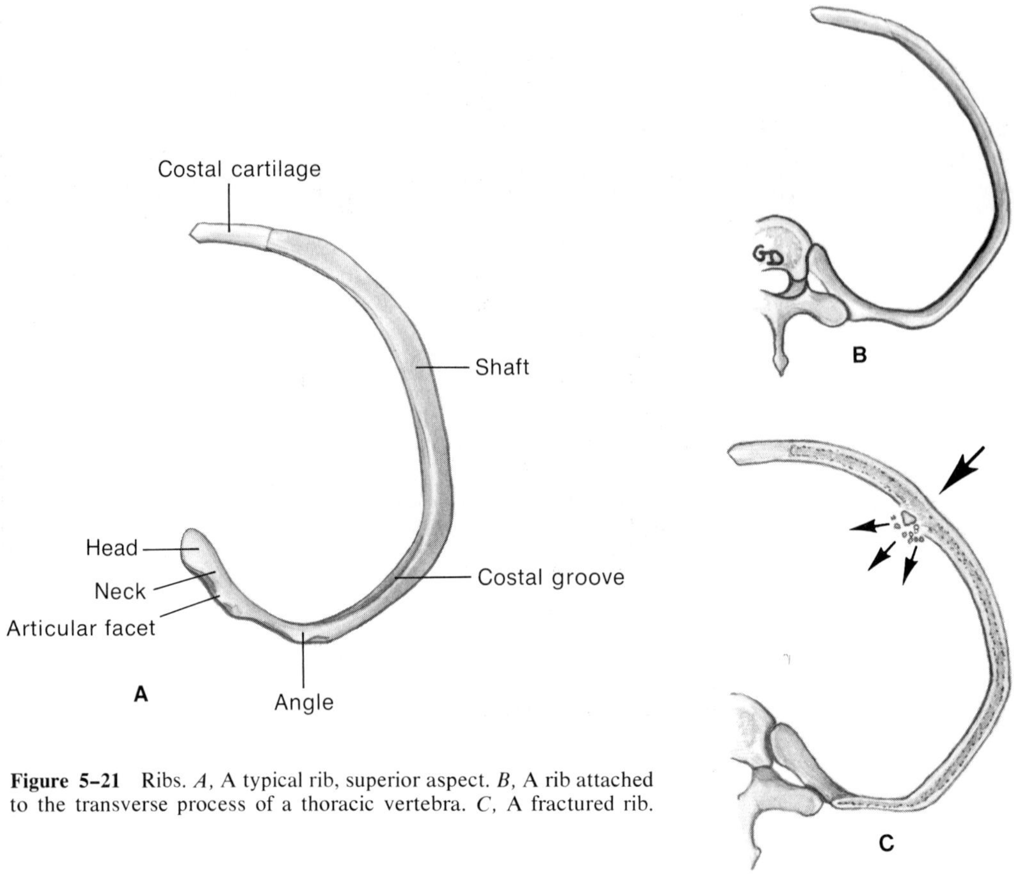

Figure 5–21 Ribs. *A*, A typical rib, superior aspect. *B*, A rib attached to the transverse process of a thoracic vertebra. *C*, A fractured rib.

united at sutures, or junction lines (Fig. 5–23). There are four of these:

1. The *coronal* suture, located between the frontal and parietal bones.

2. The *lambdoid* suture, which joins the parietal and occipital bones.

3. The *squamous* suture, located between the temporal and parietal bones.

4. The *sagittal* suture, which unites the right and left parietal bones.

Fontanels. At birth there are several places in which fusion of the cranial bones is not complete. These "soft spots" are called *fontanels* (Fig. 5–23). Ossification of these bones takes place gradually from ossification centers generally located near the center of what will become the future completed bones. Since the process of ossification is not complete in the newborn infant, the membrane-filled fontanels occupy the spaces between the bones.

1. The *anterior* (bregmatic) is the largest of the fontanels and is situated between the angles of the two parietal bones and the two segments of the frontal bones. Normally this fontanel closes at about 18 months of age.

2. The *posterior* (occipital) fontanel is a triangular space between the occipital bone and the two parietal bones. It usually closes a few months after birth.

3. The two *anterolateral* (sphenoidal) fontanels are at the junction of the frontal, parietal, temporal, and sphenoid bones. These generally close two or three months after birth.

4. The two remaining fontanels, the *posterolateral* (mastoidal), join the parie-

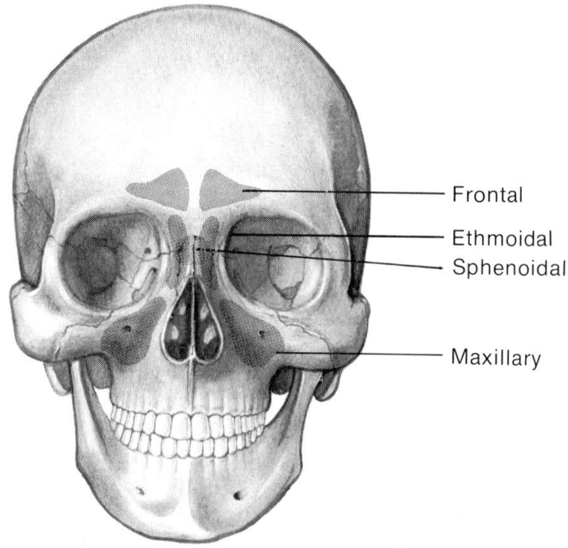

Figure 5–22 The sinuses of the skull.

tal, occipital, and temporal bones. They decrease in size as the infant grows, but do not close completely for about a year.

In abnormal conditions a fontanel may close too early or too late. In some cases, brain growth may be retarded by the extreme pressure caused by premature fusion of the bones *(microcephalus)*. When fusion

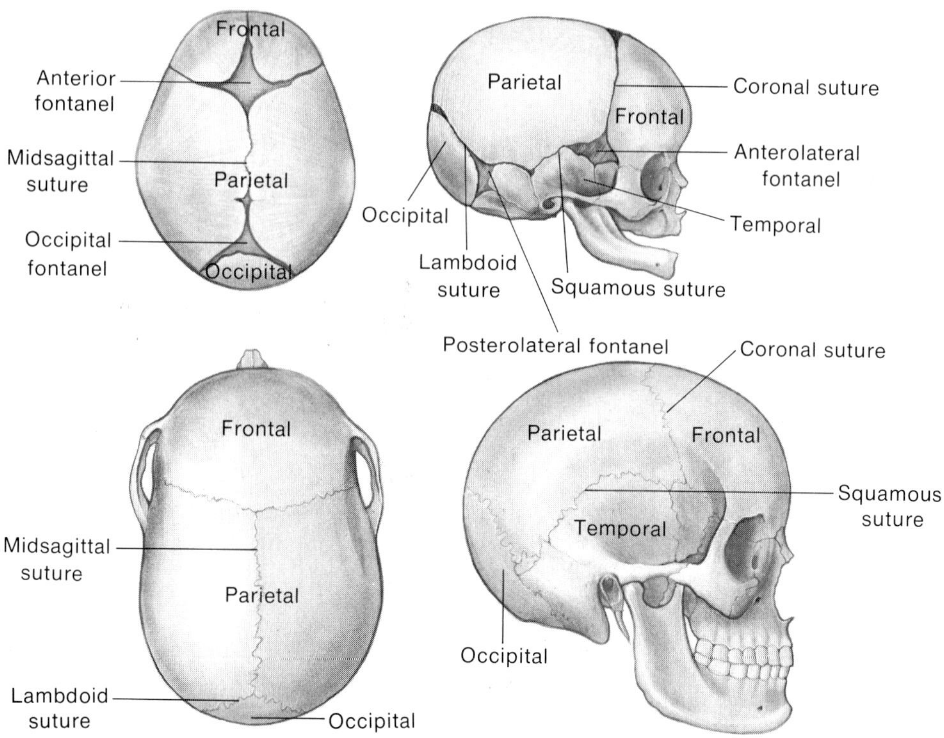

Figure 5–23 Sutures and fontanels of the fetal and adult skulls. *A*, Superior aspect. *B*, Lateral aspect.

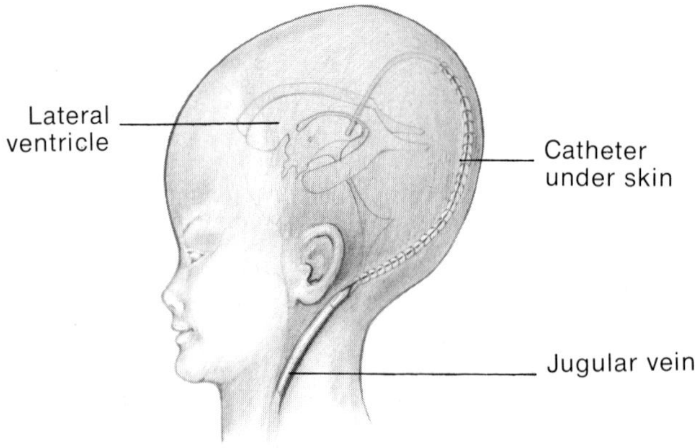

Figure 5–24 Intercranial catheterization: surgical procedure for drainage of spinal fluid from the lateral ventricle into the jugular vein to relieve excessive pressure on the brain due to hydrocephalus.

of the cranial bones is delayed because of the high internal pressure of an increased volume of spinal fluid in the brain area, the bones "float" on the stretched membranes covering the brain, giving the appearance of a large water sac *(hydrocephalus)* covered with loose cranial bones and skin. This condition can now be alleviated by drainage of the excess cerebrospinal fluid from the brain into the jugular artery of the neck by a catheter (Fig. 5–24), thereby facilitating maintenance of a normal cranial pressure while careful diagnosis and treatment of the causative factors are undertaken.

The membranous tissue between the

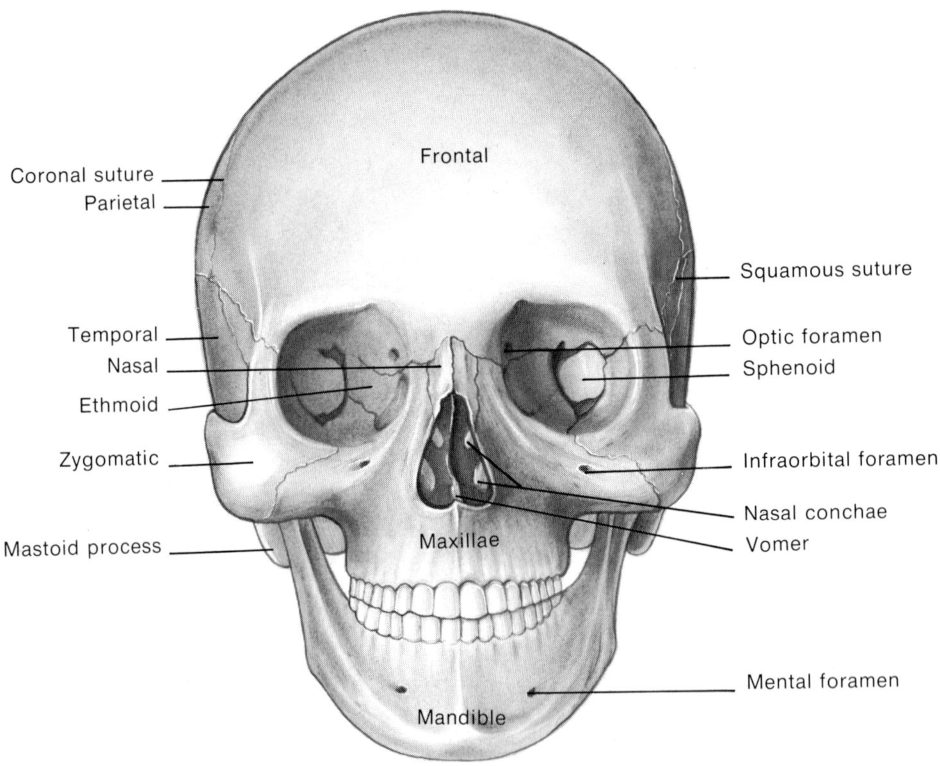

Figure 5–25 Frontal, lacrimal, zygomatic, maxillary, and nasal bones of the skull.

cranial bones at the sutures allows an overlapping during birth, thus reducing the diameter of the skull. This is called "molding," and accounts for the elongated shape of the newborn head, particularly when labor has been long or difficult.

Anterior Aspect of the Skull (Fig. 5–25). Looking at the skull from the front, one can see most of the bones of the face and some of the cranial bones.

The *frontal* bone forms the forehead and part of the roof of the orbits of the eyes and the nasal cavity. The arches formed by part of the frontal bone over each eye are sharp and prominent, and are called the *supraorbital margins* (not shown in Figure 5–25). Beneath these, inside the bone, are cavities, the frontal sinuses (Fig. 5–22), which communicate with the nasal cavity. In the upper and outer angle of each orbit are two depressions called *lacrimal fossae* (Fig. 5–28), in which lie the lacrimal glands that secrete tears.

The upper rim of the orbits, the bony sockets for the eyeballs, is formed by the frontal bone. The medial and lower rims are formed by the paired *maxillae,* which are facial bones (see below). The lateral or outer rim of the orbit is formed by the *zygomatic* bone, and the posterior wall by the *sphenoid* bone (see p. 83). The medial wall is also formed by part of the *ethmoid* bone, with the small *lacrimal* bone in the medial corner (Fig. 5–25).

The ethmoid bone (Figs. 5–25 and 5–26) is a light, cancellous bone consisting of a horizontal or cribriform plate, a perpendicular plate, and two lateral masses or labyrinths. The horizontal plate forms the roof of the nasal cavity and the base of the cranium. It is pierced by numerous openings or *foramina,* through which the olfactory nerve fibers pass from the mucous membrane of the nose to the olfactory bulb of the brain.

The *nasal* bones are two small, oblong bones side by side at the middle and upper part of the face that form the upper portion of the bridge of the nose. The lower portion is formed by the nasal cartilages.

The single *vomer* (Fig. 5–29) is a bone placed at the lower and back part of the nasal cavity, forming part of the central nasal septum. It is very thin, and varies in different individuals, often being bent to one side or the other, thus making the nasal chambers of unequal size (deviated septum).

The *inferior nasal conchae* (Fig. 5–25) are situated on the lower, lateral wall of each nostril, and consist of a layer of thin, cancellous bone curled up like a

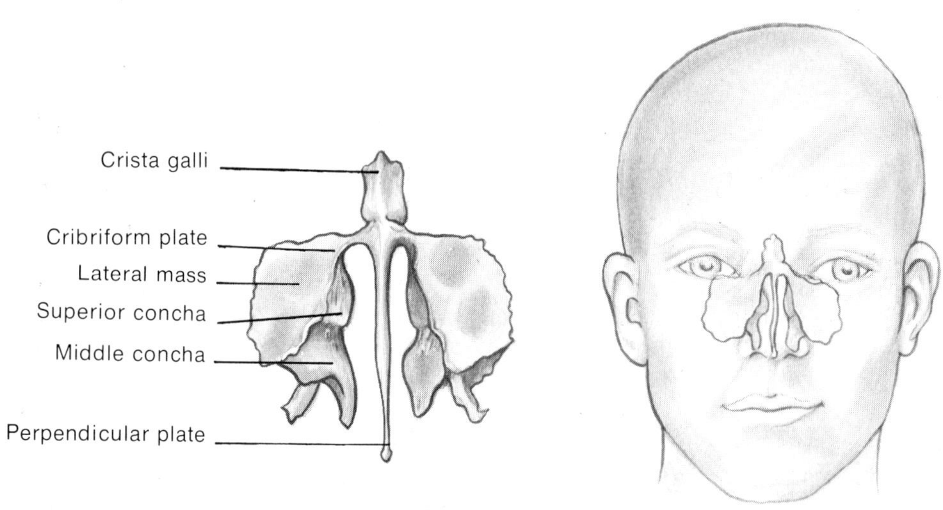

Figure 5–26 The ethmoid bone, dorsal aspect.

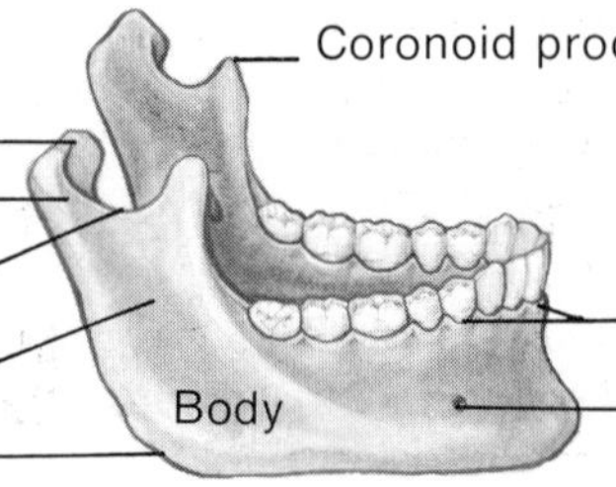

Figure 5–27 The mandible.

scroll. Structural deviations and abnormal conditions of these bones and of the vomer are involved in some of the more common facial abnormalities.

The two *maxillae,* right and left (Fig. 5–25), are joined at the intermaxillary suture in the median plane of the bone to form the upper jaw. Fusion of the maxillae generally occurs before birth, but when the two do not unite to form a continuous bone, a defect known as cleft palate (a depression of the tissue) occurs, usually associated with a cleft or "hare" lip. The maxillae also form part of the floor of the orbits, most of the roof of the mouth, the floor and lateral walls of the nasal cavity, and part of the wall of the nasolacrimal duct. Each maxilla consists of a *body*, which contains the large maxillary sinus, and four *processes*. The *alveolar* process supports the upper teeth, and the two *horizontal* (palatine) processes form the anterior and larger part of the hard palate.

The *mandible*, or lower jaw, is the largest bone of the face. It consists of a horseshoe-shaped body and two perpendicular portions, the *rami* (Fig. 5–27). The lower teeth occupy the upper margin of the body. There are two processes at the top of each ramus. The posterior process is the *condylar*; it articulates with the temporal bone to form the temporomandibular joint. The anterior *coronoid* process attaches the temporalis muscle and some of the fibers of the buccinator muscle. The deep depression between these two processes is called the *mandibular notch*. The *mental foramen,* just below the first molar tooth, serves as a passageway for the inferior dental nerve, a branch of the mandibular nerve. Branches of the nerve supply the molar and premolar teeth of the lower jaw. The mental foramen also provides a passage for blood vessels. The angle of the mandible is formed by the posterior border of the ramus and the lower border of the body.

At birth, the mandible consists of two parts that join at the *symphysis* in front to form one bone, usually during the first year of life. As the teeth erupt, the child chews, and the depth of the mandibular body increases owing to the growth of the alveolar border. In the elderly, the alveolar portion of the mandible ceases to grow. If the teeth are finally lost, the *alveoli,* or tooth sockets, become absorbed by the body of the bone, and the chin develops an angle, appearing more prominent.

Lateral Aspect of the Skull (Fig. 5–28). In this view, parts of the frontal, parietal, temporal, occipital, and sphenoid bones may be seen.

The *parietal* bones, right and left, join to form the greater part of the sides and roof of the skull. Their external surface is convex and smooth; the internal surface is concave and presents eminences and depressions that accommodate the convolutions of the brain and the numerous blood vessels that supply the meninges (Figs. 8–11 and 8–15), the membranous sheath of the brain. The two parietal bones join at the sagittal suture in the midline and at the coronal suture between the lines of articulation of the parietal and frontal bones.

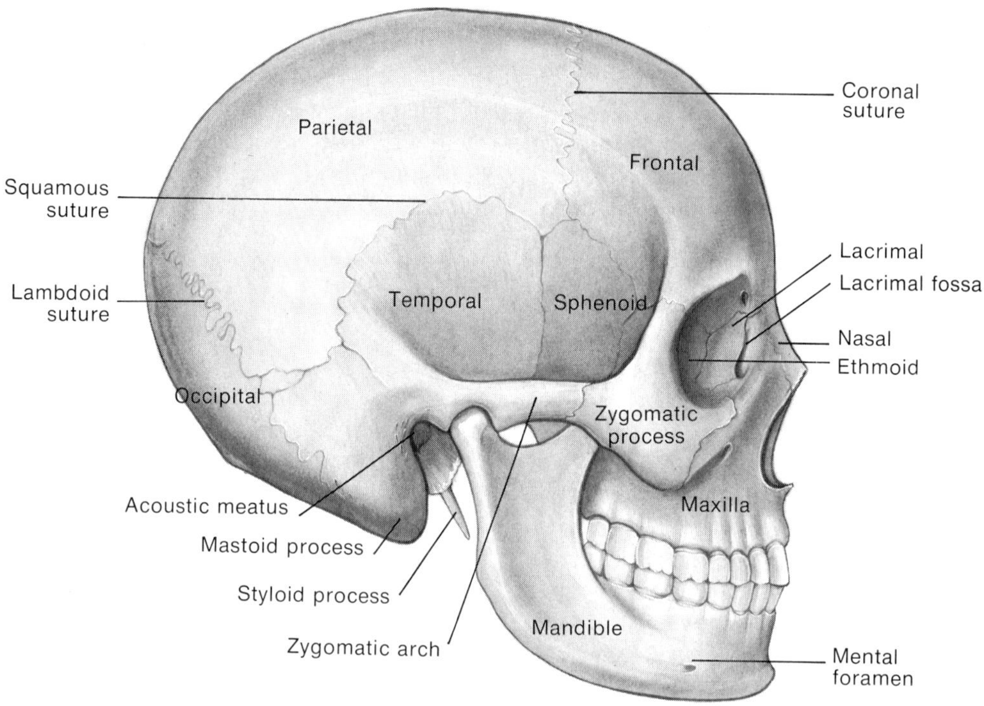

Figure 5–28 The skull, lateral aspect.

The *temporal* bones, which like the parietal bones comprise right and left components, are at the sides and base of the skull. They enclose the internal structures of the ear and articulate with the mandible. Each temporal bone is made up of four parts: squamous, petrous, mastoid, and tympanic. The *squamous* portion is the largest of the four and is in the most superior position. It is a scalelike plate of bone at the side of the skull, and forms the temple. Projecting forward from its lower part is the *zygomatic process,* which forms the lateral part of the zygomatic arch or cheek bone. Most of the interior surface of the squamous portion is occupied by the mandibular fossa, which receives the condyle of the mandible for articulation during chewing and speech.

The *petrous* portion of the temporal bone, shaped like a three-sided pyramid, contains the ear structures. On its cranial side there is a foramen, the *internal acoustic meatus,* which allows the passage of the facial, cochlear, and vestibular nerves. The petrous portion is located deep within the base of the skull between the sphenoid and occipital bones, and contains both the inner ear and parts of the middle ear within its complex cavities. The *styloid process* (Fig. 5–28) projects downward from the inferior surface of the petrous portion of the temporal bone.

The *mastoid* portion is located behind and below the *external acoustic (auditory) meatus,* the external opening of the ear (see below). The *mastoid process* (Fig. 5–28) projects downward from its inferior border. It is a rounded protuberance of the temporal bone which serves as a means of attachment for neck muscles. The mastoid portion contains air spaces called *mastoid air cells* or *sinuses*; these are separated from the brain only by thin, bony partitions. Inflammation of the cells of the mastoid (mastoiditis) is not uncommon, and when untreated is dangerous because of possible invasion of the brain or the meninges by the causative bacteria or viruses.

The *tympanic* portion is a curved plate, the upper surface of which forms the anterior and inferior walls and part of the posterior wall of the external acoustic meatus—a 15-mm canal connecting the outer with the middle ear.

The *occipital* bone is situated at the back and base of the skull. It joins the parietal bones anteriorly at the lambdoid suture. The inferior portion of the bone has a large opening, the *foramen magnum* (Fig. 5–29), through which the spinal cord passes. It is at this level that the spinal cord joins the medulla oblongata of the brain. At the sides of the foramen magnum on the external surface are two rounded projections or *condyles*, which articulate with the first cervical vertebra, the atlas. The atlanto-occipital joint between the skull and the atlas permits backward and forward movements of the head. Other obvious projections from the occipital bone are the external occipital crest and the external occipital protuberance. These can be felt at the base of the neck, and serve as a means for the attachment of muscles and ligaments.

Inferior Aspect of the Base of the Skull (Fig. 5–29). When the skull is viewed from below, the large foramen magnum, the openings of the nasal cavity, and the palatine bone (hard palate) may be seen. The bony roof of the mouth is made up of the hard palate and part of the maxillae and the palatine bones. The hard palate separates the mouth from the nasal cavity and also serves as the floor of the nasal cavity.

Superior Aspect of the Base of the Skull (Fig. 5–30). With removal of the top of the cranium and its contents, the superior structures of the base of the skull may be studied. Once again, the most obvious

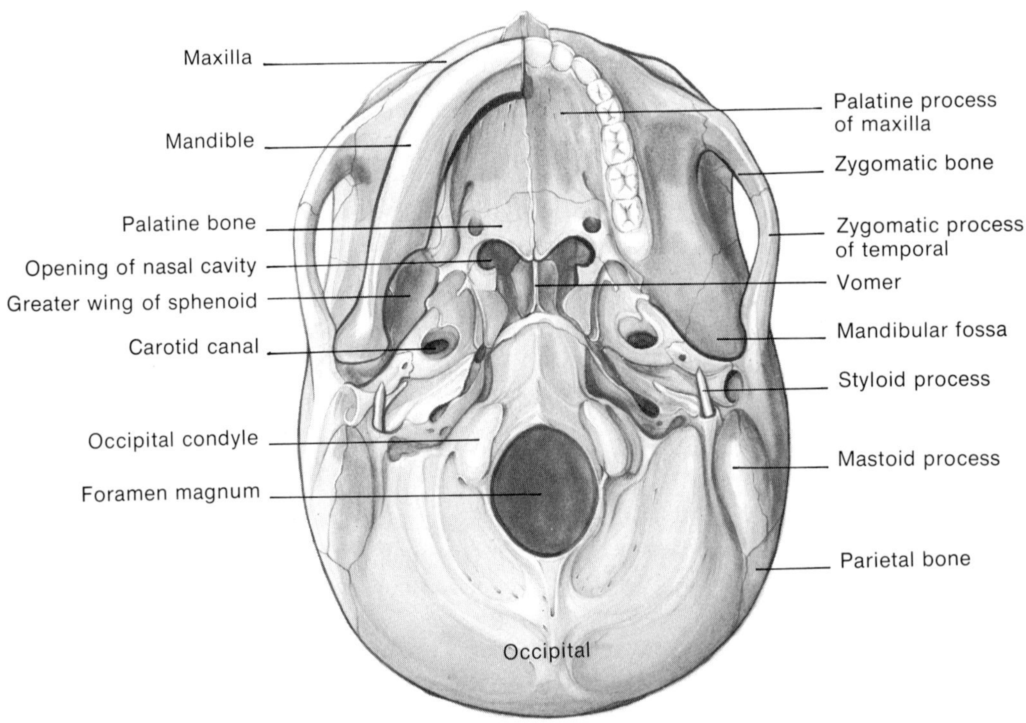

Figure 5–29 External surface of the skull, inferior aspect.

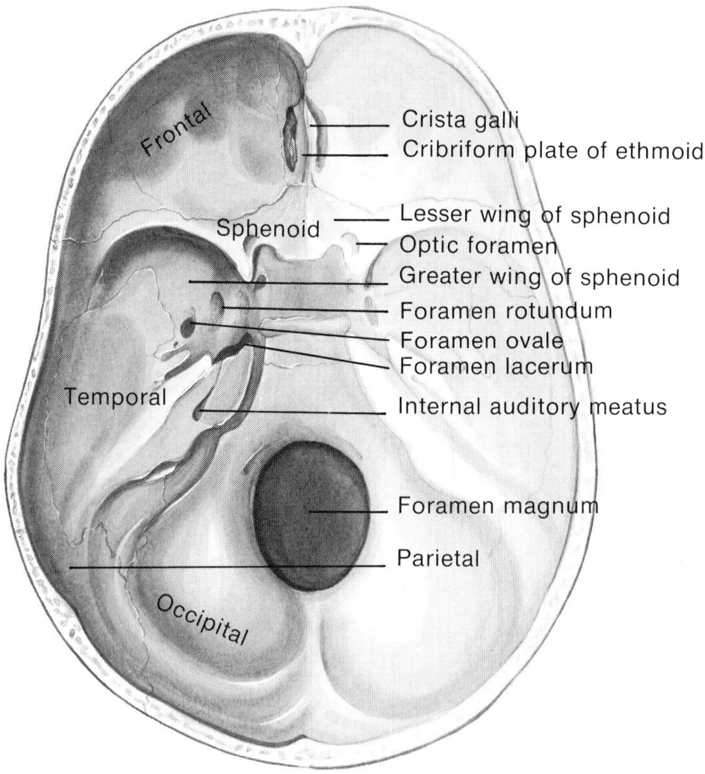

Figure 5–30 Base of the skull, superior aspect.

structure is the large foramen magnum. Anterior to the cavity is a wing-shaped bone, the *sphenoid*, mentioned earlier in this discussion (p. 79).

Anteriorly, the sphenoid bone fills the space between the orbital plates of the frontal bone; posteriorly, it separates the temporal and occipital bones. It somewhat resembles a bat with its wings extended (Fig. 5–31), and consists of a *body* (not labeled); two *greater* and two *lesser wings,* extending transversely from the sides of the body; and two *pterygoid processes,* which project downward. The body is joined to the ethmoid in front and to the occipital behind; it contains cavities, the sphenoidal sinuses (p. 75), which communicate with the nasopharynx. The upper surface exhibits a depression in the form of a saddle, the *sella turcica*. The hypophysis (pituitary gland) lies in this depression. The undersurface of the body of the sphenoid forms part of the roof of the nose.

The last bone to be mentioned is the *hyoid.* This is an isolated, horseshoe-shaped bone (Fig. 5–32) that lies in the anterior part of the neck a short distance above the larynx. It consists of a central part, the *body,* and two projections, the *greater* and *lesser horns* or *cornua,* which project backward. The hyoid bone is suspended from the styloid processes of the temporal bones, and may be felt in the neck just above the laryngeal prominence (Adam's apple). It supports the tongue and provides attachments for muscles used in speaking and swallowing.

The Appendicular Skeleton

Our discussion of the bones of the axial skeleton is complete, and we can now

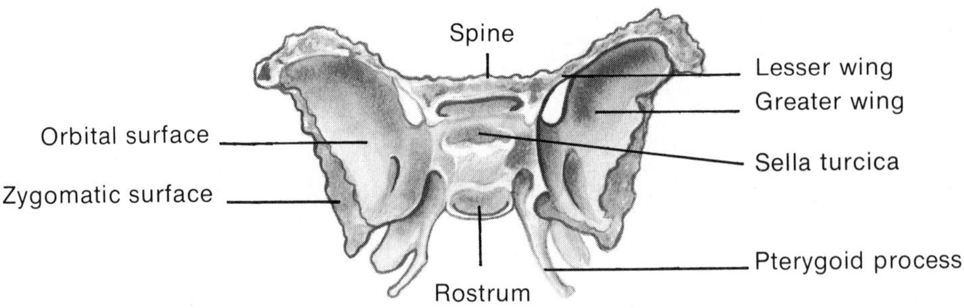

Figure 5–31 Sphenoid bone, superior aspect.

take up the second of the two divisions of the skeleton, the appendicular skeleton, which comprises the bones of the upper and lower extremities, and of the pectoral girdle (shoulder) and pelvic girdle (hip).

The Pectoral Girdle and the Upper Extremity

This division of the appendicular skeleton, beginning at the shoulder and ending at the fingers, includes 60 separate bones. The first of these, the *clavicle* or "collar bone" (Fig. 5–33), is long and has a double curvature. It is situated in a horizontal position at the upper and anterior part of the thorax, just above the first rib, occupying a space at the root of the neck between the upper end of the sternum and the scapula (shoulder blade), to which it is attached and with which it articulates. In women, the curve of the clavicle is less pronounced, and the bone generally is smoother, shorter, and more slender than in the male. The general function of the clavicle is to provide a prop for the shoulder. Because of its prominent position in the pectoral girdle, it is easily injured. Fractured clavicles, for example, are common occurrences in violent contact sports. Such an injury is easily recognized by the obvious "slumping" of the involved shoulder. Without the support of the clavicle, the arm appears to hang from the body, since it is then supported only by the scapula and the shoulder joint.

The *scapula* (Fig. 5–34), a large, flat, triangular bone, covers the dorsal aspect of the thorax between the second and seventh ribs. Its *coracoid process* is a projection originating from the anterior surface of the superior border. This process provides attachments for some of the muscles that move the arm. If you reach over your left shoulder with your right hand, you can feel the spine of the scapula as it passes across the dorsal surface and ends in the large, flat, *acromial process,* which forms the tip of the shoulder blade. The joint between the

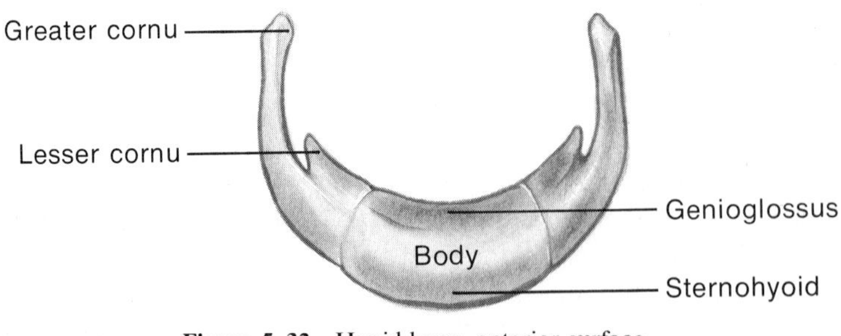

Figure 5–32 Hyoid bone, anterior surface.

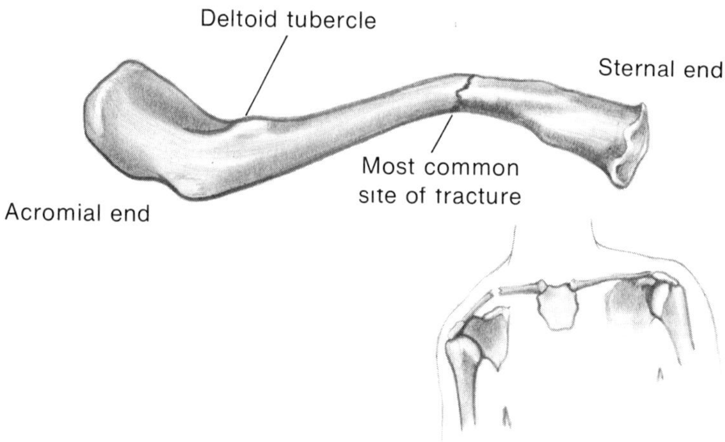

Figure 5–33 The right clavicle, inferior aspect. Inset demonstrates the "drooping" shoulder due to a fractured collar bone.

acromial process of the scapula and the lateral end of the clavicle is the only bony articulation between the upper extremity and the thorax. The projection of the acromial process can be thought of as the "roof" and the coracoid process as the upper, medial "wall"; together they form a fossa called the *glenoid cavity,* which acts as a "socket" for articulation with the humerus (Fig. 5–35).

The *humerus* of the upper arm (Figs. 5–35 and 5–36) is the longest and largest bone of the upper extremity. It extends from the shoulder to the elbow. The upper end of the humerus consists of a rounded head joined to the shaft by a constricted neck, and two eminences, the *greater* and *lesser tubercles,* which provide insertions for many of the muscles of the upper extremity. There is a depression between the two tubercles, the *bicipital groove,* which receives the tendons of part of the biceps muscle. The portion of the neck above the tubercles is called the "anatomical" neck; the portion below is called the "surgical" neck because it is so often fractured. The

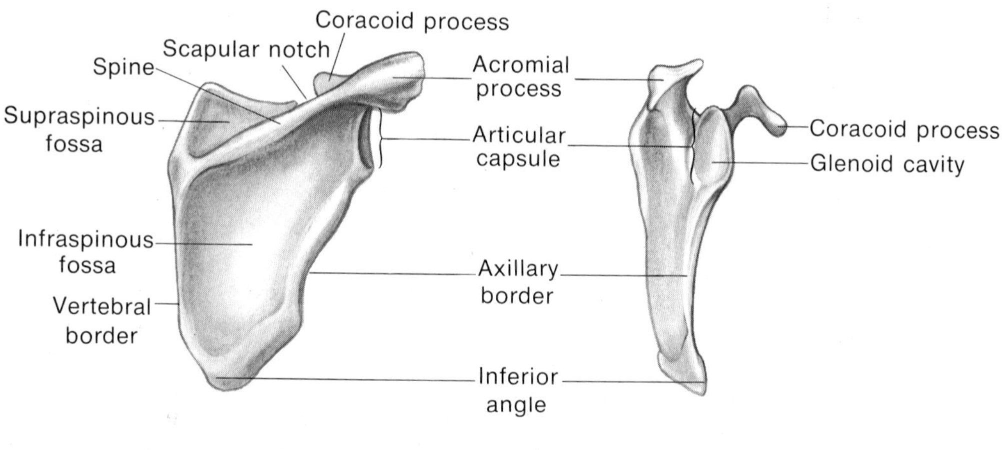

Figure 5–34 The right scapula. *A*, Posterior aspect. *B*, Lateral aspect.

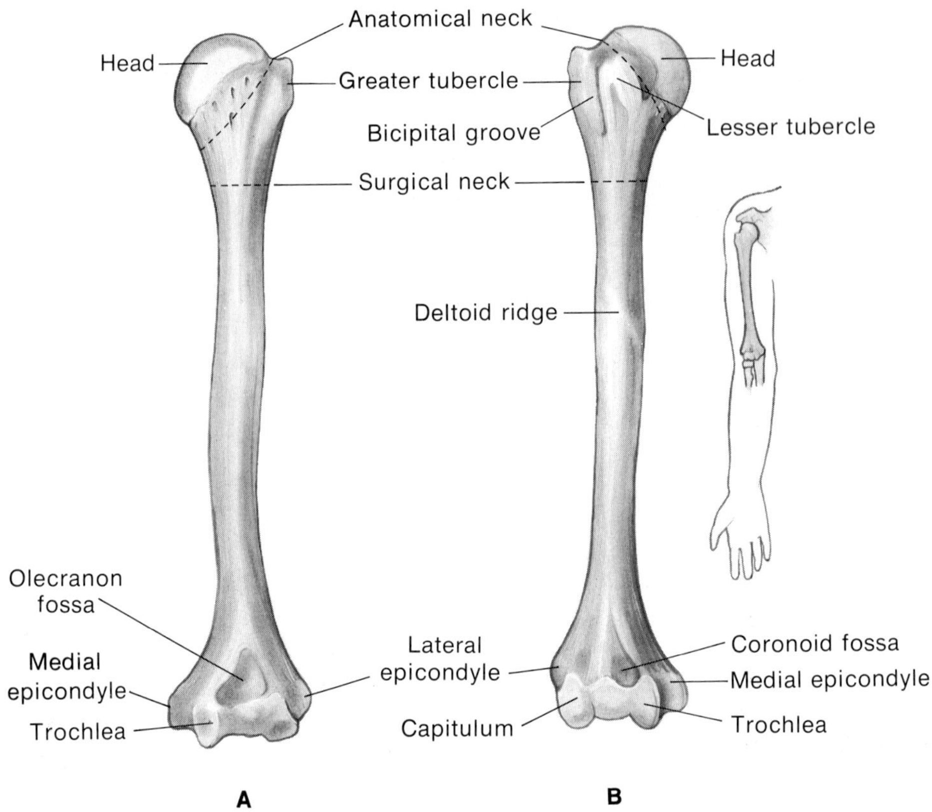

Figure 5–35 The right humerus. *A*, Posterior aspect. *B*, Anterior aspect.

head of the humerus articulates with the glenoid cavity of the scapula, forming a "ball-and-socket" joint (see Chap. 6).

The shaft of this bone is long and slender. At its lower end (distal head) the humerus widens, developing two smooth areas. On the lateral side is the *capitulum,* which articulates with the head of the radius; on the medial side is the *trochlea,* which articulates with the ulna (Fig. 5–36). Situated just above these articular surfaces are the anterior *coronoid fossa,* which receives the coronoid process of the ulna when the forearm is flexed, and the posterior *olecranon fossa,* which receives the olecranon process of the ulna when the forearm is extended. The *epicondyles* project from the medial and lateral sides just superior to the condyles (Fig. 5–35). These projections provide attachments for the tendons of both the ulna and the radius.

The *ulna* is the longer of the two bones of the forearm, and, as shown in Figure 5–37, is positioned on the medial side of the radius. Its shaft is triangular, and its enlarged upper end forms the elbow. There are two beaklike processes at this end: the *olecranon process,* which curves upward and forward; and the *coronoid process,* which, together with the olecranon process, forms a depression called the *trochlear* or *semilunar notch.* This "half-moon"–shaped depression is the articular surface for the trochlea of the humerus. At the distal (lower) end of the ulna there is a small medial projection, the *styloid process;* at both ends of the bone there are lateral facets that serve as areas of articulation with the radius.

The *radius* lies on the lateral or thumb side of the forearm (Fig. 5–37). It joins the ulna by means of a thin, fibrous sheet, or *interosseous membrane,* which crosses the area between the shafts of the two bones. At the upper (proximal) end of the radius there is a small, round head with a cuplike depression on its upper surface for articulation with the capitulum of the humerus. A prominent ridge surrounding the head allows it to rotate within the radial notch of the ulna. The head is supported by a constricted neck just superior to an eminence on the medial side of the radius, the *radial tuberosity.* There are two articular surfaces at the lower end of the radius: the *styloid process,* which articulates with the navicular and lunate bones of the wrist; and the *medial ulnar notch,* which articulates with the ulna.

A specific percussion-type fracture of the lower end of the radius is called *Colles' fracture* (Figs. 5–37 and 5–38). Such a fracture may occur in a fall when the person extends his hands to delay full body contact with the ground, thus putting too much stress on the lower third of the forearm. This kind of fall may also cause hairline fractures and bone fragmentation because of the "jamming" of the wrist bones into the lower heads of both the ulna and the radius, thereby cracking their styloid processes.

The bones of the wrist are the *carpals,* which are arranged in two rows of four each (Fig. 5–39). In the proximal row, beginning on the lateral side,. are the *scaphoid (navicular),* the *lunate,* the *triquetrum,* and the *pisiform.* In the distal row are the *trapezium (greater multangular),* the *trapezoid (lesser multangular),* the *capitate,* and the *hamate* bones. The carpals are closely joined together by liga-

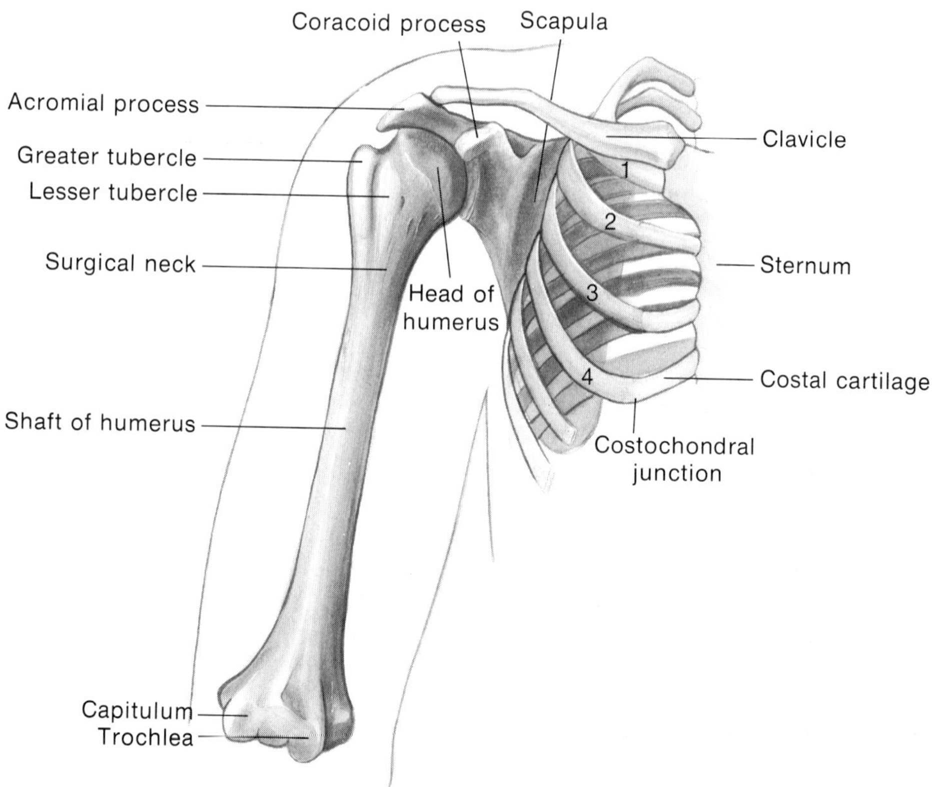

Figure 5–36 Relationship of the right humerus, scapula, and clavicle to the rib cage.

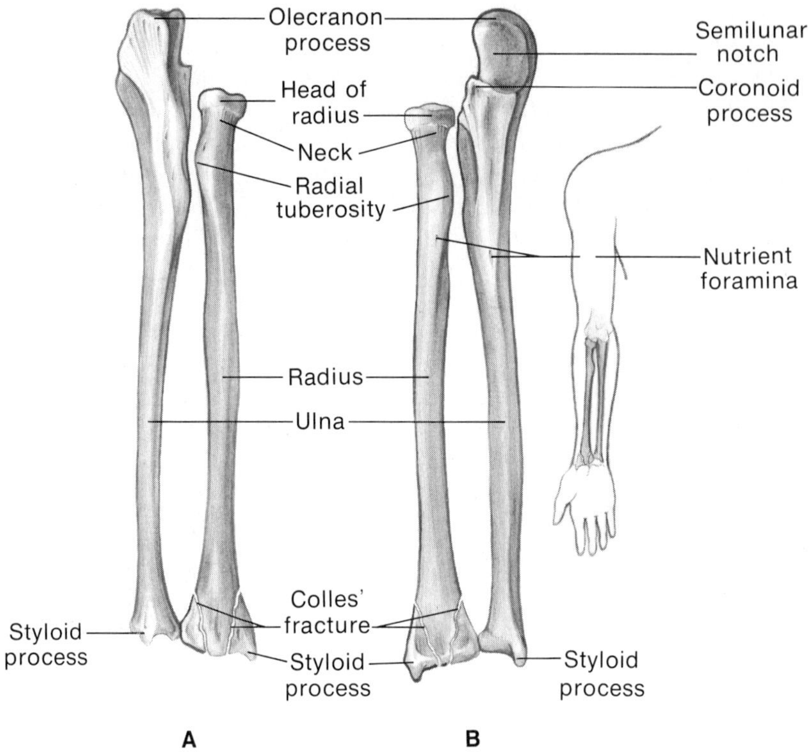

Figure 5-37 The right radius and ulna. *A*, Posterior aspect. *B*, Anterior aspect. Colles' fracture of the distal head is the most common injury to the radius.

ments; however, these ligaments are arranged to allow a certain amount of movement.

The five *metacarpal* bones form the structure of the palm of the hand; each has a base, a shaft, and a head (Figs. 5-38 and 5-39). The metacarpals are long and cylindric, and their rounded distal ends form the knuckles. At their proximal ends, these bones articulate with the carpals and with each other; at their distal ends, they are connected with the proximal phalanges (fingers).

Each hand has 14 *phalanges,* the bones of the fingers. Except for the thumb, which has only two phalanges, each finger has a proximal, middle, and terminal (or distal) phalanx (Fig. 5-39).

The Pelvic Girdle and the Lower Extremity

The pelvic girdle provides support for the trunk and attachment for the bones of the legs, which correspond in general to those of the upper extremities and bear a rough resemblance to them. Their function, however, is different. The lower extremities support the body in the erect position; therefore, they are more solidly built and their parts are less movable than those of the upper extremities.

The two *hip bones,* sometimes called the *innominate* bones (Fig. 5-40), are the broadest in the body. Each is large and irregularly shaped, and, when paired with the bone of the opposite side, forms the sides and the front wall of the pelvic cavity. Although fused in the adult, in the embryo each hip bone starts out as three separate bones—the ilium, the ischium, and the pubis (Fig. 5-40, *A*). A ring is formed by these pairs of bones, and within it are contained a part of the sigmoid colon, the rectum, the urinary bladder, and some parts of the genital system.

The female pelvis differs markedly

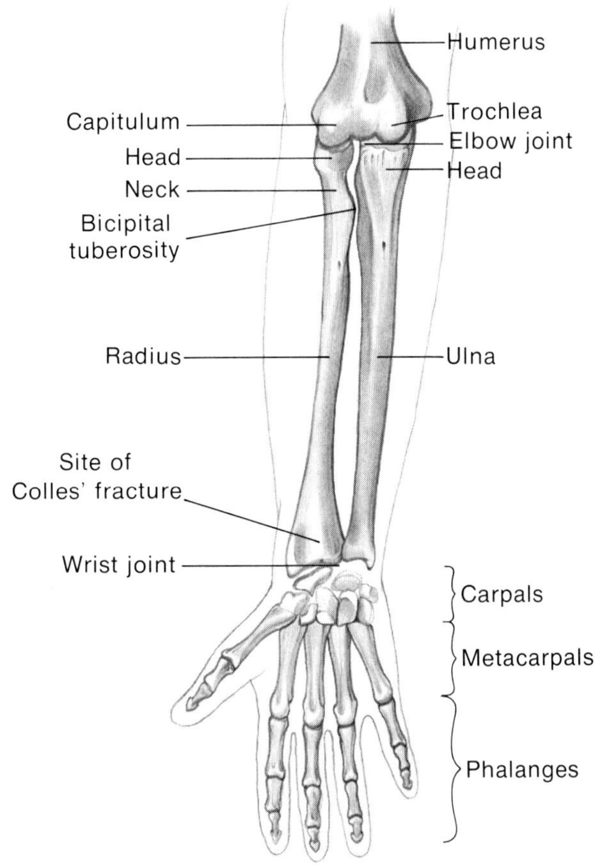

Figure 5–38 Bones of the right forearm and hand, anterior aspect.

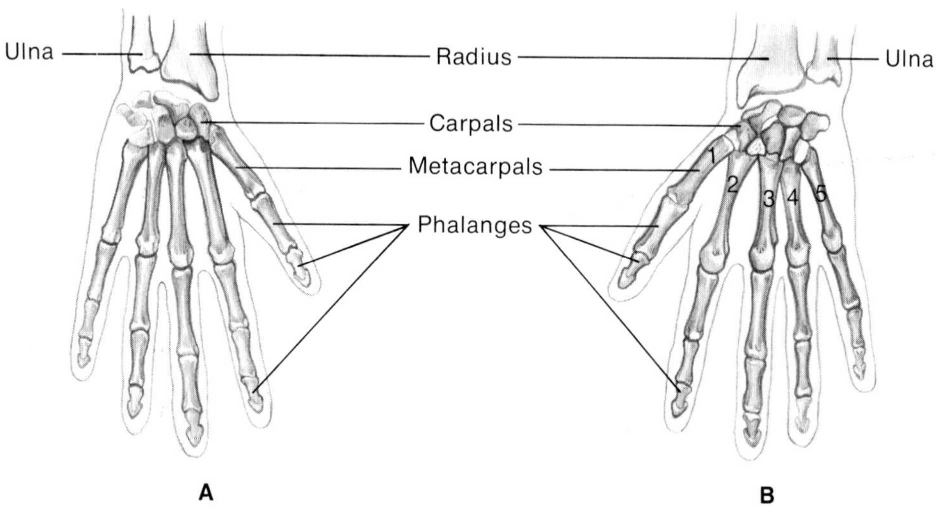

Figure 5–39 Bones of the right hand. *A*, Posterior aspect. *B*, Anterior aspect.

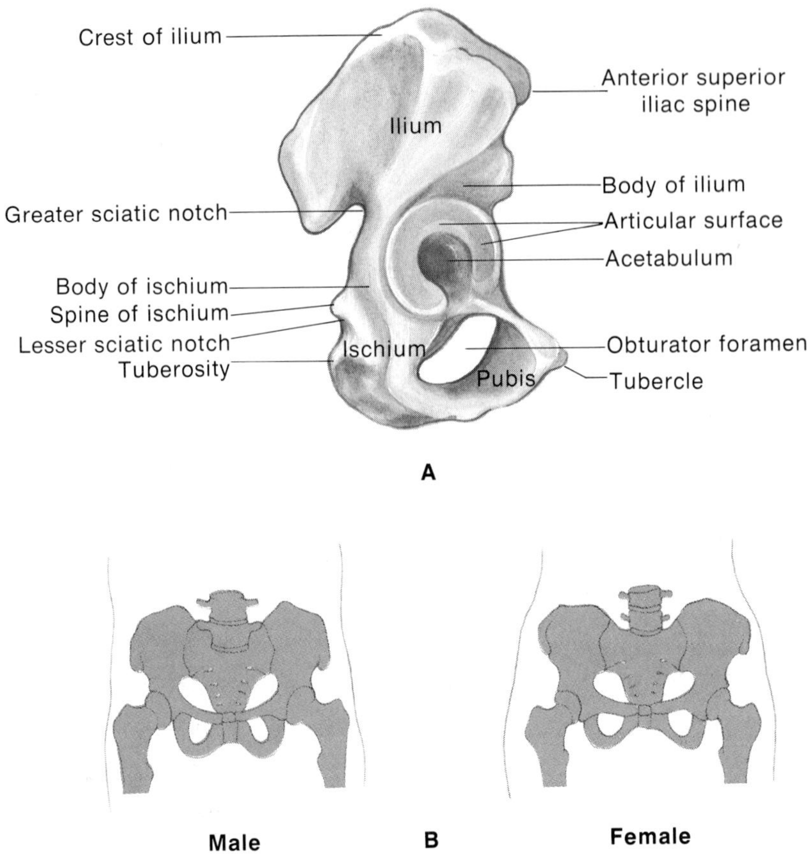

Figure 5–40 The hip bone. *A*, Right pelvic hip, lateral aspect. *B*, Comparison of proportion of male and female pelves. Note that the male pelvis is narrow and compact, whereas the female pelvis is broad and light to facilitate pregnancy and parturition.

from that of the male. These differences facilitate pregnancy and parturition (Fig. 5–40, *B*). In women, the pelvis is more shallow than in men, but it is relatively wider in every direction, especially laterally. The overall internal pelvic area of the female becomes greater during childbirth, owing to the expansion of the pubic bones at the semimovable pubic symphysis, thereby allowing more room for manipulation of the child during delivery.

The *ilium* is the largest of the three pelvic bones; it forms the superior, broad, expanded prominence of the hip. At its crest are anterior superior, posterior superior, and posterior inferior iliac spines. These anatomic landmarks provide attachments for the muscles of the abdominal wall.

The *ischium* is the lowest and strongest bone of the pelvis. It extends downward from the *acetabulum,* the socket for articulation with the head of the femur, and expands into a large tuberosity that serves as a weight bearer when the body is in a seated position.

The *pubis* consists of a body, which unites at the frontal midline with its counterpart of the opposite side to form the *symphysis pubis,* and two rami, or branches, which join, respectively, with the ilium and the ischium. A cavity, the *obturator foramen,* created by the union of all three hip bones, provides space for the passage of major vessels and nerves to the lower extremities from the pelvic cavity.

The *femur*, or thigh bone, is the longest bone in the body. Its structure is simple, but its position is not that of a vertical "pillar." Instead, it lies at a downward and inward angle from the hip joint. At the upper end of the femur is a rounded head with a constricted neck (Fig. 5–41). There are also two eminences, the greater and lesser *trochanters*. The femoral head itself articulates medially with the acetabulum to form the hip joint. The shaft of the femur, posteriorly, has a ridge for muscle attachment, the *linea aspera*.

At its lower end, the femur widens into large eminences, the lateral and medial condyles (Fig. 5–41). The intercondyloid fossa lies between these. Distally, the femur articulates with the tibia and patella at the knee joint.

The sesamoid *patella,* or knee cap, is small, triangular, and flat. It occupies the space within the tendon of the quadriceps femoris muscle at the front of the knee joint. The patella, as noted, articulates with the femur (Fig. 5–42); it is surrounded by large, fluid-filled sacs, or *bursae*. When the knee is straightened, the bone moves freely in support of its associated muscles.

The *tibia*, or shin bone, lies at the front and medial side of the leg (Fig. 5–43). Two bones form the lower leg, and the

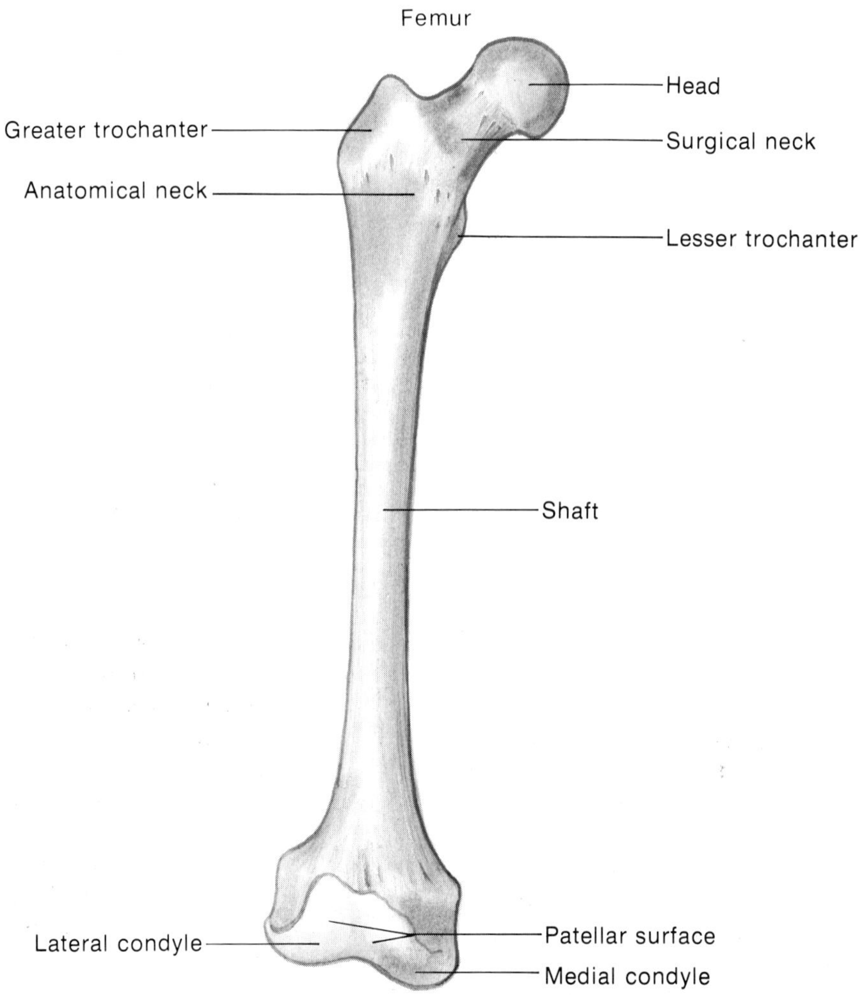

Figure 5–41 The right femur, anterior aspect.

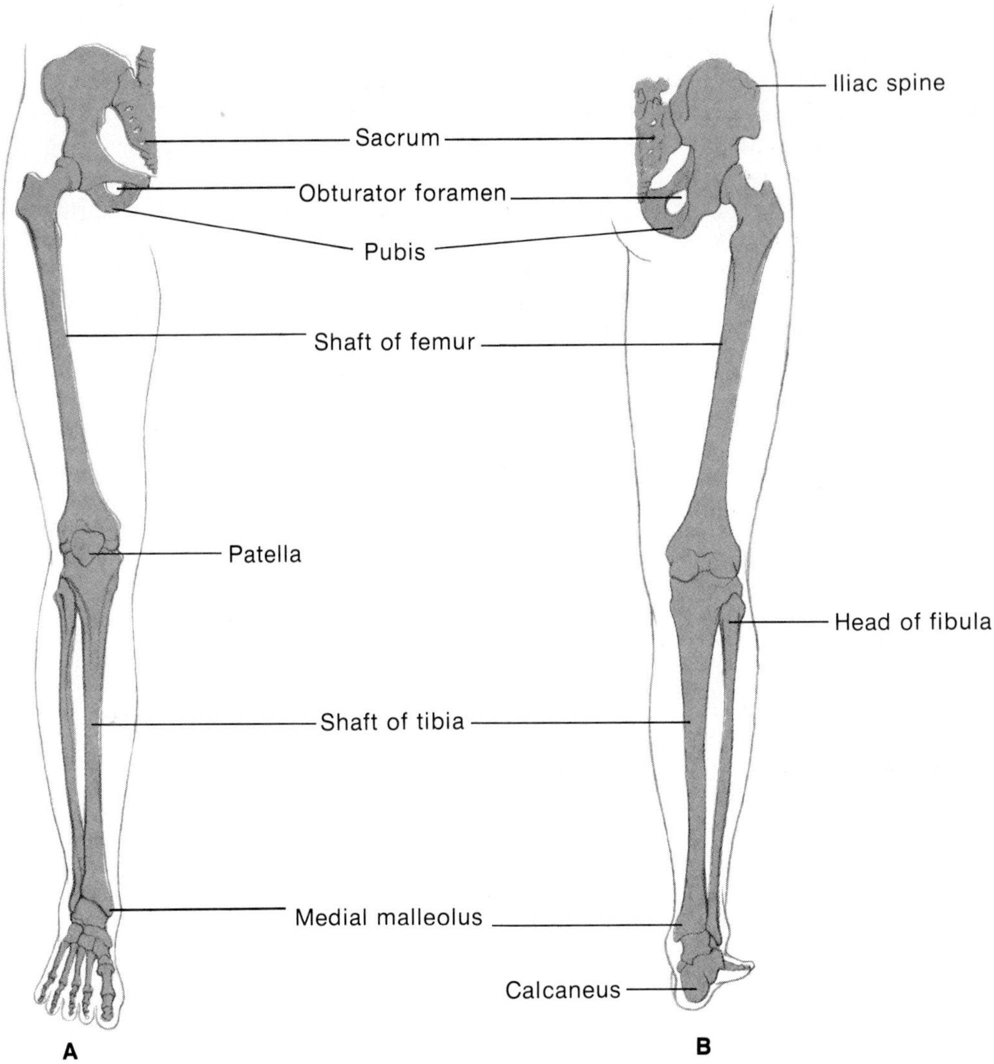

Figure 5–42 Bones of the right leg and foot. *A*, Anterior aspect. *B*, Posterior aspect.

tibia is the larger. Its upper end expands into lateral and medial condyles, which are separated by the sharp intercondyloid fossa. The concavities of these condyles articulate with the femoral condyles. At its lower end, the tibia narrows into a strong process, the *medial malleolus*, forming the inner prominences of the ankle. The *talus* (Fig. 5–44), which forms the ankle joint, articulates with the tibial surface lateral to the medial malleolus. The tibia also articulates at both of its extremities with the fibula, the other long bone of the lower leg.

The *fibula* of the calf (Fig. 5–43), the slenderest bone in the body, lies parallel to the tibia and on its lateral side. At the upper end of the fibula an irregular, quadrant head articulates with the back part of the tibia but does not reach the knee joint. The lower extremity of the fibula is prolonged downward to form the outer ankle bone, the *lateral malleolus*, which is a pointed process just beneath the skin. At this end the fibula articulates with both the tibia and the talus. A fracture of the lower end of the fibula involving injury to the lower tibial articulation is called a *Pott's fracture* (Fig. 5–43). This fracture closely resembles the Colles' fracture of the wrist, and is caused by similar trauma. Too forceful a strain or the sudden impact of too much weight from a fall will fracture the lower third of the fibula or crack the lateral malleolus of the fibula and the medial malleolus of the tibia.

The *tarsus* (Fig. 5–44, *A*) is the general name for the seven tarsal (ankle) bones that make up the articulation between the foot and the leg. They are short and resemble the carpal bones of the wrist, although they are larger. Their names are the calcaneus, talus, cuboid, navicular, and the first, second, and third cuneiforms. The largest and strongest of the tarsal bones is the *calcaneus* (Figs. 5–42 and 5–44), the heel bone, which serves to transmit the weight of the body to the ground, forming a strong lever for the muscles of the calf.

The *metatarsus,* the bony structure of the sole and instep of the foot, is formed by five bones that closely resemble the metacarpal bones of the hand. Each metatarsal is long and consists of a base, a shaft, and a head (Fig. 5–44, *A*), and each articulates at one end with the tarsal bones and at the other with the first row of phalanges (toes). The shafts of the metatarsals provide attachments for the intrinsic muscles of the toes; the heads form the joints of articulation. The first metatarsal bears the body's weight and therefore is larger than the others.

The tarsal and metatarsal bones are arranged in the form of two distinct arches: one running from the heel to the toes on the inner medial side, the *longitudinal arch* (Fig. 5–44, *B*); the other continuing across the foot in the metatarsal

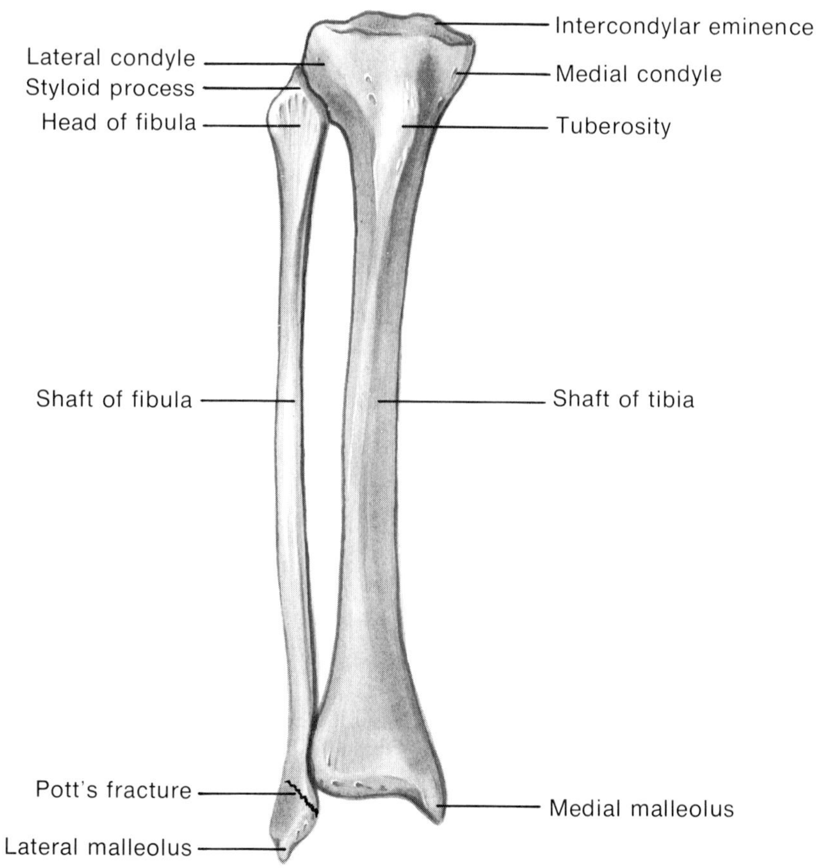

Figure 5–43 The right tibia and fibula, anterior aspect. As in the forearm, the most common fracture (Pott's) occurs to the smaller bone, at the lateral malleolus of the fibula.

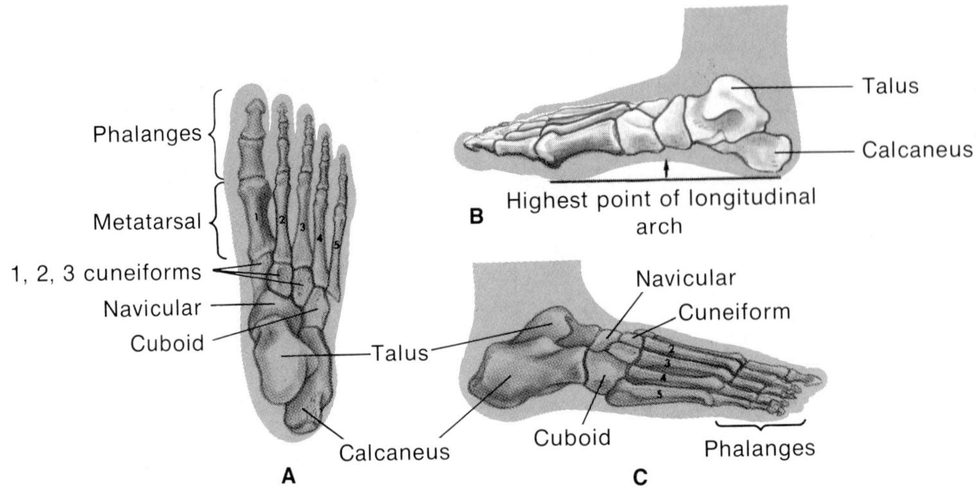

Figure 5–44 Three views of the bones of the right foot. *A*, Superior aspect. *B*, Medial aspect, indicating the longitudinal arch. *C*, Lateral aspect.

region, the *transverse arch*. These arches may become weakened and progressively broken down, a condition known as *flatfoot*. Common causes of arch collapse are prenatal, dietary, or hormonal disturbances; poor posture; and ill-fitting shoes. Flatfoot is seldom painful, but the condition does affect the function of the arches as a weight-bearing bridge when one is standing still and as a lever while one is walking.

Weakening of the transverse arch can cause pain at the base of the toes. High-heeled shoes force most of the weight of the body onto the ends of the metatarsal bones; this may enlarge them and push them sideways out of normal alignment. A painful enlargement at the joint between the first metatarsal bone and the great toe is called a *bunion*.

The long bones of the *phalanges* (Fig. 5–44, *A*) are similar in structure to the phalanges of the fingers. As in the thumb, the great toe has two phalanges; there are three in each of the other four toes.

DISEASES AND INJURIES AFFECTING THE SKELETAL SYSTEM

Several types of disease may affect the skeletal system or portions of skeletal tissue. Generalized *osteitis fibrosa cystica* is a condition characterized by a marked decrease in the amount of bone in the skeleton, the replacement of bony fibrous tissue, and the formation of cysts. Bones may be the seat of infection, as for example in *osteomyelitis* (Fig. 5–45). This condition is due to a pyogenic (pus-producing) microorganism, and it may affect the marrow, cortex (hard portion of bone), cancellous tissue, or periosteum. Degenerative diseases, such as cysts and malignant tumors (e.g., *osteosarcoma*), may originate in bone. Osseous tissues also may undergo localized death owing to insufficient circulation or bacterial invasion *(gangrene)*.

The most common injuries to which bones are subject, however, are fractures.

Types of Fractures and Their General Treatment

A broken bone is called a *fracture*. The treatment of fractures involves two distinct phases: *reduction* and *immobilization*. During reduction the bone is returned to its proper alignment, or "set." When this can be done without opening the skin, the technique is called a *closed* reduction. Manual traction, manipulation, or traction-suspension can be used to accomplish closed reduction.

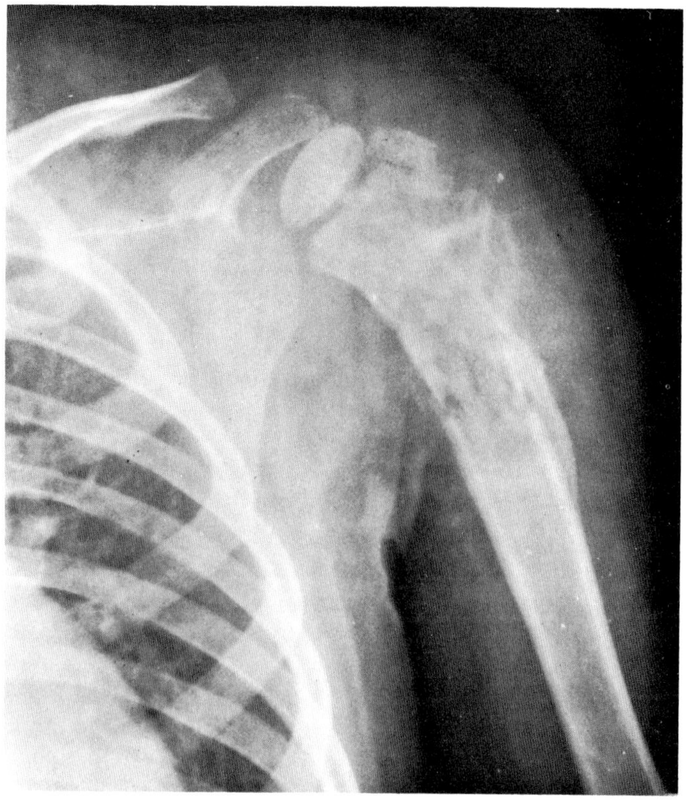

Figure 5–45 Acute osteomyelitis of the proximal humeral shaft in a young child. (From Gartland, J. J.: Fundamentals of Orthopaedics, 2nd ed. Philadelphia: W. B. Saunders Co., 1974.)

An *open* reduction requires a surgical incision at the site of the injury, and bone fragments are replaced or realigned under direct vision. When closed reduction can be used, it is preferred because there is no risk of wound infection.

Fractures are classified as *simple* if the skin is not broken; they are *compound* if the broken ends of the bone protrude through the skin (Fig. 5–46). Compound fractures present a special problem in that foreign material is likely to be found in the wound. Damage to the muscles and nerves directly under the broken skin can cause further complications. If treatment is undertaken promptly, the likelihood of bacterial infection will be minimized.

Whether the fracture is open or closed, following reduction the bone must be held securely in a stable position to promote good healing. This is called immobilization. Plaster of paris casts and splints are used most frequently to put the fracture site at rest until the bone has healed.

Several of the many different kinds of fractures are illustrated in Figure 5–46. Depending on the type, the surgeon may select the device best suited to facilitate healing. Today, the two metals most frequently employed as supports are stainless steel and cobalt; the latter is used in the form of a cobalt-chromium alloy called Vitallium. Both metals have a highly corrosion-resistant character and do not evoke blood reactions and the formation of antibodies. They are used in plates, screws, pins, heads, and the like (Fig. 5–47) to fortify or position the fracture in order to promote healing. Because the metals are noncorrosive, these devices may be left on the fracture site or removed later, depending on the device used and the location of the fracture.

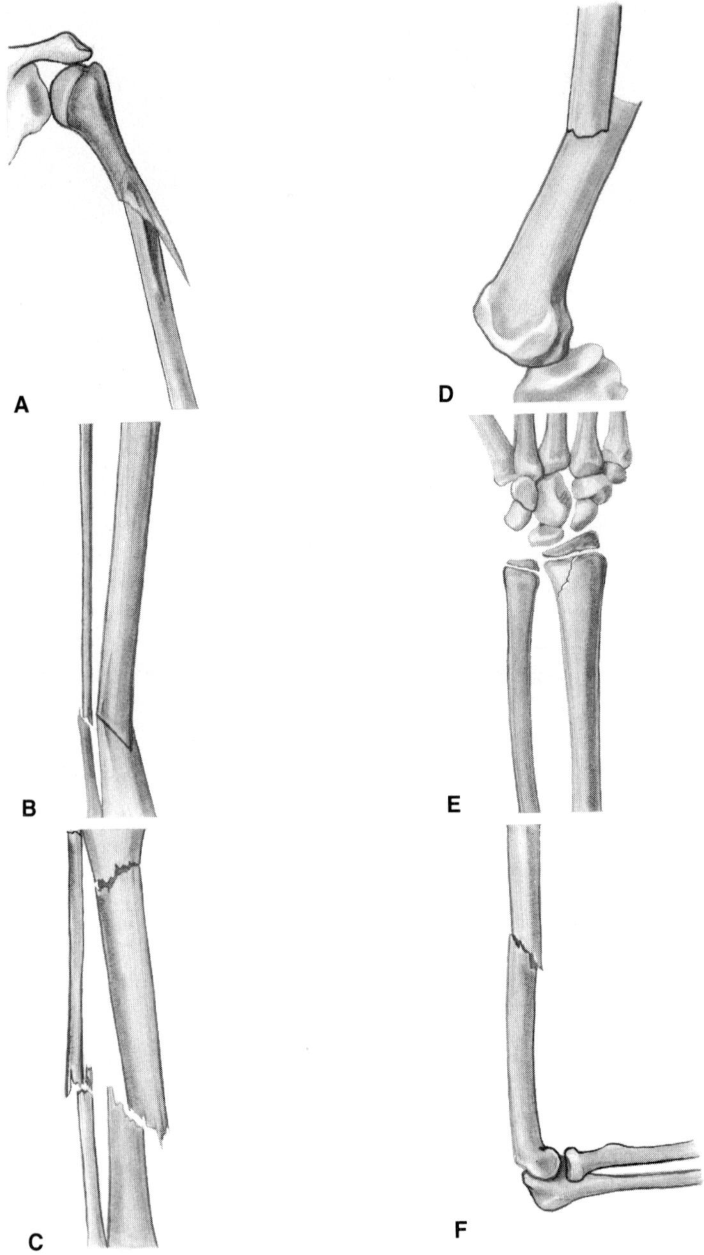

Figure 5–46 Types of fractures. *A*, Longitudinal, through upper shaft of right humerus. *B*, Spiral, of tibia and fibula. *C*, Comminuted, of tibia and fibula. *D*, Transverse, through lower shaft of femur. *E*, Impacted. *F*, Pathologic, through area of destruction secondary to metastatic carcinoma. (*A* and *D*, Redrawn from Gartland, J. J.: Fundamentals of Orthopaedics, 2nd ed. Philadelphia: W. B. Saunders Co., 1974. *C*, Redrawn from Gartland, J. J.: Fundamentals of Orthopaedics, 1st ed. Philadelphia: W. B. Saunders Co., 1965. *B*, Redrawn from Trueta, J.: Studies of the Development and Decay of the Human Frame. London: William Heinemann Medical Books, Ltd., 1968. *E*, Redrawn from Aegerter, E. E., and Kirkpatrick, J. A., Jr.: Orthopaedic Diseases, 4th ed. Philadelphia: W. B. Saunders Co., 1975.)

OUTLINE

I. Introduction
  A. Types of skeleton
    1. Exoskeleton—hard, outer shell
    2. Endoskeleton—bones covered by soft tissue
  B. Functions of endoskeleton
    1. Protection
    2. Support
    3. Muscle attachment
    4. Reservoir for minerals
    5. Hematopoiesis

II. Formation of Bone
  A. Embryonic tissue
    1. Mesenchyme
    2. Fibrous tissue containing fibroblasts
    3. Collagenous and elastic fibers

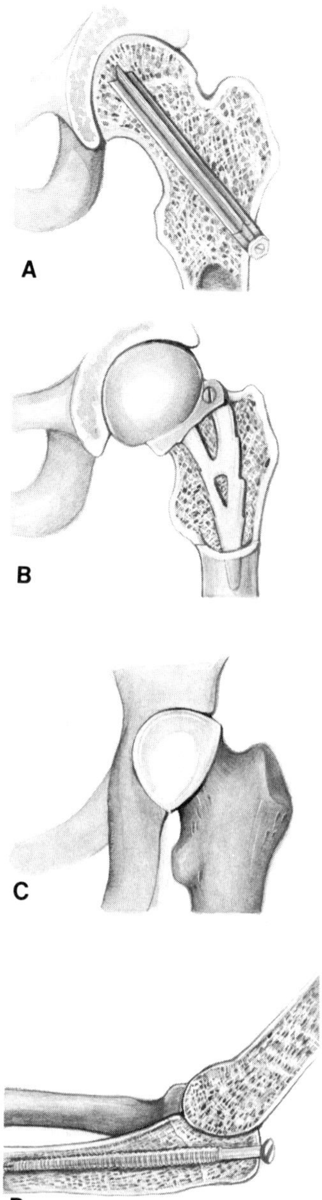

Figure 5–47 Common fractures and their repair. *A*, Fracture of the surgical neck of the femur and use of a Smith-Peterson nail to realign the head. *B*, Replacement of the femoral head by a Moore's prosthesis. *C*, Use of a Smith-Peterson cup to substitute for an acetabulum of the hip. *D*, Carpenter's screw repair of a fractured olecranon process of the ulna. (Redrawn from Trueta, J.: Studies of the Development and Decay of the Human Frame. London: William Heinemann Medical Books, Ltd., 1968.)

B. Development of bone
   1. Blood vessels bring rich contents that stimulate cell division
   2. Osteoblasts form from previous mesenchyme
   3. Osteoblasts deposit mucoprotein to form a rubbery osteoid tissue
   4. Osteoblast—a bone builder

C. Intramembranous ossification
   1. Bones of the skull
   2. Membranous covering of skull contains fibroblasts
   3. Fibroblasts change to osteoblasts, which then lay down a basis for calcium

D. Cartilaginous ossification
   1. Replacement of pre-existing cartilage by osteoblasts
   2. Hyaline cartilage forms embryonic skeleton
   3. Bone begins to form on cartilaginous model; covered by periosteum
   4. Blood vessels bring osteoblasts, which lay down collagen and mucoprotein
   5. Osteoclasts gradually dissolve old bone, while osteoblasts deposit new bone
   6. In long bones, ossification proceeds in two directions
   7. Bone growth ceases at about age 24

E. Chemical influences on bone formation
   1. Vitamin D—absorption of minerals
   2. Alkaline phosphatase—enzyme that releases calcium and phosphate ions from calcium salt solution
   3. Parathormone—hormone that regulates calcium and phosphorus levels in blood

F. Disorders of bone development
   1. Insufficiency—rickets, osteomalacia

2. Gigantism – dwarfism – acromegaly

III. Structure of Bone
   A. Microscopic structure
      1. Compact bone – osteon
         a. Haversian canals
         b. Lamellae
         c. Volkmann's canals
         d. Canaliculi
      2. Cancellous (spongy) bone
   B. Gross structure (long bone)
      1. Two epiphyses – heads
      2. Diaphysis – shaft
      3. Periosteum – covering
      4. Medullary cavity
         a. Yellow bone marrow – fat
         b. Red bone marrow – red blood cells and platelets
      5. Endosteum

IV. Types of Bones
   A. Long bones
      1. Function – act as levers
      2. Examples – femur, humerus
   B. Short bones
      1. Same basic structure as long bones
      2. Function – strength
      3. Examples – carpals, tarsals
   C. Flat bones
      1. Platelike
      2. No medullary cavity
      3. Function – protection
      4. Examples – cranium, ribs
   D. Irregular bones
      1. Odd-shaped
      2. No medullary cavity
      3. Function – articulation
      4. Examples – vertebrae, hip bones, some skull bones
   E. Sesamoid
      1. Seedlike bones found in joint capsules and in tendons
      2. Function – prevent friction
      3. Example – patella (knee cap)

V. The Skeleton
   A. Axial skeleton
      1. Vertebral column
         a. Supports trunk and neck
         b. Protects spinal cord
         c. Four curves visible from lateral aspect
         d. Twenty-six vertebrae separated by disks
            1) 7 cervical
            2) 12 thoracic
            3) 5 lumbar
            4) 2 sacral-coccygeal
         e. Typical vertebra
            1) Body
            2) Arch
            3) Four articular processes
            4) Two pedicles
            5) Lamina
         f. Vertebral injuries and diseases
            1) Kyphosis
            2) Lordosis
            3) Scoliosis
            4) Severed spinal cord – paralysis
            5) Ruptured disk
            6) Tuberculosis
      2. Thorax
         a. Sternum
         b. Costal cartilages
         c. Twelve pairs of ribs
            1) 7 true ribs
            2) 5 false ribs (last two pair are floating)
      3. Cranial sinuses
         a. Air spaces in bone
         b. Frontal and ethmoidal – nose
         c. Sphenoidal – nasopharynx
         d. Maxillary – nasal passages
         e. Inflammation – sinusitis
      4. Cranial sutures
         a. Coronal – side
         b. Lambdoid – rear
         c. Squamous – side
         d. Sagittal – middle
      5. Fontanels
         a. Bregmatic – anterior
         b. Occipital – posterior
         c. Sphenoidal – anterolateral
         d. Mastoidal – posterolateral
         e. Premature fusion – microcephalus

   - f. Abnormally large quantity of spinal fluid—hydrocephalus
 6. Bones of skull
   - a. Frontal—forehead
   - b. Ethmoid—nose
   - c. 2 nasal—nose
   - d. Vomer—nose
   - e. 2 maxillae—upper jaw
   - f. Mandible—lower jaw
   - g. Zygomatic—cheek
   - h. 2 parietal—lateral, superior
   - i. 2 temporal—lateral, inferior
     - 1) 4 parts—squamous, petrous, mastoid, tympanic
   - j. Occipital—posterior
   - k. Sphenoid—superior, inferior
   - l. Hyhoid—anterior part of neck

B. Appendicular skeleton
 1. Pectoral girdle and upper extremity
   - a. Clavicle—collar bone
   - b. Scapula—shoulder blade
   - c. Humerus—upper arm
   - d. Ulna—forearm, medial
   - e. Radius—forearm, lateral
     - 1) Colles' fracture
   - f. Carpals—wrist (8)
   - g. Metacarpals—hand (5)
   - h. Phalanges—fingers (14)
 2. Pelvic girdle and lower extremity
   - a. Hip (innominate) bones
     - 1) Ilium, ischium, and pubis
   - b. Femur—thigh, longest bone in body
   - c. Patella—kneecap
   - d. Tibia—shin, anterior
   - e. Fibula—shin, lateral to tibia
     - 1) Pott's fracture
   - f. Tarsals—ankle (7)
   - g. Metatarsals—foot (5)
   - h. Phalanges—toes (14)

VI. Diseases and Injuries Affecting the Skeletal System

A. Diseases
 1. Osteitis fibrosa cystica
 2. Osteomyelitis
 3. Osteosarcoma
 4. Gangrene

B. Fractures
 1. Repair
   - a. Closed reduction—manipulation, traction
   - b. Open reduction—surgery
 2. Classes
   - a. Simple—skin not broken
   - b. Compound—skin broken
   - c. Incomplete—bone not broken through
   - d. Complete—bone broken into at least two pieces
 3. Treatments
   - a. Metal used: Vitallium
   - b. Pins, screws, plates, cups

STUDY QUESTIONS AND PROBLEMS

1. Name five general functions of bone.
2. What are fractures? Name at least three types and a possible means of repair.
3. What is a haversian system?
4. Describe how the osteocyte receives nutrition from a vessel outside the bone.
5. Differentiate between intramembranous and cartilaginous ossification.
6. Define osteoblast, osteocyte, and osteoclast, and give the role of each in the development of bone.
7. What are the two major divisions of the skeleton?
8. List the fontanels that are in the skull at birth, and describe microcephalus.
9. Differentiate among the curvatures of the spine. What do the terms kyphosis, lordosis, and scoliosis mean?
10. Sketch a typical spinal vertebra and indicate why the lamina is its major point of weakness.
11. What roles do vitamin D, parathormone, and alkaline phosphatase play in the formation of bone?
12. What is rickets? Acromegaly? Osteomalacia?
13. List the cranial sinuses and give at least two reasons why they are important.
14. What are some of the differences between a male and a female pelvis?
15. Why is it important that we have an arch in the foot?

Chapter 6

The Articular System

The otherwise rigid bony structure of the body is made flexible by arrangements of tissues called *joints* or *articulations*. The many joints can function independently or in synchronous movements, making possible complex motions and providing support. *Arthrology* is the branch of anatomy concerned with the description and study of these articulations.

A joint may be a connection between two or more bones, or between cartilage and bone. One of the most important functions of the skeleton, namely, that of providing a system of levers, depends on the joints, which firmly secure the skeletal components for support and also make possible the movement of one component on another.

Bones can be joined together in any one of several ingenious ways, thus permitting great varieties of movement. Where free movement is essential, the articulating ends of the bones are shaped and the joint constructed so as to facilitate unhampered motion. Similarly, where only slight movement is desirable, bone shape and joint structure combine to restrict motion. Joints, where necessary, may be completely rigid.

CLASSIFICATION OF JOINTS

The degree of movement required of a joint is a useful criterion for assigning a joint to one of three classes (Table 6–1). The first class, comprising synarthroses and synchondroses, are those joints that allow virtually no movement; *amphiarthroses* allow slight movement; and *diarthroses*

Table 6–1 CLASSIFICATION OF JOINTS

JOINT	*TYPE*	*EXAMPLES*
Synarthrosis	Fibrous	
	Suture	Cranial sutures
	Syndesmosis	Tibiofibular syndesmosis
Synchondrosis	Cartilaginous	Epiphysis of growing long bone
Amphiarthrosis	Symphysis	Pubic symphysis
	Disk	Intervertebral disks
Diarthrosis	Synovial	
	Hinge	Knee, elbow, ankle
	Ball-and-socket	Hip, shoulder
	Pivot	Atlas, axis
	Ellipsoidal	Wrist
	Gliding	Intercarpal, intertarsal
	Saddle	Thumb

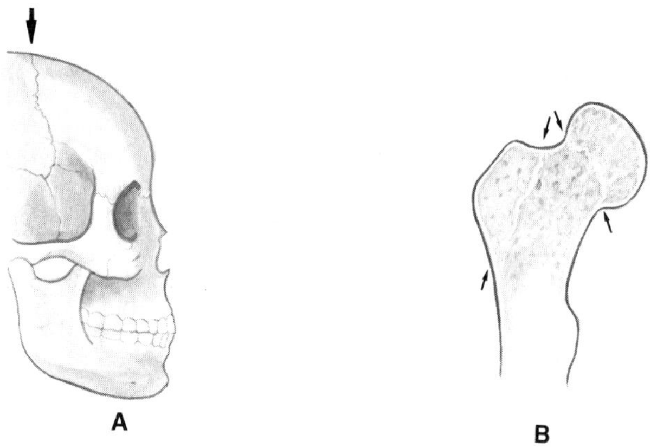

Figure 6–1 Synarthrodial joints. *A*, A suture (arrow) composed of fusing fibrous tissue. *B*, A femoral head held in position by fusing cartilage (synchondrosis) (arrows).

(also called *synovial joints*) make possible free and complex skeletal motions. In each of these groups, joints may be further classified according to their structural characteristics.

Synarthroses (Fig. 6–1). These are joints in which the bones are united by bands of fibrous tissue. They have no joint cavity, and with only a very few exceptions, they permit no movement.

1. SUTURES. The sutures of the skull are united by several fibrous layers. Growth at the cranial sutures accommodates the growth of the brain.

2. SYNDESMOSES. These joints are characterized by a greater quantity of intervening fibrous tissue than that found at the sutures. The tibiofibular syndesmosis of the leg is an example of this type of union.

Synchondroses. In these joints, two bones are held together by hyaline cartilage. This cartilage is replaced by bone in the adult. Thus an epiphysis in the body of a long bone in a growing child (Fig. 6–1, *B*) is a synchondrosis. Like the synarthroses, synchondroses have no joint cavity and allow virtually no movement.

Amphiarthroses (Fig. 6–2). In these articulations, also called *symphyses,* there is a pad of fibrocartilage between the two uniting bones. Very limited movement is permitted. The intervertebral disks, for example, have a core of pliable tissue (Fig. 6–2, *A*) that provides for some flexibility in the vertebral column. The symphysis between the pubic bones (Fig. 6–2, *B*) is also of this type, although there is practically no movement. In the adult woman the final stages of pregnancy bring some flexibility

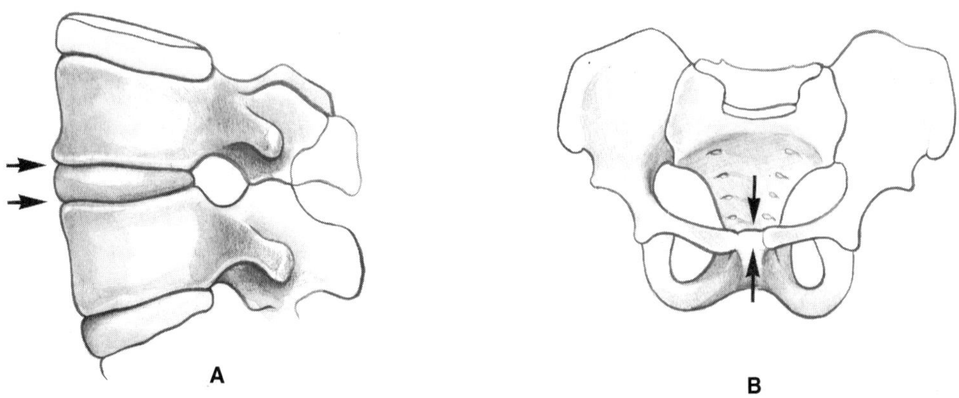

Figure 6–2 Amphiarthrodial joints. *A*, Diagram of an intervertebral disk (arrows), lateral aspect. *B*, The pubic bones are joined by a disk of fibrocartilage forming the symphysis pubis (arrow).

in the pubic articulations because of structural changes in the fibrocartilage.

Diarthroses (Synovial Joints). The joints allowing free movement necessarily are complicated structures. The bones entering into the formation of a diarthrosis have smooth articular surfaces covered with hyaline cartilage.

In all diarthrodial joints (Fig. 6–4), there is an *articular capsule* lined with a *synovial membrane*. This membrane and the articular cartilages surround the *joint cavity*, which contains *synovial fluid* (synovia). Secreted by the synovial membrane, this fluid is watery and clear and contains salts, mucoprotein, and fat. Its main functions are to lubricate the joints and to act as a "shock absorber" in preventing damage to the articular cartilages at the ends of the two adjacent bones.

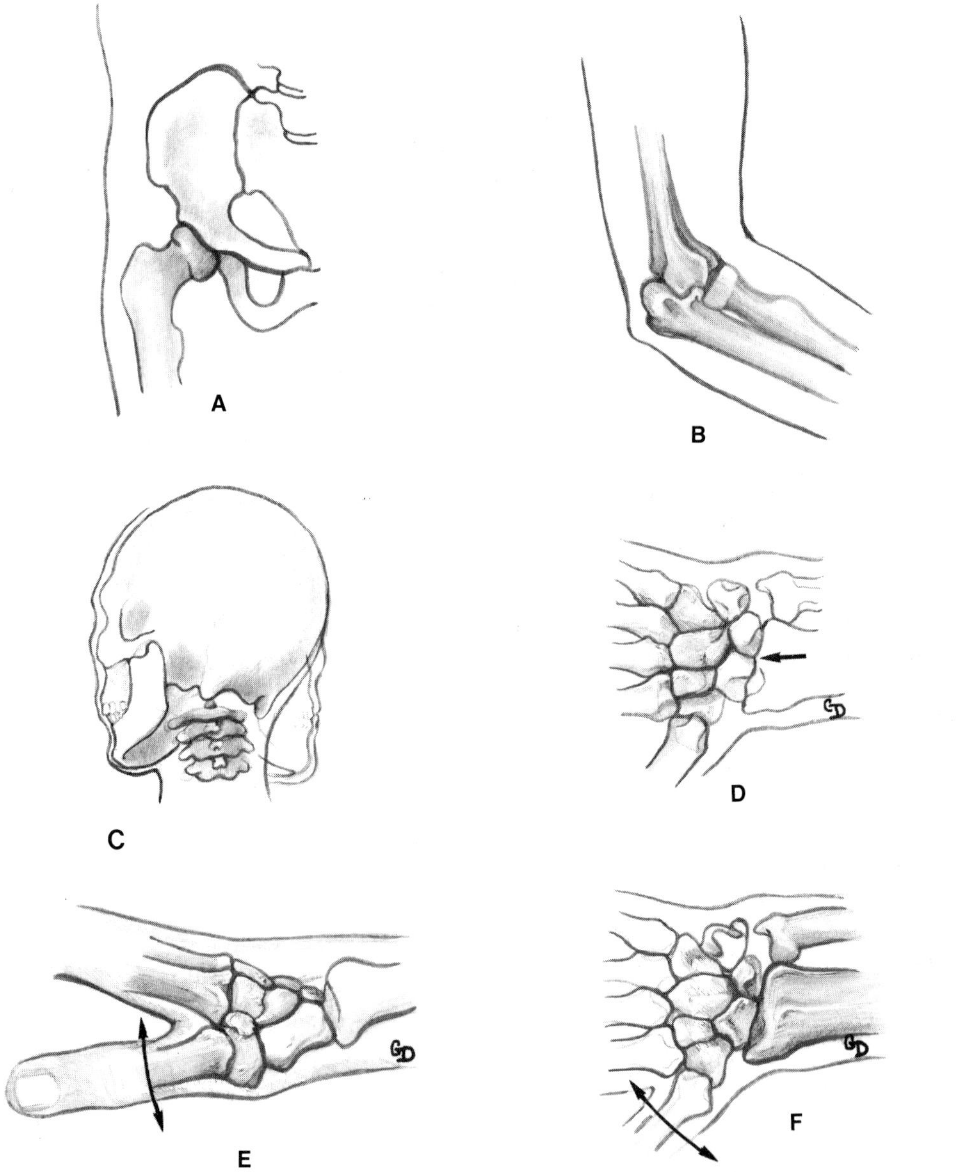

Figure 6–3 Types of diarthrodial joints. *A*, Ball-and-socket. *B*, Hinge. *C*, Pivot. *D*, Ellipsoidal. *E*, Gliding. *F*, Saddle.

The capsule of a diarthrodial joint is fibrous in nature, being made up of collagenous fibers extending from one bone to another. At certain points these bands of fibers are thickened to form capsular *ligaments,* which add strength to the joint and affect its motion.

Most joints are diarthrodial. Further classifications can be made according to the structures and configurations of the articulating ends of the bones at either side of the joint.

1. HINGE JOINT (Fig. 6–3, *A*). Precise and delicate movements are made by joints of this class. The phalanges, for example, have hinge joints. In this type, two cylindric surfaces, one concave and the other convex, meet in articulation. Only flexion and extension on one plane are possible, as in the elbow.

2. BALL-AND-SOCKET JOINTS (Fig. 6–3, *B*). The shoulder and the hip are classic examples of this type of articulation. A ball-shaped head moves within a socket that is "hollowed" to fit its contours. Such joints allow movement in all planes and rotation as well—the greatest range of any diarthroses.

3. PIVOT JOINT (Fig. 6–3, *C*). Rotating or swiveling movements, as in the atlas and axis, are made by pivot joints. A ringlike formation at the end of one bone is entered by a pivotlike process projecting from its articulating partner.

4. ELLIPSOIDAL JOINT (Fig. 6–3, *D*). An example of this type is the wrist joint formed by the radius and the carpal bones. An elliptical cavity is filled by an oval condyle, thus allowing motion in two directions.

5. GLIDING JOINT (Fig. 6–3, *E*). The joints between the carpal bones and between the tarsal bones are of this gliding type. Opposing plane surfaces are slightly convex and concave, respectively, thereby restricting movement.

6. SADDLE JOINT (Fig. 6–3, *F*). In this type, for example, the joint between the first metacarpal and the trapezium of the thumb, the end of one bone is convex in one direction and concave in the other. The end of the articulating bone is a "mirror image" so that the two fit, making possible movements in two planes at right angles.

MOVEMENT AT SYNOVIAL JOINTS (Fig. 6–4)

The movements of the parts of the body are determined by the shapes of the articular surfaces of the joints. The surfaces move around axes that are located in or near the joint. In angular movement, there is a change in the angle between bones. Movement in the joints can thus be classified as follows:

1. When the angle between two bones is increased, as in straightening the arm or leg, *extension* occurs. (Extension of the foot, however, is referred to as plantar flexion.)

2. A decrease in the angle between two bones, for instance, bending the elbow to touch the nose or moving the thigh forward and upward, is called *flexion.*

3. If the bone is moved away from the midline of the body, the movement is called *abduction.* Examples are elevating the arm or leg sideways to increase the angle between the appendage and the body.

4. In *adduction,* the reverse of abduction, the bone moves toward the midline, for example, moving the arm or leg toward the midline and increasing the angle between them.

5. *Rotation* involves turning or moving a bone on an axis without any displacement of the axis. One vertebra may pivot upon another while the plane of movement remains perpendicular to the long axis of the body.

6. A combination of angular and rotary movements in succession is called *circumduction*—for example, swinging the arms in a circle.

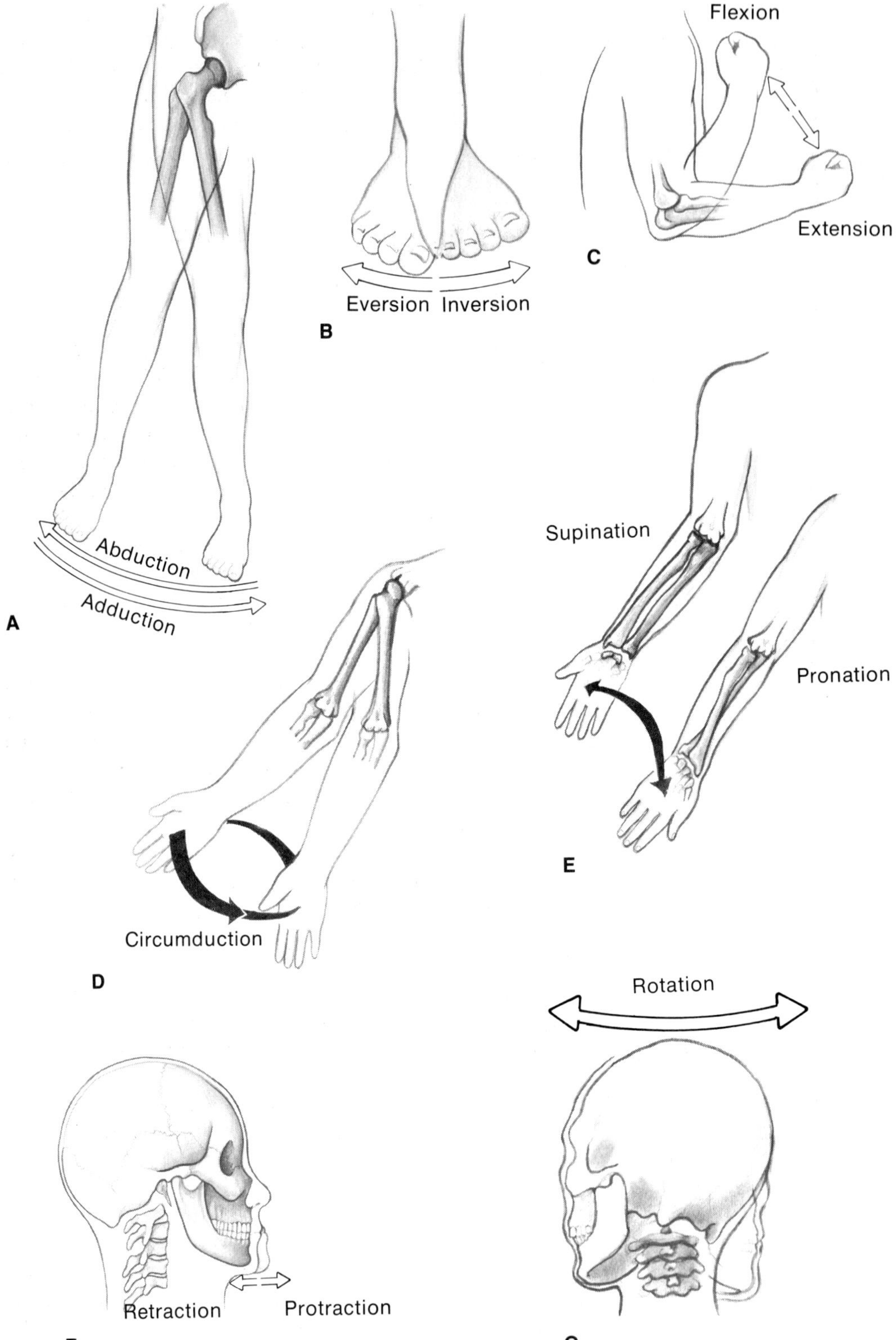

Figure 6–4 Movements of diarthrodial joints. *A*, Abduction–adduction. *B*, Eversion–inversion. *C*, Flexion–extension. *D*, Circumduction. *E*, Supination–pronation. *F*, Protraction–retraction. *G*, Rotation.

7. In *supination,* the forearm is moved so that the thumb points away from the midline. This type of movement occurs when the forearm is made to turn the palm of the hand forward or upward, resulting in parallel positions of the radius and ulna.

8. *Pronation* occurs with a movement of the forearm so that the thumb points toward the midline; that is, turning the palm inward or downward so that the radius and ulna cross each other.

9. Movement of the ankle joint so that the sole of the foot turns inward is *inversion.*

10. *Eversion,* on the other hand, is a movement of the ankle joint so that the sole of the foot turns outward.

11. Moving a part of the body forward, for example, projecting the mandible, is *protraction.*

12. In *retraction,* as the term suggests, a part of the body is moved backward, as in withdrawing the mandible.

STRUCTURE OF A SYNOVIAL JOINT

Synovial joints have three distinct features: a layer of hyaline cartilage covering the articular surfaces of the opposing bones; an articular capsule composed of bands of fibrous tissue that unite the bones and encircle the joint cavity; and a synovial membrane that lines the articular capsule and secretes a lubricating fluid, synovia, into the joint cavity (Fig. 6–5).

As previously discussed, the articular surfaces of the bones assume varying shapes according to their function. Most diarthrodial joints have congruent articular surfaces—that is, opposed surfaces that correspond to each other in shape and curvature. Some joints, however, are incongruent, and this discrepancy is sometimes compensated by the interposition of an articular cartilaginous disk that serves to cushion the surfaces, as in the knee joint (Figs. 6–5 and 6–8). Other joints have disks that entirely separate the articular surfaces, as in the sternoclavicular articulation, or the joint margins may be provided with a fibrocartilaginous lip that tends to deepen the socket, as in the glenoid cavity of the shoulder (Fig. 6–6). The surface area of most joints is increased by folds of synovial membrane, which may contain fat. These fat pads act to fill the spaces resulting from the separation of the articular surfaces as well as to reduce the friction generated by the opposing surfaces.

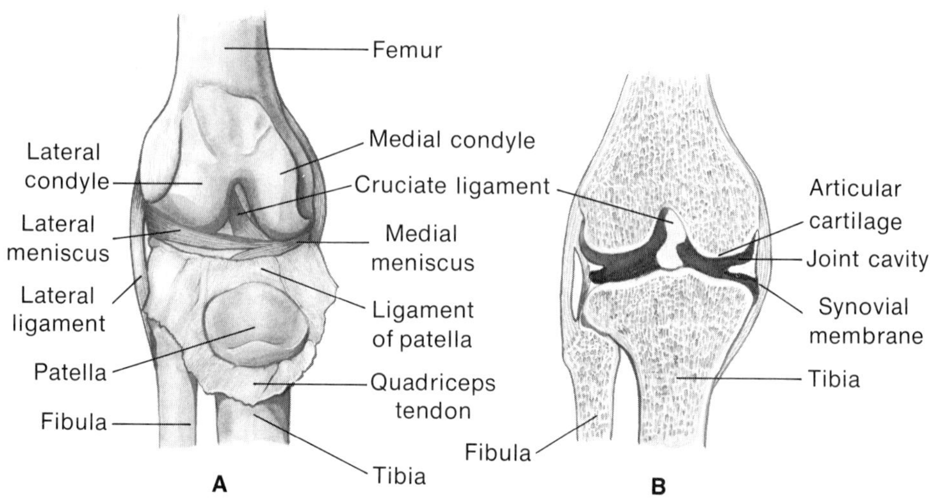

Figure 6–5 The right knee joint. *A*, Dissected from the front, with the patella resected and hanging down onto the tibia. *B*, Coronal section of the knee, exposing the joint cavity.

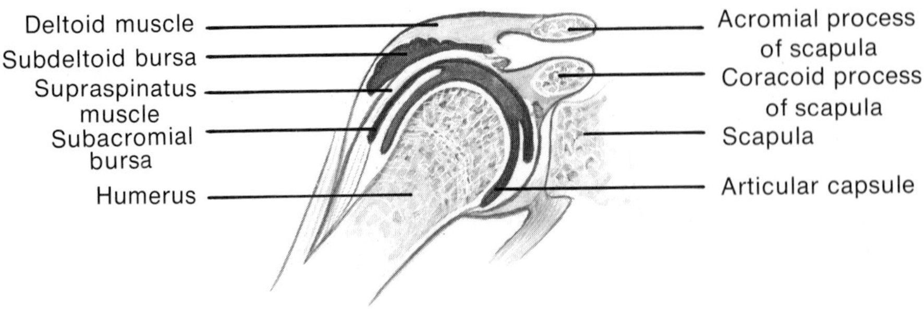

Figure 6–6 Coronal section of the shoulder, showing the relationship of the subdeltoid and subacromial bursae to the joint cavity.

The articular capsule is made up of dense fibrous connective tissue membrane continuous with the periosteum near the borders of the articular cartilage. It tends to bind the bones together, but in most joints is sufficiently loose and flexible to permit freedom of movement. In most joints it also serves to protect the joint and to support the synovial membrane.

One type of synovial membrane is frequently found outside the joint cavities where friction occurs. In such instances the membrane forms closed, padlike sacs, *bursae,* which contain a fluid similar to synovia. The major function of these specially positioned bursae is to prevent opposing surfaces from rubbing together. Therefore, bursae are found between muscles, between muscle and ligament, and in any anatomic situation where muscle and tendon glide over bony prominences or ligaments (Figs. 6–7 and 6–8).

Ligaments and muscles are also important to the function of joints. For the most part, they are fibrous bands that provide added strength. They may be incorporated into the articular capsule or may appear as independent structures. Ligaments are essentially nonelastic, but they do afford enough flexibility to permit freedom of movement. Ligaments are usually arranged so that they remain taut when the joint is in its position of greatest stability.

Muscles provide an even more important mechanism for maintaining the stability of joints. They are stronger than ligaments, maintaining the articular surfaces in firm contact at every position of the joint during relaxation and contraction (see also

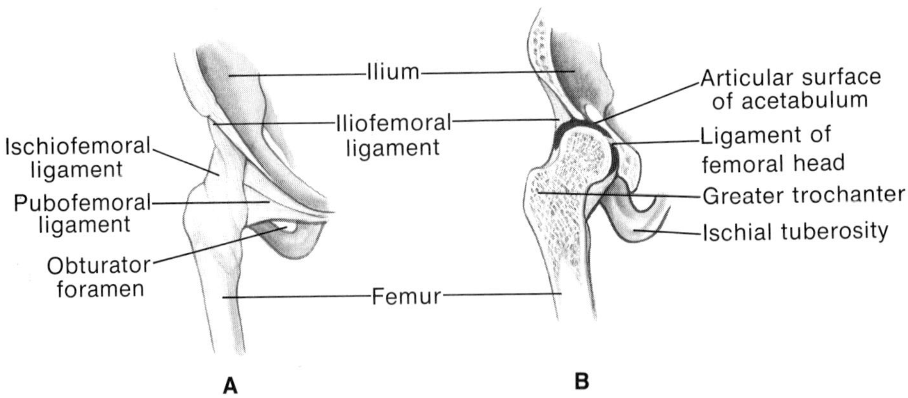

Figure 6–7 The right hip joint, anterior aspect. *A*, The articular capsule is held intact by massive groups of ligaments. *B*, Coronal section of the hip ball-and-socket joint.

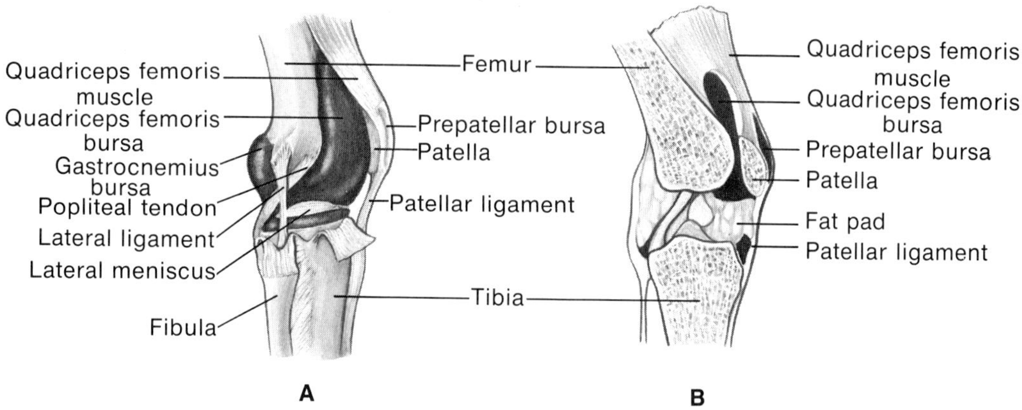

Figure 6–8 The knee, lateral aspect. *A*, Capsule of the right knee joint, indicating the position of the bursae. *B*, Sagittal section of the knee joint.

Chap. 7). In muscular paralysis a much greater than normal range of motion in the related joints is possible.

DISEASES OF THE JOINTS

Arthritis

Arthritis is a nonspecific term used to refer to any inflammatory involvement of a joint.

Suppurative arthritis is an inflammation within a joint space due primarily to invasion of bacteria, most commonly of the coccus variety. This type of arthritis is more prevalent in the middle and later years of life. Any joint may be involved, but the most frequently affected are such large joints as the knee, hip, ankle, elbow, wrist, and shoulder. The synovial membrane becomes congested and thickened, and the joint fills with a thin, cloudy fluid containing white blood cells. As the disease process advances, the inflammatory alterations in the synovial membrane grow progressively more severe, and the fluid becomes a thick pus characteristic of this disorder. The resultant synovitis (synovial inflammation) may lead to ulceration and involvement of the underlying articular cartilage. Thus extensive destruction of the joint surfaces may result in fibrous scars that seriously hamper joint function.

Gouty arthritis is the chief manifestation of a metabolic disorder that results from an alteration in the metabolism of purine, a nitrogen compound. Uric acid, a product of purine metabolism, accumulates in the blood, and urate crystals are deposited in and around the joints, leading to destruction of the articular surfaces and possible permanent loss of joint function (Fig. 6–9).

Osteoarthritis is a disease of the joints often seen in older patients. Articulations are subject to a great amount of attrition during normal function, and after years of use degenerative changes make their appearance. Most studies reveal that all persons over the age of 45 have mild and insignificant degenerative joint disease. These changes appear first in the weight-bearing joints of the spine and the lower extremities. The first pathologic change is a softening of the cartilage in the central area of the articular surfaces where the stress of weight bearing is greatest. The joints of the fingers often have small, hard nodules known as Heberden's nodes (Fig. 6–10). Subsequently a separation of the fibers occurs, followed by disintegration of the cartilage (Fig. 6–11).

Rheumatoid arthritis is the most disabling form of chronic joint disease. The condition is systemic, and widespread involvement of surrounding connective tissue usually occurs. In its early stage this

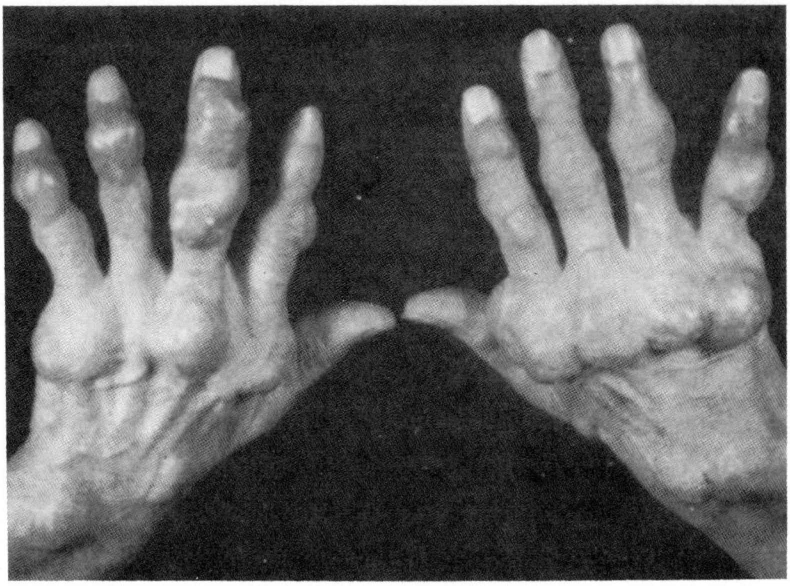

Figure 6–9 Gouty arthritis. (From Boyd, W. C.: Pathology for the Physician, 7th ed. Philadelphia: Lea & Febiger, 1965.)

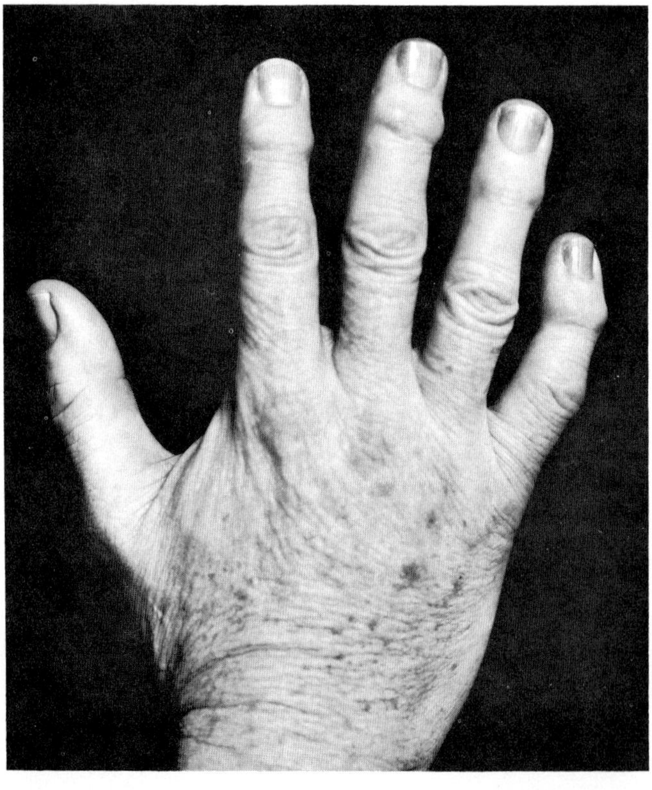

Figure 6–10 Osteoarthritis. Heberden's nodes are clearly visible. (From Gartland, J. J.: Fundamentals of Orthopaedics, 2nd ed. Philadelphia: W. B. Saunders Co., 1974.)

Figure 6–11 Osteoarthritis. Coronal sections of a developing degeneration from *A*, normal, to *C*, final fusion, when the two bones unite to become one. Note the increased width of the joint after fusion.

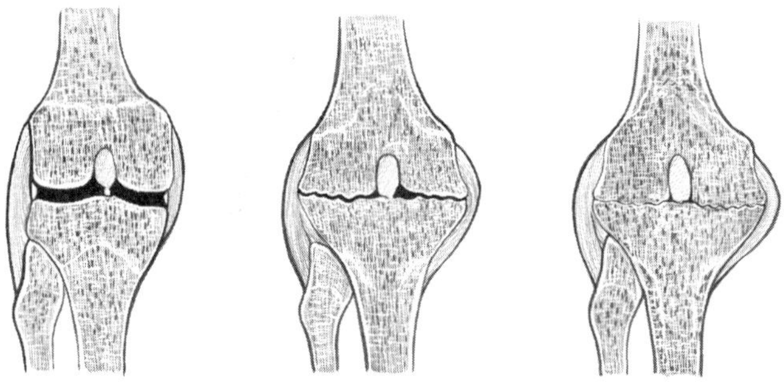

inflammation is localized in the joint capsule and synovia of such small joints as those of the hands and feet. The tissue becomes thickened from congestion and fluid accumulation, and soon invades the interior of the joint, interfering with the nutrition of the cartilage. By starvation and invasion, the articular cartilage is slowly destroyed. The changes in the synovial fluid content and inflammation in the surrounding connective tissue account for the deformities and crippling that characterize the late stages of rheumatoid arthritis (Fig. 6–12).

After extensive damage to the cartilage has occurred, the synovitis lessens, and tough, fibrous tissue prevails in the joint. This tissue adheres tightly to the opposing joint surfaces, inhibiting or preventing motion at the joint *(fibrous ankylosis).* The fibrous tissue may then become calcified and change to osseous tissue, resulting in firm bony union—an obliteration of the joint (*bony ankylosis*) which is the final stage of the pathologic process (Fig. 6–13).

Bursitis

Undue stress to or local infection of a synovial bursa may cause an inflammation called *bursitis*. Although this condition may develop in any of the periarticular bursae, some of the more common locations are the subacromial bursa close to the shoulder joint, the prepatellar bursa near the knee joint, and the olecranon bursa at the point of the elbow. Pain and limited movement of the joint result. Usually there is a varying degree of fibrositis (inflammation of fibrous tissue) at the involved bursa, and this worsens the stiffness, pain, and limitation of motion. If the condition is not corrected promptly, wasting of muscle may develop rapidly, and motion may be further restricted.

Inflammations of bursae about the ischial tuberosities and the attachment of the Achilles tendon cause similar changes.

Bunion

A bunion is a deformity of the great toe, accompanied by the formation of a bursa at the metatarsophalangeal joint. It may develop as a result of pressure or friction from badly fitting shoes. The great toe is deviated laterally toward the second toe, and the joint is made unduly prominent.

Sprains and Ruptures

A *sprain* is a mechanical derangement of the function of a joint caused by external stress or pressure. The wrenching of the joint tears some of the fibers of a supporting ligament without displacing the bones or interrupting the ligament's continuity. Many sprains, such as those of the ankle, respond favorably to rest and support with straps, splints, or casts. Complete *rupture* of the supporting attach-

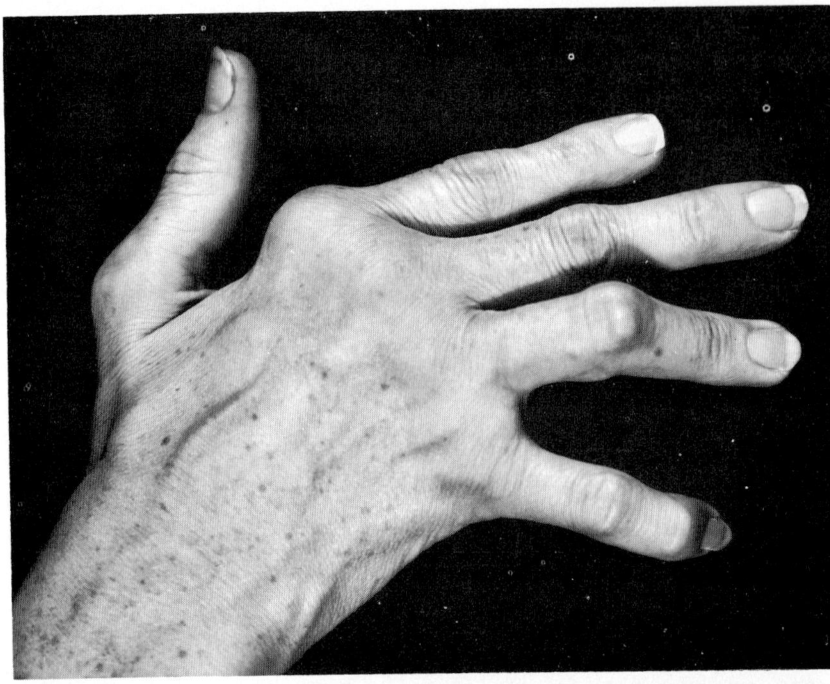

A

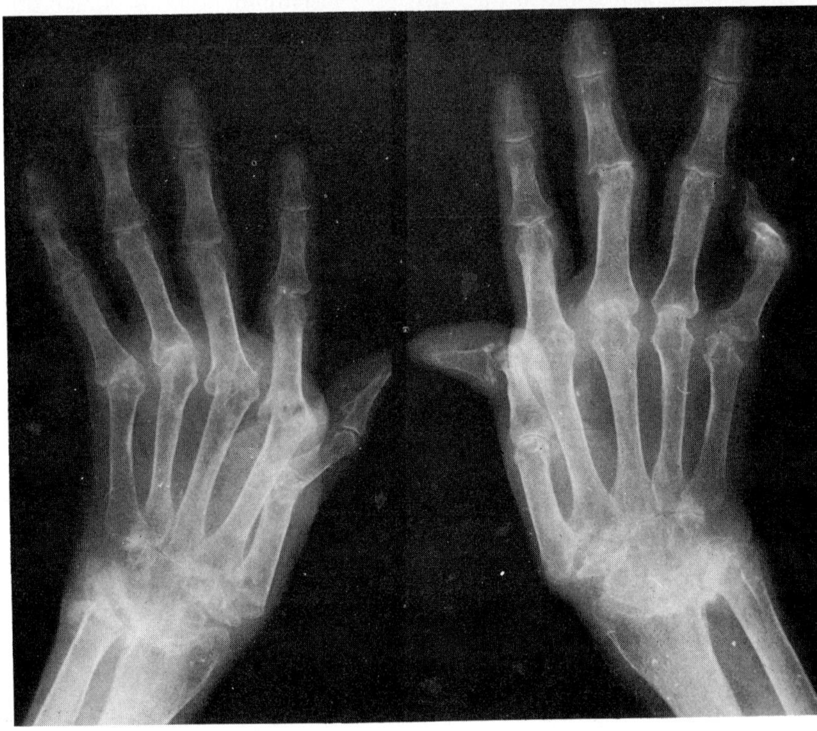

B

Figure 6–12 Rheumatoid arthritis. *A*, Changes typically seen in the hand. *B*, X-rays of hands, demonstrating bone and joint changes. (From Gartland, J. J.: Fundamentals of Orthopaedics, 2nd ed. Philadelphia: W. B. Saunders Co., 1974.)

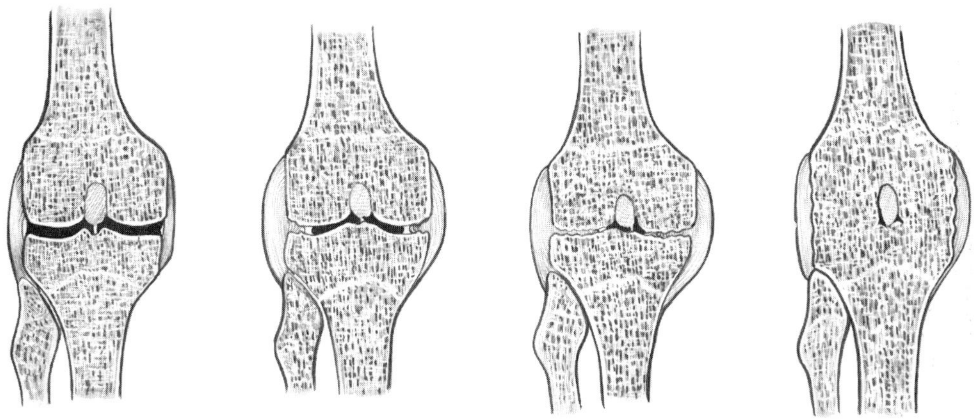

Figure 6–13 Coronal sections through joint with developing rheumatoid arthritis. As the disease progresses from *A* to *D*, fibrous tissue between the bones becomes calcified and the two bones fuse into one, resulting in ankylosis, or obliteration of the joint.

ments can be much more serious. The strong ligaments and the muscles of the thigh, for example, support the knee joint, and a twisting injury may tear the semilunar cartilage. Damage to this cartilage is a common athletic injury. Since cartilage has a poor blood supply, healing is rare. The torn cartilage or torn portion of the cartilage must be removed surgically.

OUTLINE

I. Introduction
  A. Joint—articulation between two bones
    1. Function—movement
  B. Arthrology—study of joints
II. Classification of Joints
  A. Synarthroses
    1. Fibrous
    2. Immovable
    3. Types
      a. Suture—cranial sutures
      b. Syndesmosis—tibiofibular syndesmosis
  B. Synchondroses
    1. Cartilaginous
    2. Immovable
    3. Example—epiphysis of growing long bone
  C. Amphiarthroses
    1. Semimovable
    2. Examples—intervertebral disks, pubic symphysis
  D. Diarthroses
    1. True joints—allow free movement
    2. Capsule lined with synovial membrane
    3. Synovial membrane secretes lubricating fluid—synovia
    4. Synovia fills joint cavity
    5. Types
      a. Hinge—elbow and knee
      b. Ball-and-socket—hip and shoulder
      c. Pivot—cervical vertebrae 1 and 2
      d. Ellipsoidal—radiocarpal joint
      e. Gliding—between carpal bones
      f. Saddle—thumb
III. Movement at Synovial Joints
  A. Extension
  B. Flexion
  C. Abduction
  D. Adduction
  E. Rotation
  F. Circumduction
  G. Supination

- H. Pronation
- I. Inversion
- J. Eversion
- K. Protraction
- L. Retraction

IV. Structure of a Synovial Joint
- A. Articular cartilage
- B. Articular capsule
- C. Synovial membrane
- D. Supporting and protective structures
  1. Ligaments
  2. Muscles
  3. Bursae

V. Diseases of the Joints
- A. Arthritis
  1. Suppurative—bacterial invasion
  2. Gouty—metabolic disorder
  3. Osteoarthritis—degeneration in middle or old age
  4. Rheumatoid—connective tissues as well as joints are involved
     - a. Ankylosis—fusion of two bones into one
- B. Bursitis—stress or infection of synovial bursa
- C. Bunion—inward angulation of great toe and formation of bursa at metatarsophalangeal joint
- D. Sprains and ruptures
  1. Sprain—incomplete tear of some ligament fibers
  2. Rupture—disruption of tissue continuity

STUDY QUESTIONS AND PROBLEMS

1. Of what use are bursae? What is bursitis?
2. What is a diarthrodial joint? Can it be moved?
3. Describe the structure of a synovial joint.
4. List at least five different joints and indicate where they are found.
5. What is arthritis?
6. Differentiate among the four types of arthritis described in the text.
7. Where are the semimovable joints found?
8. Define, by function, muscle, bone, joint, and ligament.

Chapter 7

The Muscles

Motion is an essential function of the body, and is made possible by special development of the property of contractility in muscle tissue. In the broad sense of the term, as used here, motion includes not only the movements of the entire body or parts of the body in their physical environment, but also those of breathing and the beating of the heart. Movements of the parts of the alimentary canal and its associated glands, of the other viscera, and of the blood and lymph vessels also must be considered.

MICROSTRUCTURE OF MUSCLE

In general, the cells and fibers of muscle tissue are specialized to possess the property of contraction. This process is accomplished by a delicate network of intercellular organoids, the muscle fibrils or *myofibrils,* which are complicated protein molecules that surround the nucleus and fill the muscle cell. When these fibrils contract, the entire fiber is made to contract or shorten.

The fibrils of muscle fiber are enclosed in a delicate but strong sheath, the *sarcolemma,* which also encloses the semifluid protoplasm of the fiber, called *sarcoplasm.* A fiber may have one or many nuclei, depending on its type and function. Individual fibers are arranged so that they lie parallel to one another, and are held together in bundles *(fasciculi)* by interlacing networks of areolar connective tissue, the *fascia.* The fibers are very small. They range from 0.001 to 0.1 millimeter in diameter (about the size of a human hair), and from 0.1 to 45 millimeters (about 2 inches) in length.

Types of Muscle

Three types of muscle can be identified by their structure, function, and location in the body: smooth muscle, skeletal muscle, and cardiac muscle.

Smooth Muscle. The fibers of smooth muscle are made up of long, slender, tapered cells grouped into thin sheets. This muscle is called smooth because of the heterogeneous arrangement of protein fibrils in its sarcoplasm. The fibrils are not grouped in large bundles as in other muscle types; therefore, they are not as readily visible under the microscope.

Each of the elongated, spindle-shaped cells of smooth muscle has a single, large, central nucleus (Fig. 7–1). The cells are usually arranged in layers surrounding viscera, forming an inner, thick, circular coat and an outer, thin, longitudinal coat. The

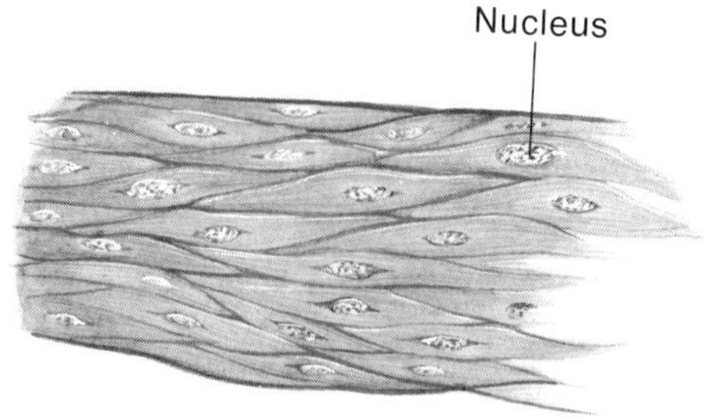

Figure 7–1 Smooth muscle. Each fiber has a single, central nucleus.

rhythmic waves of contraction of smooth muscle that pass along the walls of a hollow organ, such as the intestine, are called *peristalsis.*

Visceral muscle is most often involved in slow, sustained contractions. A single contraction may take as long as a few minutes, and the muscle tissue is capable of great extensibility. It is peculiar in its ability to be stretched considerably without a change in tension. This property is exhibited by the urinary bladder and by the uterus during pregnancy.

All smooth muscle is termed *involuntary* because its innervation originates in the autonomic nervous system, which is not under conscious control by the brain. There are nerve endings about smooth muscle cells, but motor end-plates have not been found. It is probable that impulses are transmitted by a chemical agent or by the mechanical pull exerted by one cell in contraction, which serves as a stimulus for other cells to contract.

Skeletal Muscle. The fibers of skeletal muscle are constructed to fulfill their primary purpose of facilitating skeletal movement. Each skeletal muscle can be thought of as a separate organ, since each is surrounded by its own sheath of connective tissue, the *epimysium.* Extensions of the epimysium form a tough, fibrous connective tissue, the *tendon,* which provides attachment between muscle and bone (Fig. 7-2). Because the overall action of skeletal muscles is directed by the higher brain centers, their action is voluntary, and they are therefore called *voluntary muscles.*

The striped or striated appearance of the fibers of these muscles under the microscope has led to a further designation, *striated* voluntary muscle. Each fiber consists of a long, tubular cell containing many nuclei that are positioned at the periphery of the sarcoplasm beneath the sarcolemma. Microdissection reveals cross-striped myofibrils that are closely packed together and that run lengthwise through the entire fiber. The sarcoplasm surrounds the fibrils, and has a soft, contractile consistency (Fig. 7-3). Electron microscopic studies show that a myofibril is composed of two distinct filaments, a thick filament (about 100 Å in diameter) and a thin filament (about 50 Å in diameter). These filaments are about 1.5 and 2.1 micrometers long, respectively. (An angstrom, Å, is 1/1,000,000 of a millimeter; a micrometer—formerly designated micron, μ—is 1/1000 of a millimeter). Chemical analysis has demonstrated that the thick filament contains myosin, whereas the thin filament

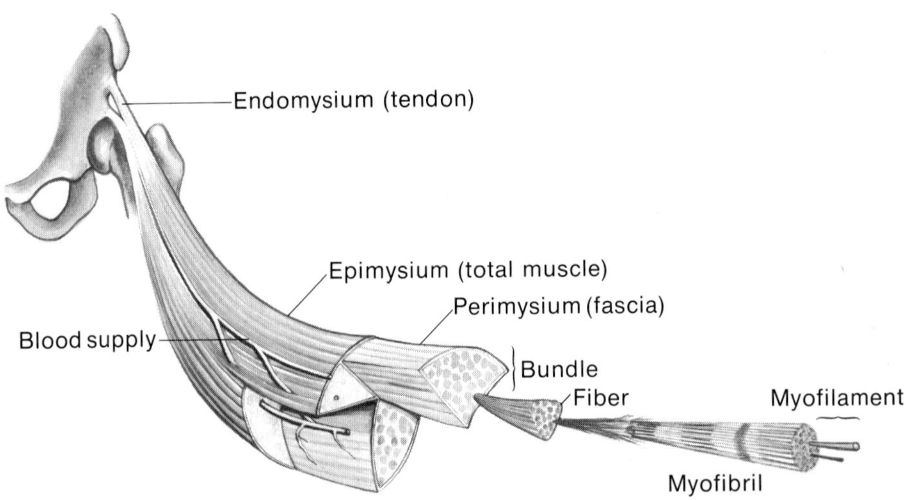

Figure 7-2 Cross-sectional detail of a skeletal muscle.

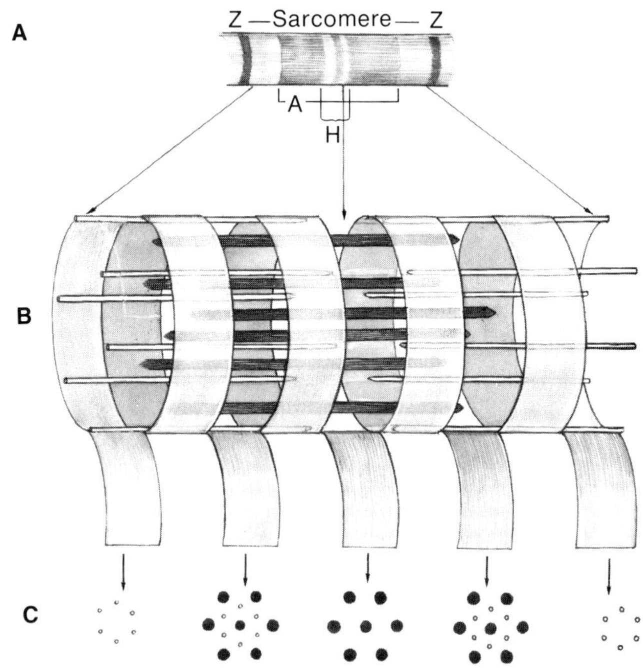

Figure 7–3 Three-dimensional transverse section through a skeletal muscle fiber. *A* and *B*, Each single muscle fiber is made up of myofibrils that appear as striations and are repeated in organized networks of bands called sarcomeres. *C*, The thick and thin filaments are arranged in a hexagonal pattern, with a thick filament surrounded by many thin filaments, except in areas where they do not overlap one another.

contains actin; actin and myosin are the large protein molecules ultimately responsible for muscle contraction (Fig. 7–3; see also p. 116). The arrangement of actin and myosin filaments contributes to the alternating light and dark areas on a striated, or striped, muscle fiber. It is the overlapping of the filaments that provides the shaded effect of the *A band* (Fig. 7–3). The light-colored area, the *I band*, is made up of thin filaments only; the narrow *H band* contains thick filaments only. A narrow band of dense material, the *Z line,* lies in the middle of the I band and divides the fiber into segments, called *sarcomeres,* from Z line to Z line. The Z lines, besides being the demarcation points of the sarcomeres, are thought to provide an area of entry into the fiber for energy materials found between the fibers.

Cardiac Muscle. Cardiac muscle, as its name suggests, is the muscular tissue of the heart. The cells are arranged end-to-end and end-to-side, forming a meshwork of branching fibers (Fig. 7–4). The dark-stained partitions or bands that separate each cell are called *intercalated disks.* Each fiber contains a single central nucleus, and is striated. This lattice arrangement of cardiac cells forms a functional

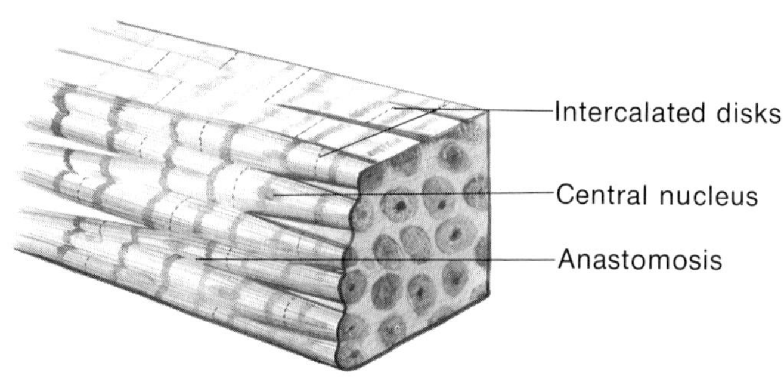

Figure 7–4 Cardiac muscle. Intercalated disks and branching fibers form an anastomosis of muscle tissue.

network, so that impulses transmitted over the membrane of any single fiber spread to all fibers.

The tissue is made up of an interconnected system of cells that spreads over the atria (upper chambers) of the heart and causes these chambers to contract as a single unit. Another set of similar muscles forms the walls of the lower chambers, the ventricles.

Cardiac muscle, like visceral muscle, is involuntary, that is, it is not under the conscious control of the brain. The outstanding characteristic of this muscle is its ability to contract and relax for a prolonged period of time—one's life.

Mitochondria are particularly abundant in the sarcoplasm of cardiac muscle. There is also an excellent blood supply, which is essential in keeping the "pump" healthy. Any disease process that interferes with the blood supply to the heart muscle may result in a "heart attack."

CHEMISTRY OF MUSCLE TISSUE

The chemical makeup of skeletal muscle has been studied extensively. Approximately 75 per cent is water; 25 per cent is solid material (20 per cent protein, 5 per cent carbohydrates, lipids, inorganic salts, and nonprotein nitrogenous compounds).

Proteins. The principal function of muscle tissue, as we have seen, is contraction. In 1942, F. B. Straub found a structural protein related to muscle contraction, which he named actin. Myosin, another protein that participates in contraction of muscle, had been isolated by W. Kühne about the middle of the last century.

Myosin is the most abundant protein in muscle; indeed, almost half the dry weight of the contractile part of a muscle is myosin. It is a complex protein molecule, water-soluble by nature, and takes the form of long fibrils with numerous links that enable it to combine with an energy source, adenosine triphosphate (ATP).

Actin, the second protein found in muscle tissue, is less abundant than myosin but equally important. It also is a fibril, and forms the second link in the actomyosin complex, a contractile fiber.

When actin and myosin are extracted from muscle tissue and mixed in solution, a new actomyosin complex is formed. The addition of ATP solution results in an immediate reduction in the size of the complex. This change in length of the fibers demonstrates the fundamental mechanism of muscle contraction.

Carbohydrates. Muscle tissue contains large quantities of carbohydrate in the form of glycogen and glucose.

Nonprotein Nitrogenous Compounds. Present in small amounts, but of great importance, are the nonprotein nitrogenous substances, adenosine triphosphate (ATP) and creatine phosphate (CP).

Energy Sources

Muscle contraction requires energy. Its immediate source is the energy-rich organic triphosphate derivatives in muscle tissue. The intermediate source of energy is the phosphate-bound creatine compound; the ultimate source is the intermediary metabolism of glycogen (glucose) performed by the cell.

ATP provides the immediate source of energy for the contractile process. This energy-rich compound contains two high energy bond locations available for immediate use. When a muscle cell is stimulated, an enzyme causes one of ATP's three phosphate radicals to split off, releasing a large quantity of energy:

$$\underset{\text{(ATP)}}{A - P \sim P \sim P} \rightarrow \underset{\text{(ADP)}}{A - P \sim P} + P + \Delta$$

Energy released by the breakdown of ATP is used in the performance of work.

ATP is resynthesized from adenosine diphosphate (ADP) by the addition of a phosphate group. Under normal conditions, the energy for this endothermic reaction is supplied by the breakdown of glucose to CO_2 and H_2O, but there also exists in muscle another energy-rich phosphate compound that can supply the necessary energy more quickly. Within a fraction of a second after the contractile process ends, ADP is converted back to ATP. Most of this intermediate energy supply comes from phosphocreatine (creatine phosphate, CP), which is hydrolyzed to creatine and phosphate groups accompanied by the release of considerable energy.

$$C \sim P \rightarrow C + P + \Delta$$

Phosphocreatine is in far greater supply in muscle than is ATP, but its energy is utilized by reconversion of ATP.

$$\text{ADP} + \text{Phosphocreatine} \rightarrow \text{ATP} + \text{Creatine}$$

All of the energy for phosphocreatine resynthesis—and much of the energy for ATP resynthesis—comes from the breakdown of glucose to CO_2 and H_2O. Glucose in the bloodstream enters cells, where it is degraded through a series of chemical reactions to pyruvic acid. Glycogen, a source of intracellular glucose, is a carbohydrate polymer found in abundance in the cells of the liver and in the skeletal muscle. When adequate oxygen is present, pyruvic acid enters the citric acid cycle and is metabolized—through this cycle and the accompanying respiratory chain—to CO_2 and H_2O. This process is called *aerobic respiration.* The metabolism of one mole of glucose to CO_2 and H_2O liberates sufficient energy to form 38 moles of ATP from ADP.

If oxygen supplies are insufficient, the pyruvic acid formed from glucose does not enter the citric acid cycle, and is reduced to lactic acid, a toxic by-product. This process is called *anaerobic respiration,* and is associated with the net production of enough energy to form the equivalent of only 2 moles of ATP per mole of glucose.

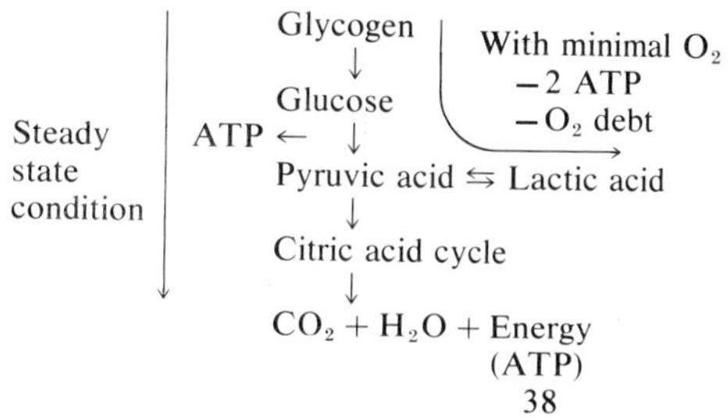

Figure 7–5 The oxygen debt mechanism.

The Oxygen Debt Mechanism

The presence of an anaerobic (oxygen-deficient) pathway for glucose breakdown has an important physiologic consequence. During muscle exercise, the blood vessels dilate and the blood flow is increased, so that the available oxygen supply also is increased. This is referred to as a "steady state" condition. However, when muscular exertion is very great, aerobic resynthesis of energy stores cannot keep pace with their utilization. Under these conditions, phosphocreatine is created by using energy released by the anaerobic breakdown of glucose to lactic acid. This use of the anaerobic pathways is limited because of the accumulation of lactic acid in the muscles. As lactic acid accumulates, the person is said to be running up an "oxygen debt" (Fig. 7–5). During the period following strenuous exercise, rapid breathing "pays" the debt owed to the body, and lactic acid is changed back to pyruvic acid, which is then oxidized to provide energy for the conversion of the remainder to glycogen, as well as for the restoration of phosphocreatine and ATP.

If, for any reason, a person is unable to repay this debt and to prevent an accumulation of lactic acid, muscle fatigue will result. In this condition the resting phase of muscle contraction lags far behind the contraction phase. The muscles are physically unable to respond to a stimulus of any type, and therefore are inefficient. Rest generally revives the muscle cells.

Rigor Mortis

Soon after death the body becomes very stiff. When dead, a muscle loses its irritability, shortens, and becomes opaque, rigid, and inextensible. These changes are attributed to the depletion of ATP and the accumulation of lactic acid. The cause of rigor is not well understood, but is related directly to the disappearance of glycogen. It begins in the muscle of the jaw and passes successively down the neck, arms, trunk, and legs. When glycogen is removed from a muscle after death, rigor sets in almost immediately.

CONTRACTION OF MUSCLES

Mechanism of Muscle Contraction

The really significant question in the cellular physiology of muscle is the manner in which the protein molecules operate to produce muscular contraction and relaxation. Although many hypotheses have been proposed, the sliding filament hypothesis presented by H. E. Huxley has been the most strongly supported.

The precise means by which actin and myosin molecules are made to react with each other—and, more specifically, to cause contraction of the myofibril—are not known; nevertheless, many attributes of myosin and actin have been discovered.

First, actin is known to combine readily with ATP.

Second, myosin is known to be an enzyme which, if adequate numbers of calcium ions are present, can release the energy carried by ATP. Thus myosin is commonly referred to as "ATPase." In the absence of calcium ions, it does not have this property.

Third, another ion, magnesium, which normally is abundant in the intercellular fluid, acts to inhibit the reaction between actin and myosin and therefore to keep the myofibril in a relaxed state.

On the basis of these facts, the sliding filament theory was proposed. As discussed earlier, each muscle fiber contains many myofibrils embedded in the sarcoplasm. Each fibril consists of myosin and actin filaments arranged longitudinally in the sarcomere (Fig. 7–3). The arrangement is more precise in cardiac and skeletal muscle than in smooth muscle, but the contractile process is essentially the same in all three types of muscle tissue. Experiments have shown that the thin filaments (actin), which extend from the ends (Z lines) of the sarcomere but do not touch the middle, will combine with the thick filaments (myosin). The latter are positioned in the middle of the sarcomere in the presence of adenosine diphosphate (ADP). When stimulated, myosin is activated by calcium ions to perform its enzyme role as ATPase. It causes the terminal phosphate radical to split from the ATP molecule:

$$\text{ATP} + \text{myosin (ATPase)} \xrightarrow{\text{Ca}^{++}} \text{ADP} + \text{P} + \Delta$$

This hydrolyzes ATP to ADP, releasing large quantities of energy. At the same time, the binding sites (active spots) on the actin filament are exposed, attracting the "cross-bridges" of the myosin filament (Fig. 7–6). Contraction of the sarcomere takes place when there is an interaction or attraction between these filaments, causing the thin actin filaments to move inward among the myosin filaments in a kind of ratchet or oscillating action. This requires the presence of active sites on the actin filaments into which the "cross-bridges" temporarily hook to pull the filament a short distance and then release it and hook onto the next site (Fig. 7–6).

In summary, the conversion of adenosine triphosphate (ATP) to adenosine diphosphate (ADP), in the presence of calcium ions, sets up a field of attraction that draws the actin and myosin filaments together, producing a shortening of distance between the Z lines of a sarcomere—a contraction. The chemical changes involved in this process are not

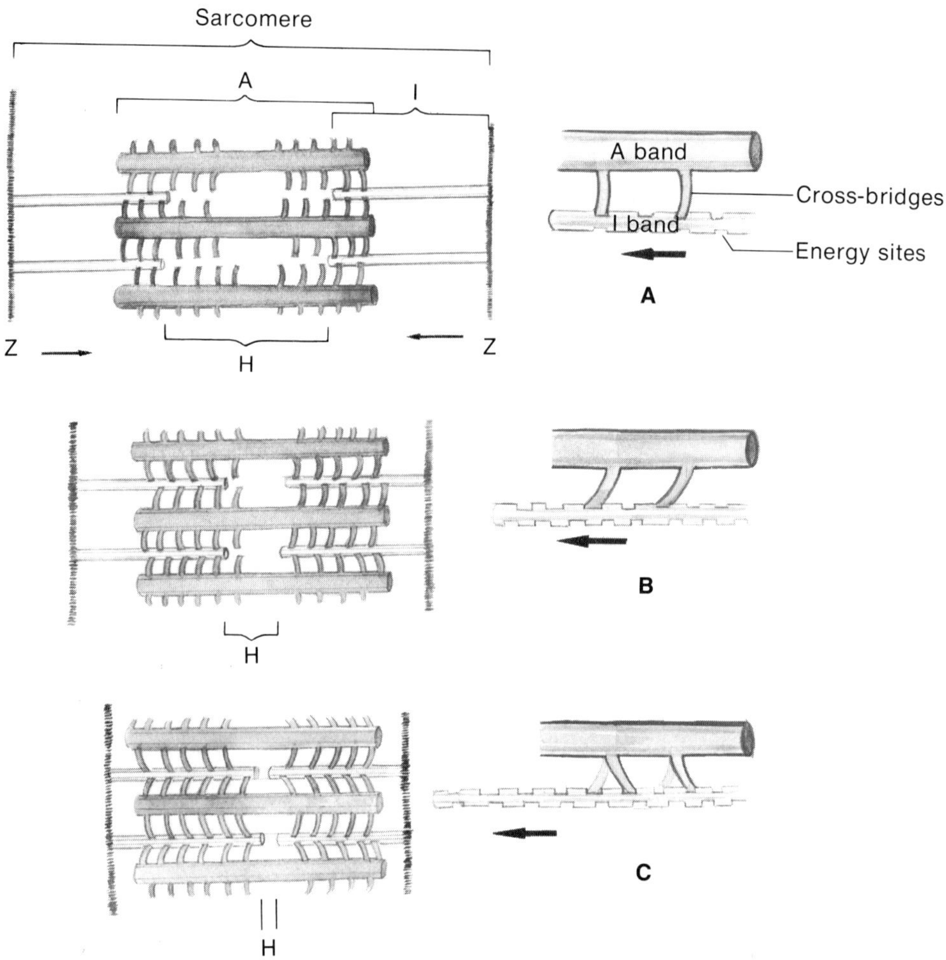

Figure 7-6 Contraction of a muscle. *A*, Huxley's sliding filament theory proposes that the A bands contain flexible cross-bridges that come in contact with energy sites on the more numerous I bands. *B* and *C*, With the availability of energy, the cross-bridges pull the active filament a short distance *(B)*, release it, and attach to another site *(C)*, resulting in a shortening of the H zone between the I bands: contraction.

yet completely understood; however, the force created by this "actomyosin complex" is so strong that it is more powerful, pound for pound, than any engine designed by man.

The "efficiency" of an engine is calculated to be the percentage of energy input that is converted into work instead of heat. The chemical energy (input) of a muscle that can perform work is less than 25 per cent, the remainder being lost as heat. This quantity can be realized only when the muscle contracts at a moderate speed—neither too slowly nor at an excessively rapid rate.

Many features of muscle contraction can be demonstrated by a kymograph (Fig. 7-7), a device used to elicit muscle twitches in the laboratory. This can be accomplished either by stimulating the nerve that conducts impulses to a muscle or by stimulating the muscle belly itself. The stimulus causes a single, sudden contraction of the muscle—a muscle twitch (Fig. 7-8)—which can be recorded on the blackened drum of the kymograph. Succeeding twitches can be produced experimentally to demonstrate such environmental and physical effects as heat, work, cold, fatigue, and so forth.

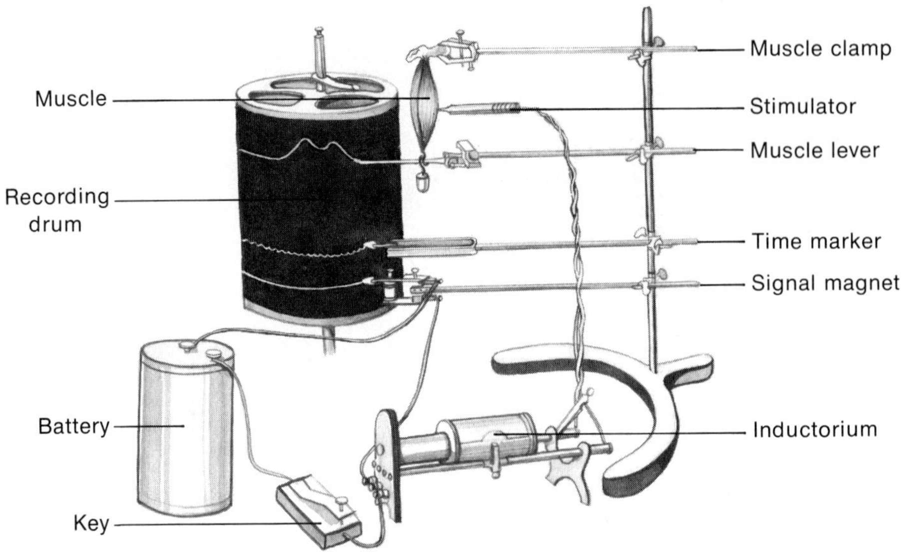

Figure 7-7 A kymograph used in analyzing muscle action.

Properties of Muscle Related to Contraction

There are four distinct properties of muscle having to do with aspects of contraction. These are *excitability, contractility, extensibility,* and *elasticity.*

Excitability

Excitability (irritability) is the property of being able to receive a stimulus and respond to it. Thanks to this property it is possible to stimulate a muscle directly with an electric shock (current) and cause it to contract or respond. This is of utmost importance in the treatment of paralysis due to motor deficiencies or to disorders caused by disease. A muscle that does not respond over a prolonged period will degenerate, or atrophy.

Contraction does not occur immediately upon the presentation of an adequate stimulus. A finite period is necessary for the chemical processes resulting in a contraction to take place (Fig. 7-8). This period, approximately 1/100 second, is called the *latent period.* At its end, the muscle begins to shorten and thicken—a period of approximately 4/100 second called the *contraction period.* It is during this period that tension is developed. After the work is done, the muscle returns to its normal size and shape, and undergoes a *relaxation period* of about 5/100 second. As an example, the frog's gastrocnemius muscle takes about 1/10 second to respond to an adequate stimulus and to return to normal.

At this point, some definitions of terms are necessary:

1. *Stimulus*—an excitant or irritant that causes a reaction, such as a movement.

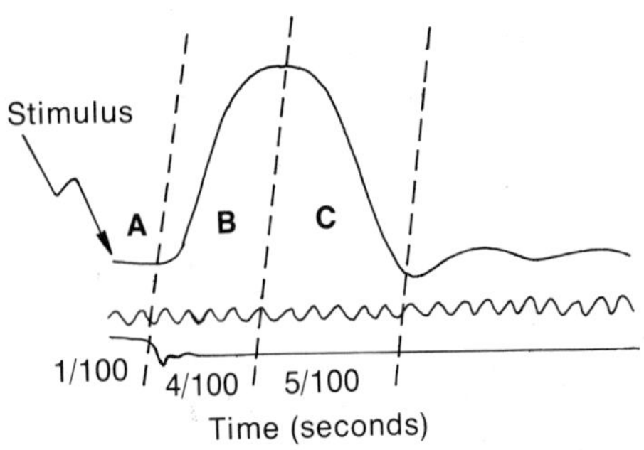

Figure 7-8 A single muscle twitch. *A*, Latent period of approximately 1/100 of a second after muscle receives an impulse. *B*, Contraction period: muscle works for 4/100 of a second. *C*, Muscle returns to normal size and shape during the relaxation period, 5/100 of a second.

(a) *Liminal* (threshold)—the weakest stimulus from a nerve fiber that will cause a response.
(b) *Subliminal*—any stimulus of lesser intensity than the liminal or threshold.
(c) *Maximal*—a stimulus that causes all the motor units of the muscle to respond, resulting in the greatest possible contraction of the muscle.
(d) *Adequate*—a stimulus of the specific form of energy to which the nerve fiber is most sensitive.

2. *Response*—the reaction, such as movement of a muscle, that results from an adequate stimulus.

Although a single subliminal stimulus cannot make a muscle respond, a series of subliminal stimuli applied to a muscle can elicit a response. This is spoken of as the "summation effect."

Contractility

Contractility is the property of a muscle that enables it to change its shape. As previously stated, the conversion of potential energy (ATP) into energy for muscle contraction is accompanied by the loss of some energy in the form of heat.

When a muscle contracts *isometrically,* it remains the same length and all the energy developed escapes as heat, since no mechanical work is performed.

When a muscle develops enough tension to overcome its work load, the muscle contracts and performs mechanical work—an *isotonic* contraction in which the tension within the muscle stays the same.

Examples of these two types of contraction can be found in two types of exercise. In exercises using the principle of opposing forces—for example, pushing against a wall—no mechanical work is done, and so the muscle works only to expend a great deal of energy as heat. Contrast this isometric exercise with the popular weight-lifting procedures, where the muscle overcomes the opposing weight and contracts, doing mechanical work as well as generating heat—an isotonic contraction.

The basic unit of muscular contraction is the *motor unit* (see also p. 118). This consists of many fibers, all innervated by the same motor axon. These fibers are actually intermeshed so that a single impulse causes a contraction throughout the entire muscle body, not only in one area of the muscle.

Motor units are subject to two basic laws of contraction:

1. The *"All-or-None Law"*—the muscle fiber will either respond entirely to a stimulus or not respond at all.

2. The *Law of Tonicity*—the healthy muscle is in a state of partial, continual contraction. Tonus is also described as the involuntary resistance to stretch *(stretch reflex).*

These two laws might seem at first to conflict with each other, but in reality they are complementary in their actions. The all-or-none principle applies strictly to the individual fiber, whereas the law of tonicity applies to the entire muscle, that is, to hundreds of fibers. The fundamental mechanism whereby muscle tone (tonus) is produced is not fully understood, but physiologically tone is known to be due to nerve impulses.

Tonus in skeletal muscle is created by the stretch reflex. It is completely dependent on the central nervous system for maintenance. Muscles that control the skeleton are stretched over the joints, for instance, the "extensor" muscles used for maintaining posture. By means of tonic contraction, posture is maintained for long periods with little evidence of exhaustion. This effect is achieved mainly by a mechanism in which different groups of muscle fibers contract in relays, giving alternating periods of rest and activity to given groups of fibers. In man, the retractors of the neck and extensors of the back are among the muscles exhibiting the highest degree of tonus.

Muscle tone is often reduced during illness or when prolonged bed rest is necessary. Muscles can lose their tone in several ways. When its nerve is cut, injured, or diseased, a muscle does not receive a constant flow of impulses and thus loses its tone. It soon becomes flaccid (soft) and then atrophies (wastes away). This process is called the *atrophy of denervation.*

Tone may also be lost by immobilization or inactivity of a part, such as in bedridden people or when limbs are held in a cast for long periods. The normal flow of nerve impulses to such muscles is reduced, and atrophy follows. This is called *atrophy of disuse,* and is not as serious as atrophy of denervation, since the nerve connection remains normal, and the muscle can recover when it is again normally active.

Tonus in smooth muscle differs from that of skeletal muscle in that it is dependent upon the peripheral rather than the central nervous system. Smooth muscle tonus can be influenced markedly when drugs are introduced into the tissue fluid that bathes the muscle fiber. The response to a particular stimulus may occur as a rhythmic contraction, an increase in tonus, or a decrease in tonus. Drugs such as epinephrine (Adrenalin) and sympathin are known to stimulate a greater activity of the muscle; other drugs, such as strong concentrations of acetylcholine and curare, lead to a general loss of tone and muscle activity.

Cardiac muscle tone is also independent of the central nervous system. Because of its special nature, it will be discussed in greater detail later in this chapter.

Extensibility

Extensibility in living muscle cells is the property that allows the muscle to be stretched or extended. All types of muscle can be stretched when force is applied. For example, this property makes possible the increase in size of the urinary bladder as urine accumulates. If the bladder could not distend, a very uncomfortable dribbling of urine would occur instead of the usual periodic emptying.

Elasticity

Elasticity is the ability of a muscle to return to its original form when a stretching force is removed. Provided muscle damage (rupture) has not occurred, on removal of the weight the muscle will return to its normal length. Since most muscles of the body are generally somewhat stretched, the two properties of extensibility and elasticity are of value in that they tend to keep the muscle in continuous readiness for contraction.

Some Phenomena of Contraction

Skeletal muscles contract quickly and relax promptly. In contrast, visceral (smooth) muscle contractions develop slowly, are maintained for some time, and fade out slowly. The contraction of a skeletal muscle is the result of a stimulus discharged by the nerve fibers innervating it. If a single contraction were to be analyzed, it would show brief periods of pause, contraction, and relaxation, as discussed previously. In general, the stronger the stimulus (up to a certain maximum), the stronger the contraction will be. Muscles do their best work at about body temperature 37° C (98.6° F). If the temperature is raised or lowered, the muscle will respond accordingly in the strength and duration of its contraction (Fig. 7–9). If the temperature is raised much above 37° C, the muscle loses excitability and becomes functionally depressed very rapidly—a condition called *heat rigor,* a permanent shortening. This effect of temperature is understood and put to use by athletes. Before strenuous muscular activity during physical contests, they use "warm up" exercises or wear jackets to keep their bodies at a constant tempera-

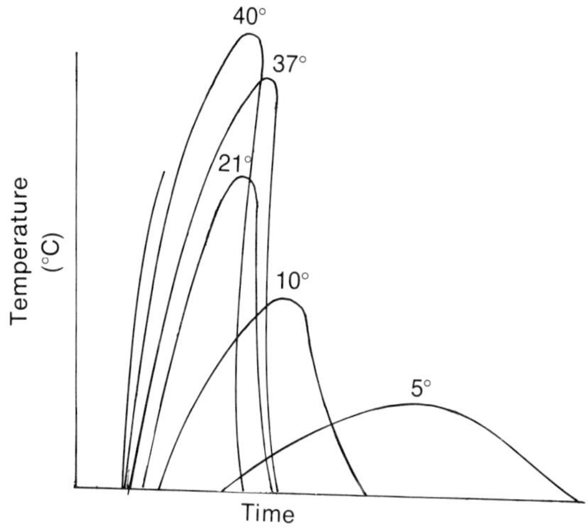

Figure 7–9 The effect of temperature on muscle action. The lower the temperature, the longer the contraction, but the strength of the contraction is lessened.

ure. This prevents "cooling off" and thereby keeps the muscle from contracting less efficiently.

As discussed above in relation to isotonic contraction, the development of tension enables a muscle to overcome resistance and, under proper conditions, to perform such mechanical work as lifting a weight. The amount of mechanical work done is determined by multiplying the weight of the load by the height to which the load is lifted. The result expresses the work done by the muscle. When a muscle contracts without a load, or when the load is too heavy to be lifted, no mechanical work is done.

Tetany occurs when a number of stimuli are applied to a muscle in rapid succession, so that little time is offered for relaxation between the successive contractions (Fig. 7–10). There is a degree of "fusion" of the twitches; the greater the frequency of the stimuli, from 3 per second to 8 per second, the more nearly complete is the fusion (incomplete tetany), until at 16 per second the recording shows no evidence of individual twitches. This steady state of contraction is called complete tetany. The tension developed during tetany may be four or more times as great as that during a single twitch.

A single muscle contraction would be of little use in coordinated movement. Those muscles that enable us to stand up, walk, run, or bend over maintain a state of physiologic tetany for a long period without ill effect. The condition of tetany can apply, however, only to a portion of a muscle rather than to the entire muscle. The slow flexing of a muscle is a coordinated movement involving many fibers in a state of contraction, and is much slower than a muscle twitch.

Tetany is closely related to muscle tone. In the maintenance of muscle tone only a few motor units are involved at one time, so that no actual movement occurs. If all the motor units in use contracted and relaxed at the same time, the result would

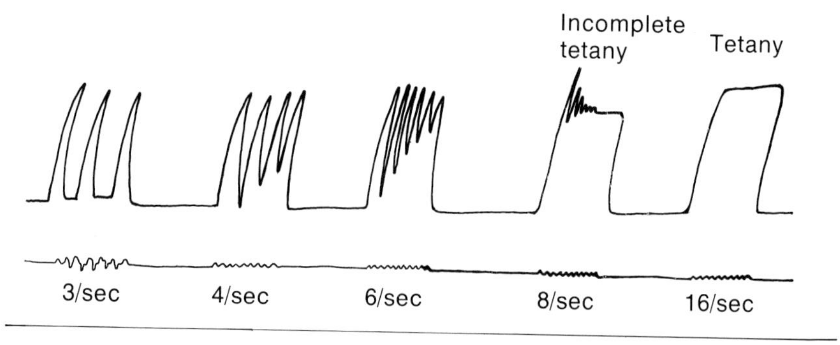

Figure 7–10 Tetanic contraction. As the number of impulses to a muscle is increased, the rest periods become less until finally the muscle cannot rest: tetanic contraction.

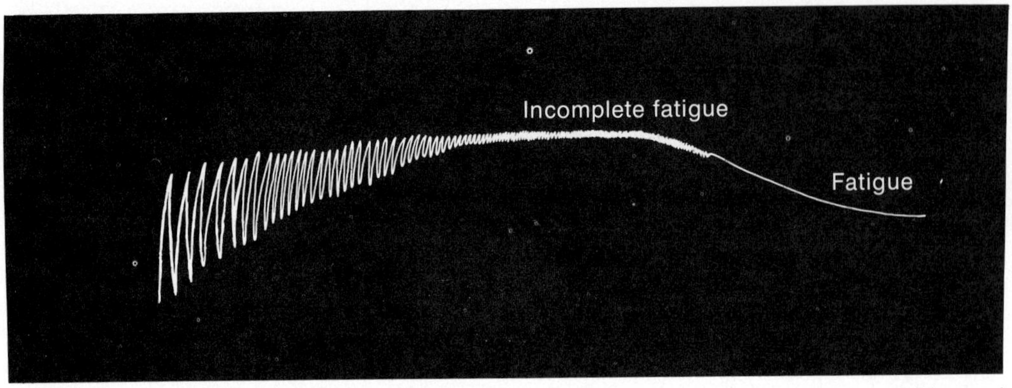

Figure 7–11 Fatigue. During stimulation at intervals of 1 second, the strength of contraction of this muscle diminished until it could not respond at all: fatigue.

be a series of jerks instead of a smooth, coordinated contraction. Such a condition, called *clonus*, sometimes occurs in response to forcible extension of some muscles.

A *spasm* is a sudden involuntary contraction of a skeletal muscle. If it persists, it is called a *tonic* spasm, or *cramp*. The characteristic "charley horse" is a spasm due to an overexertion of the muscle of the thigh. Most cramps or spasms are painful.

In muscles undergoing contraction, the formation of carbon dioxide and lactic acid, both toxic waste products of cellular metabolism, first causes an increase in the irritability of the muscle. But, if the muscle is continuously stimulated, the strength of contraction becomes progressively less until the muscle refuses to respond (Fig. 7–11). This is true *fatigue*, and is caused by the toxic effects of waste products that accumulate during exercise. Added factors are the loss of nutritive materials and the inability of the muscle to store sufficient oxygen to compensate for strenuous exertion.

It is important to observe that in fatigue not only does the muscle gradually fail to contract, but it also fails to relax (Fig. 7–11). This explains why muscle fatigue is sometimes confused with a cramp or spasm.

Exercise stimulates circulation, thereby bringing fresh blood to the tissues and altering the local pressure in the muscle. Exercise increases the size, strength, and tone of muscle fibers. Massage and passive exercise (as in hydrothermy machines) may, if necessary, be used as substitutes for active exercise. Although physical exercise is desirable to aid the body's metabolism, the continued use of fatigued muscles is harmful if their energy supply (glycogen) is exhausted and they must utilize the protein of their own cells. Normally, the sensation of fatigue protects us from such extremes.

Conduction of the Contractile Impulse at the Neuromuscular Junction

We can control with great precision the force applied by our muscles. We can perform such delicate actions as picking up a tiny insect, or, using a powerful grip, lift a hundred-pound weight. This gradient of force is regulated by the nervous system. The muscle fibers contract in response to impulses from the central nervous system, to which every skeletal muscle is attached by means of a branch of a nerve fiber. The impulse proceeds down each nerve fiber branch, ending at the neuromuscular junction, where it acts to trigger the muscle fiber and cause contraction. A nerve fiber and its associated muscle fibers together form a *motor unit*. In the large muscles of

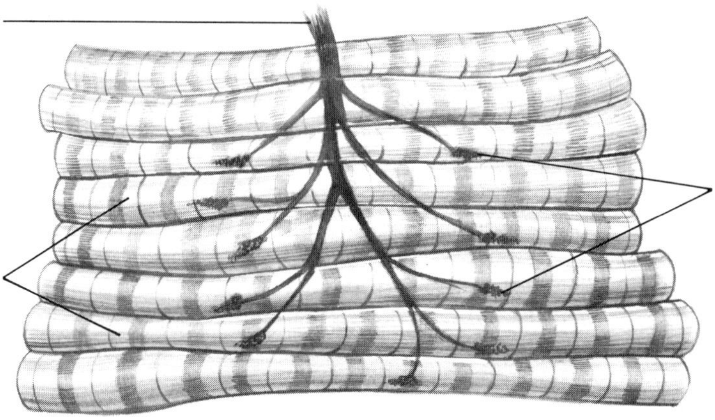

Figure 7–12 Innervation of skeletal muscle.

the legs, a single motor nerve may serve as many as one hundred muscle fibers (Fig. 7–12).

At one time it was thought that the nerve impulse crossed to the muscle fibers in the same way that an electric spark jumps across a gap. It is now known that the arrival of nerve impulses at the motor end-plate (Fig. 7–13) causes the production of acetylcholine, a chemical that alters the muscle cell membrane like an electric shock. The wave of excitation (stimulation) produced spreads along the muscle fiber in each direction until the muscle fiber contracts.

Acetylcholine is the mediator or carrier at all junctions involving muscle and nerve. Within a fraction of a second, acetylcholine is destroyed by the enzyme, cholinesterase, which prevents the re-excitation of the muscle fiber until the arrival of another impulse. If the breakdown of acetylcholine into acetyl and choline did not occur, the chemical would keep the synaptic bridge intact and allow continual contractions, which could result in exhaustion of the muscle.

The synthesis of acetylcholine at the end-plate involves a second enzyme, choline acetylase, which is found in high quan-

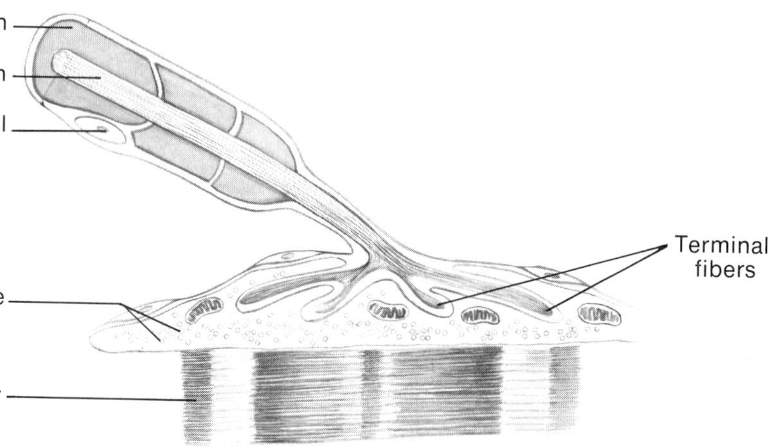

Figure 7–13 Cross-section of a motor end-plate on a skeletal muscle, showing the actual connection between the muscle fiber and the terminal branch of the nerve fiber.

tities at nerve endings where acetylcholine is the mediator. Choline acetylase catalyzes the production of acetylcholine from the available acetyl and choline compounds. It remains in an inactive state until an impulse from the motor nerve activates it again.

Events at the neuromuscular junction can be modified in many ways, such as by disuse or by drugs. Curare, a South American Indian poison, acts to bind acetylcholine and block transmission of the impulse across the myoneural junction, thus preventing the contraction of muscles. This knowledge has led to the controlled use of curare and associated drugs in conjunction with anesthesia for various types of surgery where muscular relaxation is desired. Curare has also been very useful in the emergency treatment of children who develop muscular spasms of the throat (croup), and has prevented strangulation or suffocation in such patients.

Succinylcholine also blocks transmission at the myoneural junction. Its action involves tying up the cholinesterase and preventing the muscle fibers from returning to normal after they have contracted once.

Some complex organic fluorine compounds are potent inhibitors of cholinesterase. These compounds are the "nerve gases" developed as military weapons. They can cause paralysis of the muscles and death from asphyxiation.

Another group of substances affects the release of acetylcholine from the nerve terminals. Botulinus toxin, produced by *Clostridium botulinum,* blocks the release of acetylcholine, preventing excitation of the muscle membrane. Botulinus toxin is responsible for many deaths from food poisoning. This, again, is a weapon used in germ warfare.

Myasthenia gravis, a condition characterized by weakness of the skeletal muscles, is due to a blockage of impulses at the neuromuscular junction. The beneficial therapeutic action of anticholinesterase agents would indicate that myasthenia is essentially a problem of competitive inhibition by a curare-like substance.

THE SKELETAL MUSCLES

The skeletal muscles (Fig. 7–14, *A*) are connected, for the most part, to the skeleton, and their principal function is to produce movement of one bone upon another. In order to accomplish this, they must have two attachments: one to the bone that is fixed, called the *origin* of the muscle; the other to the bone to be moved, the *insertion* of the muscle. This does not mean that muscles attach only to bones. Some attach to other muscles as well.

As mentioned earlier in this chapter, muscles are attached to bones by *tendons,* which are cords or bands of white fibrous connective tissue.

To understand the action of the skeletal muscles, you should know their names. The following list of ways in which muscles are named includes specific examples of each:

1. *Location*—intercostal (between the ribs), brachii (arm), occipitalis (head).
2. *Action*—adductor, flexor, extensor.
3. *Shape or size*—trapezius (trapezoid), maximus (largest), minimus (smallest), longus (long), brevis (short).
4. *Direction of fibers*—rectus (straight), transversus (across).
5. *Number of heads of origin*—biceps (2 heads), triceps (3 heads), quadriceps (4 heads).
6. *Points of attachment*—sternocleidomastoid (sternum, clavicle, and mastoid process of the temporal bone).

The motions produced by muscle contraction are described with reference to the standard "anatomic position" of the body—standing erect with arms straight down and palms directed forward (Fig. 7–15). Movements parallel to the longitudinal axis are *flexion,* or folding together, and *extension,* an unfolding movement in the opposite direction (Fig. 7–14, *B*). *Abduc-*

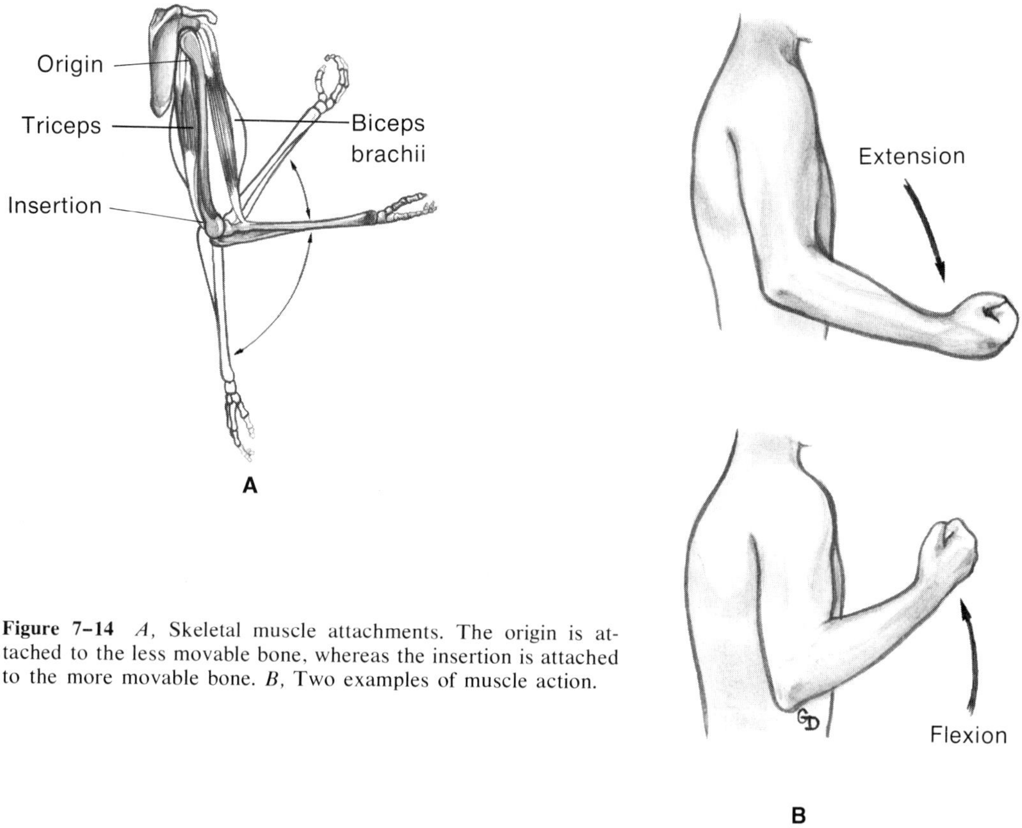

Figure 7–14 *A*, Skeletal muscle attachments. The origin is attached to the less movable bone, whereas the insertion is attached to the more movable bone. *B*, Two examples of muscle action.

tion refers to movement away from the axis; *adduction* means toward the axis of the body.

Muscles that produce movement in directions opposite to one another are said to be *antagonists* (Fig. 7–14, *A* and *B*). Muscles that work together to accomplish a particular movement are *synergists*. Two muscles that bend an arm at the elbow are synergists with one another (e.g., the biceps and brachialis flex the lower arm), but both are antagonistic to a muscle that straightens the arm (e.g., the triceps extends the lower arm).

Muscles of Facial Expression and Mastication (Table 7–1)

The muscles of facial expression enable one to express physically such feelings as disdain, pleasure, pain, disgust, fear, anger, and surprise (Fig. 7–16). The muscles of mastication aid in the digestive process by providing power for chewing food before it is swallowed.

The *epicranius* consists of two separate muscular portions: the *occipitalis* and *frontalis* muscles covering their respective cranial bones. Connecting the two is a large sheet of fibrous tissue, the *galea aponeurotica* of the scalp. The epicranius as a whole raises the eyebrows and moves the scalp, thereby expressing surprise.

The *orbicularis oculi* (Figs. 7–15, *A*, and 7–16) encircles the eyelids and closes the eye. It is a flat, elliptical muscle that arises from the nasal portion of the frontal bone and the frontal process of the maxilla. The fibers spread laterally, forming a broad, thin layer that occupies the eyelid, encircles the orbit of the eye, and spreads over part of the temple and downward on the cheek. Contraction of some of the fibers causes the eyelids to close gently, as in sleep. Contraction of most fibers creates

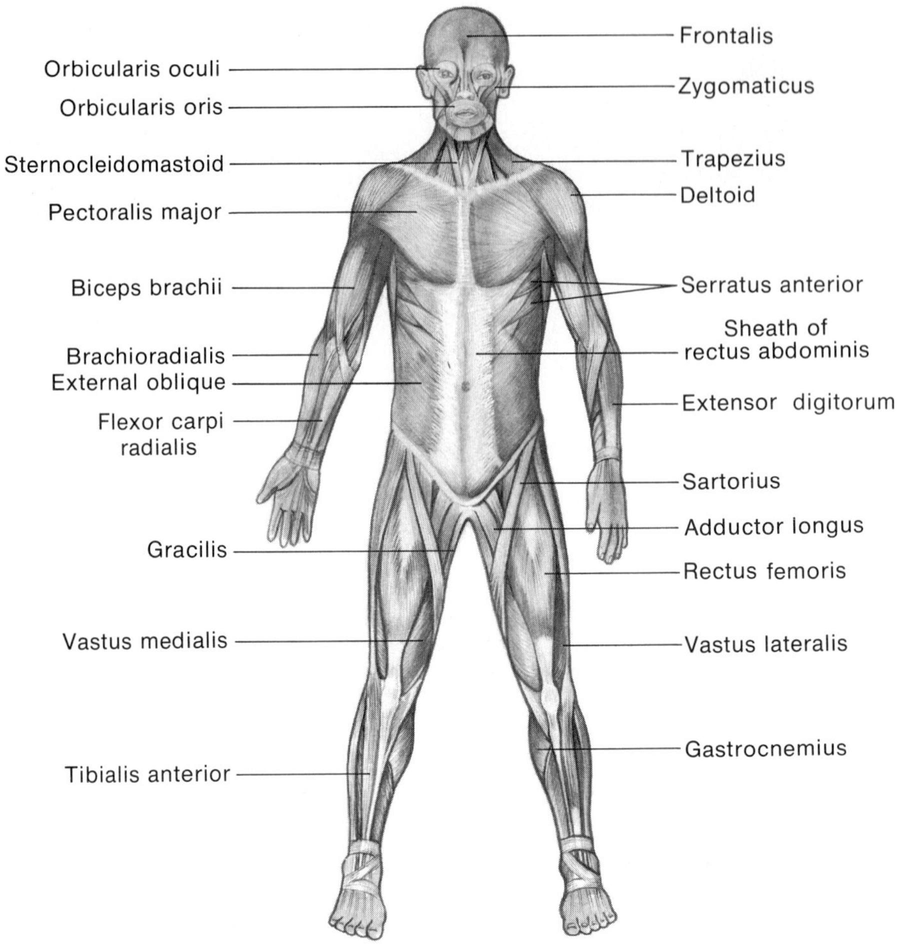

Figure 7–15 Muscles of the body. *A*, Anterior aspect.

Illustration continued on the opposite page

a squinting effect, as when looking into the sun.

The *orbicularis oris* (Figs. 7–15, *A*, and 7–16) encircles the mouth and functions in closing the lips. Some of the fibers are derived from other facial muscles that are inserted into the lips; some pass in an oblique direction from under the surface of the skin through the thickness of the lips to the mucous membrane; others connect the maxillae and septum of the nose above with the mandible below. This muscle may cause protrusion or pouting of the lips.

The *zygomaticus* (Figs. 7–15, *A*, and 7–16) comprises major and minor parts, and arises from the zygomatic bone. The major part inserts into the skin at the angle of the mouth; the minor part inserts into the orbicularis oris muscle. As a whole, the zygomaticus pulls the angles of the mouth upward, backward, and laterally, as in smiling or laughing.

The *buccinator* (Fig. 7–16) is the muscle of the cheek. It arises from the maxilla and mandible, and converges toward the angle of the mouth and inserts into the orbicularis oris. This muscle draws the corner of the mouth laterally, and flattens the cheek. It aids in swallowing, whistling, and blowing such wind instruments as the

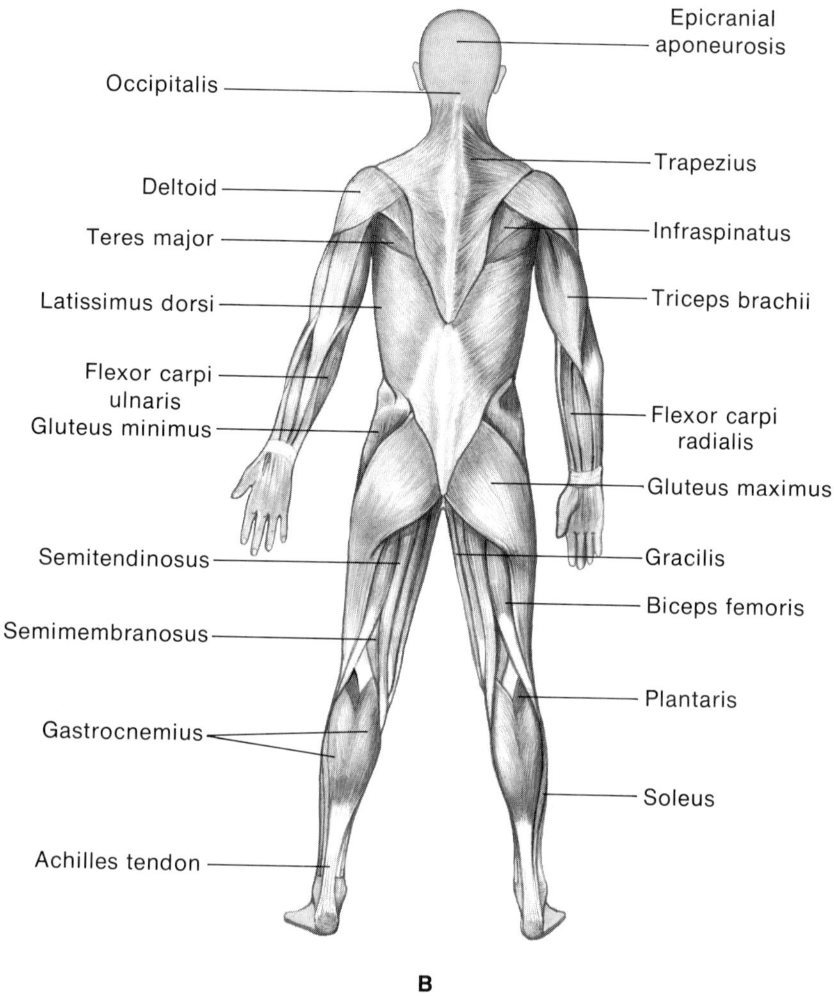

Figure 7–15 *Continued.* *B,* Posterior aspect.

trumpet, from which it received its name. It compresses the cheek during mastication, keeping food under the pressure of the teeth.

The *mentalis* (Fig. 7–16) is a short, thick muscle that originates in the mandible and inserts into the skin of the chin. The mentalis protrudes the lip and wrinkles the skin of the chin, expressing doubt or disdain.

The *platysma* (Fig. 7–16) is a broad band of muscle that originates on the skin and fascia covering the pectoralis and deltoid (shoulder) muscles. It passes upward, inserting into the mandible and the skin around the mouth. It draws the mandible and corners of the mouth downward and wrinkles the skin of the neck, expressing sadness or fright.

The *masseter* (Fig. 7–16) arises from the zygomatic arch and passes downward over the external aspect of the ramus of the mandible. It can easily be felt in action when the jaws are clenched tight, and appears as a bulging mass at the angle of the jaw.

The *temporalis* (Fig. 7–16) arises from the temporal fossa and passes downward beneath the zygomatic arch, inserting on the coronoid process of the mandible. It

Table 7–1 MUSCLES OF FACIAL EXPRESSION AND MASTICATION

MUSCLE	*ORIGIN*	*INSERTION*	*ACTION*
Epicranius			
Occipitalis	Occipital bone	Galea aponeurotica	Draws scalp backward
Frontalis	Galea aponeurotica	Skin of eyebrow	Elevates eyebrow; wrinkles forehead
Orbicularis oculi	Nasal portion of frontal bone; frontal process of maxilla	Angle of eye	Closes eyelids
Orbicularis oris	Muscle fibers around mouth	Angle of mouth	Protrudes lips
Zygomaticus major	Zygomatic bone	Angle of mouth	Draws angle of mouth up and backward (laughing)
Zygomaticus minor	Zygomatic bone	Orbicularis oris	Draws upper lip upward and laterally
Buccinator	Mandible and alveolar process of maxilla	Orbicularis oris	Draws corner of mouth laterally
Mentalis	Incisive fossa of mandible	Skin of chin	Protrudes lip; wrinkles skin of chin (doubt)
Platysma	Fascia of upper pectoral and deltoid muscles	Mandible and skin at corner of mouth	Depresses corner of mouth (sadness, fright)
Masseter	Zygomatic arch	Angle and lateral surface of mandible	Raises mandible; closes jaws
Temporalis	Temporal fossa of temporal bone	Coronoid process of mandible	Closes jaws
Lateral pterygoid	Sphenoid bone and lateral pterygoid plate	Neck of mandible	Draws jaws forward and sideward
Medial pterygoid	Maxilla and palatine bones	Ramus of mandible	Closes jaws

closes the jaw, and appears as a bulging muscle just in front of the ear at the temple.

The *pterygoid* muscles (lateral and medial) are deeply placed (Fig. 7–17). They extend from the inferior aspect of the skull to the neck of the mandible and to the medial surface of the ramus of the mandible. The lateral pterygoid is a short, thick muscle that assists in opening the jaws. The medial pterygoid assists in closing the jaws. The side-to-side movements, such as occur in grinding the teeth, are produced by alternating action of the pterygoid muscles on the two sides. Acting together, the lateral pterygoids also protrude the jaw (protraction).

The muscles of the tongue are concerned with speaking, mastication, and swallowing. They are divided into right and left paired groups by a fibrous septum that is attached below the hyoid bone. On each side there are two sets of muscles; the extrinsic have their origin outside the tongue, and the intrinsic are contained within it. Two extrinsic muscles are the *genioglossus* and *styloglossus.* The genioglossus thrusts the tongue forward, retracts it, and depresses it. The styloglossus draws the tongue upward and backward. During general anesthesia, these muscles become relaxed, along with all other muscles, and it is necessary to lift the mandible upward and forward in order to prevent the tongue from falling backward and obstructing the air passage (larynx).

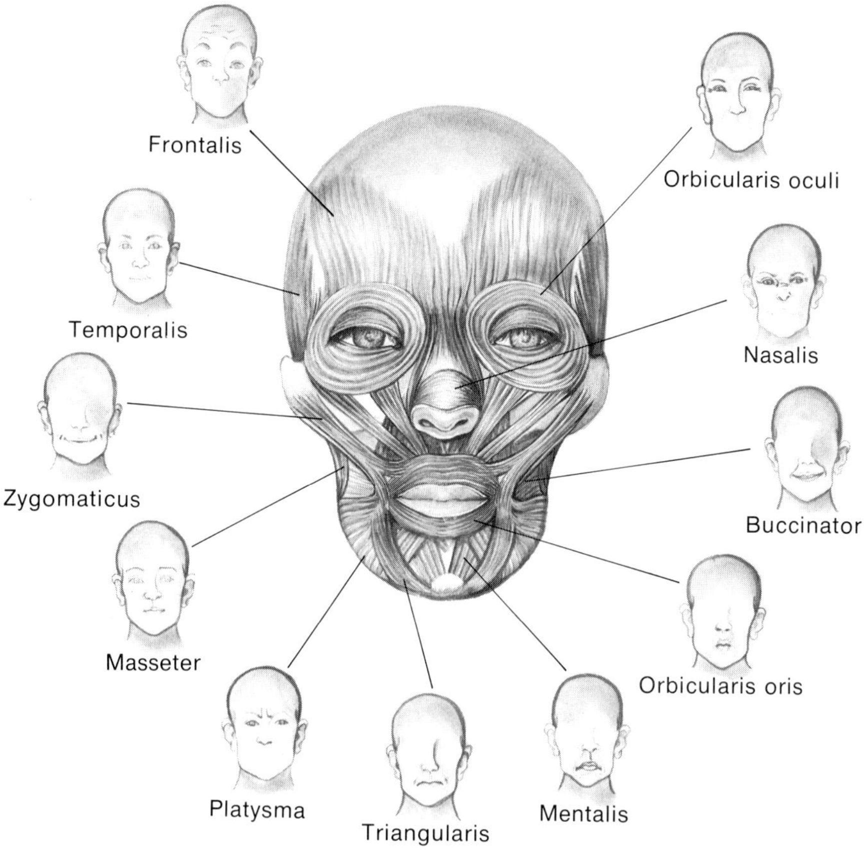

Figure 7–16 Muscles of facial expression and mastication.

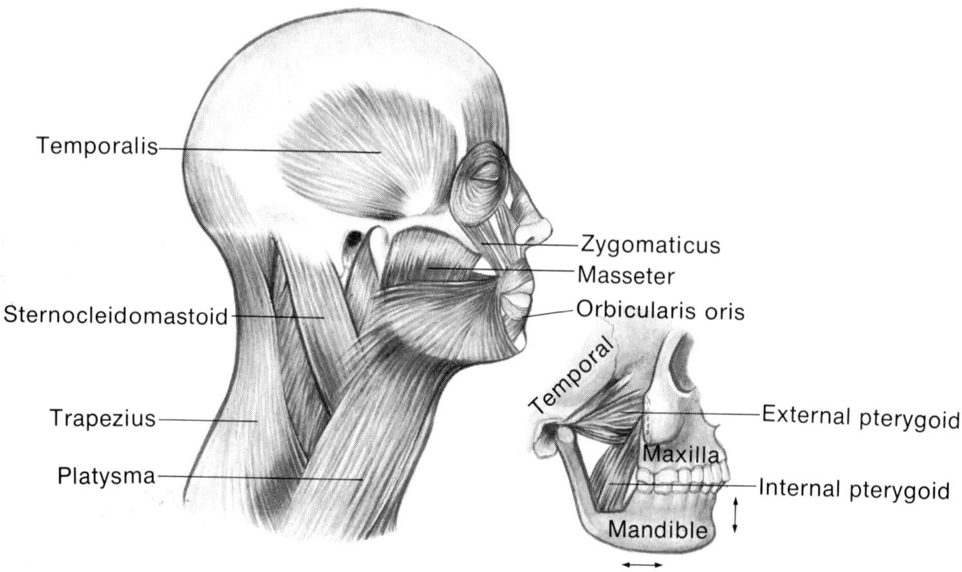

Figure 7–17 The superficial and deep muscles of mastication, lateral aspect.

Table 7–2 MUSCLES OF THE HEAD, SHOULDER, AND ARM

MUSCLE	ORIGIN	INSERTION	ACTION
Sternocleidomastoid	Two heads from sternum and clavicle	Mastoid process of temporal bone	Acting individually, each draws head toward shoulder of same side; acting together, they flex head forward
Semispinalis capitis	First six thoracic vertebrae; last four cervical vertebrae	Occipital bone	Extends and rotates head
Trapezius	Occipital bone; 7th cervical and all thoracic vertebrae	Acromial process of clavicle and spine of scapula	Draws scapula toward spine; rotates scapula to lift shoulder
Levator scapulae	Upper four cervical vertebrae	Vertebral border of scapula	Elevates scapula
Rhomboid major	Second to fifth thoracic vertebrae	Vertebral border of scapula	Elevates and retracts scapula
Rhomboid minor	Last cervical and first thoracic vertebrae	Vertebral border of scapula	Elevates and adducts scapula
Latissimus dorsi	Last six thoracic vertebrae; lumbar vertebrae; iliac crest	Anterior surface of upper humerus	Adducts, extends, and rotates arm medially
Pectoralis major	Cartilages of first six ribs, sternum, and medial half of clavicle	Crest of greater tubercle of humerus	Flexes, adducts, and rotates arm medially
Pectoralis minor	Cartilages of second to fifth ribs	Coracoid process of scapula	Pulls scapula forward and downward
Serratus anterior	Lateral surface of first nine ribs	Vertebral border of scapula	Draws scapula forward and laterally; slight rotation
Coracobrachialis	Coracoid process of scapula	Medial surface of humerus	Flexes and adducts arm; medial rotation of arms
Deltoid	Spine of scapula; clavicle; acromial process	Lateral surface (deltoid tubercle) of humerus	Abducts arm
Teres major	Lower angle of scapula	Anterior surface of humerus (lesser tubercle)	Adducts, extends, and rotates arm medially
Teres minor	Axillary border of scapula	Greater tubercle of humerus	Rotates arm laterally
Supraspinatus	Supraspinous fossa of scapula	Greater tubercle of humerus	Abducts arm and rotates it laterally
Infraspinatus	Infraspinous fossa of scapula	Middle facet of greater tubercle of humerus	Rotates arm laterally

Muscles that Move the Head
(Table 7–2)

The *sternocleidomastoid* (Fig. 7–17, *A*) is a strong, bandlike muscle named from its points of origin and insertion. It arises from two heads, one at the upper part of the sternum and the other at the inner border of the clavicle, and inserts into the mastoid process of the temporal bone (Fig. 7–18). It can easily be seen by turning the head to one side; its cordlike prominence extends obliquely down the side of the neck from the ear to the clavicle. The sternocleidomastoid draws the head toward the shoulder when one muscle works alone. When both muscles act together, the head is flexed forward, and in forced inspirations both assist in elevating the thorax. If one of the muscles is paralyzed or abnormally contracted, a deformity called *torticollis* (wryneck) results.

The *semispinalis capitis* arises from the transverse processes of the first six

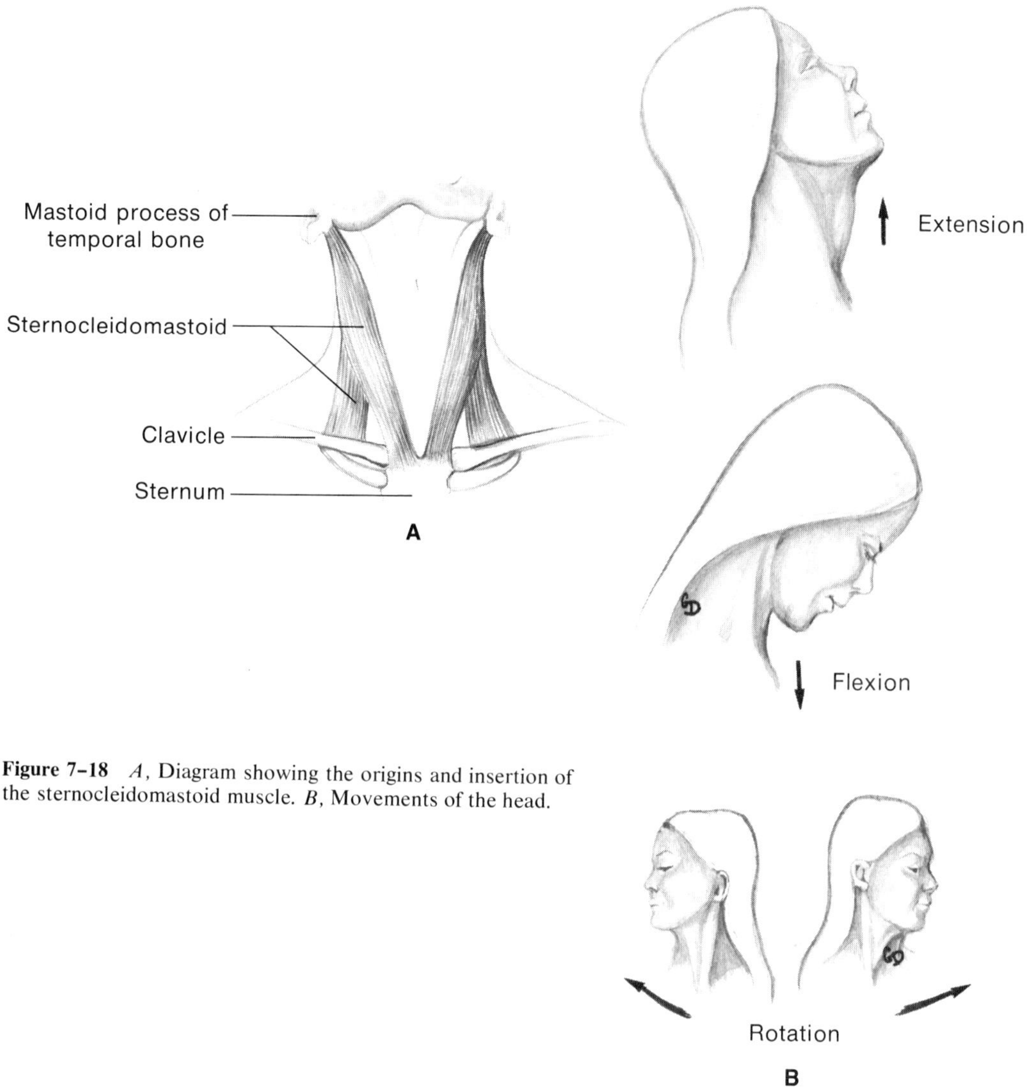

Figure 7-18 *A*, Diagram showing the origins and insertion of the sternocleidomastoid muscle. *B*, Movements of the head.

thoracic and last four cervical vertebrae and inserts on the occipital bone. When both muscles contract, the head is extended; when one acts alone, the head is rotated toward that same side.

Muscles that Move the Shoulder and Arm (Table 7-2)

Many muscles originate or insert on the scapula (Fig. 7-19), and it can be moved in many directions—forward, backward, upward, and downward (Fig. 7-20). It can also be rotated to demonstrate the vast range of movement in the shoulder joint. The head of the humerus is held firmly in place in the glenoid cavity during these movements. A dislocation of the shoulder generally involves a stretching of the ligaments and tendons to the muscles that form the shoulder cavity, thus allowing the humeral head to slip out of position. Physical manipulation followed by a period of enforced immobility generally returns the structures to their normal positions.

The *trapezius* (Fig. 7-15, *B*)—so

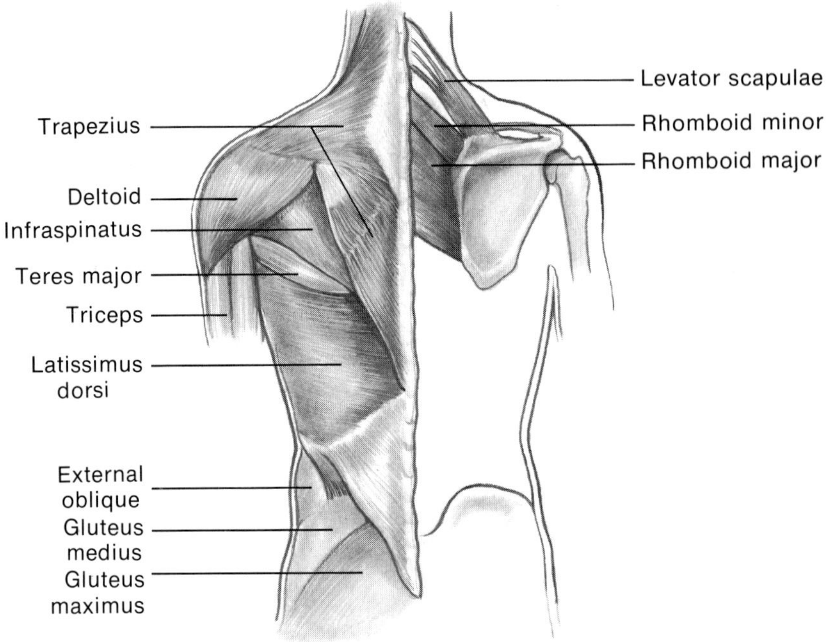

Figure 7–19 Muscles of the neck, thorax, and arm, posterior aspect. The superficial muscles have been removed on the right side to permit exposure of the deep muscles connecting the scapula to the spinal vertebrae.

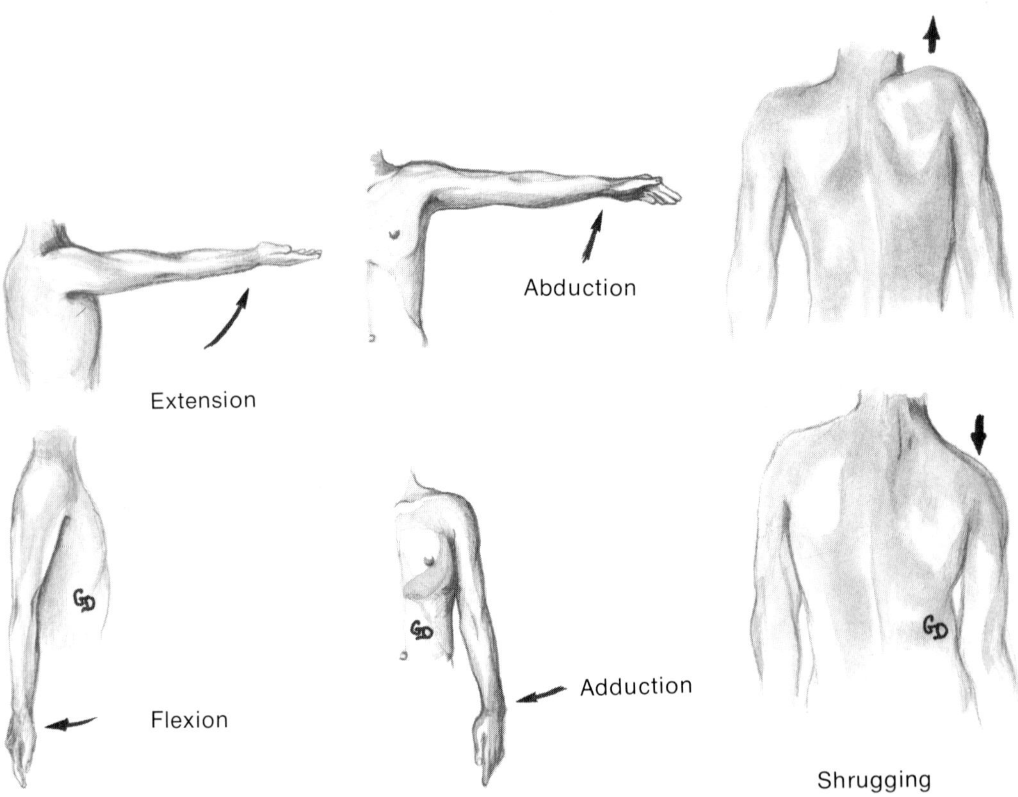

Figure 7–20 Movements at the shoulder.

called because its right and left portions together make a large diamond-shaped (trapezoid) sheet—arises from the occipital bone and the spinous processes of the seventh cervical and all thoracic vertebrae. The muscle inserts on the clavicle, the acromial process, and the spine of the scapula. In relation to the shoulder girdle, the trapezius lifts the shoulder (shrugging) if the upper end is fixed; if the shoulders are fixed, the head will be drawn backward (extension). When the entire muscle acts as a unit, the scapulae are drawn toward the spine (adduction), at the same time being rotated in a movement that allows the body to lift the arms over the head. Inflammation or abnormal cooling (sitting in a draft) of the trapezius results in a "stiff neck."

The *levator scapulae* arises from the upper four cervical vertebrae and inserts onto the vertebral border of the scapula. This muscle assists the trapezius in elevating the scapula.

The *rhomboid* muscles (major and minor) are just beneath the trapezius on the back. These originate on the spines of the last cervical and first five thoracic vertebrae and insert on the vertebral border of the scapula. The rhomboid muscles move the scapula backward and upward in a slight rotation.

The widest of all the muscles, the *latissimus dorsi* (Fig. 7–15, *B*), is triangular, covering the loin and the lower half of the thoracic region. It originates from the spines of the last six thoracic vertebrae, the lumbar vertebrae, and the iliac crest. Its fibers converge to form a flat tendon that inserts on the anterior surface of the upper part of the humerus. The latissimus dorsi forms part of the posterior axillary fold. It gives a powerful downward and backward sweep to the arm, with some inward rotation, such as occurs in swimming. This muscle is the chief muscle used in downward blows (karate) or in lifting the body (climbing).

The *pectoralis* muscles (major [Fig. 7–15, *A*] and minor) form the anterior axillary fold (Fig. 7–21). The pectoralis major originates on the clavicle, sternum,

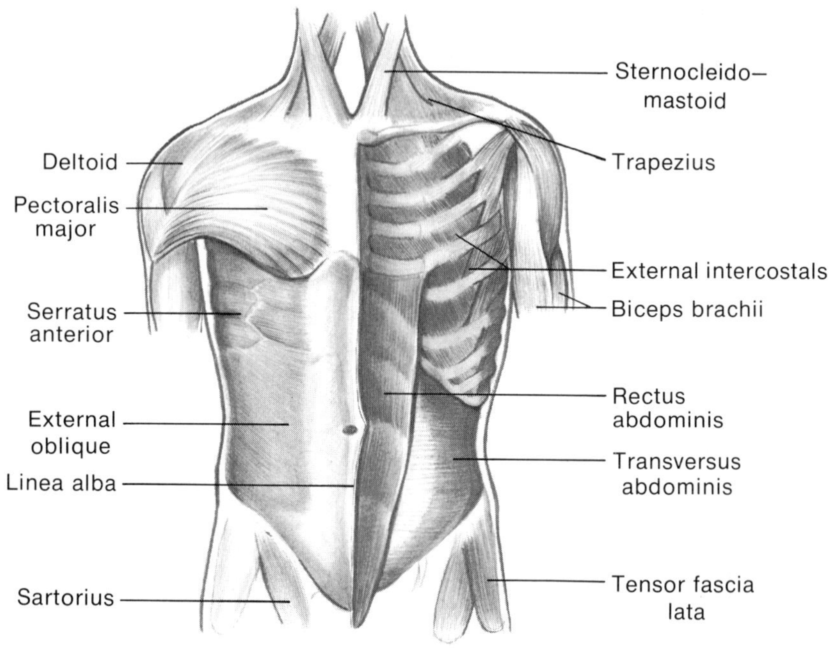

Figure 7–21 Muscles of the neck, thorax, and arm, anterior aspect. Some superficial muscles have been removed on the left side to permit exposure of the deep muscles.

and first six rib cartilages, and inserts on the greater tubercle of the humerus. This muscle adducts the arm, draws it across the chest, and rotates it medially. It draws the shoulder girdle forward and down, creating the flight motion so efficiently demonstrated by birds.

The pectoralis minor is a thin, triangular muscle, quite unlike the thick, large pectoralis major. The minor lies behind the major and arises from the costal cartilages of the second to fifth ribs. It inserts on the coracoid process of the scapula, and pulls the scapula forward and downward, assisting in flight motion.

Both the major and minor pectoralis muscles are removed surgically in a standard radical mastectomy for breast cancer.

The *serratus anterior* (Fig. 7–15, *A*) occupies the side of the chest and the medial wall of the axilla (armpit). It arises from the outer surfaces of the first nine ribs and the intercostals between them. The fibers pass upward and backward, inserting on the vertebral border of the scapula. The name of this muscle derives from the sawtooth appearance of its origins from several ribs. These origins on the chest wall permit it to move the scapula forward away from the spine (abduction) in a pushing motion.

The *coracobrachialis* is located at the upper and medial part of the arm. It arises from the coracoid process of the scapula, and inserts on the medial surface of the humerus. It carries the arm forward (flexion) and assists in its adduction.

The *deltoid* (Fig. 7–15, *A* and *B*) is a thick, powerful muscle that covers the shoulder joint. It arises from the clavicle, the acromial process, and the spine of the scapula, and inserts onto the lateral (deltoid) tubercle of the humerus. Contraction of the fibers abducts the arm from the side of the body and raises it laterally, as when reaching for something sideways. The thickness of the muscle makes it a prime area for intramuscular injections (IM).

The *teres major* (Fig. 7–15, *B*) is thick and flat. It arises near the lower angle of the scapula and inserts on the anterior surface of the humerus (lesser tubercle). It extends the arm, drawing it downward. It also helps to adduct the arm and rotate it medially.

The *teres minor, supraspinatus,* and *infraspinatus* muscles collectively form a tendon after the muscles have passed over the shoulder joint, and insert on the greater tubercle of the humerus. The teres minor arises from the axillary border of the scapula; the supraspinatus and infraspinatus originate from their respective fossae in the scapula. These three muscles protect the shoulder joint and assist in the lateral rotation of the arm. The supraspinatus also aids the deltoid in abduction of the arm.

Muscles that Move the Forearm and Hand (Table 7–3)

The biceps *(biceps brachii)* of the arm (Figs. 7–15, *A*, and 7–24) originates from two heads, one long and one short. The long head arises from the upper margin of the glenoid cavity; the short head arises from the tip of the coracoid process of the scapula (Fig. 7–22). Both heads unite and terminate in a flat tendon, which is inserted onto the tubercle on the proximal end of the radius. As the biceps contracts, the radius turns, supinating the hand. This muscle also flexes the forearm, and its long head helps to hold the head of the humerus in the glenoid cavity.

The *brachialis* (Fig. 7–22) has its origin from the distal, anterior portion of the humerus, and inserts onto the coronoid process of the ulna, covering the front part of the elbow joint. It is a strong flexor of the forearm.

The triceps of the arm *(triceps brachii)* arises by three heads of origin: the long head from the axillary border of the scapula; the lateral and medial heads from the posterior surface of the humerus (Fig. 7–15, *B*). The muscle fibers terminate in a common tendon that inserts onto the olecranon process of the ulna. The triceps is

Table 7–3 MUSCLES OF THE FOREARM AND HAND

MUSCLE	*ORIGIN*	*INSERTION*	*ACTION*
Biceps brachii	Short head from coracoid process; long head from scapula (glenoid cavity)	Tubercle on proximal end of radius	Flexes forearm; supinates forearm and hand; long head helps hold head of humerus in glenoid cavity
Brachialis	Anterior surface of humerus (distal part)	Coronoid process of ulna	Flexes forearm
Triceps brachii	Long head from axillary border of scapula; lateral and medial heads from posterior surface of humerus	Olecranon process of ulna	Extends forearm; long head extends and adducts forearm
Pronator teres	Medial epicondyle of humerus and coronoid process of ulna	Middle of lateral surface of radius	Pronates hand; flexes forearm
Supinator	Lateral epicondyle of humerus and ridge of ulna	Anterior and lateral surfaces of radius	Supinates hand and forearm
Flexor carpi radialis	Medial epicondyle of humerus	Base of second metacarpal	Flexes and abducts hand
Flexor carpi ulnaris	Medial epicondyle of humerus and proximal two-thirds of ulnar border	Pisiform bone	Flexes and adducts hand
Palmaris longus	Medial epicondyle of humerus	Fascia of palm	Flexes hand
Flexor digitorum superficialis	Medial epicondyle of humerus, coronoid process of ulna	Middle phalanges of fingers	Flexes middle fingers
Flexor digitorum profundus	Shaft and coronoid process of ulna	Distal phalanges of fingers	Flexes distal fingers
Flexor pollicis longus	Anterior surface of radius and coronoid process of ulna	Distal phalanx of thumb	Flexes thumb
Extensor carpi radius longus	Lateral supracondylar ridge of humerus	Base of second metacarpal	Extends and abducts hand
Extensor carpi ulnaris	Lateral epicondyle of humerus	Base of fifth metacarpal	Extends and adducts hand
Extensor pollicis longus	Lateral side of dorsal surface of ulna	Base of second phalanx of thumb	Extends and abducts thumb
Extensor digitorum	Lateral epicondyle of humerus	Common extensor tendon of each finger	Extends fingers and hand
Abductor pollicis longus	Posterior surfaces of ulna and radius	Base of first metacarpal	Abducts and extends thumb
Adductor pollicis	Principally second and third metacarpals	Proximal phalanx of thumb	Adducts and opposes thumb

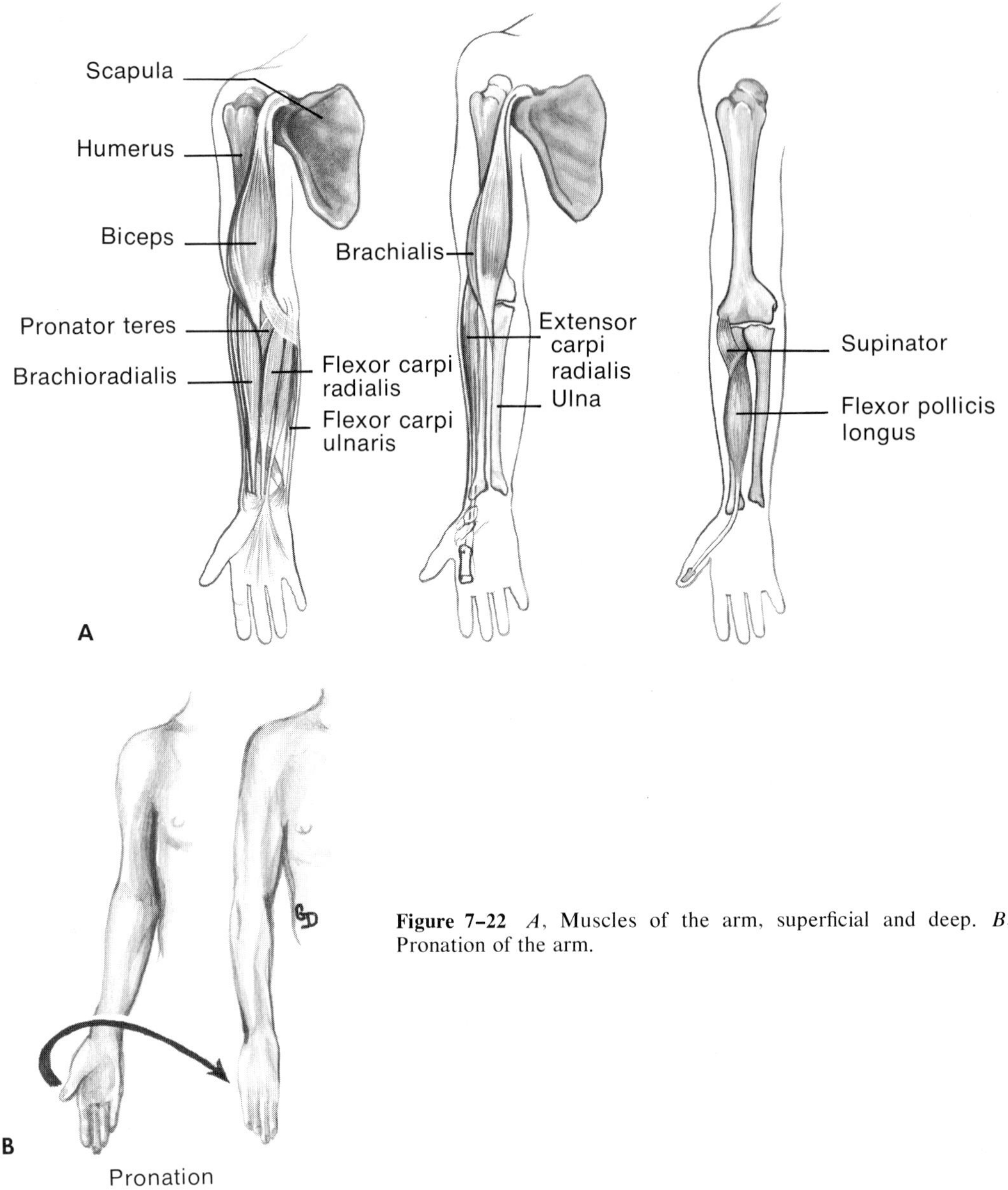

Figure 7–22 *A*, Muscles of the arm, superficial and deep. *B*, Pronation of the arm.

the great extensor of the forearm, and therefore is the direct antagonist of the biceps and brachialis.

The *pronator teres* (Fig. 7–22) has two heads of origin: the humeral, or larger head, and the ulnar. This muscle extends obliquely across the forearm and inserts as a flat tendon onto the lateral surface of the radius. Its primary function is to pronate the hand, but it also flexes the forearm.

The *supinator* (Fig. 7–22) has its origin on the lateral epicondyle of the humerus and the ridge of the ulna, and is inserted onto the lateral margin of the radius. As its name implies, it supinates the hand and forearm.

Movements of the wrist and fingers depend on several muscles of the forearm (Figs. 7–22 and 7–23). The *flexor carpi radialis* and the *flexor carpi ulnaris* both originate on the medial epicondyle of the humerus and aid in flexion at the elbow

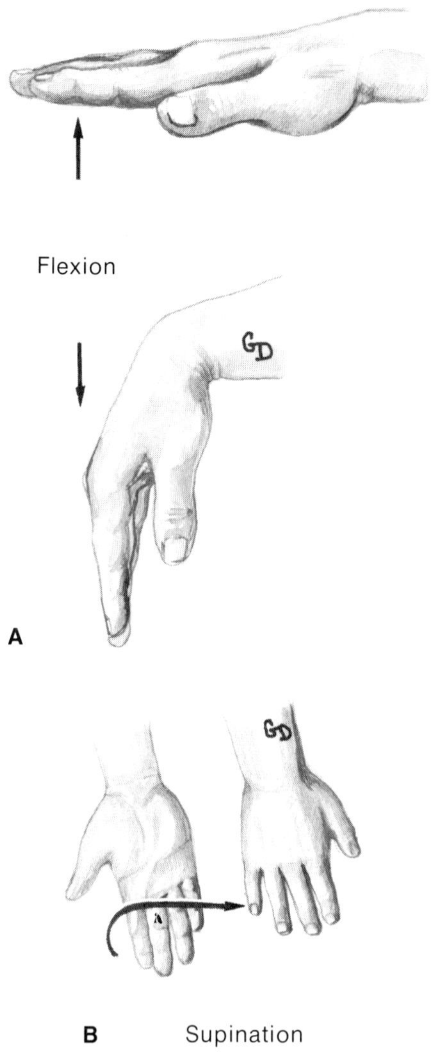

Figure 7–23 *A*, Flexion of the wrist. *B*, Supination of the hand.

and in flexion of the hand at the wrist joint. The radialis abducts the hand; the ulnaris adducts the hand.

The *extensor carpi ulnaris* and *extensor carpi radialis* originate on the lateral epicondyle of the humerus, and are synergistic with the flexor carpi muscles in their action across the elbow joint. More important, however, they extend the hand at the wrist through their insertions on the dorsal side of the second and fifth metacarpals at the ulnar and radial sides of the hand.

Muscles of the forearm also control flexion and extension of the fingers. The proximal joints of the fingers are flexed by the *flexor digitorum* muscles that arise from the medial epicondyle of the humerus and the coronoid process of the ulna. Extension of the fingers is carried out by the *extensor pollicis* muscles in conjunction with the *extensor digitorum* group of muscles (Fig. 7–15, *A*).

The thumb and fingers can be abducted (fanned out) by contraction of the *abductor pollicis* (thumb) and *interossei dorsales* (dorsal metacarpal area) muscles. Contraction of the *adductor pollicis* and *interossei palmaris* (palm) muscles adducts (moves together) the thumb and fingers (Fig. 7–24, *B*).

The *palmar aponeurosis,* or palmar fascia, is a tough, fibrous sheet of connective tissue that covers the palm just beneath the skin. Its presence allows great pressure to be exerted on the palm without injury to the nerves, blood vessels, and tendons beneath.

The muscles of the forearm and hand would be ineffective if they or their tendons could become stuck together or could rub on one another, causing a great deal of heat from friction. Therefore, numerous synovial (fluid-secreting) tendon sheaths encase muscles and tendons to reduce friction between tendons and adjacent structures (Fig. 7–24, *A*). Exact anatomic knowledge of these sheaths is a necessity when surgical repair of a tendon laceration or severance is required in order to restore normal function following injury. Suturing the wrong tendons together can result in partial paralysis of the parts involved. Occasionally an infection sets in following a laceration. Such infection can invade the tendon sheath – a condition called *tenosynovitis* (p. 151). An inadequate nerve supply or the growth of adhesions may cause stiffness of the wrist or fingers. Stiffness or impaired use of the thumb, for example, can immobilize the entire hand – compromising one of the functions that separate man from the lower animals.

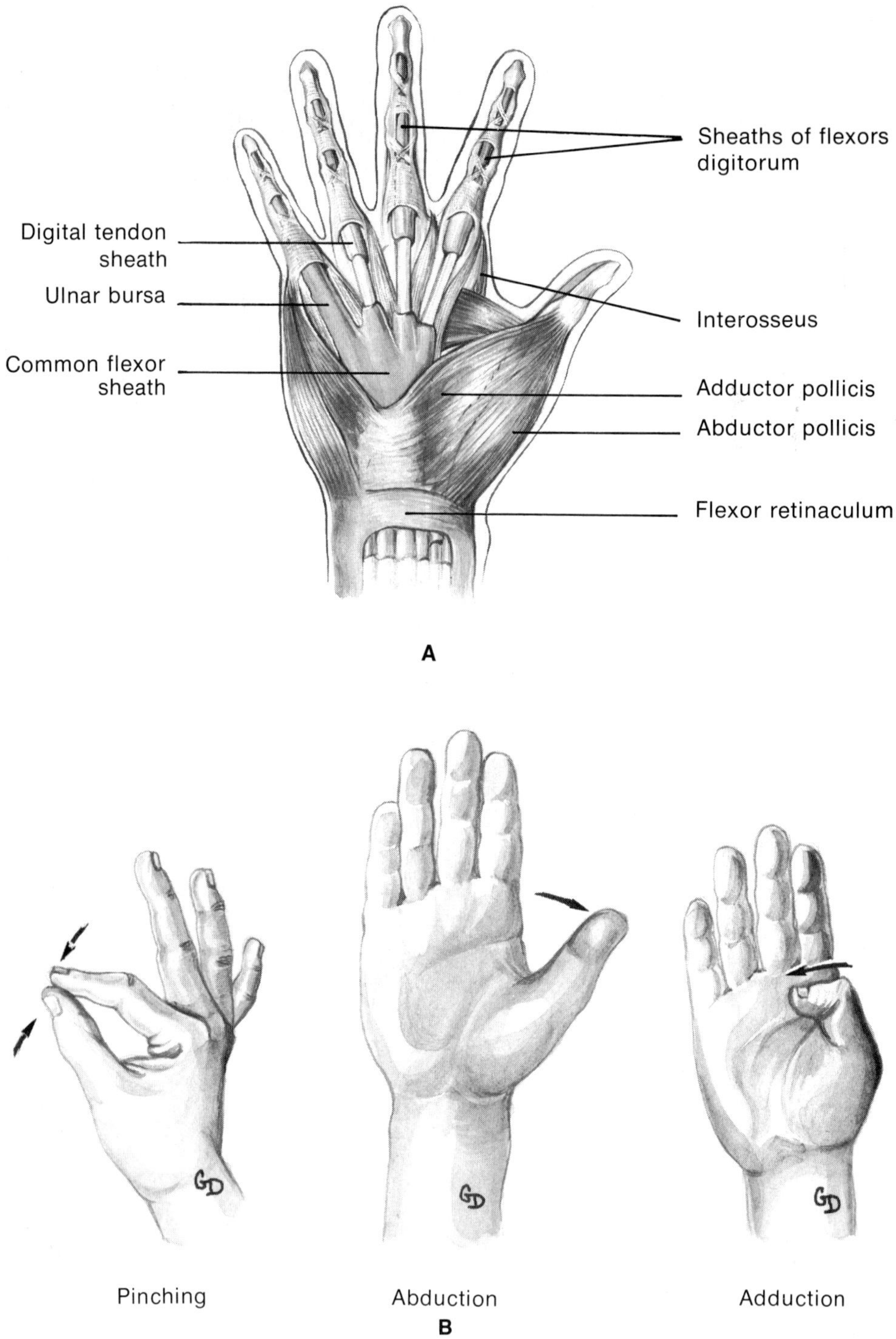

Figure 7–24 The hand. *A*, Palmar surface of the wrist and hand, showing the muscles and sheaths of the flexor tendons. *B*, Movements of the hand.

Table 7–4 MUSCLES OF RESPIRATION

MUSCLE	ORIGIN	INSERTION	ACTION
Diaphragm	Xiphoid process; lower six costal cartilages and four ribs; upper lumbar vertebrae	Central tendon of diaphragm	Expands thorax; compresses contents of abdominal cavity
External intercostals	Lower border of each rib	Upper border of next rib	Draw adjacent ribs together (inspiration)
Internal intercostals	Lower border of each rib	Upper border of next rib	Draw adjacent ribs together (expiration)

Muscles of Respiration (Table 7–4)

There are two independent mechanisms for respiration: (1) the muscles of the thoracic wall, and (2) the diaphragm and its antagonists, the abdominal muscles.

The *external intercostal* muscles, 11 on each side, occupy the spaces between the ribs (Fig. 7–21). Each arises from the lower border of a rib and inserts onto the upper border of the rib below. The muscles extend from the tubercles of the ribs behind to the costal cartilages of the ribs in front, where they end in membranes that connect with the sternum. The direction of the fibers is obliquely downward. In respiration, the external intercostals enlarge the thoracic cavity from side to side and from front to back when they contract and elevate the ribs.

The *internal intercostal* muscles, also 11 on each side, extend from the sternum to the angle of the ribs. Each arises and inserts in the same way as the external intercostals; however, the direction of their fibers, while also downward, is in an opposite direction from that of the externals. These muscles act in depressing the ribs, and therefore are concerned with expiration. The external intercostals are concerned with inspiration.

The *diaphragm* is a dome-shaped musculofibrous partition that separates the thoracic and abdominal cavities (Fig. 7–25, *A*). It forms the convex floor of the thoracic cavity and the concave roof of the abdominal cavity. Its peripheral portion arises from the xiphoid process of the sternum. The costal fibers arise from the lower six costal (rib) cartilages and the lower four ribs on either side, and the lumbar fibers arise from the tendinous arches and the bodies of the upper lumbar vertebrae. The fibers converge, to be inserted onto the central portion of the diaphragm, called the central tendon.

There are three large openings in the diaphragm: the esophageal, for passage of the esophagus and the vagus nerves; the aortic, for passage of the aorta, and frequently the azygos vein and the thoracic duct as well; and the vena caval, for passage of the inferior vena cava and branches of the phrenic nerves.

The diaphragm is the principal muscle of respiration. When the muscular portion contracts, the central tendon is pulled downward, which increases the vertical diameter of the thorax. Because of this, a vacuum is created, sucking in air from the outside (inspiration). Relaxation of the diaphragm decreases the vertical axis, forcing the air back out of the thoracic cavity (expiration).

Abnormal stretching of one of the openings in the diaphragm, generally the esophageal, can cause a weakness to develop. This weakness may allow a portion of the stomach (the cardiac portion) to slip part way into the thoracic cavity, producing a great deal of discomfort. This condition is called a *diaphragmatic* or *hiatus hernia* (Fig. 7–25, *B*).

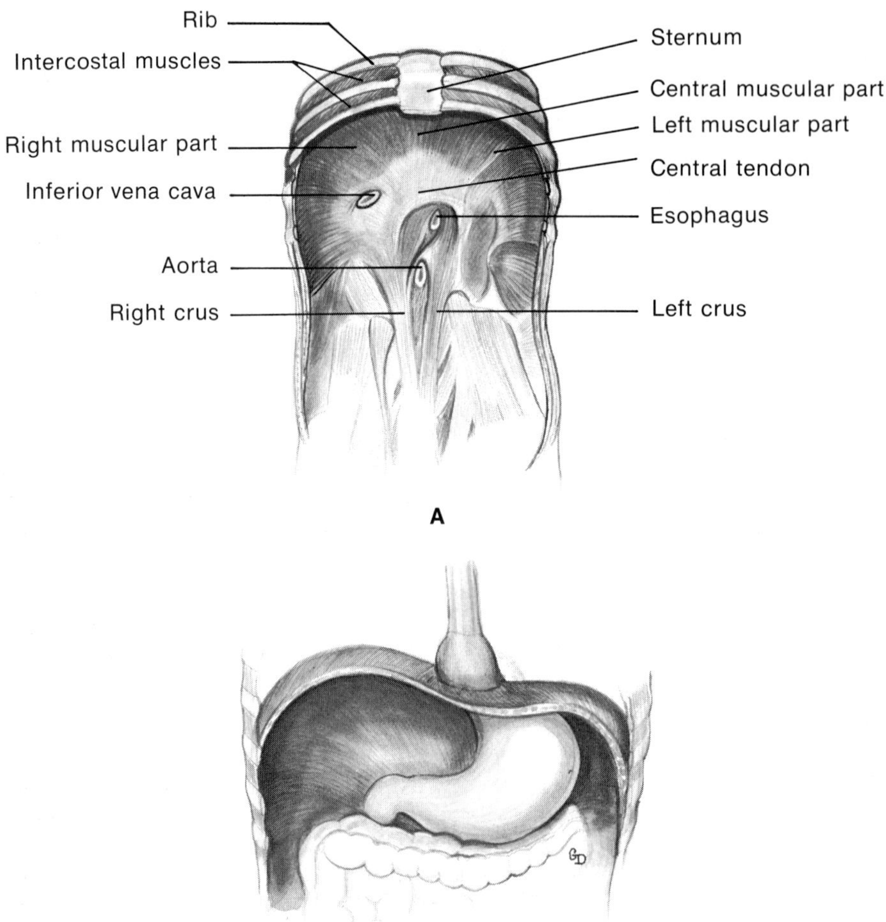

A

B

Figure 7–25 The diaphragm. *A*, Undersurface of the large muscle partition, showing the openings of digestive, nervous, and circulatory systems. *B*, Diaphragmatic or hiatus hernia.

Muscles of the Abdominal Wall
(Table 7–5, *A*)

The *external oblique* (Fig. 7–15, *A*) originates on the outer surface of the lower eight ribs. Its fibers pass forward and down, and insert on the iliac crest, the iliac spine, and the anterior rectus sheath (Fig. 7–26). The external oblique is the strongest and most superficial muscle of the anterolateral abdominal wall. Its fibers are short and broad, and do not reach the edge of the rectus abdominis (Fig. 7–21), which lies medial to it. Therefore, the anterior wall is composed of a broad aponeurosis, which inserts into the linea alba (midline region). The superficial inguinal ring is a circular opening in the aponeurosis just above and lateral to the pubic tubercle. Major vessels and nerves (and the spermatic duct in males) pass through this ring. It is a common area for weakness to occur following overstrain or stretching, and increased size of the inguinal ring in men may allow the intestines to pass from the abdominal cavity into the scrotum, causing marked discomfort (Fig. 7–27). This condition is called an *inguinal hernia*.

Table 7–5 MUSCLES OF THE ABDOMINAL WALL

MUSCLE	ORIGIN	INSERTION	ACTION
External oblique	Lower eight ribs	Iliac crest; anterior rectus sheath	Compresses abdomen; flexes and rotates vertebral column
Internal oblique	Inguinal ligament; iliac crest; lumbodorsal fascia	Lower three or four costal cartilages; pubic bone	Compresses abdomen; flexes and rotates vertebral column
Tranversus abdominis	Lower six costal cartilages; iliac crest; lumbodorsal fascia; inguinal ligament	Linea alba; crest of pubis	Compresses abdomen
Rectus abdominis	Symphysis pubis; crest of pubis	Fifth to seventh costal cartilages; xiphoid process	Flexes vertebral column and pelvis

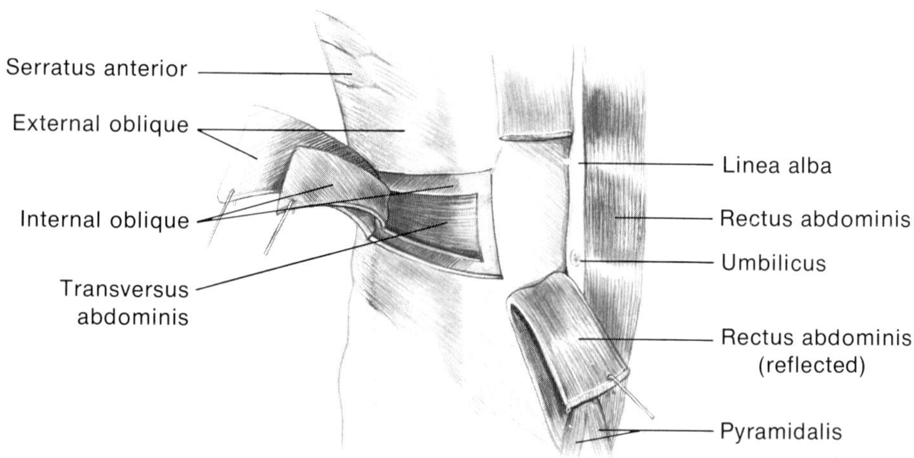

A

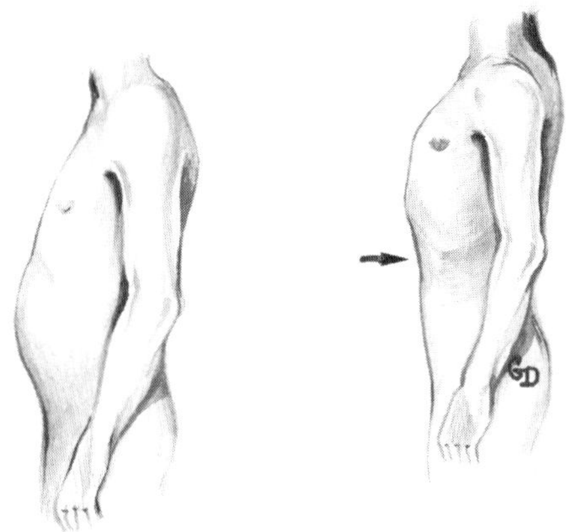

B

Figure 7–26 Muscles of the abdominal wall. *A*, The muscles have been reflected to illustrate their layered position in the formation of the abdominal aponeurosis. *B*, Abdominal contraction aids respiration.

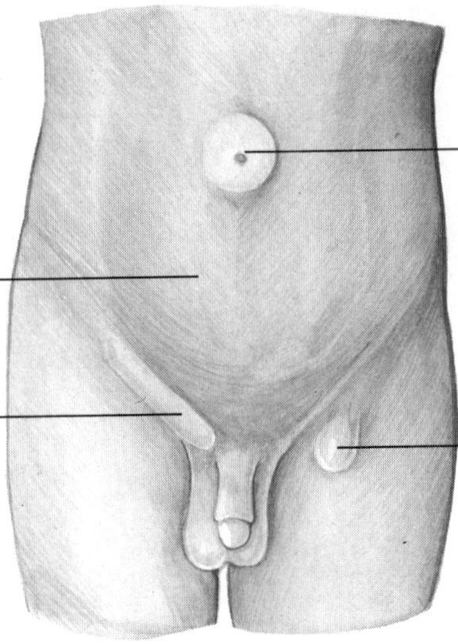

Figure 7–27 Abdominal hernias in the male.

If the ring subsequently closes, a strangulation of the intestine occurs, called an *incarcerated* or *strangulated hernia.* Immediate surgical attention is necessary, as a gangrenous condition may result from the constricted blood supply to the intestine.

The *internal oblique* lies just beneath the external oblique (Fig. 7–26, *A*). It arises from the inguinal ligament, the iliac crest, and the lumbodorsal fascia. The fibers extend upward and forward, and insert onto the lower three rib cartilages to form the aponeurosis that inserts onto the linea alba. The lower part of the aponeurosis inserts on the pubic bone.

The *transversus abdominis* is positioned deep in the lateral abdominal wall (Figs. 7–21 and 7–26, *A*). Its fibers arise from the lower six costal cartilages, the lumbodorsal fascia, and the inguinal ligament. Most of its fibers, as the name of the muscle suggests, have a horizontal (transverse) direction, and form the aponeurosis that inserts onto the linea alba and the crest of the pubis. The deep inguinal ring is found in this muscle. It provides the internal opening for the inguinal canal, whose outer opening is found in the external oblique muscle.

The external oblique, internal oblique, and transversus abdominis muscles, with their aponeuroses, form a wall or girdle for the abdomen. Their muscle fibers cross in three directions, acting to compress the abdominal contents. This action also assists in expiration. When the diaphragm relaxes, the abdominal muscles contract; when the diaphragm contracts, the abdominal muscles relax.

The *rectus abdominis* (Figs. 7–21 and 7–26, *A*) is the last of the four major abdominal muscles. It is a long, straplike muscle made up of vertical fibers, and is situated at the front of the abdomen, where it is enclosed in the fibrous sheath formed by the aponeuroses of the other three abdominal muscles. Its fibers arise from the symphysis pubis and the crest of the pubis. The muscle inserts onto the costal cartilages of the fifth to seventh ribs and onto the xiphoid process of the sternum. Its major function is to flex the vertebral column and the pelvis.

When all of the abdominal muscles are contracted they assist in parturition, defecation, micturition, and emesis. They also bend the thorax forward, and, when acting on one side only, bend the trunk of the body in that direction.

In addition to the inguinal hernia already discussed, other weaknesses of the abdominal wall include:

1. *Umbilical hernia*—protrusion of the abdominal contents in the region of the umbilicus.

2. *Femoral hernia*—an enlargement of the canal in the femoral region just behind the inguinal ligament. The inguinal canal is larger in males than in females; therefore, the inguinal hernia is more common in men and the femoral hernia more frequent in women.

Muscles that Move the Thigh
(Table 7–6)

The ball-and-socket arrangement of the hip joint permits free movement in all directions (Fig. 7–28).

The *iliopsoas* arises from the transverse processes of the lumbar vertebrae and the iliac crest and fossa. Its fibers converge and pass along the edge of the pelvis to the front of the thigh, where they are inserted onto the lesser trochanter of the

Table 7–6 MUSCLES OF THE THIGH, LEG, AND FOOT

MUSCLE	ORIGIN	INSERTION	ACTION
Iliopsoas	Transverse processes of lumbar vertebrae; iliac fossa	Lesser trochanter of femur	Flexes thigh; flexes trunk; rotates thigh medially
Gluteus maximus	Posterior part of sacrum; lateral surface of ilium	Fascia lata; gluteal ridge of femur	Extends, abducts, and rotates thigh laterally
Gluteus medius	Lateral surface of ilium	Great trochanter of femur	Abducts thigh
Gluteus minimus	Lateral surface of ilium	Great trochanter of femur	Abducts and rotates thigh medially
Tensor fasciae latae	Anterior part of iliac crest	Iliotibial band of fascia lata	Flexes, abducts, and rotates thigh medially
Adductor magnus longus brevis	Pubic bone and ischium	Linea aspera of femur	Adduct and flex thigh
Biceps femoris	Long head from ischial tuberosity; short head from lateral lip of linea aspera of femur	Head of fibula; lateral condyle of tibia	Flexes leg; extends thigh
Semitendinosus	Ischial tuberosity	Proximal part of tibia	Flexes leg; extends thigh
Semimembranosus	Ischial tuberosity	Medial condyle of tibia	Flexes leg; extends thigh
Gracilis	Body and inferior ramus of pubis	Medial surface of shaft of tibia	Flexes leg; adducts thigh
Sartorius	Anterior superior iliac spine	Medial surface of proximal end of tibia	Flexes thigh and leg; rotates thigh laterally
Quadriceps femoris	By four heads from ilium and femur	Tuberosity of tibia, through a common tendon	Extends leg; flexes thigh
Gastrocnemius	Medial and lateral condyles of femur	Calcaneus, through tendon of Achilles	Flexes leg; plantar flexes, adducts, and inverts foot
Soleus	Head and proximal third of fibula and tibia	Calcaneus, through tendon of Achilles	Plantar flexes foot
Tibialis posterior	Posterior surfaces of tibia and fibula	Tarsals (cuneiform, cuboid); second to fourth metatarsals	Plantar flexes foot
Tibialis anterior	Lateral condyle and upper portion of tibia	First cuneiform and base of first metatarsal	Dorsally flexes foot
Peroneus tertius	Distal third of fibula	Base of fifth metatarsal	Dorsally flexes foot
Peroneus longus	Head and lateral surface of fibula; lateral condyle of fibia	Lateral side of first metatarsal and medial cuneiform bone	Plantar flexes and everts foot
Peroneus brevis	Lower two-thirds of lateral surface of fibula	Tuberosity of fifth metatarsal	Everts foot; aids in extension

Figure 7–28 Movements of the hip joint.

femur. The iliopsoas is actually a compound muscle made up of the *iliacus* and the *psoas major*. The action of both is primarily to flex the thigh. The iliopsoas is a powerful muscle that also aids in flexing the trunk and in rotating the thigh medially.

The *gluteus maximus* (Fig. 7–15, *B*), so named because it is the largest of the gluteal group, is a broad, thick, and fleshy muscle that forms the major area of the buttock. It arises principally from the posterior part of the sacrum and the lateral surface of the ilium, and inserts onto the fascia lata and the gluteal ridge of the femur. When the pelvis is fixed, the gluteus maximus is the antagonist of the iliopsoas. It is a powerful extensor of the thigh, and helps to rotate the lower extremity laterally or outward. Its practical use is in climbing stairs, running, or other activities requiring greater than ordinary force. It is the prime muscle for intramuscular injections in the lower half of the body.

The *gluteus medius* and *minimus* (Fig. 7–15, *B*) lie under the maximus on the lateral side of the hip joint. They arise from the lateral surface of the ilium, and insert onto the greater trochanter of the femur. They act to abduct the thigh and rotate it medially.

The muscles of the *adductor* group (*magnus, longus* [Fig. 7-15, *A*], and *brevis*) arise from the pubic bone, and are inserted into the linea aspera of the femur. These muscles are the major adductors and flexors of the thigh, and are used extensively in such activities as horseback riding.

Muscles that Move the Leg and Foot
(Table 7-6)

The hamstring muscles are the *biceps femoris,* the *semitendinosus,* and the *semimembranosus* (Fig. 7-15, *B*). These muscles arise from the ischial tuberosity and insert onto the head of the fibula and the lateral and medial condyles of the tibia. Their functions are to flex the leg up on the thigh and to extend the thigh (Fig. 7-29). When the knee is flexed, they rotate the leg inward. This group of muscles is commonly overexerted by athletes who try to develop them too quickly or to push them to their fullest extent before the muscles are warmed up. Such basic exercises as bending the knees, high jumping, and the like call these muscles into play.

The *sartorius* and *gracilis* are the major flexors of the anterior leg and thigh (Figs. 7-15, *A*, and 7-29). The sartorius is a long, narrow muscle that arises from the anterior superior iliac spine. The gracilis, a long, slender muscle on the inner side of the thigh, arises from the body and inferior ramus of the pubis. Both muscles are inserted onto the medial surface of the tibia. The sartorius also assists in rotating the thigh laterally.

The four-headed *quadriceps femoris* covers the front and sides of the femur. Its parts are the *rectus femoris,* which arises from the ilium, and the *vastus lateralis, vastus medialis,* and *vastus intermedius,* which arise from the femur (Fig. 7-15, *A*). The fibers of these four muscles unite at the lower part of the thigh to form a strong tendon inserted on the tuberosity of the tibia, which is closely applied to the patella, the knee bone. The patella develops within the tendon. The quadriceps femoris extends the leg, aiding greatly in walking and running. The rectus femoris portion flexes the thigh.

The ankle is a hinge joint. In the foot, upward movement is called *dorsal flexion;* that is, the foot is bent toward the anterior part of the leg. Downward movement is called *plantar flexion* (extension): the foot is bent backward and downward away from the body. *Inversion* (turning the sole of the foot inward) and *eversion* (turning the sole of the foot outward) also take place at the ankle joint.

The *gastrocnemius* and *soleus,* together known as the *triceps surae,* form the calf of the leg (Figs. 7-15, *B*, and 7-30). The gastrocnemius has two heads that arise from the medial and lateral condyles of the femur. The soleus arises from the head of the fibula and upper end of the tibia, underneath the gastrocnemius. Both muscle fibers are directed downward, and are inserted into a common tendon—the *tendon of Achilles,* the thickest and strongest tendon in the body. These two powerful muscles are used extensively in walking, standing, and leaping, as they extend or plantar flex the foot at the ankle joint and flex the femur on the tibia.

The *tibialis posterior* arises from the posterior surfaces of the tibia and fibula, and inserts onto the undersurfaces of the tarsal bones (cuneiform, cuboid) and the second to fourth metatarsal bones. This muscle plays an important role in supporting the longitudinal arch of the foot. It also extends the foot at the ankle joint in conjunction with the soleus and gastrocnemius.

The *peroneus longus* is found on the lateral aspect of the leg. It arises from the lateral condyle of the tibia and from the head and lateral surface of the fibula, and inserts onto the first cuneiform and first metatarsal bones after crossing obliquely

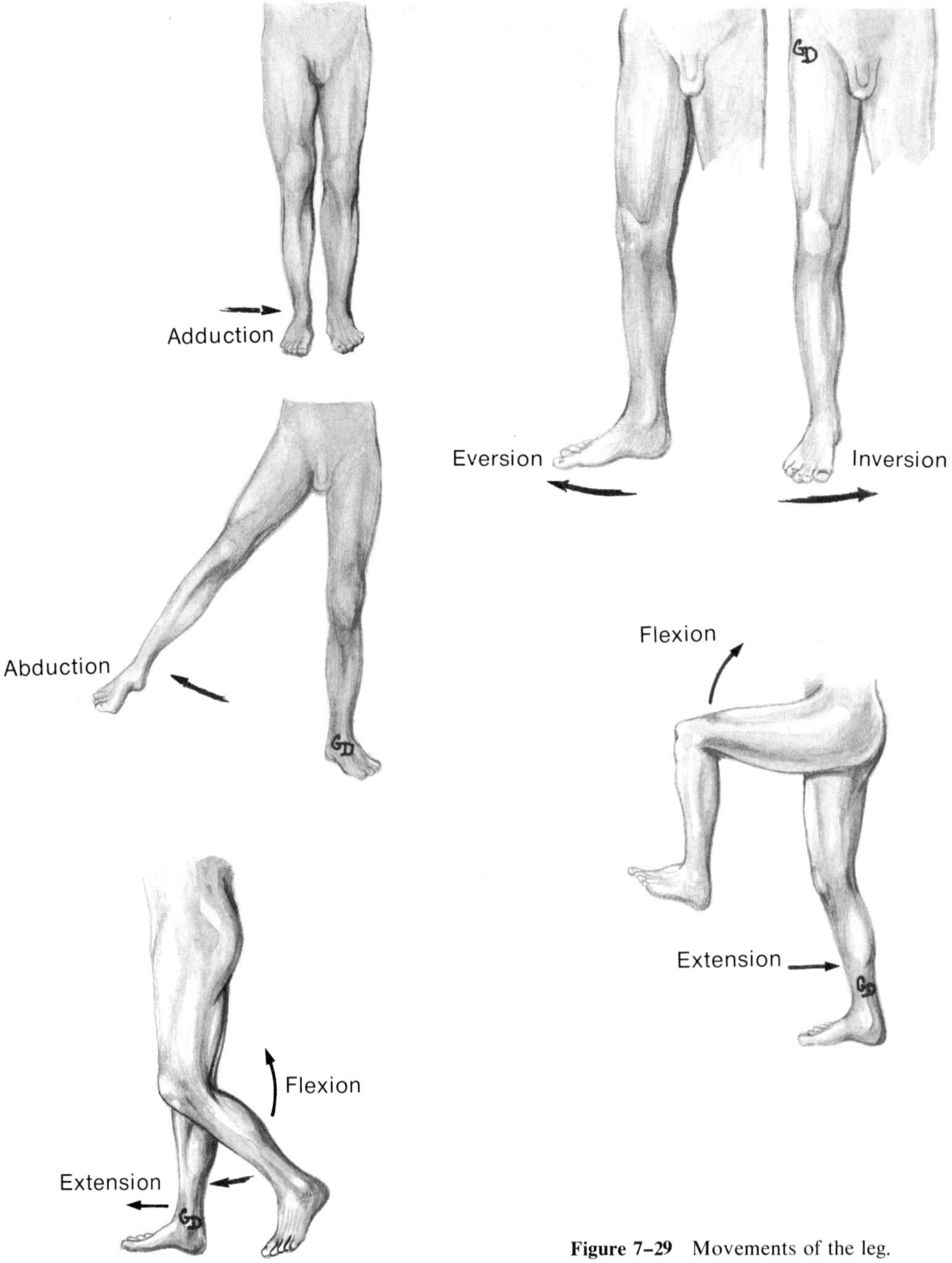

Figure 7–29 Movements of the leg.

over the sole of the foot and terminating in a long tendon. It extends and everts the foot and also helps to maintain the transverse arch.

The antagonists to these extensors are the *tibialis anterior* and the *peroneus tertius* and *brevis* (Fig. 7–30; peroneus tertius not shown). The tibialis anterior arises from the lateral condyle and upper portion of the tibia. Its terminal tendon crosses over the tibia and inserts into the first cuneiform and the base of the first metatarsal bones. The peroneus tertius arises from the distal third of the fibula and inserts onto the base of the fifth metatarsal bone. These two muscles flex the foot dorsally. The tibialis anterior also inverts the foot at the ankle. The peroneus brevis is a short muscle that arises from the lower two-thirds of the fibula and inserts into the fifth metatarsal bone. The peroneus brevis acts with the longus to extend and evert the foot upon the leg. Nerve involvement of the paralyzing variety in the region below the knee generally affects the ability of the tibialis anterior to flex the foot. If this function is impaired, the foot tends to "hang" from the lower leg, a condition called "foot drop." As the name suggests,

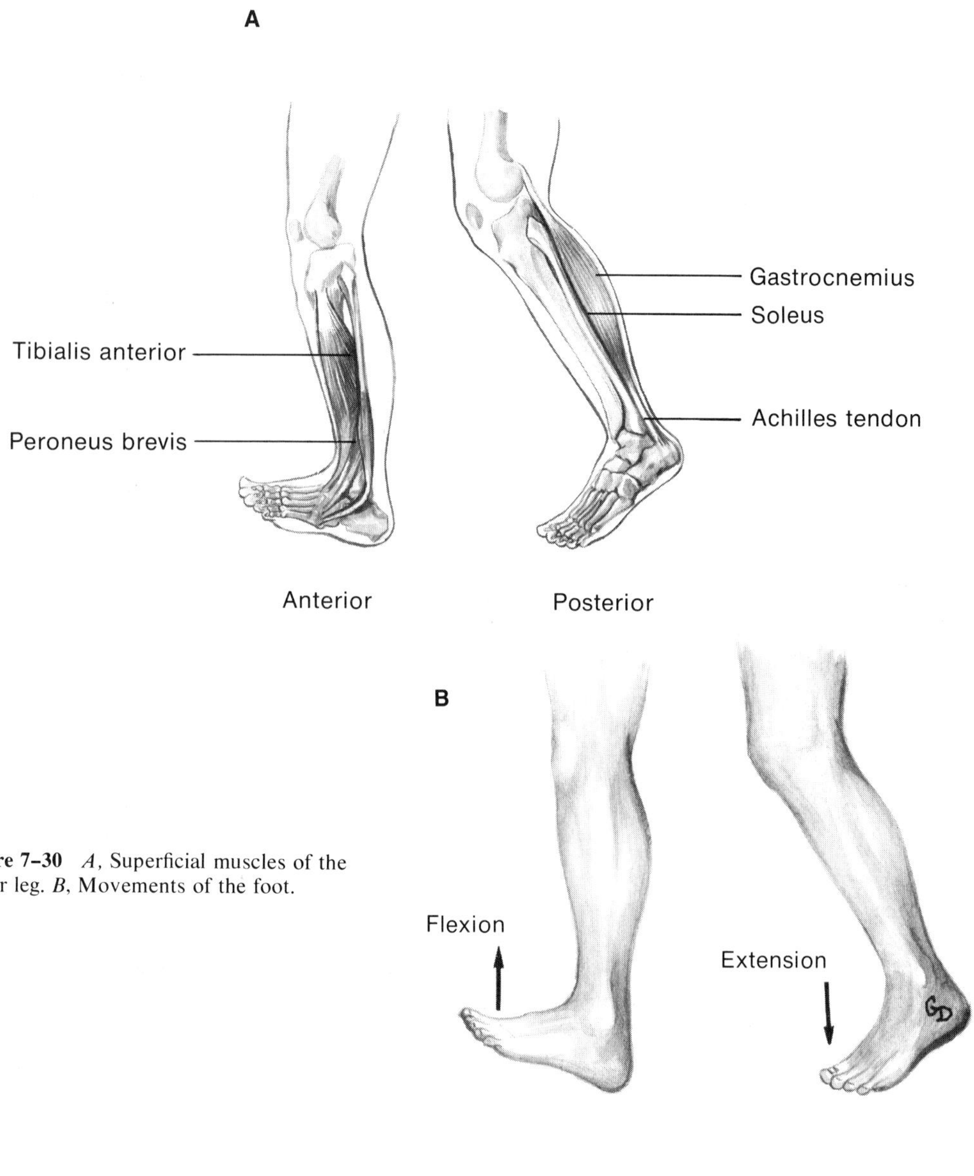

Figure 7–30 *A*, Superficial muscles of the lower leg. *B*, Movements of the foot.

there are marked problems in walking or standing erect. A person with this abnormality has to lift his affected leg high and "throw" his foot forward to take a step.

DISORDERS AFFECTING MUSCLES AND RELATED CONNECTIVE TISSUE

The major symptoms of disorders affecting muscles are paralysis, weakness, atrophy, pain, and spasm.

Myalgia means muscular pain. *Myositis* is inflammation of muscular tissue, especially of voluntary muscles. *Fibrositis* is inflammation of white fibrous tissue, as in muscle sheaths. Tendons, tendon sheaths, and fascia are also subject to inflammatory disorders.

Paralysis. Examples of paralysis, a common pathologic condition of muscle resulting from nerve injury, are known to all. Such muscles cannot contract normally and may, with disuse, show indications of atrophy. The failure of some muscles to contract may prove fatal; for example, after abdominal operations, the muscles of the ileum (the distal portion of the small intestine) may become paralyzed, resulting in a potentially lethal obstruction. Paralysis of the muscles of respiration makes it necessary for a mechanical breathing device to be employed.

Spastic paralysis, a taut, tetanic contraction, is experienced when a group of neurons connecting the brain and the spinal cord is injured. *Flaccid paralysis* (a limp, flabby form) can follow injury to the group of neurons connecting the spinal cord to muscle. If paralysis continues, the lack of use of the affected muscles will result in atrophy (wasting). In addition, muscles may undergo spontaneous atrophy of unknown cause that first appears in the lower limbs.

Chemical elements, as we have seen, play an important role in muscle metabolism. Either a deficiency of potassium or too much potassium (potassium intoxication) can cause paralysis. The enzyme transaminase is present in great concentrations in both cardiac and skeletal muscle. If such muscle tissue is destroyed, the levels of transaminase in the blood will increase, providing a valuable test for determining the presence and extent of such injury, especially in the heart. The test is used in the diagnosis of myocardial infarction.

Fibrillation. Sometimes muscle fibers will contract only individually or in small groups. This phenomenon, called a fibrillation, gives the appearance of quivering. Immediate mechanical stimuli or electric shocks stop the muscle, allowing it time to start contracting again in rhythm. This is the procedure used to stop the heart from fibrillating.

Muscular Dystrophy. The dystrophies are actually a group of related muscle diseases that are apparently inherited, at least in part. Women are usually the carriers but males are most often affected. The shoulder and hip muscles become enlarged, but this is not a true hypertrophy because the muscle is gradually replaced by fat and connective tissue. A blood test will reveal elevated levels of creatine (a nonprotein substance stored in muscle) and abnormal amounts of transaminase.

Myasthenia gravis. Myasthenia gravis, characterized by easy exhaustion of muscles, is a combined neuromuscular problem in that its cause is an impairment of impulse conduction from nerve to muscle. It is thought to be a deficiency of the conducting neurohumor, acetylcholine, or an increase in cholinesterase, the enzyme that readily inhibits acetylcholine activity.

Primary Fibrositis. This is a disease process of fibrous connective tissue. It may occur in acute attacks, causing painful stiffness of the neck or lower back (lumbago), or it may begin insidiously and become a nagging, chronic ailment affecting many parts. Its cause is unknown, but current research suggests that chemical

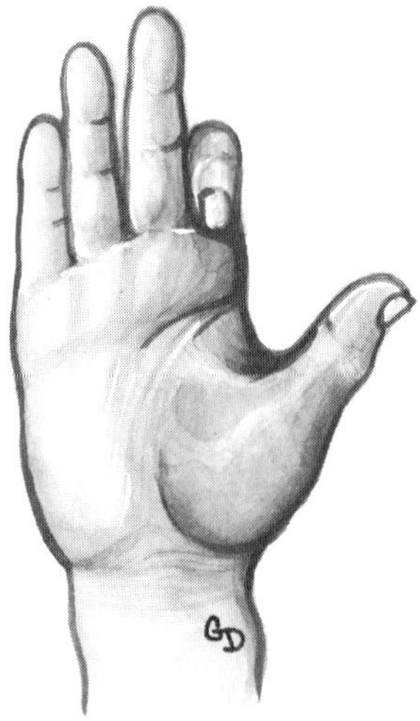

Figure 7–31 Trigger finger.

changes may be the chief etiologic factor. The involved portions become stiff and painful. Irritation of the periarticular connective tissue and muscles leads to restriction of movement.

Tenosynovitis. Tenosynovitis is an inflammation of the tendon sheaths. It may interfere with free passage of the enclosed tendons, leading to dysfunction of the associated joints. Weakness and atrophy result if the infection becomes chronic. Since the joints themselves are not diseased, no deformities result, and when the infection is arrested function returns to normal.

The area along the flexor tendon sheaths of the fingers is commonly affected. The finger frequently cannot be extended by the "frozen flexor" apparatus, and a "trigger finger" results (Fig. 7–31).

Ganglion. This is a small cystic swelling that develops in connection with a tendon sheath close to a joint. The commonest location is the back of the wrist. The swelling is attached to the outer surface of the tendon sheath and eventually undergoes mucoid degeneration, with numerous small cysts forming in the area. Eventual impairment of wrist movement warrants surgical removal of the ganglia.

Clubfoot. Clubfoot is a congenital malformation of one or both feet in which the front part of the foot is inverted and rotated. It is accompanied by shortening of the Achilles tendon and contracture of the fascia in the sole of the foot.

OUTLINE

I. Introduction
  A. Motion
    1. Made possible by contractility of muscle tissue
    2. Includes breathing, movement, beating of heart, digestion, and visceral movements
II. Microstructure of Muscle
  A. General Structure
    1. Myofibrils—protein molecules that accomplish contraction
    2. Sarcolemma—strong sheath enclosing myofibrils
    3. Sarcoplasm—muscle fiber protoplasm
    4. Fasciculi—bundles of muscle fibers
    5. Fascia—areolar connective tissue that holds together the fasciculi
  B. Types of muscle
    1. Smooth
      a. Long, slender fibers
      b. Fibers grouped in thin sheets
      c. Each fiber has a single, central nucleus
      d. Found in blood vessels and surrounding viscera
      e. Involuntary control
    2. Skeletal
      a. Voluntary control of the skeleton
      b. Surrounded by epimysium
      c. Each fiber cell has many nuclei
      d. Protein bands appear as striations on outside of fiber
        1) Thick filament—myosin (H band)
        2) Thin filament—actin (I band)
        3) Arranged in rows that interlock (A band)
        4) Fibers divided into seg-

ments (sarcomeres) by 2 lines
5) Availability of energy affords a contraction
3. Cardiac
a. Cells form network of branching fibers
b. Involuntary
c. Intercalated disks—bands separating individual cells
d. Many mitochondria to supply abundant energy

III. Chemistry of Muscle Tissue
A. Makeup of skeletal muscle
1. 75 per cent water
2. 25 per cent solid material
a. 20 per cent protein (actin and myosin)
b. 5 per cent carbohydrates, lipids, inorganic salts, and nonprotein nitrogenous compounds
B. Energy sources
1. ATP—immediate
2. CP—intermediate
3. Glycogen—ultimate
C. Oxygen debt mechanism
1. High energy derived from aerobic respiration
2. Deficiency of oxygen to tissues prevents energy production
3. Lactic acid accumulates
4. Body rest resupplies the muscles with the needed oxygen to pay back the energy need

IV. Contraction of Muscles
A. Mechanism of muscle contraction
1. Available ATP combines with actin
2. Myosin, in the presence of calcium, acts as an enzyme to free energy from ATP
3. Magnesium inhibits reaction between actin and myosin—myofibril remains relaxed
4. Actomyosin complex theory
a. Conversion of ATP to ADP binds actin and myosin together at energy "bridges" and "sites," causing a shortening of the segments of a muscle fiber—a contraction
B. Properties of muscles related to contraction
1. Excitability
a. Muscle receives an impulse and responds to it
b. Muscle contraction
1) Latent period
2) Contraction period
3) Relaxation period
c. Stimulus—excitant or irritant that causes a reaction
1) Liminal
2) Subliminal
3) Maximal
4) Adequate
d. Response—reaction
2. Contractility
a. Permits muscle to change its shape
b. Isometric contraction—no mechanical work is done
c. Isotonic contraction—muscle overcomes workload
d. All-or-none law—muscle fiber responds totally to a stimulus or not at all
e. Tonicity (tonus)—muscle is in a state of partial, continual contraction
f. Atrophy—lack of tonus
g. Epinephrine—muscle stimulant
3. Extensibility
a. Muscle can be stretched
4. Elasticity
a. Muscle returns to its original size
C. Some phenomena of contraction
1. Skeletal muscles contract and relax more quickly than smooth muscles
2. Muscles do their best work at body temperature
3. Tetany—too many stimuli without allowing the muscle to relax

4. Spasm—sudden involuntary contraction of a skeletal muscle
5. Fatigue—inability of a muscle to respond to a stimulus
6. Exercise stimulates circulation and strength in a muscle

D. Conduction of the contractile impulse
  1. Motor unit
    a. Nerve body
    b. Nerve fiber
    c. Motor end-plate
    d. Associated muscle fibers
  2. Chemical conductor
    a. Acetylcholine
  3. Acetylcholine cycle
    a. Activated in the end-plate by an impulse
    b. Diffuses into the muscle with the impulse
    c. A muscle contraction occurs
    d. Cholinesterase inactivates the acetylcholine
    e. The separate acetyl and choline radicals diffuse back into the end-plate
    f. Choline acetylase re-establishes acetylcholine, which awaits a new impulse
  4. Inhibitors
    a. Curare
    b. Succinylcholine
    c. Nerve gases
    d. Botulinus toxin
    e. Myasthenia gravis

V. Skeletal Muscles
  A. How muscles are named
    1. Location—intercostal
    2. Action—flexor
    3. Shape—trapezius
    4. Direction of fiber—rectus
    5. Number of heads of origin—biceps
    6. Points of attachment—sternocleidomastoid
  B. Muscle action
    1. Antagonists—muscles working opposite to each other
    2. Synergists—muscles working together
  C. Muscles of facial expression and mastication
    1. Epicranius—surprise
    2. Orbicularis oculi—squinting
    3. Orbicularis oris—pouting
    4. Zygomaticus—smiling
    5. Buccinator—chewing
    6. Mentalis—doubt
    7. Platysma—sadness
    8. Masseter—chewing
    9. Temporalis—sternness
    10. Pterygoid—consternation
    11. Genioglossus, styloglossus—speaking, chewing, swallowing
  D. Muscles that move the head
    1. Sternocleidomastoid—forward
    2. Semispinalis capitis—extension, rotation
  E. Muscles that move the shoulder and arm
    1. Trapezius—lifts shoulder (shrugging)
    2. Levator scapulae—elevates scapula
    3. Rhomboid (major and minor)—elevate and retract scapula
    4. Latissimus dorsi—sweeps arm downward
    5. Pectoralis (major and minor)—draw shoulder forward, medially, and downward
    6. Serratus anterior—moves scapula forward (pushing)
    7. Coracobrachialis—moves arm forward (flexion)
    8. Deltoid—abducts arm
    9. Teres (major and minor)—rotate arm medially and laterally
  F. Muscles that move the forearm and hand
    1. Biceps—flexes forearm
    2. Brachialis—flexes forearm
    3. Triceps—extends forearm
    4. Pronator teres—pronates hand
    5. Supinator—supinates hand and forearm
    6. Flexor carpi (radialis, ulnaris)—flex, abduct, adduct hand
    7. Extensor carpi (radialis, ul-

naris)—extend, abduct, adduct hand
8. Palmaris longus—flexes hand
9. Flexor digitorum—flexes fingers
10. Extensor digitorum—extends fingers
11. Adductor and abductor pollicis—adduct and abduct thumb

G. Muscles of respiration
1. Intercostals—enlarge thoracic cavity
2. Diaphragm—enlarges thoracic cavity
a. Weakness—diaphragmatic hernia

H. Muscles of the abdominal wall
1. External oblique, internal oblique, transversus abdominis—form the abdominal aponeurosis or girdle
a. Weaknesses—umbilical and femoral hernias
2. Rectus abdominis—flexes vertebral column and pelvis

I. Muscles that move the thigh
1. Iliopsoas—flexes thigh
2. Gluteus (maximus, medius, minimus)—extend, abduct, rotate thigh
3. Adductor (magnus, longus, brevis)—adduct, flex thigh

J. Muscles that move the leg and foot
1. Biceps femoris, semitendinosus, semimembranosus—flex leg, extend thigh
2. Sartorius, gracilis—flex leg and thigh
3. Quadriceps—extends leg; flexes thigh
4. Gastrocnemius, soleus—extend foot; flex leg
5. Tibialis posterior—extends foot
6. Peroneus (longus, tertius, brevis)—extend and evert foot
7. Tibialis anterior—dorsally flexes foot
a. Foot drop

VI. Disorders Affecting Muscles and Related Connective Tissue
A. Paralysis
1. Spastic
2. Flaccid
B. Fibrillation
C. Muscular dystrophy
D. Myasthenia gravis
E. Primary fibrositis
F. Tenosynovitis
G. Ganglion
H. Clubfoot

STUDY QUESTIONS AND PROBLEMS

1. Give a general description of the three types of muscle, indicating their anatomy and function, and the nature of their contractions.
2. What is complete tetany? Why doesn't it occur in cardiac tissue?
3. What is rigor mortis? What is its cause?
4. What part do actin and myosin play in muscle contraction?
5. Can a muscle contract without calcium? Magnesium?
6. Sketch a single muscle twitch and define the anatomy of a muscle contraction.
7. What is a threshold stimulus? What is meant by "summation effect"?
8. Can we create an oxygen debt? Describe it.
9. What do the intensity of stimulus and number of stimuli have to do with a muscle's action?
10. Define tonus and the all-or-none law. Do they complement each other or contradict each other? Why?
11. Distinguish between an isotonic and an isometric contraction.
12. Of what value are ATP, CP, and glycogen to muscle? Describe their roles.
13. What are the origin and the insertion of a muscle?
14. What are flexors, extensors, rotators, supinators, and depressors?
15. Differentiate between fibrillation and spasm.
16. Define extensibility and elasticity. Why are they important?
17. How is a nerve impulse transmitted to a muscle?
18. State at least four ways in which muscles are named. Give examples.
19. Name some inhibitors of acetylcholine and show how they achieve this effect.
20. What is peristalsis?

Chapter 8

Patterns of Control

Part One: The Nerve Cell; The Voluntary Nervous System

Once some of the body systems have been studied, it becomes fairly obvious that not one of these systems is capable of functioning alone. The systems are interdependent. All must work together as one functioning unit in order that homeostasis may be maintained within the body. The mechanism that insures that the organs and systems operate in smooth coordination is the nervous system. Conditions within and outside the body are constantly changing, and one purpose of the nervous system is to respond to these internal and external changes so that the body may adapt itself.

THE NEURON

The basic unit of nervous tissue is the nerve cell or *neuron* (Fig. 8–1; not all component parts shown). Neurons are composed of a *cell body* containing a large *nucleus* with one or more *nucleoli, mitochondria, Golgi apparatus, Nissl bodies* (which are associated with the conduction of a nerve impulse), and numerous *ribosomes*. Extending from this cell body are threadlike projections of cytoplasm known as *nerve fibers*. There are two types: *dendrites*, which conduct impulses to the cell body; and *axons*, which conduct impulses away from the cell body.

There are three main types of neurons:

1. Motor Neurons. These pass messages from centers in the brain and spinal cord to effector organs, such as muscles and glands.

2. Association Neurons. These, also called internuncial or intercalated neurons, relay messages between neurons within the brain and spinal cord.

3. Sensory Neurons. These receive messages from the environment and pass them to centers in the brain and spinal cord.

Neurons vary in shape and size and have one or more fibers or processes. Thus they may also be classified according to the number of processes they exhibit (Fig. 8–1):

1. Monopolar. All neurons must have at least one axon; they may have no dendrites, one, or more than one. In monopolar neurons, there is a single process that divides shortly after leaving the cell body; one branch conveys the impulse toward the central nervous system, and is, therefore, technically the axon. The other branch goes toward the periphery and conveys impulses from sense organs to the cell, acting as the dendrite. Such unipolar cells have only a sensory function.

2. Bipolar. These neurons have two processes, one at each of the opposite poles of the cell. One dendrite carries impulses to the cell; one axon conveys impulses from the cell to centers in the central nervous system. As an example,

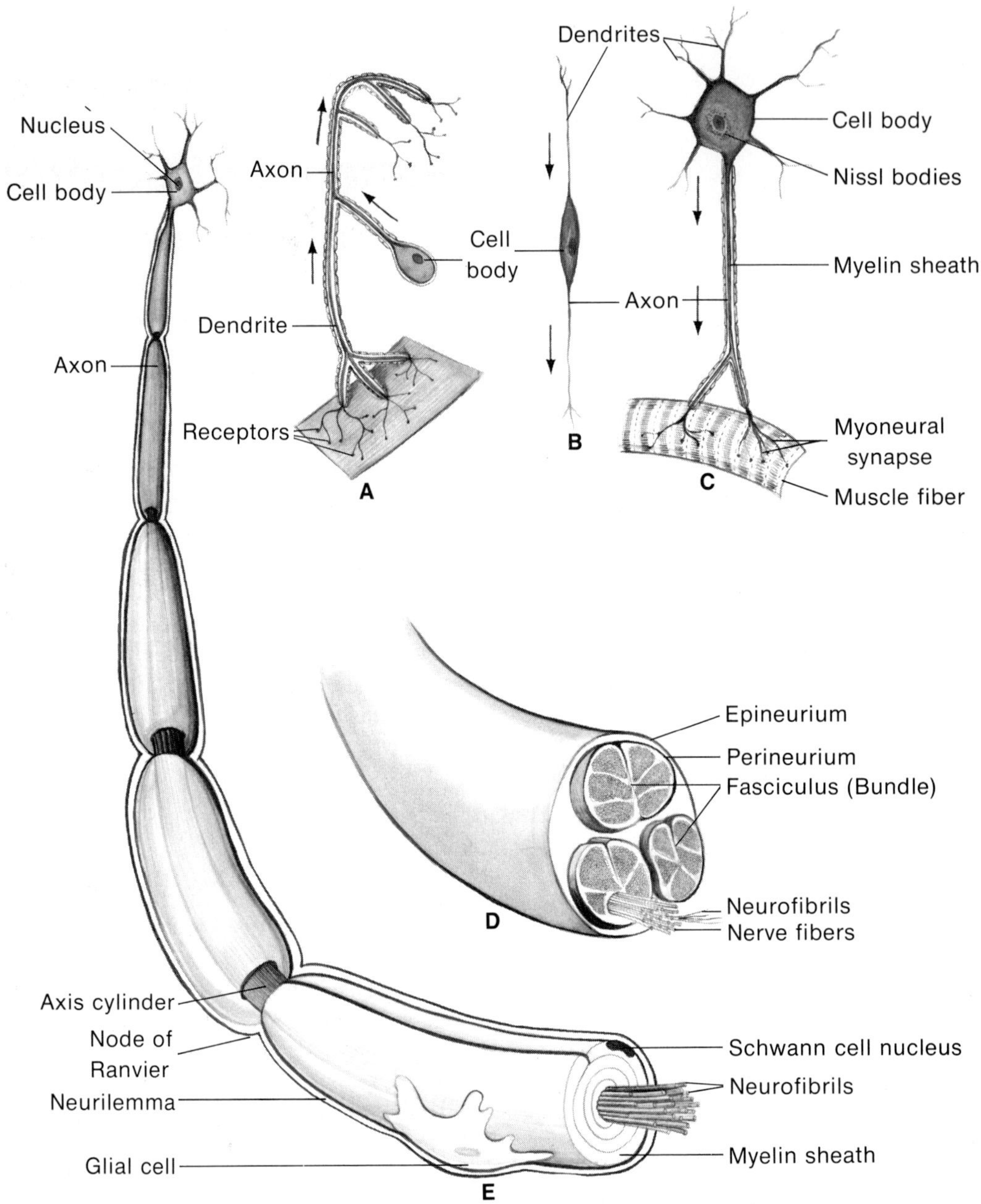

Figure 8–1 The structure of a neuron. *A*, Monopolar sensory neuron. *B*, Bipolar retinal neuron in the eye. *C*, Multipolar motor neuron, showing a muscle-nerve synapse. *D*, Structure of a typical nerve. *E*, An enlarged extension of an axon, showing its protective myelin sheath and neurilemma.

neurons associated with the auditory nerves are of the bipolar type and are sensory in function.

3. Multipolar. These are the commonest type of nerve cell. They are usually star-shaped (stellate), and have numerous branching processes. Short processes—dendrites—convey ingoing impulses to the cell; a large, single process—the axon—carries outgoing impulses from the cell body. Most multipolar neurons are motor or associative in function.

As indicated above, the dendrites of sensory neurons differ markedly from those of other neurons. They are usually single, and may be either very long (up to 3 feet) or short; in any case, they do not have the treelike appearance typical of other dendrites. Each sensory nerve fiber (dendrite) has a special structure, the *receptor* or *end-organ* (Fig. 8–1), where the stimulus is propagated. Such sensations as pain, touch, hearing, and seeing which involve these sensory neurons will be discussed in a later chapter.

The sensory nerve fibers that are connected with receptors conduct impulses to the brain and spinal cord. When grouped together they form *afferent* nerves. The motor fibers that carry impulses from the centers out to the muscles and glands form *efferent* nerves. Some nerve structures contain a mixture of afferent and efferent nerve fibers, and often are referred to as "mixed nerves."

Structure of Nerve Fibers

The nerve fiber is a long, thin tube consisting of a cell membrane enclosing a core of cytoplasm. The end of a fiber may be a few feet away from the cell nucleus. As in all cells, the nucleus is necessary for the prolonged life of the cell, but if a fiber is severed from its cell body and kept in a suitable salt solution it can carry impulses for several days.

Satellite cells that probably help nourish and protect the nerves are attached to nerve cells. *Schwann cells* are satellites wrapped in thin concentric layers around the nerve fiber of certain nerve cells, forming a myelin sheath or coat *(white matter)*. On these *myelinated* fibers the satellite cells are wound many times in a spiral, forming thick sleeves of fatty substance, the *myelin* (Fig. 8–1). Each sleeve is separated from the next by a depression, the *node of Ranvier*. At the nodes, the nerve fiber has free access to the surrounding fluids, allowing exchange of nutrients and waste materials. Most of the peripheral nerves of the body and also the nerves found in the outer periphery of the spinal cord are myelinated. Nonmyelinated nerves *(gray matter)* do not need the protection afforded by the laminated myelin; these generally are found in the organs themselves, where the organ tissue provides protection and nourishment.

The axon, when it leaves the central nervous system (that is, the brain and spinal cord), usually acquires a covering of myelin. This, in turn, is encased in a *neurilemmal sheath* (a thin membrane) continuous with the cell body. Bundles of axons with their coverings make up the peripheral nerves. Each bundle has a connective tissue sheath, the *perineurium* (Fig. 8–1). A number of bundles are encased in *epineurium*.

All accessory or supporting nerve cells of the central nervous system that neither receive nor conduct impulses are grouped within the category *neuroglia*. The neuroglia are comparable to the Schwann cells of the peripheral nervous system.

PHYSIOLOGY OF NERVOUS TISSUE

Regeneration of the Fiber

Although nerves do not reproduce themselves, they can regenerate new parts under special circumstances. When a nerve fiber is cut, the severed end of the axon degenerates. If an injury destroys a number of axons supplying a muscle block, the muscle can no longer function and ultimately atrophies. Frequently, it is possible to repair the damage surgically when the injury is localized. The connection between the cut ends of the fibers is re-established by carefully bringing the severed ends of tissue together; when this is done the nerve fibers very often grow into the sheaths remaining in the severed tissue. This delicate operation is based on the dis-

Table 8–1 INTRA- AND EXTRACELLULAR IONS

ION	INTRACELLULAR IONS mEq/L	EXTRACELLULAR IONS mEq/l
Potassium	155	4
Sodium	12	135

covery that, provided the sheath of myelin is intact, growth of the existing fiber can take place and new myelin can be generated.

Transmission of the Nerve Impulse

In all living cells there is a difference in electrical potential between the inside of a cell and the area just outside the cell wall. Consequently, if these two areas are connected by a conductor, a current will flow. The potential difference comes from the unequal distribution of ions between the inside and outside of the cell (Table 8–1).

It is now known that, although the nerve impulse is electrical in nature, its *transmission* along the fiber is the result of *chemical* reactions involving potassium and sodium ions. The impulse relies on a series of changes in electrical polarity as it moves along the cell membrane. The charge itself is the result of the movement of potassium and sodium ions (charged atoms) across the differentially permeable membrane of the fiber cells (see pp. 51–53).

The operation of nerve impulse transmission depends on the separation of the positively charged sodium and potassium ions on either side of the cell membrane. During the resting stage, the differentially permeable membrane is polarized, with the inner surface negative with respect to the outer surface. This difference in electrical potential is due to the greater concentration of sodium ions outside the cell and of potassium ions inside. Membrane polarity is maintained by the neuron through energy derived from cellular respiration. Large numbers of mitochondria, the cellular bodies where energy is generated, are found in nerve fibers.

The excitation of a nerve by a stimulus produces a momentary change in the permeability of the membrane, resulting in sodium ions passing into the fiber and potassium ions moving out. The exchange of ions causes a drop in voltage, and that part

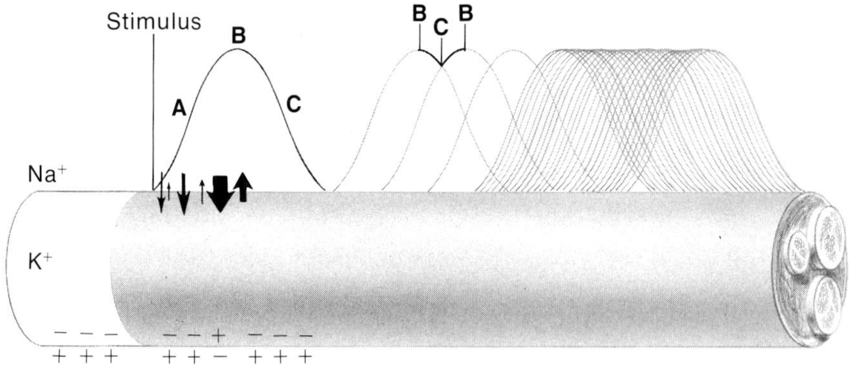

Figure 8–2 Conduction of a nerve impulse. Size of the arrows suggests the ease with which K^+ and Na^+ pass through the membrane during the transmission of the impulse along the fiber. The + and – signs indicate the quantity of positive ions outside and inside the nerve membrane. *A*, Polarized zone with more sodiums outside the membrane than inside, and more potassiums inside than outside. *B*, Depolarized zone: after the membrane has been stimulated, the sodium ions are allowed to flow in at a greater rate than the potassium ions flow out, creating a wave of negativity on the outside of the membrane. *C*, Repolarized zone. The sodium ions are forced back out (sodium pump) and the potassium ions diffuse, restoring the greater quantity of positive ions on the outside.

of the fiber becomes depolarized (Fig. 8–2), creating the nerve impulse.

The depolarization lasts for only an instant; however, many more sodium ions continue to flow into the cell. As an excess of positive ions is created inside the cell, an excess of negative ions occurs outside the cell wall. The potential, therefore, has reversed.

The negative, stimulated point of the membrane sets up a local current, with the positive point adjacent to it acting as a stimulus, thus creating a depolarization of that part of the membrane by reversing its polarity from positive to negative (Fig. 8–2). This cycle of reversal of potential goes on over and over in rapid succession, and the impulse travels from point to point over the full length of the fiber, much as a ripple of water moves toward the shoreline. This movement of an impulse or "wave of negativity" is called *conduction.*

Repolarization results partly from the "sodium pump" of the active transport theory (pp. 52–53), but is due mainly to the membrane's decreased permeability to sodium, resulting in an increased permeability to potassium. The latter diffuses outward, creating an excess of positive ions on the outside, thus restoring the resting potential of the membrane.

In the nervous system, the impulse begins at the dendrite or cell body and is then conducted throughout the length of the axon.

When myelin covers the neuron, the mechanism of impulse conduction is a bit different. Myelin acts as an insulator, but this insulation is broken at the nodes of Ranvier (Fig. 8–1). Thus, the positive charges must flow from one node to the next by passing over the myelin. Because the charges leap from node to node, this form of conduction is termed *saltatory,* meaning dancing or leaping.

Velocity of Conduction

The velocity at which a neuron conducts an impulse is a function of its diameter. In man, large motor neurons conduct impulses at a velocity of 80 to 100 meters per second. Sensory nerves generally have a similar diameter but transmit more slowly. Nonmyelinated fibers conduct impulses more slowly than do myelinated ones. Their velocity is about 0.5 meter per second.

All-or-None Response

Nerves operate on the "all-or-none" principle. Once started, the impulse travels with the same intensity and speed, no matter how strong the stimulus that initiated it. Every neuron has a specific threshold of excitation, but once that threshold is exceeded, the response is unchanged by increasing the intensity of the stimulus.

Summation of Action Potentials

A nerve is composed of many neurons (Fig. 8–1). The action potential recorded from a nerve depends on the number of neurons firing. Because a nerve usually contains neurons with different activation thresholds, it is possible to alter the nerve action potential by increasing the intensity of stimulation. A weak stimulus activates only a few neurons; a strong one, more. The total nerve action potential represents the summation of the action potentials of all the active neurons.

Conduction of an Impulse across Synapses

Nerves are interconnected at junctions called *synapses.* The word means "contact." These junctions, where two or more neurons are in contact, consist of finger-like processes whose ends are sufficiently close to one another to allow certain chemicals to diffuse rapidly across the intervening space.

A close look at a terminal knob of a presynaptic neuron (Fig. 8–3) discloses

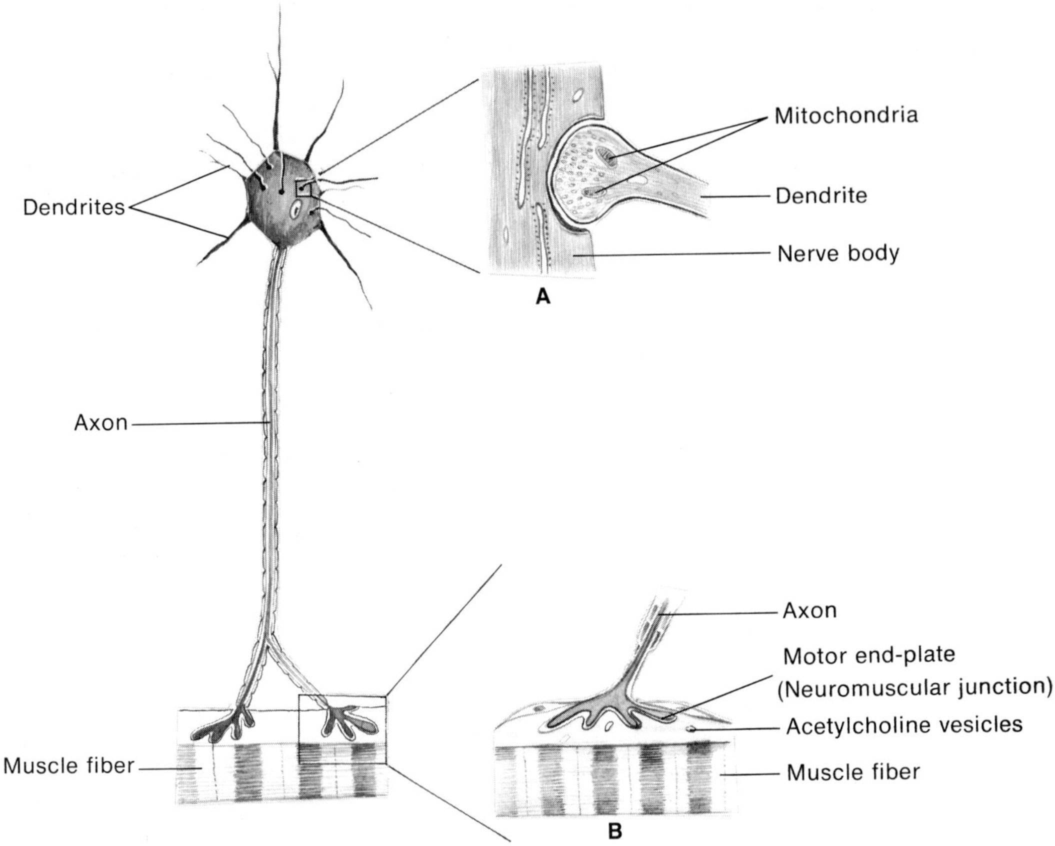

Figure 8–3 Types of synapses. *A*, Nerve-nerve body synapse. *B*, Nerve-muscle synapse (myoneural junction).

that it contains many mitochondria and vesicles. Note also that there is a space of approximately 250 Å between the knob and the neuron. It is now generally believed that when an impulse reaches the terminal knob, it causes the excitatory chemical, acetylcholine, to be released by the vesicles. The acetylcholine then flows across the gap and comes in contact with the membrane of the neuron, causing an increase in membrane permeability. As a result, sodium ions rush in and permit the passage of the impulse from the terminal knob to the neuron.

Once the impulse has been sent across the synapse, the acetylcholine is broken down, as in muscle, into two components, acetyl and choline, by the enzyme cholinesterase. If this did not occur, the chemical would keep the synaptic resistance at such a low level that there would be a serious interference with the nerve paths.

As you will recall from Chapter 7, the main features of the point at which motor fibers enter a skeletal muscle—the *neuromuscular junction* or *motor end-plate* (Fig. 8–3)—include a terminal branch of an axon and a gap similar to the neuronal synapse. Acetylcholine is the common transmitter for this gap. This type of synapse is commonly called a *myoneural junction*.

As noted earlier, the body has a means of resupplying the necessary acetylcholine for synaptic transmission. After the acetylcholine has been broken down by cholinesterase, the two free compounds, acetyl and choline, diffuse back across the gap and are reunited by a second enzyme, choline acetylase, in the terminal knob vesicles to await a new impulse.

THE CENTRAL NERVOUS SYSTEM

The central nervous system (CNS) is made up of the brain and spinal cord. In the embryo, the central nervous system develops from a neural tube. The anterior end of the tube enlarges and, greatly modified, forms the brain, which reaches its greatest complexity in man. The original tubular structure still may be seen in the intercommunicating ventricles of the brain and in the central canal of the spinal cord. To facilitate understanding of the CNS, we will first consider the anatomy and function of the spinal cord and the nerves arising from it.

The Spinal Cord

The spinal cord occupies the entire spinal canal of the embryo and thus extends down into the tail portion of the vertebral column. However, the column of bone grows much more rapidly than the nerve tissue of the cord; therefore, the end of the cord soon is too short to extend into the lower part of the spinal canal. This disparity in growth increases during development, and in the adult the cord ends between the first and second lumbar vertebrae in the region just below the area to which the last rib attaches.

Examination of the spinal cord reveals that it has a small, irregularly shaped internal section of gray matter (nerve cell bodies) and a larger surrounding area made up of white matter (nerve fibers). A cross-section of the cord shows that the gray matter is arranged in columns of cells ex-

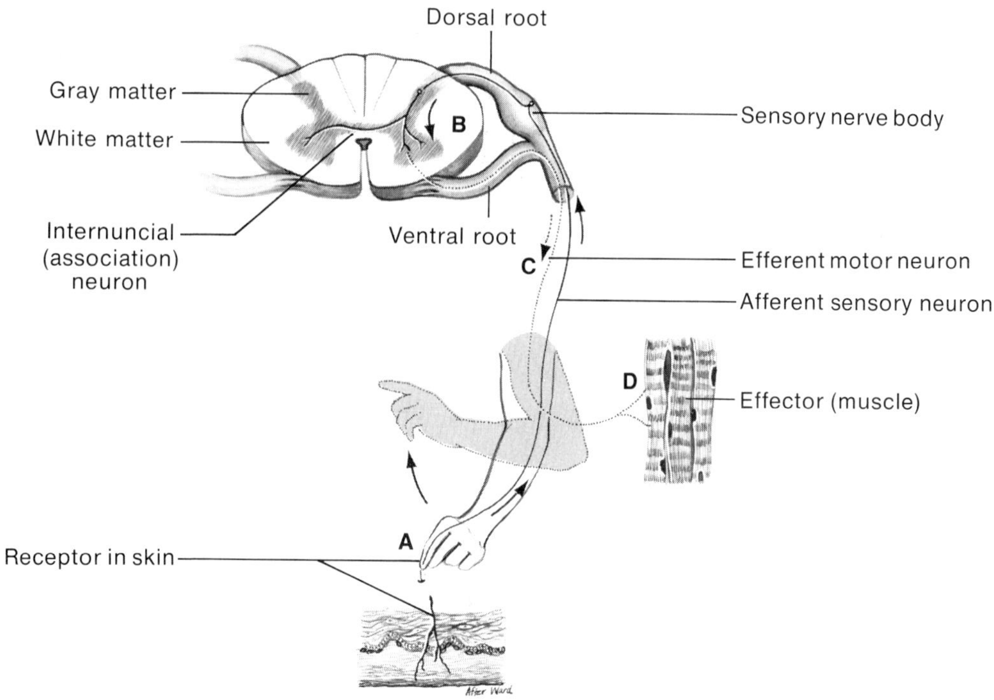

Figure 8–4 Diagram demonstrating a simple reflex arc. An impulse is initiated at the finger by a receptor *(A)* in the skin. The impulse travels over the sensory afferent neuron to the spinal cord, where it is transmitted via the internuncial neuron *(B)* to the efferent motor neuron. The impulse then travels via the efferent motor neuron (*C*) to the muscle (*D*), which effects a response.

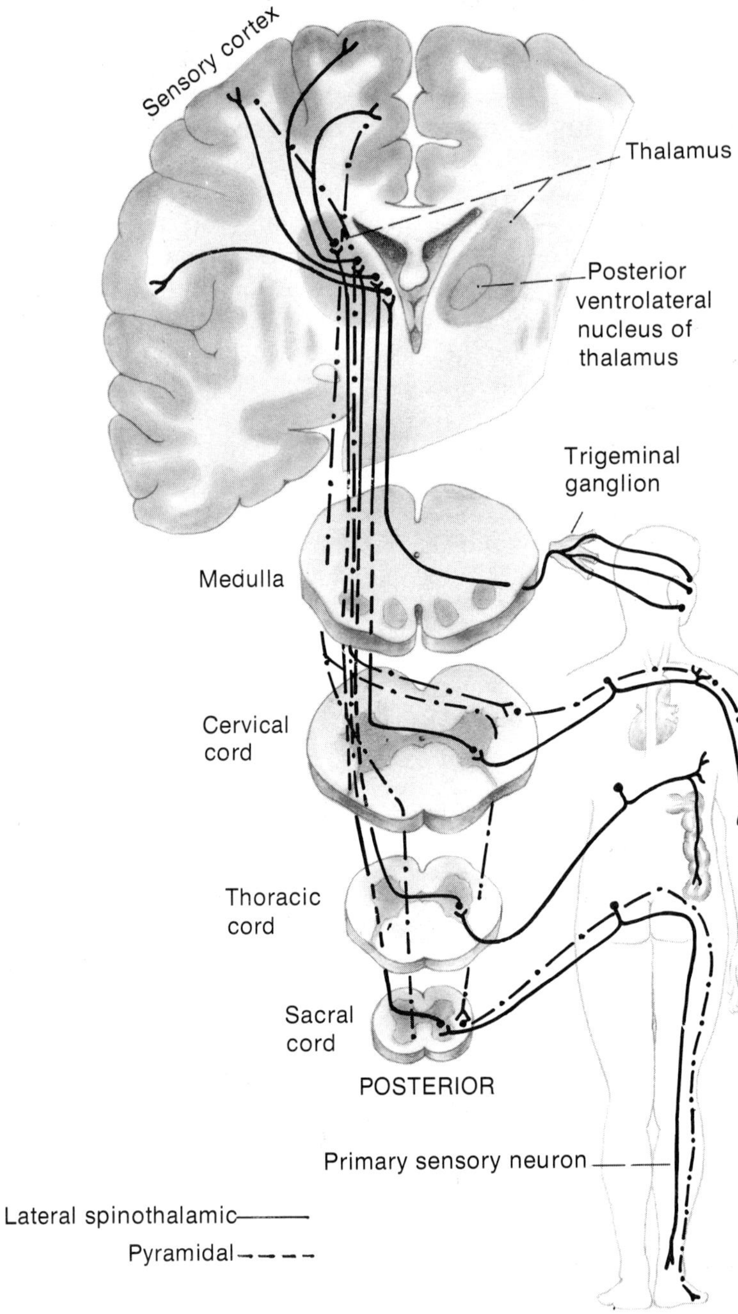

Figure 8–5 Elements of ascending (sensory) and descending (motor) tracts of the spinal cord. The lateral spinothalamic tract is sensory for pain and temperature. The pyramidal tract conveys motor impulses primarily for voluntary muscle movements. (Modified from Jacob, S. W., and Francone, C. A.: Structure and Function in Man, 3rd ed. Philadelphia, W. B. Saunders Co., 1974.

tending up and down dorsally, one on each side. Another pair of columns is found in the ventral region; a third, less conspicuous pair is situated between the ventral and dorsal columns. These three pairs of columns of gray matter give, in cross-section, an H-shaped appearance (Fig. 8–4).

The white matter, consisting of thousands of myelinated nerve fibers, is arranged in three columns outside the gray matter on each side. Its function is to convey information up and down the cord. At each small space between the vertebrae (intervertebral space), spinal nerves arise from each side of the cord and distribute themselves in the tissues. There are about as many pairs of spinal nerves as there are vertebrae, but at the lower end of the spine at the level of the first lumbar vertebra, the cord divides. All the spinal nerves which emerge from the vertebral column below that point continue as separate strands. Above this point, the spinal nerves arise at each vertebral level, and dorsal and ventral roots (Fig. 8–4) are present on each side. The dorsal roots contain the afferent (sensory) fibers that bring information into the CNS from the receptor, where the impulse is propagated.

Each column (or funiculus) is subdivided into tracts—large bundles of nerve fibers arranged in functional groups (Fig. 8–5). Ascending (sensory) tracts transmit impulses to the brain, and descending (motor) tracts transmit impulses from the brain to the various levels of the spinal cord. Information can thus be conveyed up and down the cord (see also p. 165).

At each of the intervertebral spaces, the fibers of the dorsal roots may or may not be myelinated. Their cell bodies are located outside the cord proper in a swelling called the dorsal root ganglion. The ventral roots carry the efferent motor fibers, and their cell bodies are inside the spinal cord. These motor fibers carry impulses from the CNS to the muscle or gland, and are heavily myelinated. A third neuron, the association or internuncial neuron, is found exclusively within the brain and spinal cord, and transmits impulses from the afferent to the efferent neurons.

Spinal Reflexes

The least complicated example of how the sensory mechanism operates is the simple reflex. If a man pricks his finger, he pulls back his hand. The basic events that occur here are shown in the diagram of the reflex arc in Figure 8–4. The receptors at *A* are stimulated. An impulse travels over the sensory afferent nerve to the spinal cord, where it synapses at *B* with an internuncial neuron. This, in turn, synapses with the motor efferent fiber at *C*, and an impulse travels out to stimulate the muscle at *D*, which contracts and withdraws the hand.

In a very few cases (patellar and pupillary reflexes), the sensory fiber synapses directly with the motor fiber. Other associative fibers may convey the impulse to other motor fibers at different levels in the spinal cord (even to the brain) to deliver the information that a stimulus has been applied.

The Spinal Nerves

Strictly speaking, the spinal nerves are part of the peripheral nervous system rather than of the CNS, but we are considering them here as part of our discussion of the spinal cord. Each of the 31 pairs of spinal nerves that arise from successive levels of the spinal cord begins as a series of fibers from both the dorsal and ventral roots of the cord. The dorsal root contains only sensory fibers; the ventral root contains only motor fibers and acts as a final pathway for all motor impulses leaving the spinal cord. The first cervical nerve leaves the vertebral canal by passing between the first cervical vertebra and the skull. Below this, successive cervical nerves pass between adjacent vertebrae.

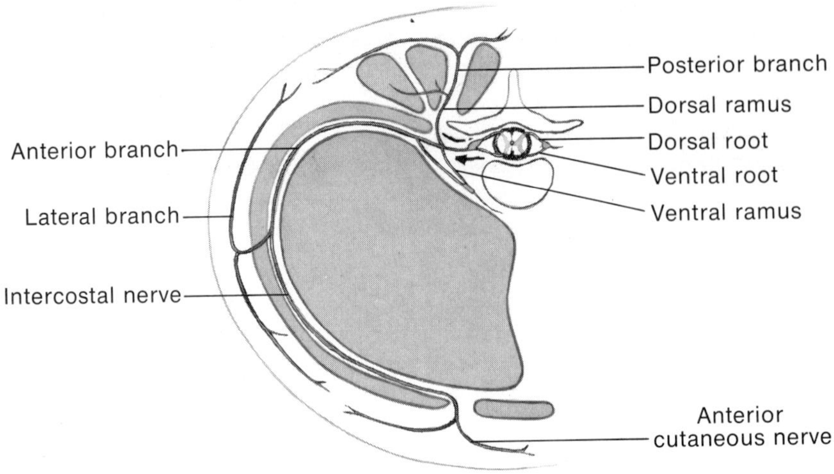

Figure 8–6 Diagram of the course and branches of a typical intercostal nerve.

Below the cervical region the spinal nerves likewise leave the vertebral canal below their respective vertebrae: for example, the first thoracic nerve leaves the canal between the first and second thoracic vertebrae.

As each spinal nerve emerges from between its respective vertebrae, it divides into two small branches. The first of these, the *dorsal,* or *posterior, ramus,* passes posteriorly (dorsally) to supply a specific segment of the skin, bones, joints, and longitudinal muscles of the back (Fig. 8–6). The other *ventral,* or *anterior, ramus* supplies a portion of the anterior or lateral aspect of the trunk, or of the limbs. The ventral rami are larger than the dorsal rami and behave quite differently. In the cervical, lumbar, and sacral regions, ventral rami form extensive networks *(plexuses)* of nerves. The ventral rami in the thoracic region course in the intercostal spaces to supply primarily the intercostal muscles and the skin overlying them. Most nerves arising from these plexuses carry fibers of neurons from more than one segment of the spinal cord (Fig. 8–7).

Cervical Plexus. Ventral rami of the first four cervical nerves form the cervical plexus. Peripheral branches innervate the skin and certain muscles of the back of the head, the neck, and the shoulder. The *phrenic nerve,* which is distributed to and controls the diaphragm, arises from this plexus. Crushing damage to the phrenic nerve at its origin will result in respiratory paralysis and death. Severe reflex disorders, such as hiccups, can be arrested by pinching the phrenic nerve to paralyze the diaphragm temporarily.

Brachial Plexus. This large plexus of the neck and shoulder originates from ventral rami of the fifth to eighth cervical nerves and the first thoracic nerve. It supplies the neck, shoulder, arm, forearm, wrist, and hand. Among the nerves that supply the arm are the axillary, median, radial, and ulnar nerves. Injuries to the nerves of the upper extremities result in the loss of use of the part of the extremity that the specific nerve supplies. Damage to the radial nerve, for example, results in an inability to extend the hand at the wrist, called *wristdrop*. Median nerve damage results in an inability to flex and abduct the thumb and index finger, a distinguishing function of the hand. Loss of the ability to oppose the thumb and any of the fingers prevents a person from picking up small objects. Damage to the ulnar nerve impairs wrist function and adduction, and the fingers cannot be spread apart, resulting in a *clawhand.*

Lumbar Plexus. Ventral rami of the

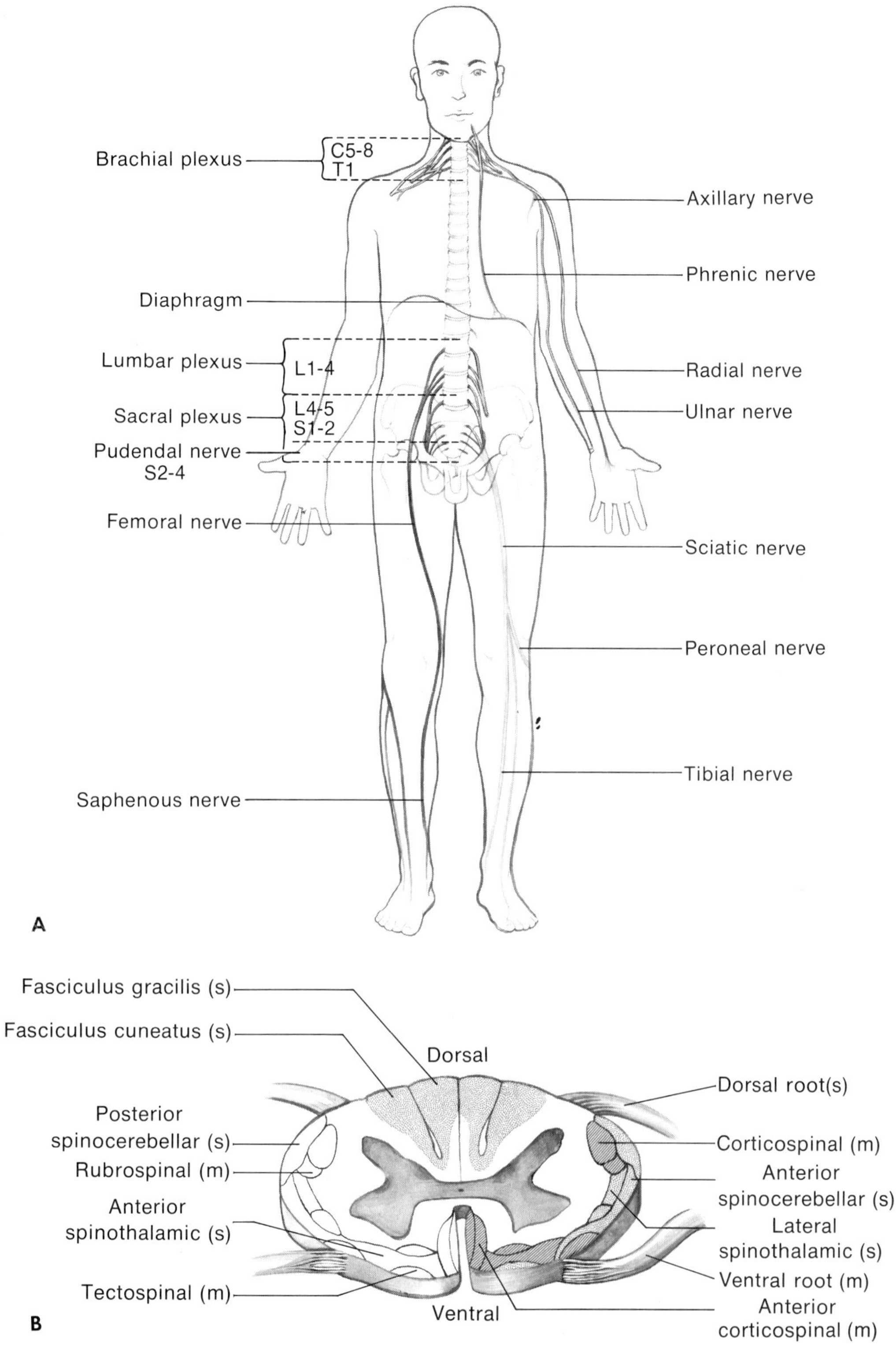

Figure 8–7 *A*, Spinal plexuses. Branches of the spinal cord, anterior aspect. *B*, Sensory and motor tracts from the spinal cord to the brain.

second to fourth lumbar nerves divide into anterior and posterior sections to form the lumbar plexus, located on the inside of the posterior abdominal wall. Its nerves contribute to the supply to the muscles and skin of the abdominal wall, buttocks, medial thigh, and front part of the leg and foot. The major nerve arising from this plexus is the *femoral nerve,* which supplies all the flexor muscles of the thigh, as well as portions of the hip and leg. Injury to the femoral nerve may result in loss of the ability to extend the leg and to flex the hips. "Lumbago" is characterized by an inflammation of the nerves at the base of the spine and buttocks area.

Sacral Plexus. Ventral rami of the fourth and fifth lumbar and first four sacral nerves form the sacral plexus. This group of nerves combined with the nerves of the lumbar area is commonly referred to as the *lumbosacral plexus,* since both groups supply the lower extremities. The major nerve of this plexus is the *sciatic,* the largest nerve in the body. It supplies the flexor muscles of the thigh and divides into the *peroneal* and *tibial nerves,* which supply all the muscles of the lower leg and foot.

Dislocated hips and herniated intervertebral disks are common causes of damage to the sciatic nerve. Injury to the nerve itself results in an impairment of extension at the hip and flexion at the knee. Inflammation of this nerve is characterized by sharp, shooting pains along the course of the nerve, and, if the condition persists, the muscles supplied by the nerve may deteriorate or atrophy. Damage to the peroneal nerve results in loss of flexion of the foot and toes—*footdrop.*

The *pudendal nerve,* formed by the rami of the second, third, and fourth sacral nerves, is regarded by some authorities as being the major nerve of a separate plexus. However, it is generally considered part of the sacral plexus. It supplies the muscles and skin of the perineum, that is, the floor of the pelvis. It is this nerve that is blocked in local anesthesia for childbirth, a procedure known as "saddle block" anesthesia.

Conduction Pathways

The spinal cord, as we have seen, carries signals in both directions between the

Table 8–2 MAJOR NERVE PLEXUSES OF THE BODY

PLEXUS	*ORIGIN*	*PRINCIPAL NERVE(S)*	*RESULT OF DAMAGE*
Cervical	C1–4	Phrenic	Respiratory paralysis; death
Brachial	C5–8; T1	Axillary	Weakened abduction and rotation of arm
		Median	Impaired flexion and abduction of hand; loss of flexion and abduction of thumb and index finger; loss of thumb apposition
		Radial	Wristdrop
		Ulnar	Clawhand
Lumbar	L2–4	Femoral	Loss of extension of leg and flexion at hip
			Lumbago—inflammation
Sacral	L4–5; S1–4	Sciatic	Loss of extension at hip and flexion at knee
			Sciatica—inflammation
		Peroneal	Footdrop

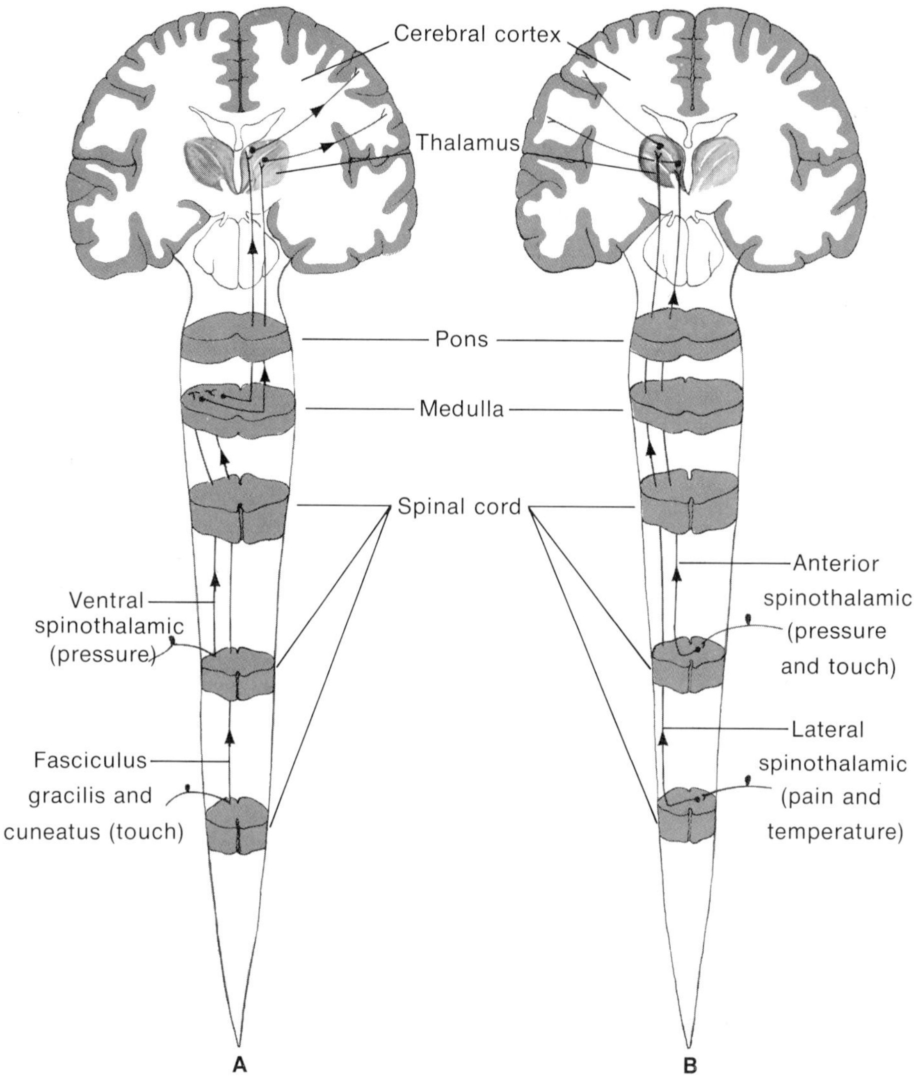

Figure 8–8 Sensory pathways to the brain. *A*, Touch and pressure pathways from the trunk and limbs via the fasciculus gracilis and cuneatus nerves. *B*, Pressure and touch, and pain and temperature pathways from the trunk and limbs.

brain and the remainder of the body. This takes place along the ascending and descending tracts (Figs. 8–5 and 8–8). In most cases, the name of the tract will indicate the funiculus in which it travels, the location of its cells of origin, and the level of location of its axon termination.

Ascending Tracts (Figs. 8–5 and 8–8). The ascending tracts carry messages along the spinal cord in the direction of the brain. They include:

1. The *fasciculus gracilis* and *fasciculus cuneatus* are two major afferent tracts that form the first link in the pathway for the sensations of motion, movement, touch, and pressure. The cell bodies of the neurons composing these tracts lie in the dorsal ganglia of spinal sensory

nerves, and their fibers extend upward to the medulla. Other neurons connect the medulla with the thalamus and to the sensory interpreting areas of the cerebral cortex.

2. The lateral and anterior *spinothalamic* tracts lie in the white matter of the spinal cord, and convey stimuli of pain, temperature, pressure, and touch. They originate in the posterior cell column, and ascend to the thalamus, where they cross to the opposite side in the anterior region of the cord. Damage to the lateral spinothalamic tract results in loss of the sensations of pain and temperature on the side of the body opposite the injury.

3. The anterior and dorsal *spinocerebellar* tracts convey impulses from the muscles of the legs and trunk between the sixth cervical and second lumbar regions. They carry stimuli from the proprioceptors in muscles, tendons, and joints, and terminate in the cerebellum. These transmit impulses concerned with unconscious muscle sense, and play an important role in posture and muscle tone. Damage to these tracts results in uncoordinated voluntary movements due to poor skeletal muscle tone.

Descending Tracts (Fig. 8–5). The descending tracts travel from the brain down the spinal cord and terminate in the gray matter of the cord. They include:

1. The *rubrospinal* tract (*extrapyramidal* tract) conveys impulses from the cerebellum to the motor cells of the anterior column. The fibers cross immediately from the red nucleus of the midbrain, and descend to various levels of the cord where they make connections with spinal motor nerve roots. This pathway is concerned with reflexes that aid in righting the body and with the tone of muscles affecting posture.

2. The *corticospinal* tract (*pyramidal* tract) forms the great motor pathways between the cerebral cortex and spinal nerves, and carries impulses for voluntary movement—especially those concerned with skilled movements of the fingers and hands. Most of the corticospinal fibers cross in the medulla and descend as the large tract in each side of the spinal cord. Owing to the crossing of the fibers, the left half of the brain controls the skeletal muscles on the right side of the body, and the right side of the brain controls those on the left side of the body.

Injuries and Disorders of the Spinal Cord

Injuries to the Cord. The spinal cord may be partially injured or entirely transected. Partial injury to the ascending (sensory) tracts causes lack of coordination, since spinal reflexes are disturbed. Walking becomes uncoordinated, movements are jerky, and loss of balance occurs. Partial injury can also produce paresthesias (abnormal sensations), weakness, exaggerated reflexes, and a decreased ability to sense pain and temperature.

Complete transection of the spinal cord results in paralysis of the skeletal muscles below the level of injury. Sensation is abolished, since all descending and ascending pathways are interrupted. If the injury occurs in the pyramidal tract, or if the motor neurons of the cord are injured, paralysis may be complete. It then would be of the *flaccid* type. *Spastic* paralysis is caused by certain extrapyramidal tract injuries of the brain or cord, and is indicated by an exaggerated tonicity and uncoordinated reflexes.

Disorders Involving the Cord. *Poliomyelitis,* an acute viral disease most commonly found in children, affects both the spinal cord and the brain. The virus may destroy the motor nerve cells in the spinal cord (Fig. 8–9), in which case paralysis of one or more limbs results.

Tabes dorsalis, a disease that represents the late form of syphilis, is characterized by progressive degeneration of the posterior funiculus, resulting in ataxia (loss of muscle coordination) and weakness of the extremities.

Multiple sclerosis is a disease of the

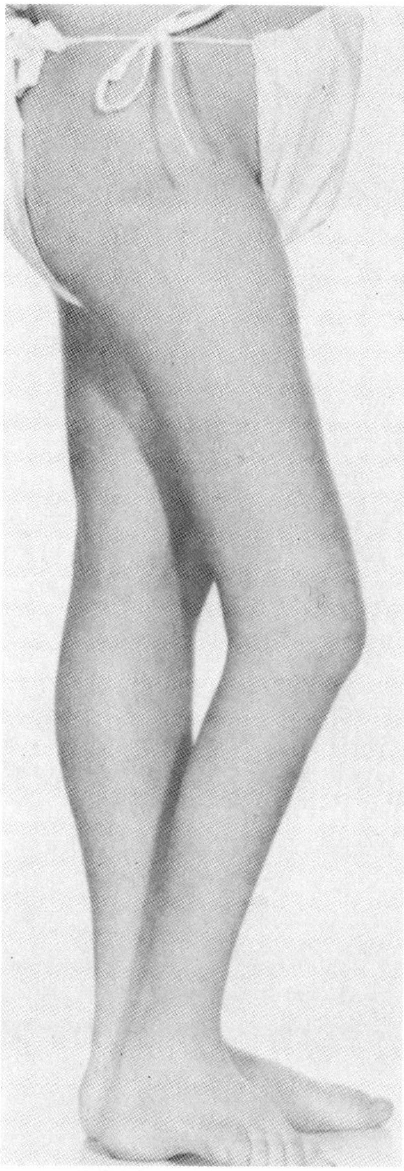

A

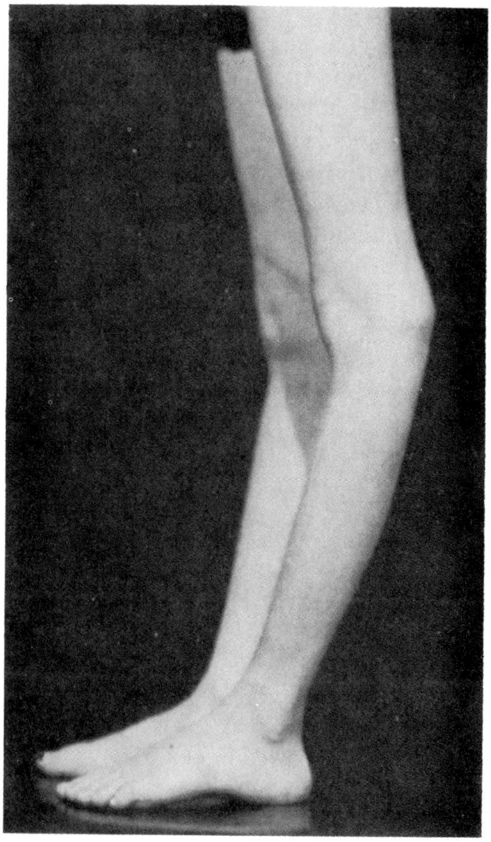

B

Figure 8–9 Deformities of the knee in patients with poliomyelitis. *A*, Flexion deformity. *B*, Hyperextension. (From Tachdjian, M.O.: Pediatric Orthopedics, Vol. 2. Philadelphia: W. B. Saunders Co., 1972.)

central nervous system characterized by a demyelination of the ascending and descending tracts of the spinal cord, and thus involving both sensory and motor functions. There is no specific treatment. The disease is chronic and marked by periods of absence of symptoms.

Congenital spinal cord defects are most frequent in the lumbosacral region, but also may be found in the cervical or thoracic areas. These include *spina bifida,* a defect in the bony covering of the spinal cord that is sometimes associated with *meningocele,* a sac containing malformed cord and roots and covered by a thin, vascular membrane (Fig. 8–10). The trunk and limbs below the level of the deformity are paralyzed and often underdeveloped; there is no control over the excretory organs.

Disorders of Spinal Nerves. *Neuritis* in the strict sense means "inflammation of a nerve," but the term is also used to refer to degenerative and other disorders that may involve nerves. Actually, neuritis is a symptom, not a disease. A lack of vitamin B (thiamine) is thought to be important in chronic alcoholism in which neuritis is fairly common. Sciatica is another form of neuritis.

Shingles (herpes zoster) is characterized by numerous blisters along the course of certain nerves. Its cause is believed to be a virus that attacks the sensory cell bodies inside spinal ganglia. Recovery is usual within a few days, but neuralgic pains may persist for years.

Coverings of the Brain and Spinal Cord

The *meninges* are three layers of connective tissue, or membranes, that surround the brain and spinal cord to form a complete enclosure (Fig. 8–11).

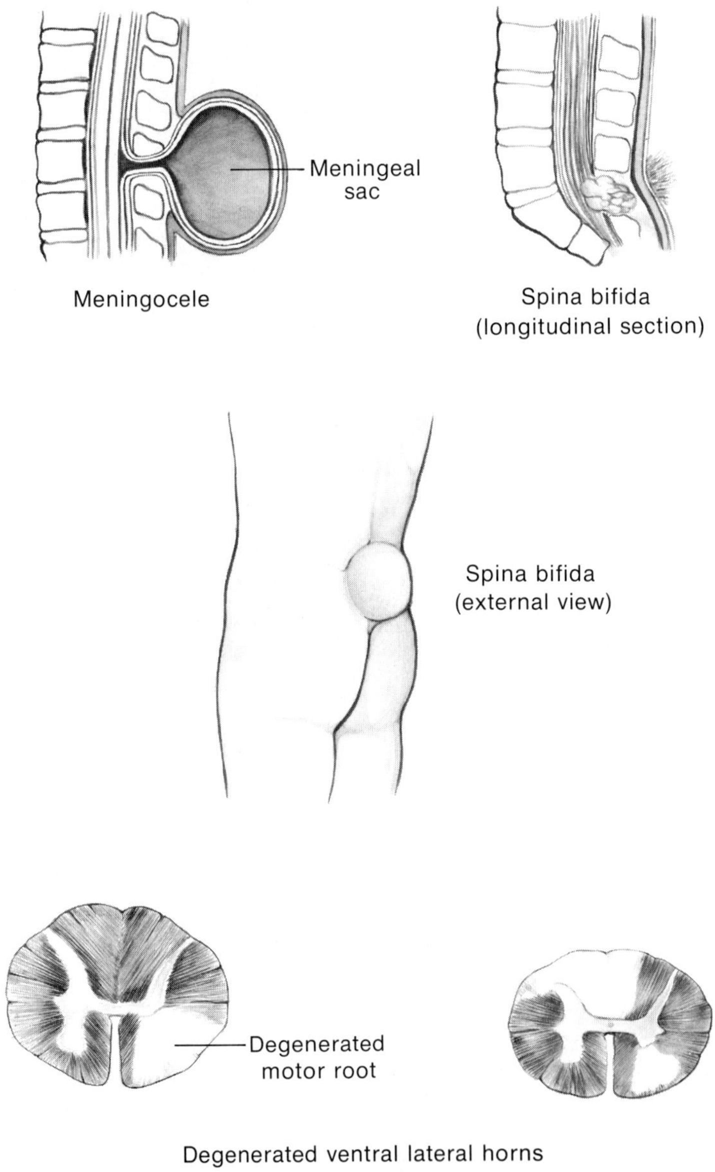

Figure 8–10 Disorders of the spinal cord.

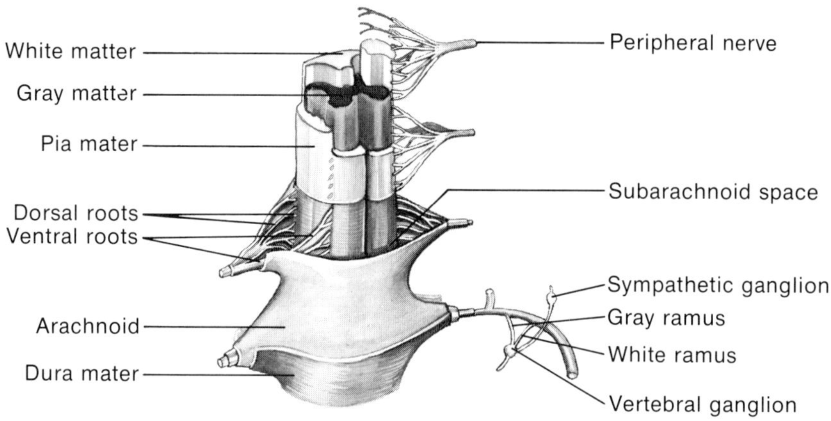

Figure 8–11 Diagram of the spinal cord, showing its three meningeal layers and its association with the sympathetic trunk.

1. The *dura mater* is the outermost layer. It is thick and tough, and provides the greatest protection directly under the bony skull and inside the vertebrae.

2. The *arachnoid* is a delicate covering attached to the deepest of the meninges (the pia mater) by weblike fibers, so that a space for cerebrospinal fluid is created between the arachnoid and the innermost membrane—the *subarachnoid space.*

3. The *pia mater* is a thin, highly vascular membrane. It adheres to the brain and spinal cord, and dips into all depressions, unlike the other two meninges. The blood supply to the brain is carried to a large extent by the pia mater.

Meningitis is an inflammation of the meninges caused by pathogenic bacteria. This organism (meningococcus) may attack the membranes around the cord or may attack the entire membranous enclosure (spinal meningitis). If the inflammation is of prolonged duration and extends to the meninges above the cervical vertebrae, it may be termed *meningoencephalitis* because of its close proximity to the brain.

Spinal Puncture and Anesthesia

Cerebrospinal fluid may be removed from the subarachnoid space at a level beyond the termination of the spinal cord. To avoid injury to the cord, a spinal puncture (also called spinal tap or lumbar puncture) is usually done between the third and fourth lumbar vertebrae, about at the level of the top of the hip bone. The patient is positioned flat, lying on his side, or sitting on a table, and is bent forward so that the spinal vertebrae will be in a flexed position (Fig. 8–12, *A*). If the puncture is performed without trauma, a clear fluid is obtained. This cerebrospinal fluid may then be studied in the laboratory for evidence of disease or injury (see p. 175).

In spinal anesthesia, anesthetics such as cocaine or its procaine derivatives may be injected into the subarachnoid space below the level of the cord, thereby removing temporarily all sensation from the lower part of the body. The patient remains awake, but feels nothing in the lower half of his body.

Intracranial tumors can produce a distortion of the ventricles (cavities) of the brain where cerebrospinal fluid is produced. Such distortions can be detected by an injection of air into the lumbar subarachnoid space (pneumoencephalography).

Myelography is the x-ray examination of the spinal cord and vertebral canal by the injection of a suitable contrast medium (Pantopaque) into the subarachnoid space. The opaque medium permits the visualization of the subarachnoid space. Since the

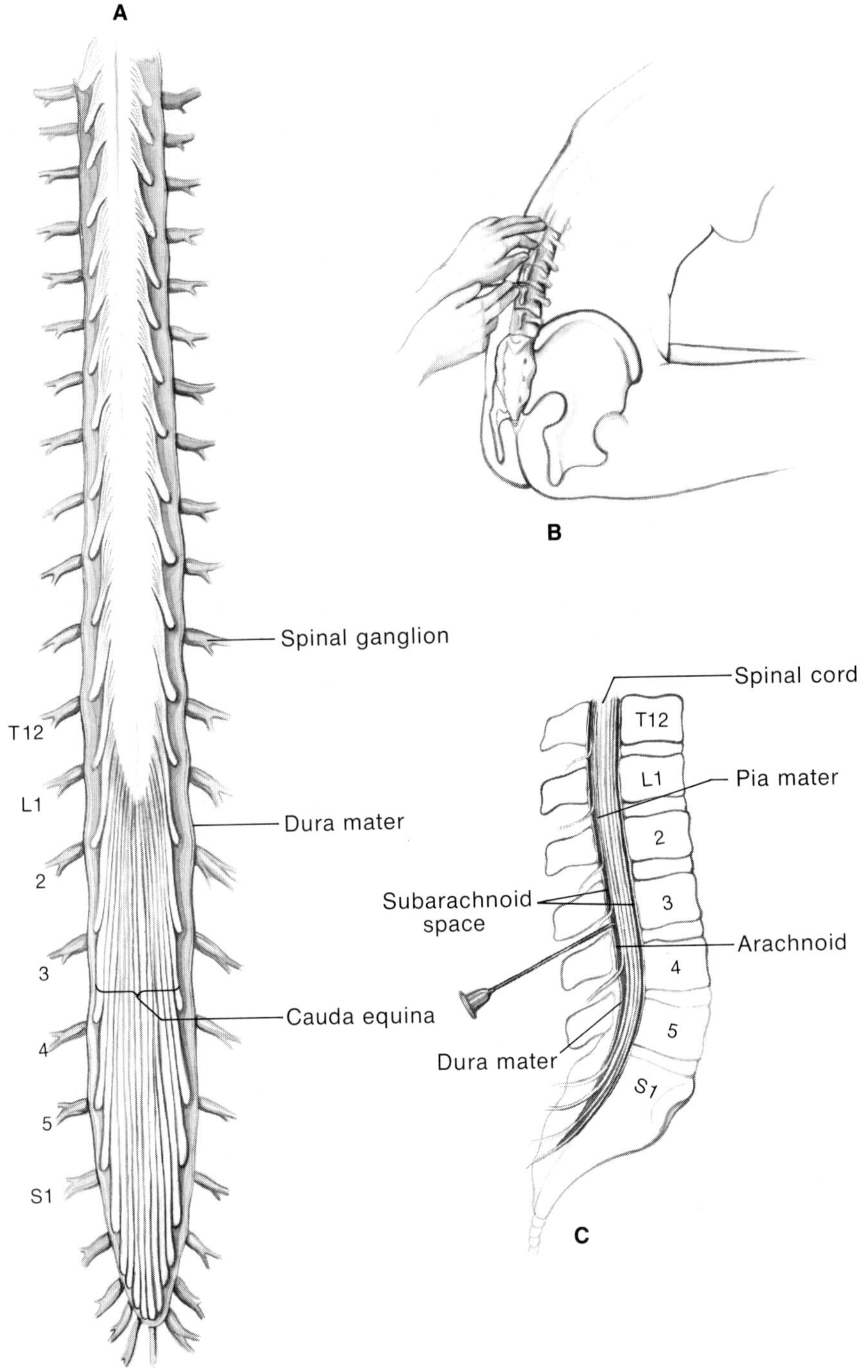

Figure 8–12 Spinal puncture. *A*, Dorsal aspect of the spinal cord, showing the ganglia, spinal nerves, and end of the spinal cord (cauda equina). *B*, Flexed position in a sitting posture to allow passage of the spinal needle between the vertebral spinous processes. *C*, Longitudinal section through the lower half of the vertebral column, showing the position of the spinal needle in the space between the arachnoid and pia mater where spinal fluid is stored.

lumbar area is the usual site of intervertebral disk rupture and herniation, this diagnostic method is of great value in such cases. Myelography is also helpful in locating spinal tumors. Any space-occupying lesion or tumor within the spinal canal of sufficient size will displace and depress the spinal cord (Fig. 8–12, *B*). As a rule, a myelogram is one of many tests done to diagnose the condition.

The Brain

Divisions of the Brain

From a developmental standpoint the brain can be divided into three large areas: the forebrain, the midbrain, and the hindbrain. These areas can be further subdivided as shown in the following outline:

- I. Forebrain (prosencephalon)
  - A. Telencephalon (cerebrum)*
    - 1. Cerebral cortex (gray matter)
    - 2. White matter (core)
  - B. Diencephalon
    - 1. Thalamus
    - 2. Hypothalamus
- II. Midbrain (mesencephalon)
  - A. Corpora quadrigemina
  - B. Cerebral peduncles
- III. Hindbrain (rhombencephalon)
  - A. Pons
  - B. Medulla
  - C. Cerebellum

*The term "cerebrum" has also been used to designate the forebrain and midbrain together, and as a synonym for the entire brain.

The largest part of the brain is the *cerebrum,* which is divided into two *cerebral hemispheres,* right and left. The *brain stem* is the lower, stemlike portion of the brain connecting the cerebral hemispheres with the spinal cord. It comprises, from above downward, the diencephalon, midbrain, pons, and medulla. The *midbrain* contains a massive number of cerebral nerve fibers. Below, it is connected with the *pons,* which in turn is connected with the *medulla.* Finally, the medulla connects with the spinal cord through a large opening in the base of the skull (Fig. 8–13).

Next in size to the cerebral hemisphere is the *cerebellum,* which is located behind the brain stem. It is connected with the pons, medulla, and midbrain by the three cerebellar peduncles (see p. 178).

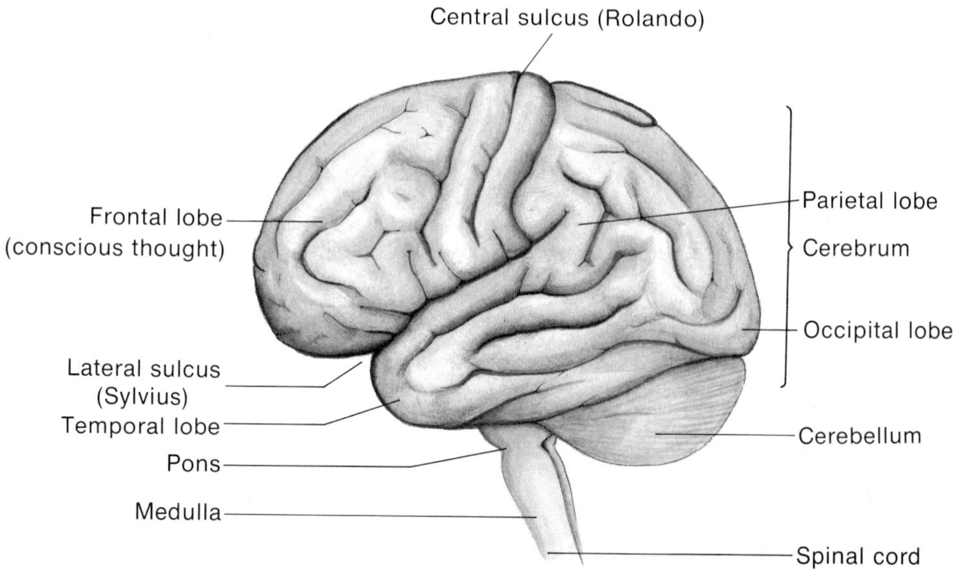

Figure 8–13 Left side of the brain, showing the cerebrum, cerebellum, pons, medulla, and spinal cord.

Collectively, the cerebellum, pons, and medulla make up the major parts of the hindbrain..

The Ventricles and Cerebrospinal Fluid

The four *ventricles* or cavities of the brain communicate with one another and are continuous with the central canal of the spinal cord (Fig. 8–14). The ventricles, the central canal of the cord, and the subarachnoid spaces are filled with cerebrospinal fluid. The largest cavities are the two lateral ventricles of the cerebral hemispheres. Each lateral ventricle communicates with the third ventricle by way of an interventricular foramen, or passage *(foramen of Monro)*. The third ventricle connects with the fourth by way of the cerebral aqueduct *(aqueduct of Sylvius)*, which traverses the midbrain. The fourth ventricle, found at the opening of the cerebellum, is continuous with the central canal of the spinal cord (Figs. 8–14 and 8–15).

Each ventricle is lined with a specialized cell layer called the *ependyma*. At a certain point a network of capillaries projects from the pia mater into the ventricle. The capillaries and ependyma together form the *choroid plexus*. The choroid plexuses are the source of cerebrospinal fluid. Fluid is secreted in each plexus, filters out through the membrane by osmosis, and circulates slowly through the ventricles.

The greatest amount of fluid is produced in the choroid plexus of each lateral ventricle. The fluid flows through the foramen of Monro into the third ventricle, then through the aqueduct of Sylvius into the fourth ventricle. The fluid may then flow directly through the central canal of the spinal cord, or it may mix with the spinal fluid produced by the choroid plexus of the cerebral cortex (Fig. 8–15, *inset*) that has flowed posteriorly on the outside of the cerebrum. The opening in the spinal canal that affords this interchange, the *foramen of Magendie*, is found just below the cerebellum. The direction of flow in the subarachnoid space is downward from the brain, and the fluid is returned to the bloodstream through arachnoid villi, which absorb it. Because production of fluid is more rapid than absorption, there is a constant flow out of the ventricles into the subarachnoid space. Should the openings of the fourth ventricle become blocked, the fluid will accumulate and distend the ventricles, leading to the condition of hydrocephalus or "water on the brain" (Fig. 5–24).

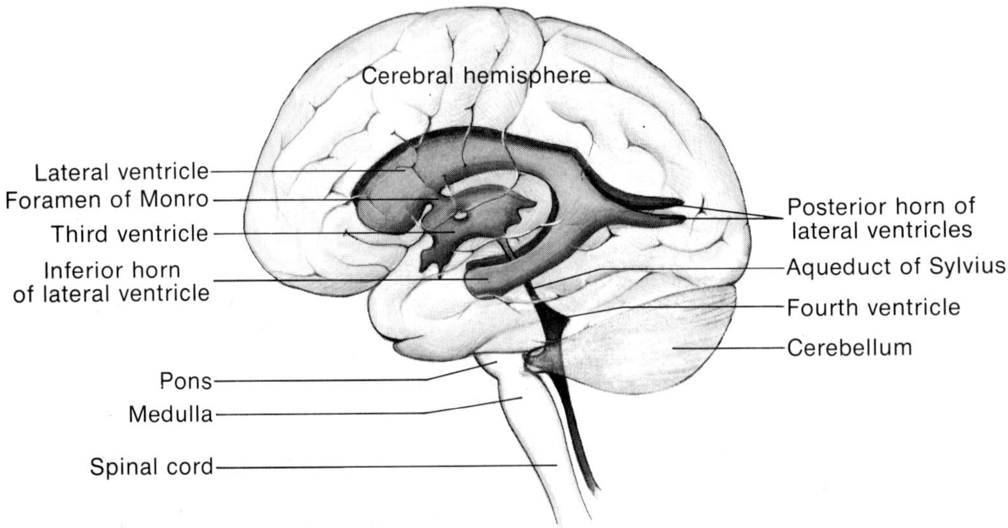

Figure 8–14 The brain ventricles, lateral aspect.

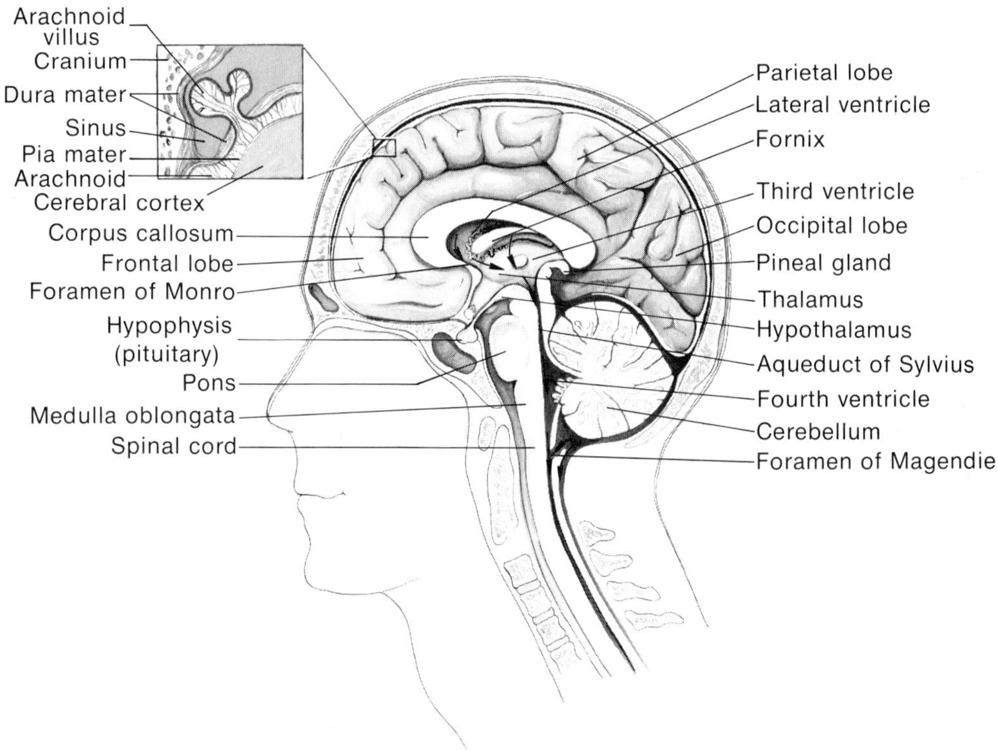

Figure 8–15 Midsagittal section of the human brain, showing circulation of cerebrospinal fluid in the brain and spinal cord. Inset: the cranial meninges.

Cerebrospinal fluid is a clear, colorless, watery blood filtrate. It contains most of the blood chemicals but not the large molecules or formed elements (blood cells). Some of its constituents are water, glucose, sodium chloride, potassium, traces of protein, and a few white cells. The average quantity of cerebrospinal fluid in an adult varies from 100 to 150 ml.

The function of cerebrospinal fluid is to protect the brain and spinal cord from mechanical injury and shock by acting as a water barrier to absorb the effects of trauma. It also serves to supply the nervous tissue with the nutrients necessary to life and to remove waste products from the nerve cells to the bloodstream.

Cerebrospinal Fluid as a Diagnostic Aid. The changes that cerebrospinal fluid undergoes in disease have proved to be of major diagnostic value in the treatment of adults and particularly infants, who are not able to describe their ailments. In acute meningitis, the vessels of the inflamed meninges allow protein and white blood cells to proliferate in the spinal fluid. The fluid is therefore purulent (pus-filled) and turbid. Tuberculous meningitis is detected by a marked decrease in the quantity of chloride and by the presence of tubercle bacilli in the fluid. The condition of the fluid in poliomyelitis is similar, except that the white cell count is higher, the chloride quantity is normal, and no bacilli are present. Precipitation tests, such as the colloidal gold test, assist in the detection of such diseases as aseptic and tuberculous meningitis, encephalitis, poliomyelitis, multiple sclerosis, and syphilitic paresis. There is a change in the ratio of the proteins albumin and globulin in spinal fluid in disease. Thus it is evident that examination of cerebrospinal fluid obtained through a lumbar puncture can be of great value in the diagnosis of suspected spinal cord diseases.

Brain tumors can sometimes be detected by compressing large veins in the neck, creating a back-up of the blood into the cranial sinuses, and a resultant increase in cerebrospinal fluid pressure. The normal pressure ranges from 70 to 200 mm of water; pressure will increase sharply and remain elevated until the venous blood is allowed to flow normally again. If the pressure does not increase or fall rapidly when released, it is possible that a tumor or lesion exists in the area. A pneumoencephalogram, or ventriculogram, is an x-ray picture of the skull after some of the fluid has been removed and replaced by air from a lumbar puncture. Intracranial pathologic states can be detected in this manner.

The Cerebrum

The cerebrum, the largest part of the human brain, accounts for up to four-fifths of the brain's total mass and weight (Fig. 8–13). It fills the entire upper portion of the cranial cavity, and is said to contain billions of neurons and synapses. These form an amazing network of pathways. The outer nerve tissue of the cerebral hemispheres is gray matter, and is called the *cerebral cortex*. This gray matter is arranged in folds or convolutions (gyri), separated by depressions *(sulci)* or deep grooves called *fissures*. Internally, the cerebral hemispheres are largely white matter with a few areas of gray matter.

Fissures of the Cerebral Cortex. The major longitudinal fissure extends from the posterior to the anterior of the cerebrum, dividing it into the two cerebral hemispheres. The cerebral hemispheres with their coverings constitute the true brain. Each has a set of controls for sensory and motor activities of the body. The controls for the right side of the body are located in the left cerebral hemisphere, and vice versa. When one area of the brain is damaged, the corresponding area of the other hemisphere often can take over the functions of the damaged region.

The *corpus callosum* (Fig. 8–15) is located in the longitudinal fissure. It is made up of nerve fibers that connect the two hemispheres, and is concerned with learning and memory.

The *lateral sulcus* (fissure of Sylvius) spreads dorsally and separates the temporal from the frontal and parietal lobes (Fig. 8–13).

The *central sulcus* (fissure of Rolando) starts at the midpoint of the upper border of the cerebrum, and extends downward toward the lateral sulcus, separating the frontal and parietal lobes. The central sulcus also separates centers for sensory and motor functions.

Localization of Cerebral Function. As a result of numerous experiments on animals and close observation of humans, physiologists have been able to localize certain areas of the brain that control motor, sensory, and other activities. In no case, however, is the control of a given function limited to a single area. All parts of the cerebrum are connected, and changes in nervous activity of any one part alter the excitability of the entire brain.

MOTOR AREA. The motor area is the thickest part of the cerebral cortex, and contains cells from which most motor impulses originate. It lies in a band on each hemisphere in the frontal lobe just anterior to the central sulcus (fissure of Rolando), and extends down to the lateral sulcus on each side (Fig. 8–16). The many kinds of cells in the motor cortex include those of the pyramidal tracts. The impulses for voluntary muscular movement originate from the motor area of the cortex, and when a small electric shock is administered to this area, contractions are observed in muscles on the opposite side of the body. By varying the location of the stimuli, we find that the regions of the motor cortex relate to muscles at various levels of the body in a distinct, but reverse, order.

A knowledge of the location and function of the motor area has proved of great value in the diagnosis of brain injuries. Motor paralysis of a given part can result

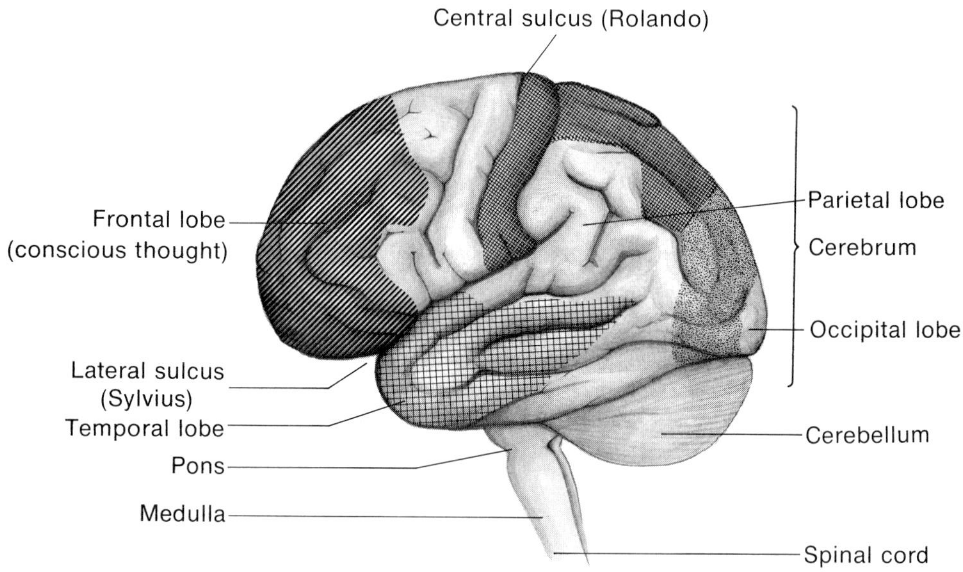

Figure 8–16 Centers for cerebral function, lateral aspect.

from injury or pressure involving the motor area of the brain. Cerebral hemorrhage (stroke or apoplexy), a rupture of a major vessel in the base of the cerebrum, has the same effect as injury or trauma. If a blood clot forms or a vessel is ruptured, preventing nervous impulses from reaching the muscles, a motor paralysis of the opposite side of the body occurs.

The motor area for speech (Broca's area) is located at the base of the frontal lobe, and is thought to be concerned with the formation of words both in speaking and writing. As before, the speech area is located on the opposite side of the brain from the activity involved; that is, left hemisphere for a right-handed person, and vice versa.

Dysfunction of the frontal lobe, if the entire motor area is involved, results in hemiplegia (paralysis of the opposite half of the body). Generally the paralysis is more severe in the upper extremities than in the lower. Irritation of the precentral gyrus creates convulsive seizures of the jacksonian type. These are seizures that begin in one part of the body and extend to another part.

SENSORY AREAS. The sensory cortex is concerned with the interpretation of sensory impulses, and is located mainly in the postcentral region (Fig. 8–16). The area is divided into a large number of sensory regions like those of the motor area just in front of the central sulcus. Each of these regions, when stimulated, produces sensory impressions in a special segment of the body.

The somatosensory or somesthetic area interprets the acute sensations of pain, temperature, and touch. Recognition of the size and shape of objects is dependent on this area.

The visual area is situated at the posterior of the occipital lobe, and is responsible for interpreting optic stimuli.

The auditory or hearing area is located in the superior part of the temporal lobe, and localizes sound stimuli. The anterior part of the temporal lobe serves as a projection area for taste and smell.

Damage to the centers for hearing, vision, or speech can result in *aphasia.* Motor aphasia, loss of the ability to speak, is associated with an injury to the frontal gyrus. Failure to comprehend spoken or written words is sensory or receptive aphasia. Injury to the posterior portion of the superior temporal gyrus produces an auditory receptive aphasia, whereby the patient has unimpaired hearing but still cannot understand the significance of the spoken word.

ASSOCIATION AREAS. The association areas of the cortex make up the greater part of the lateral surface of the cerebral lobes. These areas are concerned with the emotional and intellectual processes involved in memory, reasoning, and judgment (Fig. 8–16).

1. In the *temporal association area* are found the factors that involve recognition of body image, individuality, and personality, as well as the inhibitory mechanism of behavior prompted by instinct.

2. The *parietal association area* is concerned with organization of thought, understanding spoken speech, and musical memory.

3. The *occipital association area* is concerned primarily with visual image interpretations.

4. In the *frontal association area* the association patterns involved in such conscious activities as writing and walking develop. This area functions to conceive and will motor acts.

Injury to the association areas affects many activities. Temporal lobe disturbances include auditory aphasia, occasional visual hallucinations, imagined bad odors, and dreamy states. Occipital injury or disease creates numerous visual problems, including hallucinations and generalized convulsive seizures.

The Thalamus

The thalamus is a large oval-shaped structure made up of gray matter; it bounds the third ventricle (Fig. 8–15). It receives all sensory impulses, either directly or indirectly, from all parts of the body. Some of the impulses are associated, some are sorted out, and others are relayed to specific areas of the cerebral cortex. The thalamus is a sensory integrating center where body malaise, crude touch, pain, and temperature are interpreted. For this reason, it is believed to be a center for primitive, crude sensation. In addition, the thalamus is interconnected with the motor system, and can expedite or inhibit motor impulses from the cerebral cortex.

The Hypothalamus

The hypothalamus is the principal center integrating the autonomic or peripheral functions. It is located directly below the thalamus, and forms the floor of the third ventricle (Fig. 8–15). Because of its significance in involuntary actions, the hypothalamus will be discussed in detail in the next chapter.

The Cerebellum

The cerebellum occupies the lower and posterior part of the skull cavity (Figs. 8–13, 8–15, and 8–17). It lies below the posterior portion of the cerebrum and behind the pons and the upper part of the medulla. It is oval and constricted in the center. When cut in two, it resembles a tree, and thus sometimes has been called the arbor vitae, the tree of life. Like the cerebrum, it has a cortex of gray matter and an interior of both gray and white matter.

The cerebellum is made up mainly of masses of nerve tissue called *peduncles*. These are connecting pathways between the cerebellum and the rest of the central nervous system. The structure functions as a reflex center through which coordination and refinement of muscular movement are brought about. Coordination of impulses from sensory organs and muscle contractions facilitates equilibrium. It is believed that motion sickness is partly due to a disorganization of the part of the cerebellum that coordinates posture and balance through the medulla. Damage to the cerebellum leads to jerkiness of voluntary movements and to poor balance.

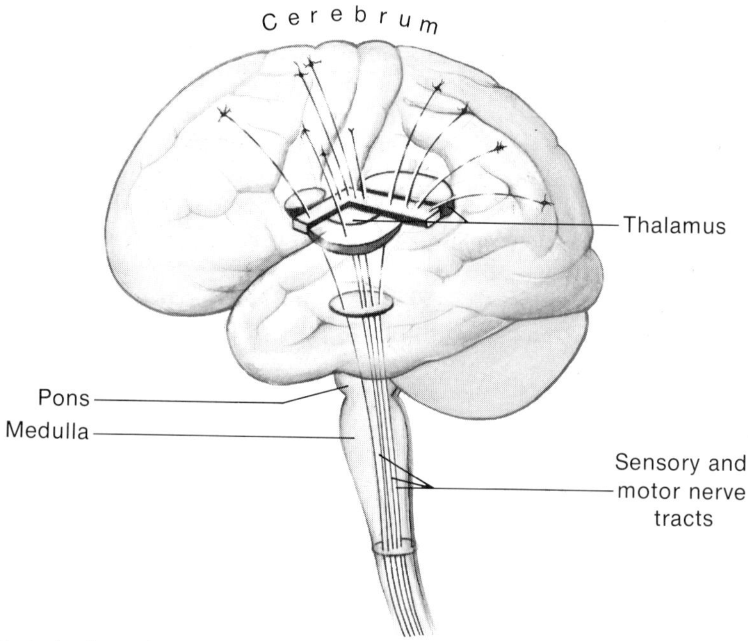

Figure 8–17 Reticular formation of nerves in the spinocerebrum. Note the function of the thalamus as a relay center from the spinal cord to the cerebral cortex.

The Brain Stem

Pons. The pons is situated anterior to the cerebellum between the midbrain and the medulla (Figs. 8–13, 8–15, and 8–17). It consists of interlacing transverse and longitudinal white fibers intermixed with gray matter. The pons, as its name suggests, forms bridges that unite the two cerebellar hemispheres and also the medulla and the midbrain. The fifth, sixth, seventh, and eighth cranial nerves emerge from the pons to serve their respective areas. There is a pneumotaxic center in the pons that participates in the regulation of respiration.

Medulla. The medulla (sometimes incorrectly called the "spinal bulb") is continuous with the spinal cord, and, together with the pons, is a major component of the reticular formation of nerves that constitutes the upper part of the extrapyramidal pathway (Figs. 8–13, 8–15, and 8–17). Stimuli traveling along the pathway reach the reticular formation, which has the capacity to arouse the brain to wakefulness. Although the cortex is the actual seat of the power to think—to respond to stimuli with anything more than a reflex reaction—it must itself be stimulated into action. Such stimulation is supplied by the reticular formation and relayed to the cerebrum through the thalamus.

All of the afferent and efferent tracts of the cord are represented in the medulla, and many of them cross from one side to the other. Others terminate in the medulla. The nerve cells of the medulla are grouped to form nuclei, in some of which the ninth through twelfth cranial nerves originate. Other nuclei serve as centers for control of bodily functions, for example, cardiac, respiratory, and vasomotor centers. In addition to control of respiration and circulation, many other reflex activities are effected through the medulla by means of the vagus (tenth cranial) nerve. Such reflex activities as sneezing, coughing, vomiting, and alimentary tract movements, as well as the major system reflex controls, make the medulla the vital reflex center of the body.

Disorders of the Brain

Stroke—more correctly called "cerebrovascular accident" and sometimes known as "apoplexy"—is by far the most common kind of brain disorder. Rupture of a blood vessel (cerebral hemorrhage; Fig. 8–18, *C*) and an ensuing blood clot (thrombus or embolus) may cause permanent destruction of brain tissue. Such disorders are more frequent in patients with arterial wall disease (arteriosclerosis) and generally occur after the age of 40. The effects of a stroke depend on the extent and location of arterial involvement. Various forms of paralysis may result, depending on the depth of the affected white matter in the lower part of the cerebrum. Categories of paralysis include:

1. *Monoplegia*—paralysis of a single extremity.
2. *Hemiplegia*—paralysis of half the body and its extremities.
3. *Paraplegia*—paralysis of both

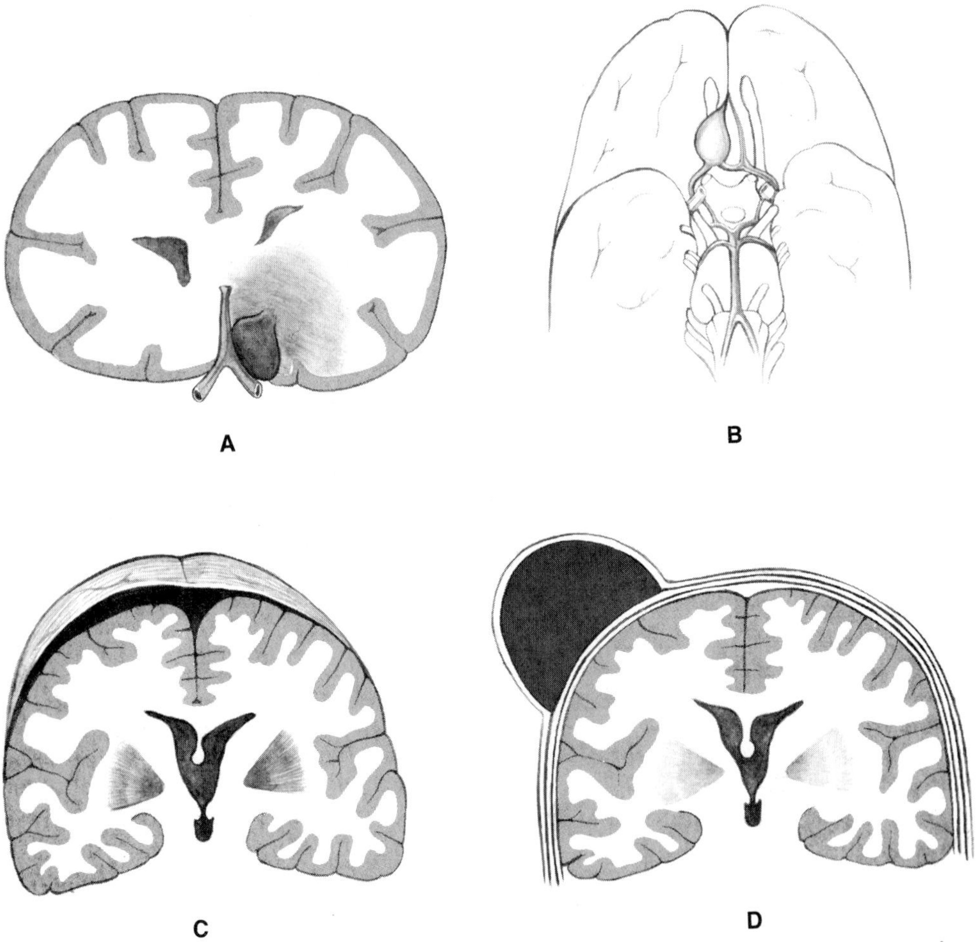

Figure 8–18 Disorders of the brain. *A*, A ruptured aneurysm intracerebrally. *B*, Aneurysm of the anterior cerebral artery. *C*, Hemorrhage of the pia mater. *D*, Subdural hematoma.

lower extremities and the lower part of the body.

4. *Diplegia*—paralysis of both upper or both lower extremities.

5. *Quadriplegia*—paralysis of all four extremities.

Parkinson's disease (paralysis agitans) affects principally the elderly. It is characterized by rigidity without other evidence of corticospinal tract involvement, and by tremors of the extremities. Muscle weakness and stooped posture are common. Until recently, little treatment was available for the disease. Today, however, surgery and the use of certain new drugs are effective in many patients.

Cerebral palsy is a congenital brain disorder. It is characterized by diverse disorders of muscles that vary in degree from weakness to complete paralysis. The disease may involve the muscles of the lower extremities and speech functions as well. Continuous muscle re-education, speech training, and other physio-occupational therapies have helped cerebral palsied patients considerably.

Epilepsy, once popularly called "falling sickness," is a disease of the CNS characterized by recurring convulsions and loss of consciousness that may be brief or prolonged. In most cases the cause of the disorder is not known. Severe seizures are termed "grand mal"; if the patient suffers no more than brief periods of dizziness or loss of memory, they are called "petit mal." Most of those afflicted may lead normal, active lives if they use appropriate medications.

Tumors of the brain may develop at any age, but are more common in the young and in middle-aged adults. The severity, growth, and destructiveness of the lesion depend on the degree to which it compresses the brain tissue. Early surgery offers hope in some cases.

Chorea (St. Vitus' Dance) is a convulsive nervous disease characterized by involuntary jerking movements. It afflicts young women most often, and is believed to be hereditary.

THE CRANIAL NERVES

Twelve pairs of cranial nerves emerge from the lower surface of the brain and pass through the foramina in the base of the cranium. These nerves form part of the peripheral nervous system. They are classified as motor, sensory, and mixed nerves. The origin of the cranial nerves is comparable to that of the spinal nerves. The motor fibers of the spinal nerves arise from cell bodies in the ventral column of the cord, and the sensory fibers arise from cell bodies in the ganglia outside the cord. The motor cranial nerves arise from cell bodies within the brain, their nuclei of origin. The sensory cranial nerves arise from groups of nerve cells outside the brain. These cells may form ganglia on the trunks of the nerves, or may be located in the sense organs. The central processes of the sensory nerves run into the brain and end around nerve cells that form their nuclei of termination. The nuclei of both motor and sensory nerves are connected with the cerebral cortex.

The cranial nerves are numbered according to the order in which they arise from the brain (Fig. 8–19); they are also named to describe their nature, function, or distribution:

I. *Olfactory* (sensory)—carries impulses of smell to the brain.

II. *Optic* (sensory)—special nerve of the sense of sight.

III. *Oculomotor* (motor)—concerned with contraction of most of the eye muscles (the extrinsics).

IV. *Trochlear* (motor)—supplies motor fibers for movement of the superior oblique muscle of the eye.

V. *Trigeminal* (mixed)—the largest cranial nerve and chief sensory nerve of the face and head; motor fibers extend to the muscles of mastication (chewing).

VI. *Abducent* (motor)—supplies motor fibers to the lateral rectus muscle of the eye.

VII. *Facial* (mixed)—sensory fibers are supplied to the anterior two-thirds of

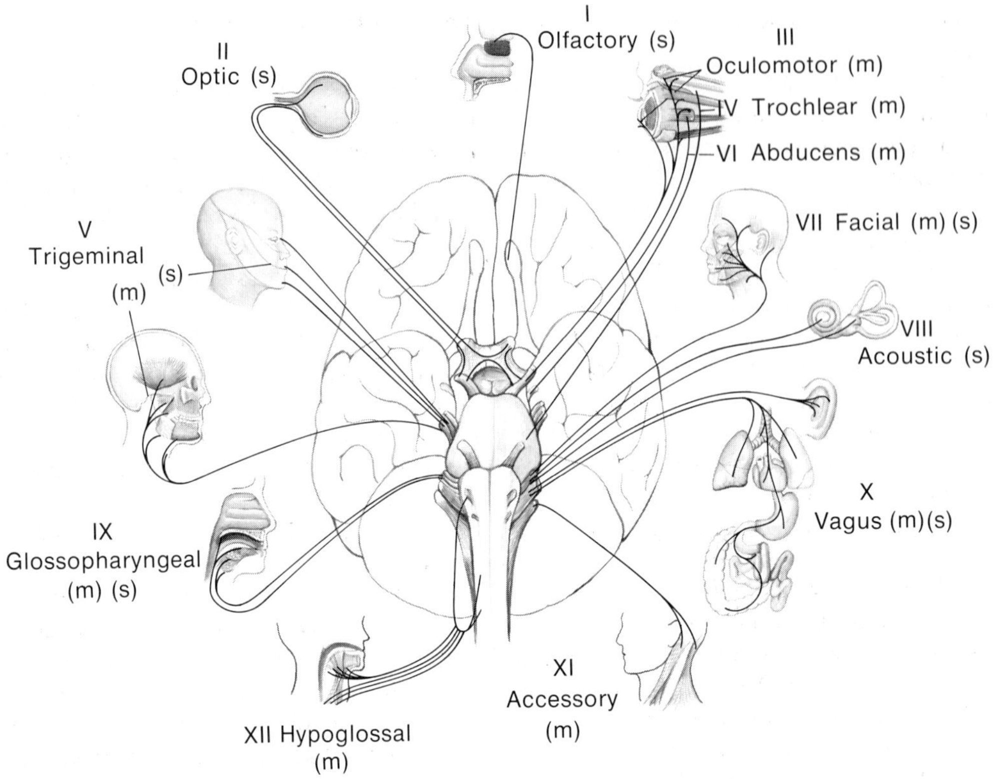

Figure 8–19 Cranial nerves. Twelve pairs of nerves arise from the undersurface of the brain to supply the head and neck and most viscera. They may be sensory (s), motor (m), or mixed in function.

the tongue (taste) and to the lacrimal and salivary glands; motor fibers supply the muscles of facial expression.

VIII. *Acoustic* or *vestibulocochlear* (sensory)—contains sense fibers for hearing and balance (semicircular canals of the internal ear).

IX. *Glossopharyngeal* (mixed)—sensory fibers are supplied to the posterior third of the tongue (taste), tonsils, and pharynx; motor fibers control the swallowing muscles in the pharynx.

X. *Vagus* (mixed)—the longest cranial nerve, and the only one to leave the cranial region; motor and sensory fibers innervate most of the organs of the thoracic and abdominal cavities; the nerve acts as a cardiac inhibitor and bronchial constrictor.

XI. *Accessory* (motor)—motor nerve fibers control the major trapezius and sternocleidomastoid muscles of the neck.

XII. *Hypoglossal* (motor)—motor nerve fibers control movements of the tongue.

Disorders Involving the Cranial Nerves

Damage to the optic nerve is an important cause of blindness. Eye fluid pressure changes (glaucoma), poisons (e.g., wood alcohol), or certain infections (e.g., syphilis) may cause destruction of optic nerve fibers.

Damage to the oculomotor nerve will cause a paralysis that usually interferes seriously with eye function.

Bell's palsy is a facial paralysis caused by involvement of the facial nerve. It is generally unilateral, resulting in distortion of the face because of one-sided paralysis of the muscles of facial expression.

Neuralgia is nerve pain. Its most common form is a severe spasmodic pain affecting the fifth cranial nerve (trigeminal neuralgia). Such pain occurs at relatively long intervals at first, but the intervals shorten as time goes on.

OUTLINE

I. The Neuron
  A. Anatomy of the neuron
  B. Classification of neurons
    1. According to function
      a. Motor neurons
      b. Association (internuncial) neurons
      c. Sensory neurons
    2. According to number of processes
      a. Monopolar
      b. Bipolar
      c. Multipolar
  C. Structure of nerve fibers
    1. Schwann cells
    2. Myelin
    3. Nodes of Ranvier
II. Physiology of Nervous Tissue
  A. Regeneration of the fiber
  B. Transmission of the nerve impulse
    1. Conduction theory
      a. "Wave of negativity"
      b. Saltatory conduction
    2. Velocity of conduction
    3. All-or-none response
    4. Summation of action potentials
    5. Conduction of an impulse across synapses
      a. Acetylcholine cycle
III. The Central Nervous System
  A. The spinal cord
    1. Spinal reflexes
      a. Simple
      b. Complex
    2. Spinal nerves
      a. 31 pairs
      b. Dorsal and ventral rami
      c. Plexuses
        1) Cervical
        2) Brachial
        3) Lumbar
        4) Sacral
    3. Conduction pathways—tracts
      a. Ascending (sensory)—transmit impulses to brain
      b. Descending (motor)—transmit impulses from brain to spinal cord
    4. Injuries and disorders of the spinal cord and spinal nerves
      a. Flaccid and spastic paralysis
      b. Poliomyelitis
      c. Tabes dorsalis
      d. Multiple sclerosis
      e. Congenital defects
      f. Neuritis
      g. Shingles
  B. Coverings of the brain and spinal cord
    1. Meninges
      a. Dura mater
      b. Arachnoid
      c. Pia mater
    2. Spinal puncture and anesthesia
  C. The brain
    1. Divisions of the brain (see p. 173)
    2. Ventricles and cerebrospinal fluid
      a. 4 ventricles and foramina
      b. Cerebrospinal fluid production, composition, and function
      c. Cerebrospinal fluid as a diagnostic aid
    3. Cerebrum
      a. Fissures of cerebral cortex
      b. Localization of cerebral function
        1) Motor area
        2) Sensory areas
        3) Association areas
    4. Thalamus and hypothalamus
    5. Cerebellum
      a. Peduncles
      b. Muscle coordination
    6. Brain stem
      a. Pons } Reticular
      b. Medulla } formation
    7. Disorders of the brain

- a. Stroke
  - 1) Paralysis
- b. Parkinson's disease
- c. Cerebral palsy
- d. Epilepsy
- e. Tumors
- f. Chorea

D. The cranial nerves

1. Motor, sensory, or mixed
2. Location in reference to brain
3. Listing according to number, type, and function
4. Disorders involving cranial nerves

STUDY QUESTIONS AND PROBLEMS

1. Name the three types of neurons according to their structure, function, direction, and myelinization.
2. Define a reflex arc. Compare it with a complex reflex.
3. Explain an action potential with reference to the electrical theory of conduction.
4. What is acetylcholine? What is its main function in a synapse?
5. What do we mean by "localization of function" of the cerebrum?
6. List the three major sulci of the cerebrum and describe their anatomic and physiologic significance.
7. Relate the function of the thalamus to that of the parietal association area of the cerebrum.
8. What is cerebrospinal fluid? Where is it produced and for what functions?
9. How would you obtain a sample of cerebrospinal fluid? Once the sample is obtained, what significant procedures can be followed?
10. What are the meninges? Describe their anatomic and pathologic significance.
11. Trace the reception and interpretation of pain from the toe to the brain.
12. Compare the pons and medulla, as lobes of the brain stem, with a railroad terminal.
13. John Jones is a watchmaker who has lost the ability to use his hands as precisely as before. Relate the loss of manual dexterity to the brain and conduction pathways.
14. Mary Smith fell on her back and felt a twinge run down her left leg. Can you trace the possible impairment of nerve function from the nerve plexus down to her muscles?
15. Why is the medulla called the "vital reflex center" of the body?
16. Differentiate between the symptoms caused by a crush injury at the level of the fourth lumbar vertebra and at the level of the tenth thoracic vertebra.
17. What part of the brain is responsible for all motor activities of a conscious nature?
18. Is a neuron capable of regeneration? Explain.

Chapter 9

Patterns of Control

Part Two: The Involuntary Nervous System

If we consider the entire animal in relation to its environment, we see that it becomes aware of external changes through external receptors such as the eyes and ears, and that it uses its skeletal muscles to adapt to these changes. In like manner, internal receptors signal information about the state of the internal environment to the vital centers of the brain.

Although the internal organs, such as the heart, lungs, and stomach, contain sensory nerve endings and nerve fibers for conducting sensory messages to the brain and cord, most of these impulses do not reach the individual's consciousness. Sensory neurons from the organs are grouped with those that come from the skin and voluntary muscles. On the other hand, the efferent motor neurons that supply the glands and the involuntary, or smooth, muscles are arranged very differently from those that supply the voluntary muscles. This variation in location and arrangement of the visceral efferent neurons has led to their being classified as part of a separate division—the *autonomic* or *involuntary* nervous system.

STRUCTURE OF THE AUTONOMIC NERVOUS SYSTEM

The autonomic nervous system (ANS) has many special parts, particularly *ganglia* that serve as relay stations. Instead of a neuron alone transmitting an impulse from the spinal cord to a muscle, an autonomic pathway has a relay ganglion between the cord and the effector. A visceral efferent pathway has two types of components: *preganglionic fibers* and *postganglionic fibers* (Fig. 9–1). Many fibers pass to and from each ganglion, and each of the structures is able to discriminate in its relaying. In the case of voluntary muscle cells, each nerve fiber extends all the way from the spinal cord to the muscle, with no intervening relay station.

The principal parts of the autonomic nervous system are as follows:

The Sympathetic Division

The pathway of the sympathetic division of the autonomic nervous system begins in the thoracolumbar area of the spinal cord. Preganglionic fibers emerge from the cord and extend to spinal nerves by way of ventral roots (Fig. 9–2). They then leave the spinal nerves and branch off into white communicating branches *(rami communicantes)* that lead to the adjacent sympathetic (or *lateral*) ganglia. The bead-like lateral ganglia form two chains, the *sympathetic trunks,* one on each side of the spinal cord. The lateral ganglia contain the cell bodies of the second set of neurons (postganglionic), whose fibers emerge as gray rami communicantes and rejoin the spinal nerves for distribution to glands,

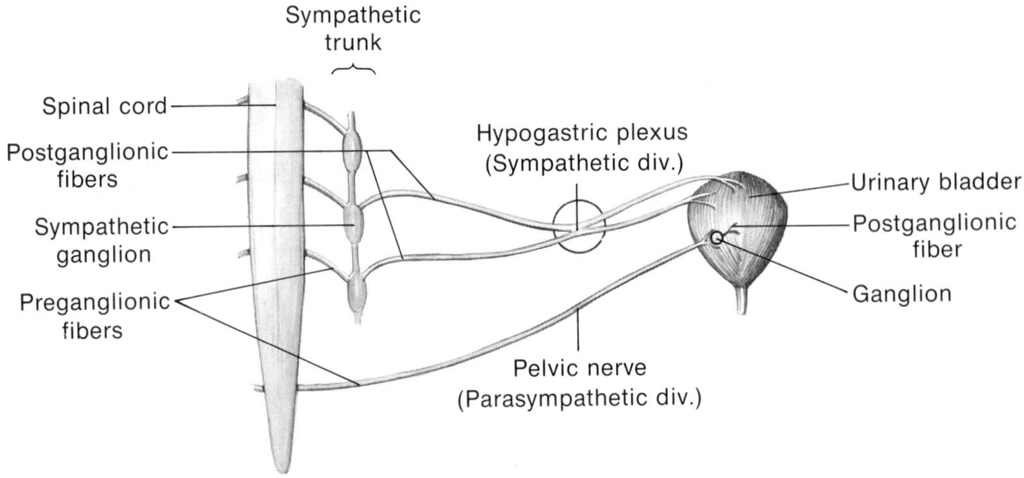

Figure 9–1 Pre- and postgânglionic fibers. Diagram of the lumbosacral segment of the autonomic nervous system, showing the arrangement of ganglia, pre- and postganglionic fibers of both the sympathetic and parasympathetic divisions to the spinal cord and visceral organs.

blood vessels of the limbs, and involuntary muscle tissues (Fig. 9–1).

The arrangement of the lateral ganglia in the sympathetic trunk is related to the segmentation of the spinal column; however, the ganglia of the cervical region are fused to make three larger ganglia (ciliary, sphenopalatine, and otic; Fig. 9–3). Each lateral ganglion is joined to those above and below it, and there are many more

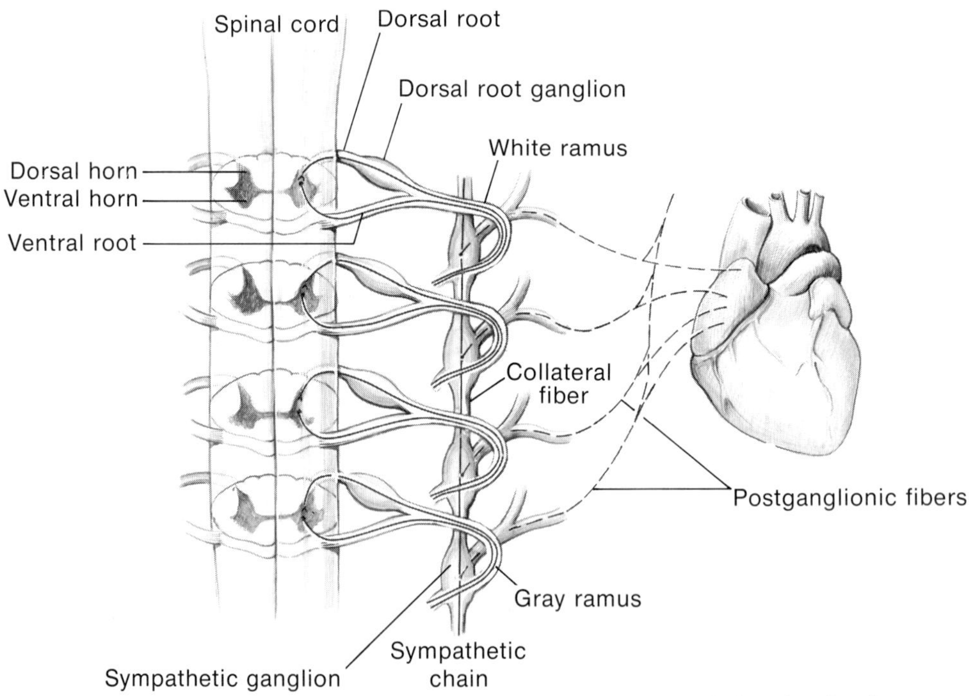

Figure 9–2 Pathways for distribution of sympathetic fibers from the spinal cord to the viscera.

postganglionic fibers than preganglionic. Therefore, a visceral efferent impulse arising from the brain has a more diffuse effect than a comparable somatic motor impulse.

The Parasympathetic Division

The parasympathetic division pathway begins in the cell bodies of the midbrain, the medulla, and the sacral part of the spinal cord. From these centers the first set of fibers (preganglionic) extends to ganglia that are usually located within the organ. The pathway then continues (postganglionic) along a second set of neurons that stimulates the tissues of the various body organs or viscera (Fig. 9–3).

The Autonomic Plexuses

The autonomic, or sympathetic, plexuses are intricate networks of nerve fibers found in the thoracic, abdominal, and pelvic cavities. They include the cardiac, celiac (solar), and hypogastric plexuses. Although they are regarded as essentially sympathetic, they also receive fibers from the parasympathetic system.

The *cardiac* plexus lies under the arch of the aorta, at the base of the heart. It receives branches from the cervical sympathetic ganglia and from the right and left vagus nerves (parasympathetic), and has a regulatory effect on the heart.

The *celiac* or *solar* plexus is the largest network of cells and fibers of the autonomic system. It lies behind the stomach and receives fibers from the sympathetic system as well as branches from the vagus nerve of the parasympathetic system. A blow to this region may slow the heart, reduce the blood flow to the head, and lead to unconsciousness.

The *hypogastric* plexus (Fig. 9–1) forms a connection between the celiac plexus above and the two pelvic plexuses below. It lies in front of the fifth lumbar vertebra and continues downward in front of the sacrum. These plexuses supply the organs and blood vessels of the pelvis.

The *enteric* plexuses, which contain both sympathetic and parasympathetic fibers, are located between the longitudinal and circular muscles of the digestive tract. Peristalsis is maintained, however, even when both the sympathetic and parasympathetic nerve supplies are cut. The nerves of the autonomic system exert a regulatory effect.

CENTRAL CONTROL OF THE AUTONOMIC NERVOUS SYSTEM

The centers controlling heart rate and vasomotor tone are located in the medulla oblongata. Following transection of the brain of an experimental animal between the pons and the thalamus, the autonomic functions controlling heart rate, vasomotor tone, respiration, and the urinary bladder remain intact. In this state, however, the animal is unable to react to environmental changes, and cannot regulate its body temperature. This is because the autonomic functions that permit adaptation depend on the thalamus, which is the principal sensory receiving center. It relays sensory impulses to higher centers, including the cerebral cortex, and also to the hypothalamus, which in turn contributes largely to the sympathetic and parasympathetic outflow regulating the visceral organs.

The posterior portion of the hypothalamus sends out impulses over sympathetic fibers; the anterior portion over parasympathetic fibers. If the environment is cold and the body temperature tends to fall, sympathetic impulses from the hypothalamus promote vasoconstriction in the skin and other mechanisms that tend to conserve heat. Conversely, warm conditions provoke parasympathetic discharges, promoting sweating and skin vasodilation (Table 9–1).

In general, the hypothalamus acts as a coordinating center for the autonomic ner-

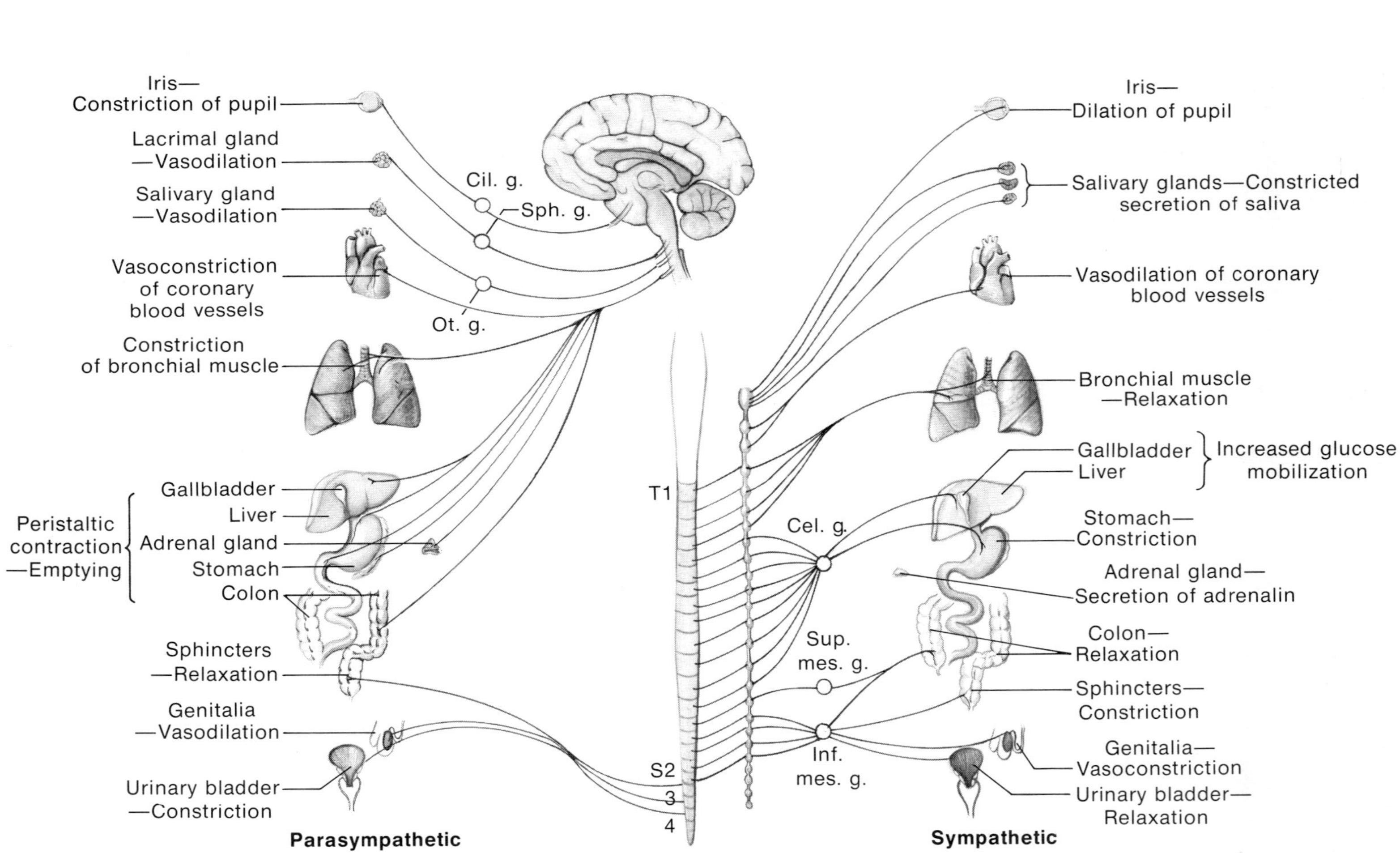

Figure 9–3 Diagram of the autonomic nervous system, showing the chain of parasympathetic ganglia on the left and the sympathetic ganglia and their nerves on the right. Abbreviations: Cil. g., ciliary ganglion; Sph. g., sphenopalatine ganglion; Ot. g., otic ganglion; Cel. g., celiac ganglion; Sup. mes. g., Superior mesenteric ganglion; Inf. mes. g., inferior mesenteric ganglion. The outer columns indicate functions of the sympathetic and parasympathetic divisions.

Table 9–1 FUNCTION OF THE AUTONOMIC NERVOUS SYSTEM

STRUCTURE OR FUNCTION	*PARASYMPATHETIC EFFECTS*	*SYMPATHETIC EFFECTS*
Pupil of eye	Constricted	Dilated
Lacrimal, salivary glands	Vasodilated	Vasoconstricted
Circulation		
Heart rate and force of beat	Decreased	Increased
Blood pressure	Decreased	Increased
Blood vessels of skin and abdominal viscera	--------	Constricted
Blood vessels in heart and skeletal muscles	Constricted	Dilated
Respiration		
Bronchioles of lung	Constricted	Dilated
Breathing rate	Decreased	Increased
Digestion		
Peristalsis	Increased	Decreased
Stomach	Increased	Inhibited
Intestine	Increased	Inhibited
Adrenal medulla	--------	Increased
Integument		
Sweat gland secretion	--------	Increased
Blood vessels of skin	Dilated	Constricted
Genitourinary		
Bladder	Constricted	Dilated
Genital organs	Vasodilated	Vasoconstricted

vous system. In this way it aids in regulating the functions of the viscera.

CHEMICAL TRANSMITTERS

Adrenergic Fibers. The terminal filaments of most sympathetic postganglionic neurons produce chiefly norepinephrine. This substance is also secreted by the medulla of the adrenal gland, and therefore these fibers are classified as adrenergic. Exceptions are the sympathetic fibers to sweat glands, to the blood vessels of the skin, and to the arrector muscles that elevate the hair. These postganglionic fibers enter the spinal nerves through the gray rami communicantes, and reach the skin incorporated in peripheral nerves.

The effects of epinephrine, also secreted by the adrenal medulla, and norepinephrine can be very widespread. The chemical substance resulting from excitation of the sympathetic postganglionic fibers is carried out by the bloodstream; it acts to stimulate the viscera controlled by the sympathetic division of the autonomic nervous system.

Cholinergic Fibers. Parasympathetic fibers also produce a chemical substance, acetylcholine, which is promptly converted to choline and acetic acid by the action of the enzyme cholinesterase. This depolarizing substance, acetylcholine, is an aid to synaptic transmission at the preganglionic terminals in both sympathetic and parasympathetic ganglia. At the postganglionic terminal of the parasympathetic or craniosacral division, secretion of acetylcholine acts to inhibit the organ. In the heart, for example, parasympathetic stimulation is followed by hyperpolarization of the cardiac cells and inhibition of the heart beat. The enzyme cholinesterase quickly destroys acetylcholine, which therefore has effects only locally, where it is secreted. Unlike norepinephrine, it probably is not carried by the bloodstream.

THE FUNCTIONS OF THE AUTONOMIC NERVOUS SYSTEM

When we speak of the sympathetic and parasympathetic nerves as being "antagonistic," we mean that the nerves from the two divisions can produce opposite effects. However, they also provide a correlated adjustment to meet any physiologic demand. Generally speaking, the effects of *parasympathetic* stimulation are most obvious during conditions of calm and rest. For example, fibers in the oculomotor nerve stimulate the smooth circular muscle of the iris, thus protecting the eye from excess light; fibers in the facial nerve stimulate the secretion of saliva, thus aiding digestion. Digestion is most suitably carried out during rest, and it is aided further by the vagus parasympathetic fibers, which stimulate secretion of digestive juices and increase the activity of the smooth muscle of the digestive tract. Other vagus nerve fibers decrease the rate and force of the heart beat during rest—a reasonable adjustment inasmuch as the skeletal muscles are inactive and the body's requirement for blood supply is lower at that time (Table 9–1).

In summary, the functions of the parasympathetic division can be thought of as conservative. Narrowing the pupil to protect the eye from excessive light; promoting the digestive processes in a restful environment; slowing the heart beat to allow the heart muscle and the skeletal muscles to recover their tone; and emptying the bowel and bladder to remove the body's waste materials: these are a few of the many ways the parasympathetic division contributes to the maintenance of the constant internal environment called homeostasis.

The *sympathetic* division is greatly influenced by changes in the external environment. It acts to prepare the body for stressful situations, and in so doing has effects mainly opposite to those of the parasympathetic division. These are commonly referred to as "fight or flight." Fear, pain, anger, and other forms of emotion are a strong influence on stimulation of the sympathetic system. This effect is made even more general because sympathetic preganglionic fibers innervate the adrenal gland, which secretes the hormone epinephrine into the circulation. Epinephrine and norepinephrine both have the effect of stimulation. For this reason, the sympathetic division might be called the emergency system, since it prepares us to remain and fight or to take flight.

In acting to prepare the body for stressful situations, the sympathetic division produces effects primarily opposite to those noted during our examination of the parasympathetic chain. Sympathetic fibers innervate the smooth muscles of the eye that contract and increase the size of the pupil, thus increasing its sensitivity in dim light. Muscular exertions require an increased blood supply, and this is accomplished by sympathetic impulses that stimulate the smooth muscles of tiny blood vessels within the digestive tract, cutting down the flow of blood to the stomach and intestine and thereby increasing the blood flow to skeletal muscle. This augmented blood supply is also aided by cervical sympathetic impulses which, arriving at the heart, cause an increase in the rate and force of the heart beat.

When an adjustment of autonomic activity occurs in response to stress, the effectiveness of the adjustment is increased by the presence of reciprocal innervation in the central connection between the sympathetic and parasympathetic systems. Increase in parasympathetic activity (inhibition) to a particular organ may be accompanied by a decrease in frequency of sympathetic (stimulation) impulses, and vice versa. Once the crisis has passed, the parasympathetic system normally acts as a balance for the sympathetic system. The parasympathetic system brings about constriction of the pupil, slowing of the heart beat, and constriction of the bronchial tubes. Most important, this system stimulates peristalsis and increases the quantity and fluidity of secretions. The saliva, for example, that seemed nonexistent during the stress situation ("dry mouth") now flows more easily and profusely.

SYMPATHETIC SYNDROMES

Certain syndromes are characteristic of diseases of the sympathetic nerve trunks. Dilation of the pupil of the eye on the same side as a penetrating wound of the neck shows evidence of a disturbance in the cervical sympathetic cord; temporary paralysis of the bowel following fracture of a lumbar vertebra can indicate loss of peristalsis and distention of the intestine by gas; compression fractures of the upper thoracic vertebrae can produce marked variations in pulse rate and rhythm.

OUTLINE

I. Introduction
II. Structure of the Autonomic Nervous System
  A. Visceral efferent fibers
    1. Preganglionic
    2. Postganglionic
  B. Ganglia
    1. Sympathetic
    2. Parasympathetic
  C. Divisions
    1. Sympathetic
      a. Begins in thoracolumbar area of spinal cord
      b. Rami communicantes – communicating branches
      c. Sympathetic trunks – lateral ganglia
    2. Parasympathetic
      a. Begins in craniosacral region
      b. Ganglia usually located within organ innervated
  D. Autonomic
    1. Cardiac – heart
    2. Celiac – stomach
    3. Hypogastric – pelvis
    4. Enteric – digestive tract
      a. Contains both sympathetic and parasympathetic fibers
III. Central Control of ANS
  A. Medulla oblongata
    1. Heart, vasomotor tone, respiration
  B. Thalamus
    1. Sensory receiving center
  C. Hypothalamus
    1. Regulates viscera
    2. Coordinating center
IV. Chemical Transmitters
  A. Adrenergic fibers
    1. Sympathetic fibers
    2. Norepinephrine
    3. General stimulation
  B. Cholinergic fibers
    1. Parasympathetic fibers
    2. Acetylcholine
    3. General inhibition
V. Functions of ANS
  A. Sympathetic and parasympathetic nerves produce opposite effects – antagonism
    1. Sympathetic – response to stress
    2. Parasympathetic – conservative; resting conditions
  B. Over-all homeostasis
VI. Sympathetic Syndromes
  A. Pupil dilation
  B. Bowel paralysis
  C. Pulse and heart rate irregularities

STUDY QUESTIONS AND PROBLEMS

1. What is the main function of the autonomic nervous system?
2. Name the two divisions of the ANS, and show how they work during and following a moment of extreme fear.
3. Explain how it is possible to stimulate the vagus nerve and yet observe that the heart rate slows or stops altogether.
4. Discuss the role of the ANS in the control of body temperature.
5. Define a plexus. What is a ganglion?
6. Distinguish between pre- and postganglionic fibers.
7. Discuss the antagonism of the ANS in reference to:
   a. Chemical mediation
   b. Anatomic variation
8. What is homeostasis? How does the ANS play a part in this physiologic phenomenon?

Chapter 10

Patterns of Control

Part Three: The Endocrine System

The activities of the body are controlled and integrated by two means: the nervous system, which was discussed in Chapters 8 and 9, and the *endocrine system*. The endocrine glands are considered a "system" only for convenience. Actually, we are concerned with different specialized glands having various embryonic origins, widely different functions, and locations in different regions of the body (Fig. 10–1). The endocrines differ from other (exocrine) glands in that they have no ducts. They liberate their secretions, known as *hormones,* directly into the bloodstream. For this reason, the endocrines are referred to as "ductless glands."

It should be noted that not all endocrine functions are associated with the immediate regulation of organs—just as the nervous system has important functions beyond its role in homeostasis. For example, the nervous system controls the skeletal muscles by means of which the animal responds to its external environment—its behavioral response. Similarly, some endocrine glands exert a slower, longer-lasting control: the hormones secreted by the reproductive glands, for instance, determine patterns of sexual and maternal behavior.

CHEMICAL NATURE OF HORMONES

Hormones are organic compounds of varying structural complexity that are carried by the blood to other parts of the body where they exert their specific effects. They are, therefore, produced in one gland or part of a gland and effective in another gland or in another region of that same gland. In simplest terms, hormones are "chemical messengers" that pass via the bloodstream to the target organ or process. They may either stimulate or inhibit a function, but in general they do not initiate a process. Usually, hormones are either steroids or such amino acid derivatives as proteins, peptides, and modified amino acids.

Hormone Function

Hormones are effective in remarkably small quantities. For example, the injection of a few micrograms of epinephrine into a dog causes a definite increase in the rate of heart beat. There is little doubt that each hormone has some effect on the fundamental metabolism of its target cells or tissues. Hormones, therefore, have a marked influence on such basic life processes as growth, development, reproduction, energy utilization, and cell permeability.

Although the endocrine and nervous systems share the common function of integration and control of body activities, there are several obvious distinctions in the manner in which they work. In general, the action of hormones occurs more slowly and lasts longer than that of the nervous system, which effects high-speed responses of short duration. The endocrines utilize the blood to carry hormones to the effector tissues, whereas the nervous sys-

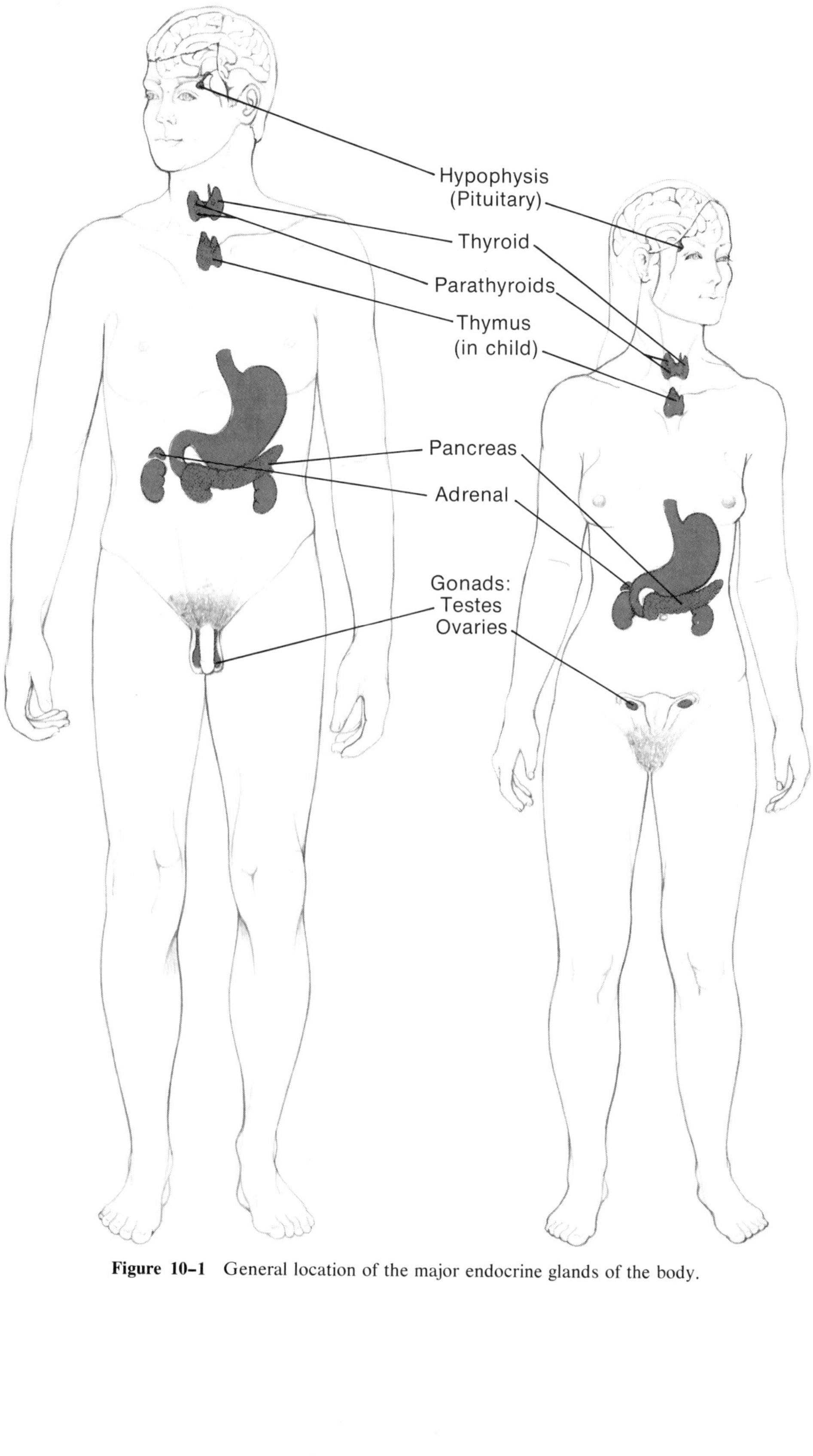

Figure 10-1 General location of the major endocrine glands of the body.

tem uses a sequence of neurons as a system of communication that finally achieves secretion of acetylcholine to elicit a response in the effector.

A closer comparison of the nervous and endocrine systems, however, indicates that they are more similar than we might think. Both secrete chemicals that act on a target organ or tissue. Some endocrine tissues, for example the adrenal medulla, secrete the same hormones or chemicals (e.g., norepinephrine) as the nerve cells. The endocrine system, like the nervous system, tends to maintain the dynamic internal balance of the animal, and it contributes to the response of the animal to its external environment. Most effector organs, except muscle, are regulated by both endocrine and nervous system secretions. Finally, each hormone plays an important role in the growth, development, and differentiation of tissues of an animal, but does so only in an interrelationship with the other hormones.

THE PITUITARY GLAND

The pituitary gland, also known as the hypophysis, is about the size of a small

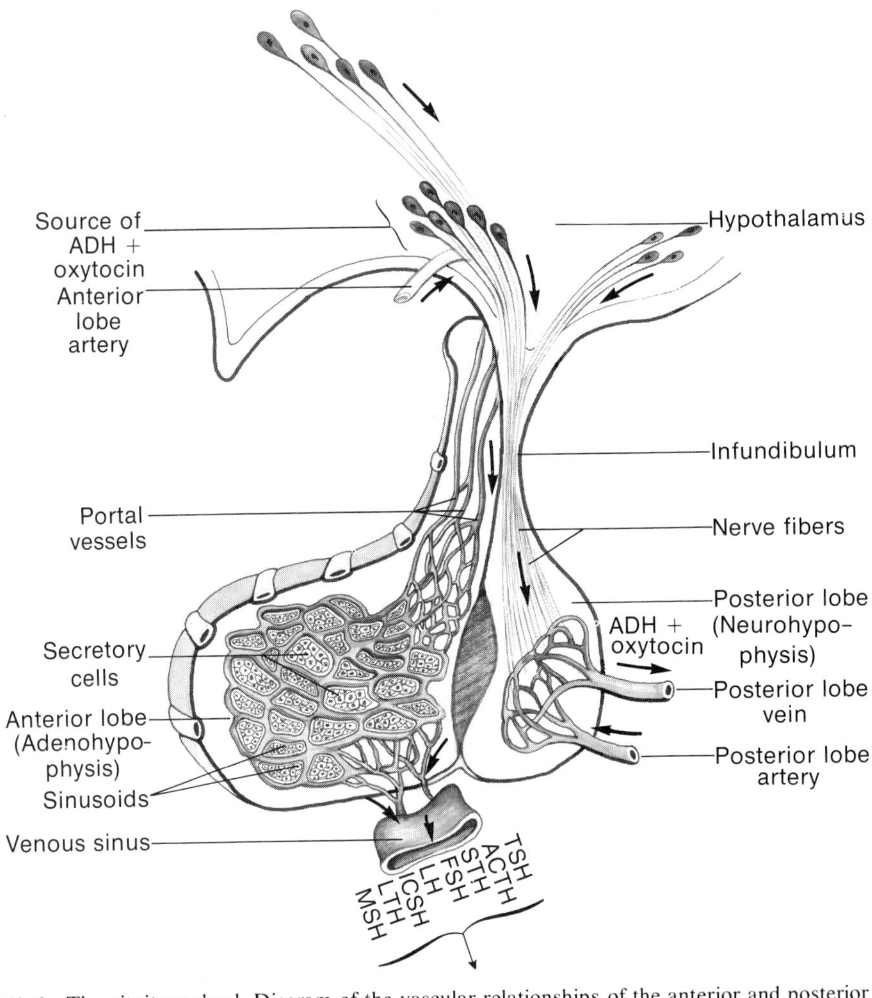

Figure 10–2 The pituitary gland. Diagram of the vascular relationships of the anterior and posterior lobes of the pituitary with the hypothalamus.

marble. It is attached by a stalk to the floor of the third ventricle in the base of the brain (Fig. 10–2).

The pituitary has two lobes, each with a specialized function. The anterior lobe is derived from the epithelial tissue of the mouth. Strictly an endocrine gland, this lobe is called the adenohypophysis, which means "glandular hypophysis." The posterior portion is derived from the brain, and therefore is composed of nervous tissue. This lobe, the neurohypophysis, is connected directly to the hypothalamus (Fig. 10–2). If it has any functional relationship with the adenohypophysis, this is extremely tenuous. The two lobes are essentially two different endocrine glands, and thus are commonly referred to as the anterior and posterior pituitary glands.

The Anterior Pituitary Gland
(Table 10–1)

The anterior lobe of the pituitary gland is often referred to as the "master gland" of the endocrine system because of its important influence in regulating and maintaining the development and function of the other endocrine glands. It has a key role in the integration of the body's growth and metabolism and in the development and function of the reproductive system, including the secondary sex characteristics.

The following hormones are produced by the adenohypophysis (Fig. 10–2):

1. *Thyrotropic* or *thyroid-stimulating* hormone (TSH), which controls various aspects of thyroid gland function. TSH stimulates the accumulation of iodine in the thyroid for conversion into thyroid hormones; it also governs their release from the thyroid into the circulatory system.

2. *Adrenocorticotropic* hormone (ACTH), which regulates the development, maintenance, and secretion of the cortex of the adrenal gland. Whereas the other hormones secreted by the anterior pituitary gland are proteins, ACTH is a polypeptide.

3. *Somatotropic* hormone (STH), which stimulates the growth of body weight and regulates the rate of skeletal growth. This hormone controls the rate of growth in epiphyses and other ossification centers of the skeletal system. A deficiency of growth hormone (hypoactivity of the pituitary) causes early closure of the epiphyses and a short stature—*dwarfism* (Fig. 10–3). Oversecretion (hyperactivity), on the other hand, which is caused by diseases such as tumors, results in excessive growth. If the epiphyseal plates have not closed, giantism occurs because of excessive growth of the long bones (Fig. 10–4). After the time of epiphyseal closure at about the age of 20 to 24, the bones thicken and coarsen, producing *acromegaly*, which is characterized by enlargement of the mandible, broadening of facial contours, and a spadelike thickening of the hands and feet (Fig. 10–5).

Hyposecretion after puberty produces a state of general ill health called *cachexia*, or *Simmonds' disease*. This condition is not entirely understood, but it is believed to be the result of a pituitary lesion that causes not only the hypofunction of all the elements of the anterior pituitary, but also, superimposed on this, a marked degree of malnutrition.

Somatotropin encourages retention of protein by the body. There is a decline of nitrogen products in the urine when the concentration of this hormone in the blood rises. Chemically, it acts to promote fat breakdown (catabolism) and conversion to protein. The rate of fat deposition is thereby decreased.

4. *Follicle-stimulating* hormone (FSH). In the female, starting at puberty and continuing throughout the years of sexual maturity, FSH promotes the growth of the ovarian follicles, in which the egg and one of the female sex hormones, estrogen, are produced. In the male, FSH induces spermatogenesis (production of sperm) in the testes. Any pituitary hor-

Table 10–1 THE PRINCIPAL ENDOCRINE GLANDS AND THEIR HORMONES

GLAND	HORMONE	PRIMARY FUNCTION	CLINICAL EFFECT(S)
Pituitary, anterior lobe (adenohypophysis)	Thyrotropin (TSH)	Regulation of thyroid	–
	Adrenocorticotropin (ACTH)	Regulation of adrenal cortex	–
	Somatotropin (STH)	Growth	Dwarfism Giantism Acromegaly Cachexia
	Follicle-stimulating hormone (FSH)	Stimulation of ovarian follicle	–
	Luteinizing hormone (LH)	Stimulation of corpus luteum	–
	Interstitial cell-stimulating hormone (ICSH)	Stimulation of interstitial cells in testes	Eunuchism
	Lactogenic hormone (LTH)	Milk production	–
	Melanocyte-stimulating hormone (MSH)	Skin pigmentation	–
Pituitary, posterior lobe (neurohypophysis)	Oxytocin	Uterine contraction	–
	Antidiuretic hormone (ADH) (vasopressin)	Kidney water retention; vasoconstriction	Diabetes insipidus
Thyroid	Thyroxine (thyroglobulin)	Oxidative catalyst	Goiter Myxedema Cretinism
Parathyroid	Parathormone	Regulation of calcium and phosphorus	Osteoporosis Tetany Kidney stones
Adrenal cortex	Aldosterone	Regulation of minerals (sodium and potassium)	Acidosis
	Cortisone (cortisol)	Carbohydrate metabolism	Addison's disease
	Androgens	Maintenance of male reproductive system	Homosexuality Masculine characteristics Cushing's syndrome
	Estrogens	Maintenance of female reproductive system	
Adrenal medulla	Epinephrine (Adrenalin)	Stimulation; alarm reaction	Inability to cope with stress
Pancreas	Insulin	Glucose transport and regulation	Diabetes mellitus
	Glucagon	Glycogenolysis	Arteriosclerosis Hypoglycemia
Testis	Testosterone	Spermatogenesis; maintenance of male reproductive system	Sterility and loss of secondary sex characteristics
Ovary	Estrogen (estradiol, estriol and estrone)	Oogenesis; maintenance of female reproductive system	Sterility and loss of secondary sex characteristics

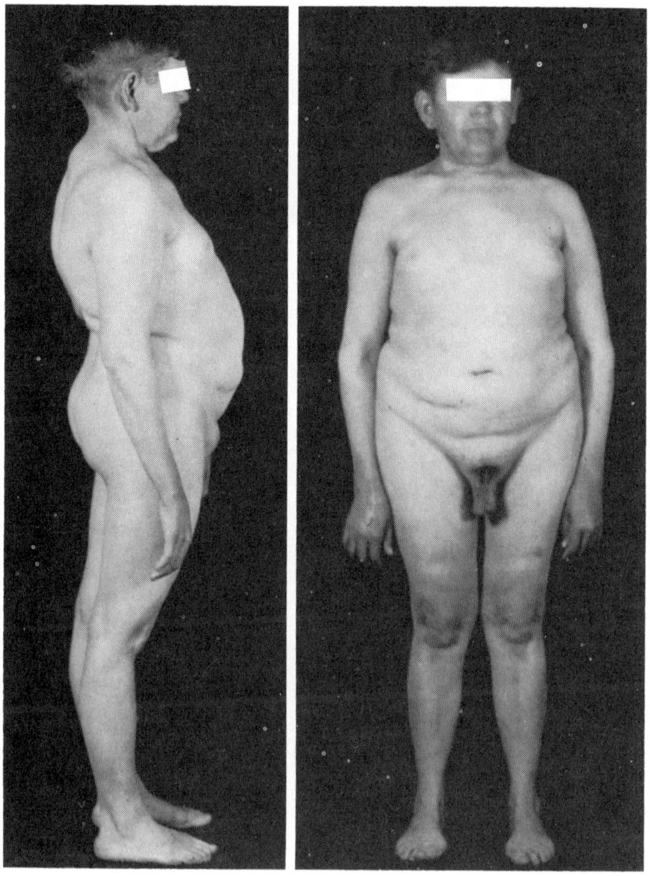

Figure 10–3 Dwarfism. (From Grollman, A.: Clinical Endocrinology and its Physiologic Basis. Philadelphia: J. B. Lippincott Co., 1964.)

mone affecting the gonads, such as this one, is called a "gonadotropic" hormone. In the male, sex hormones are called androgens; in the female, estrogens.

5. *Luteinizing* hormone (LH). In the female, LH stimulates further development of the egg as well as its release from the ovarian follicle. LH is also responsible for stimulating the formation of the corpus luteum, which, in turn, will produce progesterone. Progesterone and estrogen are the two principal sex hormones in the female, responsible for sexual drive and development of the secondary sexual characteristics (e.g., mammary gland development).

6. *Interstitial cell-stimulating* hormone (ICSH). In the male, a hormone comparable to LH stimulates the formation and secretion of the male sex hormone, testosterone, by specialized cells in the testes. ICSH and testosterone are the two hormones primarily responsible for development of masculine secondary sexual characteristics (e.g., deep voice, hair, etc.). As would be expected, hyposecretion of the gonadotropins in adults leads to atrophy (degeneration) of the gonads and a recession of secondary sex characteristics. In male children, hyposecretion leads to eunuchism, a state in which the penis and testes retain their infantile dimensions and the secondary sexual characteristics do not appear. The lack of sex hormones, however, allows growth hormones to promote bone growth; therefore, eunuchs usually are tall, but without distinctively masculine features. Hormonally deficient female chil-

Figure 10–4 Giantism. This giant, aged 13 years, is shown with his brother, aged 9, and his father. He measured 7 feet 2 inches and weighed 278 pounds. (From Behrens and Barr: Endocrinology, *16*:120, 1932.)

Figure 10–5 Acromegaly. Oversecretion of the pituitary growth hormone in the adult results in an overgrowth of those parts of the skeleton that are still able to respond. Note the enlargement of the lower jaw and hands and the thickening of the nose and ridges above the eyes. (From Turner, C. D.: General Endocrinology. Philadelphia: W. B. Saunders Co., 1948.)

dren grow tall, have no breast development, and frequently are masculine in appearance.

7. *Lactogenic* hormone or *prolactin* (LTH). This hormone seems to function only in females. It promotes the production of milk by the mammary glands and the general development of the breasts during pregnancy. LTH also helps maintain the corpus luteum, a temporary endocrine organ vital in early pregnancy. It does not produce these effects if injected into a nonpregnant woman, because it requires the levels of estrogen and progesterone secretion characteristic of pregnancy.

8. *Melanocyte-stimulating* hormone (MSH) produces the deposition of melanin in the skin, providing its pigmentation.

The central role of the anterior pituitary gland in controlling several endocrine glands and other body tissues is summarized in Figure 10–6.

The Posterior Pituitary Gland

The neurohypophysis releases at least two known and distinct hormones, *oxytocin* and *vasopressin* (antidiuretic hormone—ADH), which exert certain significant physiologic actions. It is important to note that oxytocin and vasopressin are not actually produced in the neurohypophysis. Instead, they are manufactured in the neurons of the hypothalamic region of the brain, and migrate through the axons to the posterior pituitary, where they are stored (Fig. 10–2).

1. *Oxytocin* promotes a generalized contraction of the smooth muscles of the body. While almost all such muscles are affected, the greatest response is from the uterine muscle, especially in late pregnancy. The hormone is frequently injected postpartum to promote the contraction of the empty uterus, thereby slowing bleeding

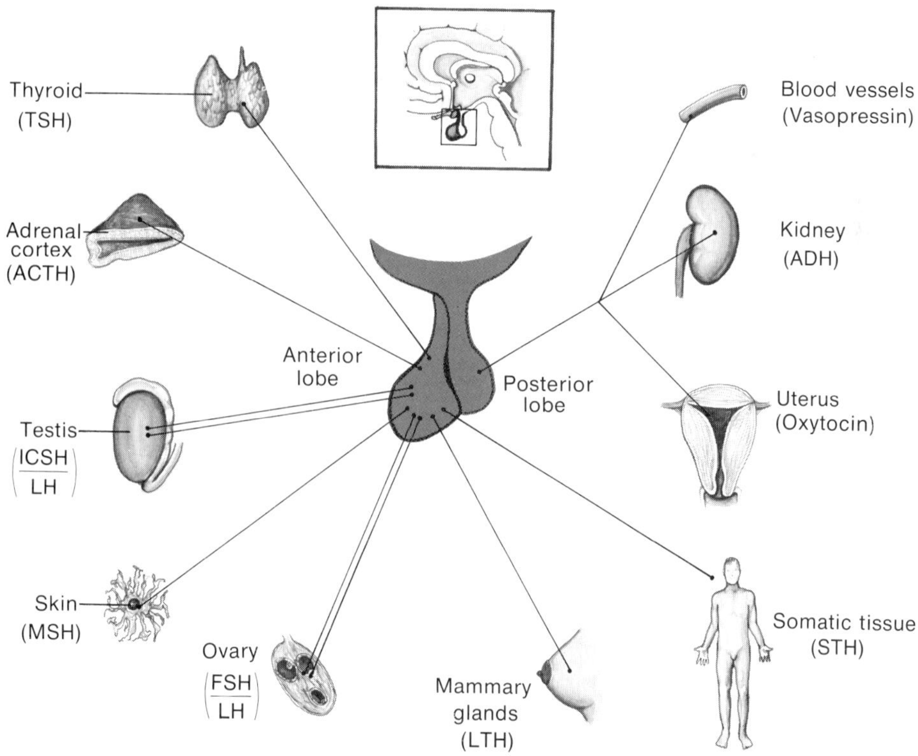

Figure 10–6 Schematic representation of the principal actions of the pituitary hormones. Inset: Relationship of the pituitary to the base of the brain.

from any minor lacerations that may have occurred during childbirth.

In the lactating woman, the infant's sucking at the nipple sends impulses to the hypothalamus, leading to oxytocin release. Because of the increase of the level of the hormone in the blood, the smooth muscle of the mammary gland contracts and ejects the milk. The response to the "sucking reflex" is called the "letdown" of the milk.

2. *Vasopressin* or *antidiuretic* hormone (ADH). Changes in the osmotic concentration of the blood and body fluids stimulate particularly sensitive neurons in the hypothalamus, resulting in the release of this hormone from the neurohypophysis. It travels by way of the blood to the distal tubules of the kidney and promotes the active resorption of water into the bloodstream for retention by the body. Lower concentrations of ADH in the blood cause a less efficient resorption of water, resulting in an increased urine output—*diabetes insipidus*.

A second, less important, function of this hormone is its role as a vasoconstrictor. It can produce a mild increase in blood pressure, and for this reason is called vasopressin. In man, the vasoconstrictor effect occurs mainly in the coronary vessels, where it tends to lower the cardiac output; therefore, no dramatic increase of systemic blood pressure is experienced.

THE THYROID GLAND

The thyroid gland is located in the anterior aspect of the neck, just below the larynx (voice box). It consists of right and left lateral lobes connected by an isthmus that lies upon the upper part of the trachea (Fig. 10–7). Internally, the gland contains numerous secretory cells responsible for the production of thyroid hormones (Table 10–1).

The principal thyroid hormone is *thyroxine*. The gland synthesizes this hormone from an amino acid, tyrosine, and from iodine, both of which must be present in the diet. Food does not contain free iodine, but rather iodide, which is converted to iodine by the thyroid gland itself.

The thyroxine molecule is relatively small, and, after synthesis, is stored in the gland as thyroglobulin, a protein complex in the form of a large colloid. Since its principal constituent is iodine (60 per cent), a determination of the protein-bound iodine (PBI) is an index of the circulating thyroid hormone.

The thyroid hormone regulates the metabolism of the tissues and is called an oxidative catalyst. In its absence, tissue oxygen consumption falls to about half its normal resting value; conversely, an excess of the hormone speeds up energy production in the cells.

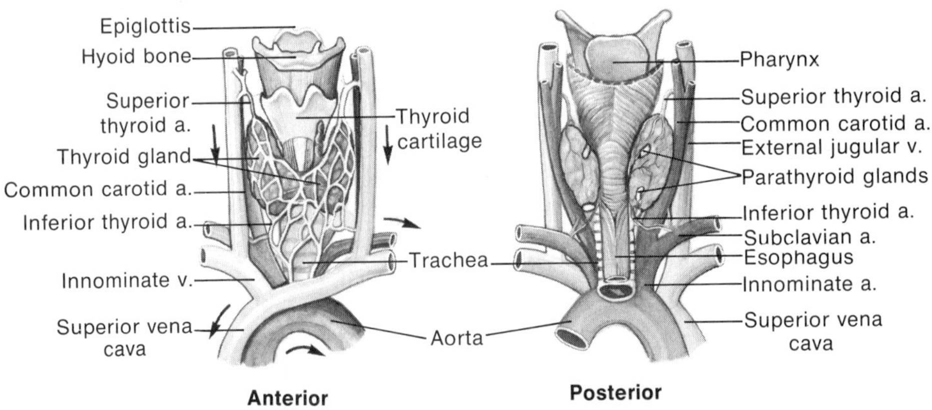

Figure 10–7 The thyroid and parathyroid glands, and their relationship to the blood supply of the trachea. a., artery; v., vein.

Insufficient secretion, hypothyroidism, may be due to a variety of factors, including atrophy of the gland, deficient dietary supply of iodine, or inadequate development of the glandular tissues. Hypothyroidism at birth gives rise to a dwarfed condition, *cretinism* (Fig. 10–8), typified by retarded physical, sexual, and mental development and a significantly lowered basal metabolic rate (BMR). (This condition should not be confused with dwarfism, the result of pituitary problems. A dwarf is physically underdeveloped, but his mental and sexual capacities are normal.) In developing hypothyroidism, early administration of thyroid hormones is necessary to prevent permanent damage.

Hypothyroidism in adults produces a condition called *myxedema* (Fig. 10–9), characterized by a low BMR, thickness and puffing of the skin, coarse and brittle hair and fingernails, and a general mental

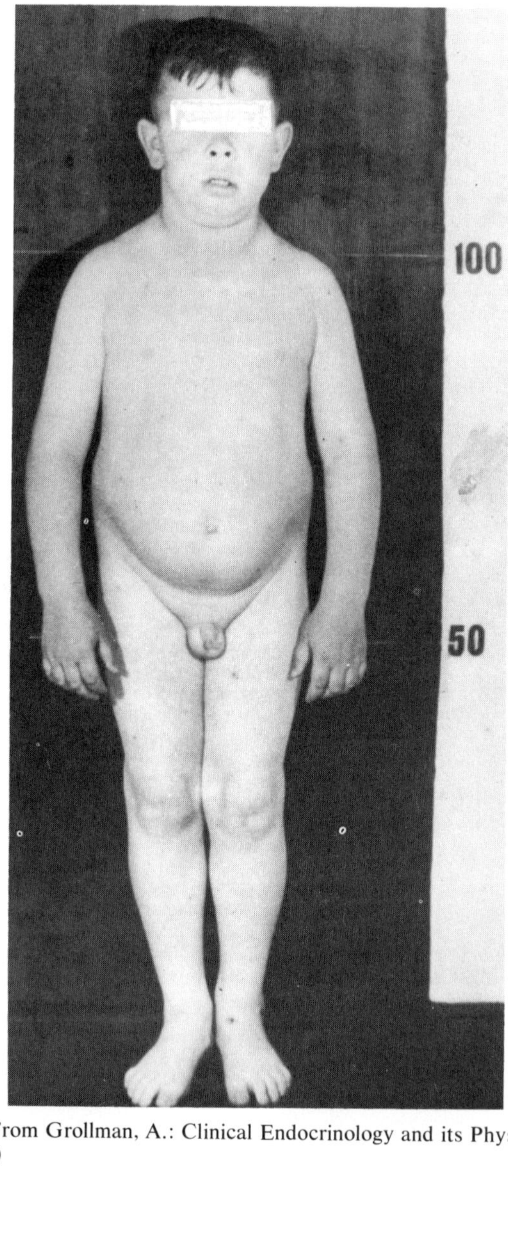

Figure 10–8 Cretinism. (From Grollman, A.: Clinical Endocrinology and its Physiologic Basis. Philadelphia: J. B. Lippincott Co., 1964.)

and physical lethargy. Again, thyroid extract administration will correct the situation.

Hyperthyroidism is accompanied by physiologic effects and symptoms that are the opposite of those observed in hypothyroidism. One of the most severe conditions is *Graves' disease,* or *exophthalmic goiter* (Fig. 10–10). This disorder is characterized by an enlargement of the entire gland, the "goiter," an elevated BMR, weight loss, and a typical protrusion of the eyeballs—hence the adjective "exophthalmic." The condition can be relieved by surgical removal of the thyroid gland, administration of radioactive iodine, or use of thyroid-inhibiting drugs. In most cases, a goiter is a tumor or growth of the thyroid gland, and can be treated surgically with much success.

Figure 10–10 Exophthalmic hyperthyroidism (goiter). Notice the protrusion of the eyes and retraction of the upper eyelids. (From Guyton, A. C.: Basic Human Physiology: Normal Function and Mechanisms of Disease. Philadelphia: W. B. Saunders Co., 1971. Courtesy of Dr. Leonard Posey.)

A simple goiter is caused by a dietary deficiency of iodine. The gland enlarges in an attempt to compensate for the lack of iodine. The goiter is not particularly injurious to health until it becomes large enough to cause undue pressure upon the trachea and other nearby structures. Simple goiter can be treated successfully by administration of iodine. If it grows too large, surgery is indicated (Fig. 10–11).

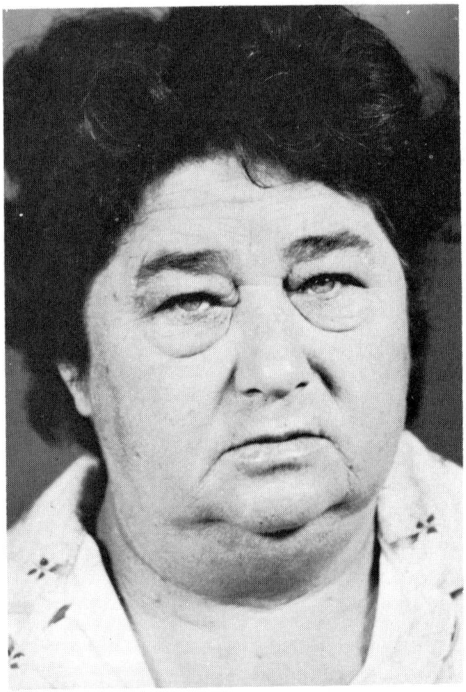

Figure 10–9 Myxedema. (From Guyton, A. C.: Basic Human Physiology: Normal Function and Mechanisms of Disease. Philadelphia: W. B. Saunders Co., 1971. Courtesy of Dr. Herbert Langford.)

As we have seen, secretion of thyroid hormones is controlled by TSH produced by the pituitary gland. The rise of thyroxine in the bloodstream inhibits, by a negative feedback mechanism, the production of TSH, thus regulating the control of the thyroid.

THE PARATHYROID GLANDS

Two small, brownish-yellow parathyroid glands lie on the capsule of the posterior aspect of each thyroid lobe. The four minute glands consist of densely packed secretory cells held in a close association with capillaries by a fibrous meshwork (Fig. 10–7). These glands are essential to life, for they regulate the supply, distribution, and metabolism of calcium and phosphorus in the body.

The secretion of the parathyroid

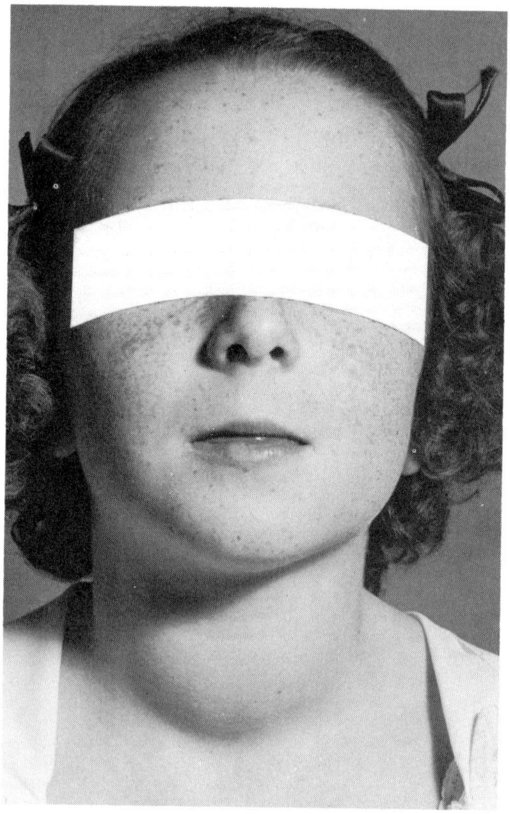

Figure 10–11 Goiter. (From Grollman, A.: Clinical Endocrinology and its Physiologic Basis. Philadelphia: J. B. Lippincott Co., 1964.)

glands, *parathormone,* is not a single compound like thyroxine, but rather a complex protein of undetermined structure. The concentration of ionized calcium in tissue fluid is an important factor in the excitability of tissues. Thus in hypoparathyroidism, when the ionized calcium concentration is reduced, an excessive excitability in muscles and nerves, known as *tetany,* occurs, leading to toxic contractions of the muscles that result in convulsions.

The mineral portions of bone are in a state of constant renewal, even in the adult, with new mineral being laid down and old being reabsorbed. The deposition of bone mineral depends on the product of the concentrations of calcium and phosphorus in the fluid around the bone. This solubility product normally is high enough for bone deposition only if both elements are present in adequate amounts. A low concentration of either element gives a low solubility product, and leads to net bone resorption.

Parathormone is secreted in response to a decrease in inorganic calcium in the blood. The hormone promotes the excretion of phosphates by the kidney tubular cells, thereby lowering the extracellular phosphate concentration. The solubility product therefore falls, and a net dissolution of bone results; the dissolved bone mineral raises the calcium level back to normal, and replaces the excreted phosphate. Such bone losses are considered serious only if protracted indefinitely, as occurs with calcium-deficient diets. The depletion of the bone mineral then leads to osteoporosis, or porous bones (Fig. 10–12).

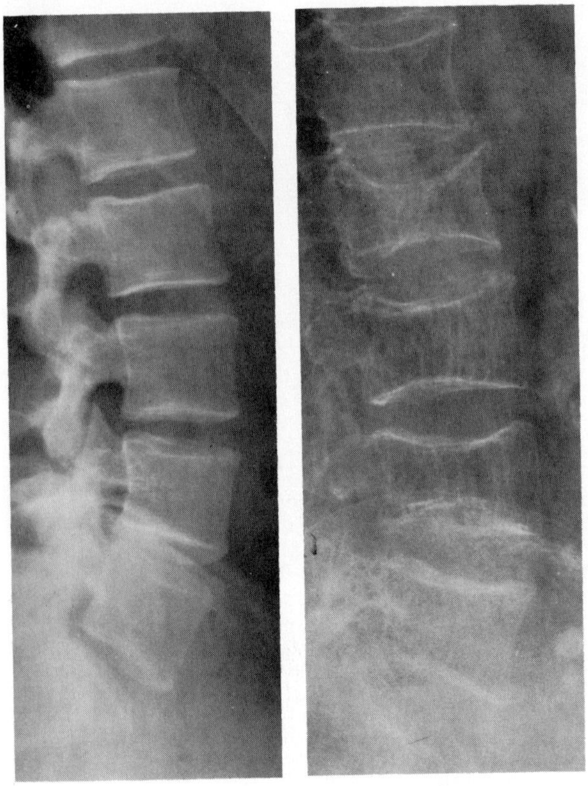

A B

Figure 10–12 Osteoporosis. *A*, Normal spine. *B*, Extreme osteoporosis of the spine. (From Squire, L. F., Colaiace, W. M., and Strutynsky, N. S.: Exercises in Diagnostic Radiology. Vol. III: Bone. Philadelphia: W. B. Saunders Co., 1972.)

An increased calcium level, hyperparathyroidism, leads to greater urinary secretion of calcium, but also causes decreased parathormone secretion. This decrease diminishes the excretion of phosphorus and raises the solubility product, causing a net deposition of bone. Kidney stones, in particular, frequently occur.

Administration of calcium salts by mouth or by injection temporarily improves the condition, as do injections of parathyroid gland extracts. An overabundant supply of parathormone is generally due to growths or tumors of the gland, which can be removed surgically.

THE ADRENAL GLANDS

The two adrenal (or suprarenal) glands lie, one on each side, just above the kidneys, and usually are covered by perirenal fat. Each appears to be a compact unit, but actually comprises two endocrine glands (Fig. 10–13). The outer part, the *cortex*, secretes a whole range of steroid hormones; the core of the gland, the *medulla*, has an entirely separate function, the secretion of epinephrine and norepinephrine (Table 10–1).

The Adrenal Cortex (Table 10–1)

The adrenal cortex secretes so many hormones that the classic experiment of removing the glands led to a bewildering series of results. In general, however, it was noted that after the cortex had been removed the animals showed poor performance and a low tolerance to infection, temperature changes, or any form of stress. They also suffered profound water and electrolyte disturbances. These animals usually died after a few weeks, but their life could be extended, for a short

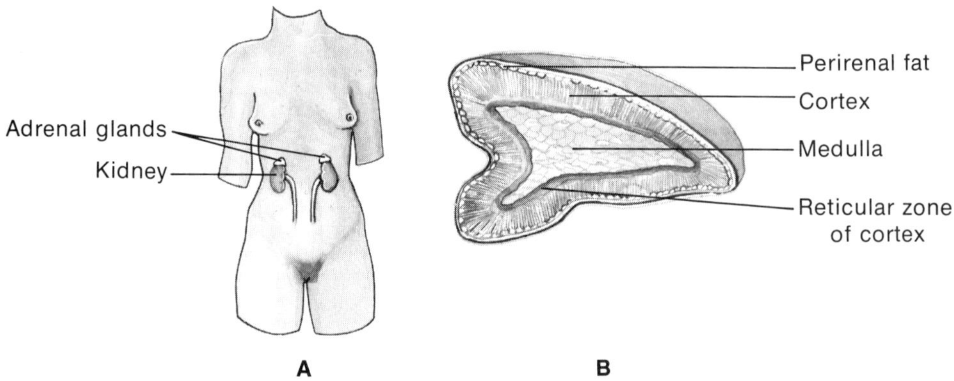

Figure 10–13 The adrenal glands. *A*, Location of the adrenal glands. *B*, Cross section, showing the cortex and medulla.

time, by giving them unlimited access to salt and water.

The many steroids secreted by the adrenal cortex fall into three main categories:

1. *Mineralocorticoids* are the steroids that regulate electrolyte metabolism. *Aldosterone* is the principal mineralocorticoid, and has a regulatory effect on the relative concentrations of the mineral ions (especially sodium and potassium) in the body fluids, and therefore on the water content of the tissues. An insufficiency of the mineralocorticoid results in an increased excretion of sodium and chloride ions and water into the urine, accompanied by a fall in sodium, chloride, and bicarbonate concentrations in the blood, resulting in a lowered pH (acidosis).

2. *Glucocorticoids* have a principal role in carbohydrate metabolism. They promote the deposition of liver glycogen from proteins (gluconeogenesis) and inhibit the utilization of glucose by the cells, thus increasing the blood sugar level. *Cortisone* and *hydrocortisone* (cortisol) are the primary glucocorticoids, and a deficiency in either results in depression of the glyconeogenic potential. It may be noted that they diminish allergic response, especially the more serious inflammatory types, such as rheumatoid arthritis and rheumatic fever.

3. *Sex hormones* are the androgens (male hormones) and estrogens (female hormones), and their effects are similar to those of the hormones produced by the testes and ovaries. However, the adrenal cortex secretes both male and female hormones regardless of the sex of the individual.

In the normal physiology of the human being, the adrenal androgens and estrogens are secreted in minute quantities and have almost insignificant effects. But when the adrenal gland oversecretes, the quantity of adrenal sex hormones has a marked effect. Adrenal tumors in women can promote an aggressive homosexuality and secondary masculine characteristics (e.g., the bearded lady of the circus). Hypersecretion of adrenal sex hormones in prepubertal boys accelerates adult masculine development and the growth of pubic hair. As with the thyroid, the adrenal cortex is controlled by the ACTH hormone from the pituitary. A diminished glucocorticoid blood concentration stimulates the secretion of ACTH. Consequent increase in adrenal cortex activity inhibits further ACTH secretion.

Hypofunction of the adrenal cortex in man is called *Addison's disease*. Symptoms and signs of this disease include hypotension, general weakness, loss of appetite and weight, and a characteristic "bronzing" of the skin. Treatment must be given promptly, for a person with Addison's disease will die within a few days because of an electrolyte imbalance. Small quantities of mineralocorticoids (cortisol) and glucocorticoids (cortisone or prednisolone) can be given by mouth daily. Patients treated in this manner have been known to live for years, although in a weakened condition.

Oversecretion of the adrenal cortex may be caused either by an overproduction of ACTH due to a pituitary tumor or by a tumor in the cortex itself. Symptoms include hypertension, salt retention, edema, demineralization of the bones, and loss of sexual function. One type of oversecretion causes a complex of hormonal effects called *Cushing's disease* or syndrome (Fig. 10–14). In this disease, sodium retention is increased, and this leads to tissue edema, increased plasma volume, and a mild alkalosis. Muscle and bone weakness are common in adrenocortical hyperactivity. The adrenal androgens are also increased to significant levels, causing masculinizing effects in women. Treatment of Cushing's disease usually consists of decreasing levels of ACTH, if possible, by surgical removal of the pituitary tumor.

The Adrenal Medulla (Table 10–1)

The adrenal medulla makes up the pulpy center of the gland (Fig. 10–13), in which groups of irregular cells are located amidst veins that collect blood from the sinusoids. This gland produces two very similar hormones, epinephrine and norepinephrine, also known as Adrenalin and Noradrenalin. Their molecular structures are relatively simple: both are amines, and they are sometimes referred to as catecholamines. Both hormones elevate the blood pressure, the former working as an accelerator of the heart rate and the latter as a vasoconstrictor. The two hormones together promote glycogenolysis (glucogenesis), the breakdown of liver glycogen to glucose, causing an increase in blood sugar concentration.

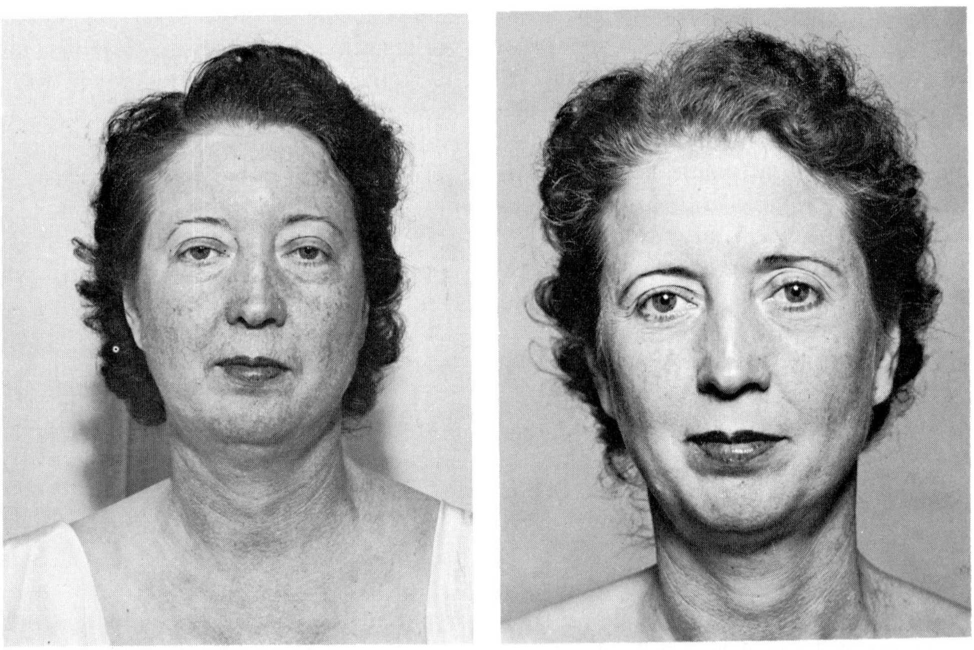

Figure 10–14 Cushing's disease. Patient before *(left)* and after *(right)* almost total removal of the adrenal glands. (From Guyton, A. C.: Basic Human Physiology: Normal Function and Mechanisms of Disease. Philadelphia: W. B. Saunders Co., 1971. Courtesy of Dr. Leonard Posey.)

Of the two, epinephrine functions most closely as the "emergency" hormone for the body. Its functions of increasing the blood sugar level and the blood flow to the skeletal muscles, heart, and viscera serve to shunt or direct the circulation of blood where necessary during exertion or increased stress.

The adrenal medulla, unlike the thyroid and other endocrine glands, is not essential for life. It can be removed surgically without causing untreatable damage.

THE PANCREAS

The pancreas lies close to the posterior abdominal wall behind the stomach (Fig. 10–15). It functions as both an exocrine and endocrine gland. Its exocrine functions involve the secretion of digestive enzymes into the small intestine; this will be discussed in the chapter on metabolism and digestion. The endocrine function of the pancreas is carried on by "islets" of special tissue scattered throughout the pancreatic tissue—the *islets of Langerhans.* These vary in number from several hundred thousand to a few million, and are scattered throughout the pancreatic tissue. Each islet consists of several types of cells in which the alpha and beta cells predominate. We now know that insulin is secreted by the beta cells and another hormone, glucagon, by the alpha cells. Both hormones are polypeptide in nature and can be synthesized commercially (Table 10–1).

The earliest studies on the effect of insulin on carbohydrate metabolism showed three basic effects of the hormone:

1. *An increased rate of glucose metabolism.* Glucose cannot pass through the cell pores, but instead must enter by some transport mechanism through the membrane matrix. Glucose probably combines with a carrier substance (not yet known), and is then transported to the inside of the cell. This process can occur in either direction. In the absence of insulin, the rate of glucose transport into the cell is about one-fourth the normal value. Conversely, an excess of insulin multiplies the normal rate; therefore, in an unknown way, insulin increases the rate of glucose transport through the membranes into most cells in the body.

2. *Regulation of blood glucose concentration.* In the absence of insulin, glucose absorbed from the gastrointestinal tract cannot be transported into the tissue cells; consequently, the normal blood glucose level of 90 mg per 100 ml may rise to as high as 1000 mg per 100 ml. In the

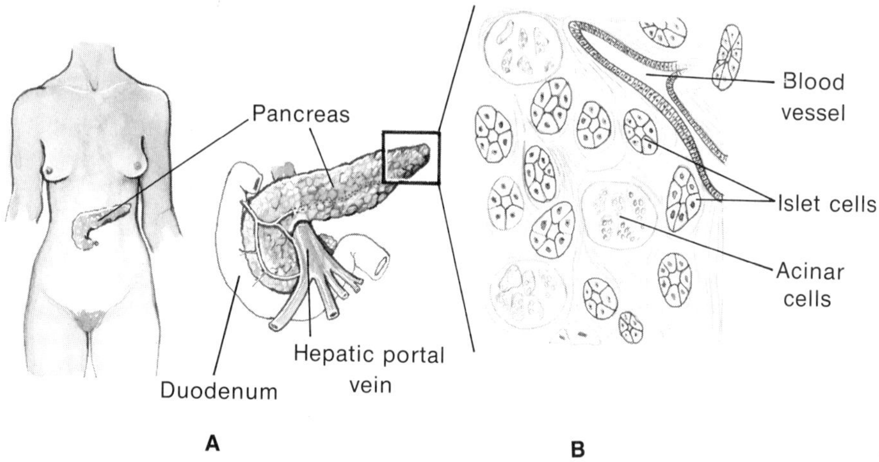

Figure 10–15 The pancreas. *A*, Location of the pancreas. Its close proximity to the duodenum facilitates transfer of digestive enzymes from the pancreas, where they are produced, to the duodenum, where they are utilized. *B*, Microscopic section, showing the arrangement of the islet cells of Langerhans.

presence of insulin, glucose is transported into the tissue cells so rapidly that the glucose level may drop as low as 20 mg per 100 ml. Therefore, the rate of insulin secretion by the pancreas must be regulated to facilitate a constant and normal glucose value in the blood.

3. *An increased glycogen storage in the tissues.* Glucose is preferentially metabolized by the cells to supply all the energy needed by the body. If there is inadequate glucose, the major share of energy is then supplied by fats. Large deposits of fat are broken down (lipolysis), with resultant release of free fatty acids into the blood. Acetoacetic acid (a ketone body), a product of fatty acid metabolism, is released by the liver into the blood, causing the condition called *ketosis.*

In a deficiency of insulin, hypoinsulinism, the glucose uptake by the cells from the blood is reduced, although absorption from the gut and breakdown of liver glycogen (glycogenolysis) both continue. The blood glucose level rises (hyperglycemia) and exceeds the renal threshold (approximately 200 mg per 100 ml), causing glucose to be lost in the urine (glycosuria). This disease, *diabetes mellitus,* is caused by hyposecretion of insulin by the pancreas. The result of the cells' inability to utilize glucose is the accumulation of ketone bodies (fatty acid breakdown), leading to acidosis. Acidosis, along with an electrolyte derangement, may result in paralysis, coma, and death. Diabetics may be safely treated through injections of insulin and a controlled diet.

Increased insulin production (hyperinsulinism) occurs occasionally, generally as the result of a pancreatic tumor. As previously noted, the central nervous system acquires essentially all its energy from glucose. If the blood glucose level drops below normal (hypoglycemia), the CNS metabolism becomes depressed, and the person falls into an "insulin shock" syndrome. If glucose is not administered immediately, permanent damage occurs.

Effective diagnosis of glucose blood levels is carried out by performing a glucose tolerance test. This three- to four-hour test can assess the body's handling of an increased quantity of glucose (Fig. 10–16). The patient drinks a prescribed quantity of glucose. Blood samples are taken before the test (fasting) and every hour for three or four hours after the drink, and the glucose contents of these samples are determined. If the blood glucose does not return to normal within two hours, there is indication of insulin deficiency.

Glucagon, like insulin, is a protein produced by the islets of Langerhans, particularly the alpha cells. Its most dramatic effect is its ability to cause glycogenolysis in the liver, which increases the blood glucose level. When the blood glucose concentration falls as low as 70 mg per 100 ml, the pancreas secretes large quantities of glucagon, which rapidly mobilize glucose from the liver. Glucagon has commonly been called the "hyperglycemic" factor.

THE FEMALE AND MALE SEX HORMONES

In addition to producing the sperm and egg, the ovaries and testes are also the primary sources of the sex hormones that control the maturation and function of the reproductive system (Fig. 10–1).

Female Sex Hormones (Table 10–1)

The principal function of the female sex hormones is the maintenance of the reproductive tract for development of the fertilized egg into a new individual. The ovary secretes two specific types of hormones, estrogen and progesterone, both of which are steroids.

1. *Estrogen* is a collective name for three different but closely related steroids. The most common of these is estradiol; the others are estrone and estriol. Starting shortly before puberty, and continuing throughout the sexually mature life of the female, secretion of estrogen from the developing follicle is periodically stimulated by FSH from the pituitary. The main re-

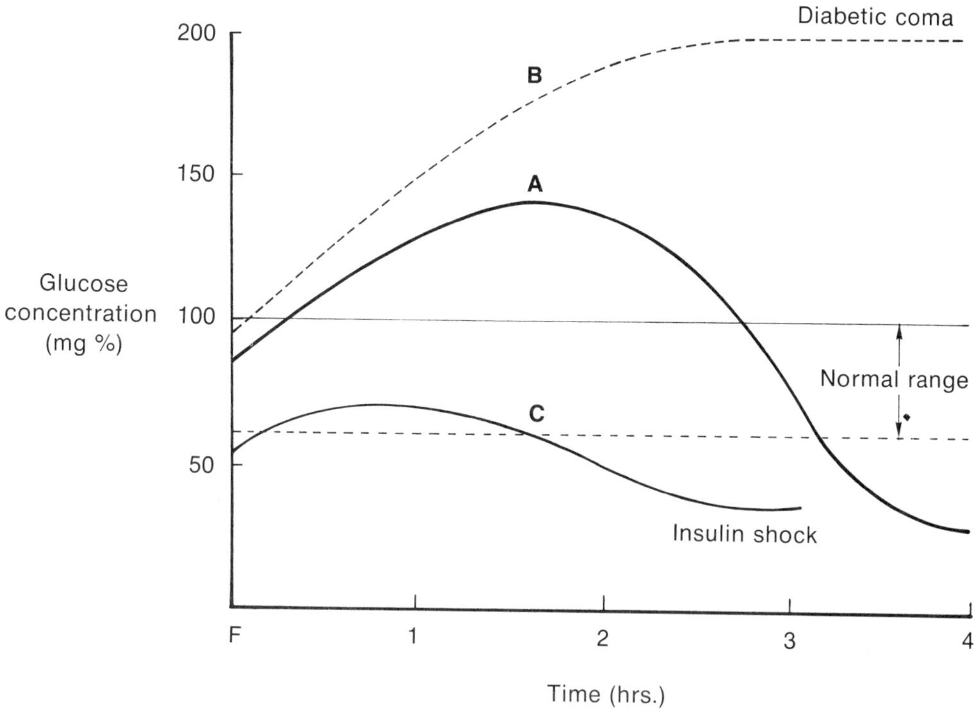

Figure 10–16 Glucose tolerance curves. *A*, In a normal person. *B*, In a potential diabetic. *C*, In a person with hypoglycemia.

sponsibilities of estrogen are to develop and maintain the female reproductive tract, the secondary sex organs (breasts, vagina, uterus), and the secondary sex characteristics.

2. *Progesterone* is the second principal ovarian sex hormone. It is secreted by a temporary endocrine gland, the corpus luteum, during the second half of the menstrual cycle, and promotes development of the uterine lining (already initiated by estrogen in the first half of the cycle) in preparation for implantation of the fertilized egg. Progesterone, like the estrogens, helps to maintain the female reproductive system. During pregnancy, progesterone influences the development of the fetal membranes and mammary glands.

Secretion of estrogen and progesterone is controlled by the level of pituitary secretions, follicle-stimulating hormone (FSH) and luteinizing hormone (LH), in the blood.

Male Sex Hormones (Table 10–1)

The steroid testosterone is the principal male sex hormone, and is secreted by the interstitial cells in the testes. It is responsible for maturation at puberty and for the maintenance of the male reproductive system and the secondary sex characteristics (hair, deep voice, and greater masculine development).

Removal of the testes (castration) in the human male before onset of puberty results in failure of the secondary sex characteristics to develop. After puberty, castration causes an attenuation of the sexual characteristics.

Production of testosterone is under the influence of the gonadotropic hormone from the pituitary (ICSH). The production of ICSH is depressed or arrested by a certain concentration of testosterone in the blood.

OTHER HORMONES

Some organs other than the ones previously mentioned produce or are governed by hormones. These include the stomach, intestine, kidney, and placenta. The physiologic importance of these hormones is best described when their respective organs are considered in detail.

OUTLINE

I. Introduction
   A. Endocrine versus exocrine
II. Chemical Nature of Hormones
   A. Chemical messengers
   B. Steroids or amino acid derivatives
III. Hormone Function
   A. Effect on cellular metabolism
   B. Comparison with central nervous system
IV. Individual Glands and Hormones
   A. Pituitary gland (hypophysis)
      1. Adenohypophysis—anterior pituitary gland
         a. Thyroid-stimulating hormone (TSH)
         b. Adrenocorticotropic hormone (ACTH)
         c. Somatotropic hormone (STH)
         d. Follicle-stimulating hormone (FSH)
         e. Luteinizing hormone (LH)
         f. Interstitial cell-stimulating hormone (ICSH)
         g. Lactogenic hormone (LTH)
         h. Melanocyte-stimulating hormone (MSH)
         i. Hyperfunction
            1) Giantism, acromegaly, and overstimulation of most glands
         j. Hypofunction
            1) Dwarfism, cachexia, and depressed activity of most glands
         k. Master gland of body due to vast systemic controls
      2. Neurohypophysis—posterior pituitary gland
         a. Vasopressin (antidiuretic hormone, ADH)
         b. Oxytocin
         c. Hyperfunction
            1) Excessive water retention
         d. Hypofunction
            1) Diabetes insipidus
   B. Thyroid gland
      1. Thyroxine
         a. Protein-bound iodine (PBI)
         b. Basal metabolic rate (BMR)
         c. General oxidative catalyst
         d. Hyperfunction—tumor
            1) Graves' disease (exophthalmic goiter)
         e. Hypofunction
            1) Cretinism—in children
            2) Myxedema—in adults
            3) Simple goiter—iodine deficiency
   C. Parathyroid glands
      1. Parathormone
         a. Regulator of calcium and phosphorus ions in blood
         b. Hyperfunction
            1) Increased calcium in blood
            2) Kidney stone production
         c. Hypofunction
            1) Depletion of bone minerals—osteoporosis
            2) Increased neuromuscular sensitivity—tetany
   D. Adrenal glands
      1. Adrenal cortex
         a. Mineralocorticoids—aldosterone
            1) Regulate concentration of sodium and potassium in blood
         b. Glucocorticoids—cortisone
            1) Promote glyconeogenesis; increase blood sugar level
         c. Sex hormones—androgens, estrogens
            1) Male and female hormones secreted by both sexes

2) Assist reproductive functions
d. Hypersecretion—tumors
1) Loss of sexual function
2) Hypertension
3) Cushing's syndrome
e. Hyposecretion—disease
1) Addison's disease
2. Adrenal medulla
a. Epinephrine (Adrenalin)
1) General stimulant—emergency hormone
b. Norepinephrine (Noradrenalin)
c. Not essential for life

E. Pancreas
1. Exocrine and endocrine functions
2. Islets of Langerhans
a. Insulin—beta cell product
1) Increased rate of glucose metabolism
2) Regulation of blood glucose concentration
3) Increased glycogen storage in tissues
b. Glucagon—alpha cell product
1) Glycogenolysis
2) Increased blood glucose concentration
c. Hypersecretion—tumor
1) Hyperinsulinism—hypoglycemia
2) Insulin shock
d. Hyposecretion—disease
1) Hypoinsulinism—hyperglycemia
2) Diabetes mellitus
e. Glucose tolerance test

F. Female and male sex hormones
1. Female sex hormones
a. Estrogens (estradiol, estriol, estrone) and progesterone
b. Development and maintenance of female reproductive system and secondary sex characteristics
2. Male sex hormones
a. Testosterone
b. Development and maintenance of male reproductive system and secondary sex characteristics
3. Hyperfunction—tumor
a. Increased production
4. Hypofunction
a. Sterility and loss of secondary sex characteristics

G. Other hormones
1. Stomach, intestine, placenta, and kidney hormones
a. To be discussed with their respective systems

STUDY QUESTIONS AND PROBLEMS

1. What is a hormone?
2. Explain the anatomic relationship between the hypothalamus and the neurohypophysis.
3. Giving examples, explain the feedback mechanism for control of the secretion of a hormone.
4. What is the physiologic response to stress?
5. What hormone enhances and prolongs sympathetic effects?
6. Which gland is called the "master gland"? Why?
7. Describe the problem leading to cretinism. Differentiate between cretinism and dwarfism.
8. What is tetany? Is it the same as muscle tetany?
9. Differentiate between the terms exocrine and endocrine.
10. What is the relationship between suckling and oxytocin?
11. Why is thyroxine sometimes prescribed for obese patients?
12. How do glucocorticoids elevate the blood sugar?
13. Compare diabetes mellitus with diabetes insipidus in terms of pathology and character of urine output.
14. What is the relationship between the parathyroid gland and kidney stones?
15. Describe the homosexual in reference to the adrenal cortex.
16. Vasectomy and castration both render the male sterile; describe the difference between the two.
17. Explain the relationship between hyperthyroidism and the BMR and PBI.
18. Explain the "self-regulation" of the adrenal cortex by the interaction of ACTH and the glucocorticoids.

Chapter 11

Radiologic Health

Radiologic technology provides a practical application to many of the biological sciences. It has been introduced here, and not earlier in this text, because a broad and basic understanding of the structure and function of tissues, of skeletal anatomy, and of basic control mechanisms is necessary if the significance of radiography and radiation therapy is to be appreciated.

NATURE OF RADIATION

You may have wondered why certain electrical instruments are harmless, whereas others are extremely dangerous. Everyone knows that radio and television waves exist, and no one is afraid to stand in front of a radio or a television set. However, too long an exposure to the radiation of an x-ray machine can cause serious injury. Electrical waves are used for the operation of radio, television, and x-ray—so why should there be such a difference?

Everyone is exposed daily to a wide variety of radiations, whether he is aware of it or not. Some of these can be felt, such as the radiant energy through which heat is transferred. Others can be seen as light, and this phenomenon can be further subdivided into all the colors of the visible light spectrum. All of these radiations are within the electromagnetic spectrum and are spoken of in terms of waves of energy.

The electromagnetic spectrum (Fig. 11–1) is a grouping of all energy waves (photons) according to their wavelengths. The amount of energy of a photon is determined by its size (wavelength) and its frequency. The quantum theory:

$$E \propto \frac{1}{\lambda} \quad \text{where } \lambda = \text{wavelength}$$

sums up the entire energy concept by stating that the amount of energy emitted by a photon is inversely proportional to its wavelength (λ). This means that the shorter the wavelength, the greater the power that it imparts.

Notice in Figure 11–1 that the entire scale can be divided almost in half. The left side of the spectrum includes electric power, heat, television, and the infrared rays, all of which are visible. The right side contains the dangerous waves, such as ultraviolet rays, x-rays, and cosmic radiation. The reason for this division is quite simple. In the first chapter of this text we saw that compounds are not just thrown together. They are made up of elements held together by an energy bond. If this bond is broken, the compounds become ionized, and this process may cause injury if the compound is water.

It takes approximately 13 ev (electron volts) to ionize a molecule of H_2O:

$$H_2O \longrightarrow H^+ + OH^-$$

This alone is not significant, but the fact that the free hydroxyl radicals can com-

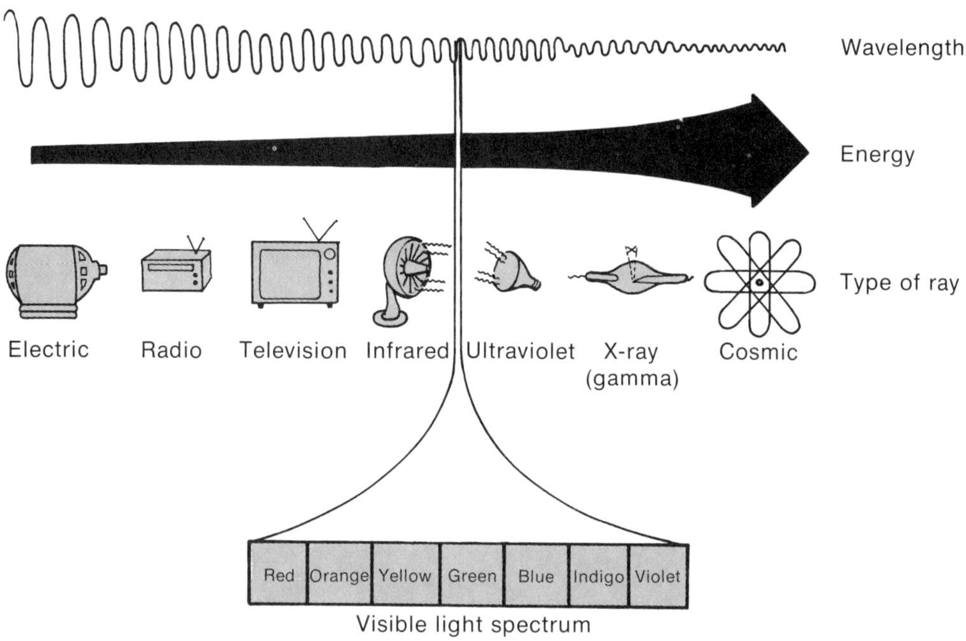

Figure 11–1 The electromagnetic spectrum.

bine to form hydrogen peroxide, a cellular poison, *is* significant:

$$2OH^{-} \longrightarrow H_2O_2$$

The body is prepared to adapt to a normal amount of hydrogen peroxide production, but it cannot accommodate massive quantities produced during a short period. This occurs in *radiation sickness.* A person exposed to a large quantity of high-energy radiation over a short period produces large quantities of hydrogen peroxide, and the resultant cellular disruption may cause permanent physical defects or death.

To return once again to our initial question, why are you not afraid to stand in front of a radio while you are afraid to stand in front of an x-ray machine? We have seen that the energy spectrum is set up so that waves are arranged in order of decreasing size (Fig. 11–1). As the energy level increases, the quantity of radiation needed to destroy body tissues appears just after the visible light range—in the ultraviolet range. Therefore, all energy waves, beginning with the ultraviolet range and proceeding from there, are ionizing and dangerous. All energy waves below the ultraviolet range are nonionizing, and are not considered dangerous. Thus, to answer our question, a radio or television wave is not small enough to be an ionizing wave; but x-rays must be ionizing to penetrate the body when used in radiographic examinations.

TYPES OF RADIATION

Ionizing radiations may be classified according to either their origin or their physical properties. They fall into two general categories: those that have a mass or weight (called *particles*), and those that are energy only (*electromagnetic*).

The following experiment was performed to study the radiations produced by a radioactive element. A piece of radium was placed at the bottom of a thick lead well. The escaping radiation was allowed to fall on a photographic plate (Fig. 11–2). When these radiations were passed through a strong electrical field, three different areas showed up on the photographic plate. These were called alpha, beta, and gamma rays:

1. *Alpha rays* are attracted to the negative field, which indicates that they are positively charged. These rays are particles, and are relatively harmless because they do not penetrate the skin.

2. *Beta rays* are attracted toward the positive field, which indicates that they are negatively charged particles. They have only a slight penetrating power, and will penetrate the skin or an organ to a depth of only a few millimeters. Most of the radioisotopes used in medicine today emit beta rays.

3. *Gamma rays* are not affected by an electrical field because they have no charge. Strictly speaking, they are not particles, but rather a form of electromagnetic radiation similar to x-radiation. Gamma rays penetrate tissue readily and will cause cellular damage as they travel through the body.

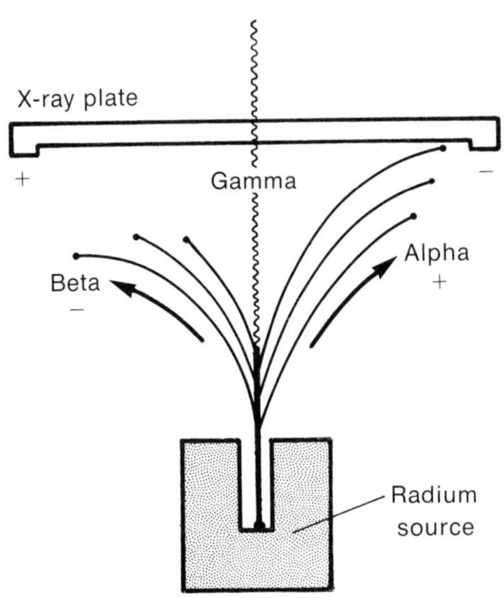

Figure 11–2 Electrical properties of radioactive emissions. The alpha particles are attracted to the negative field; therefore, they are positively charged. Conversely, the beta particles are negatively charged. The gamma emissions are not affected by either electrical field and are neutrally charged.

A fourth type of radiation involves the emission of a neutron from the nucleus of an atom. Such changes are called nuclear reactions and result in the formation of either an isotope or an entirely new element.

NATURAL AND ARTIFICIAL RADIOACTIVITY

Some elements, such as uranium and radium, are naturally radioactive; that is, they emit energy continually. Natural radioactivity may be defined as the spontaneous change of one element into another. Most of the heavier elements are naturally radioactive. Nonradioactive elements, generally the lighter ones, can be made radioactive if the element is bombarded with protons, neutrons, electrons, or alpha particles. The radioactive substance thus produced in a nuclear reactor (Fig. 11–3) is artificially radioactive, and is called a *radioisotope*. Radioisotopes have the same chemical properties as nonradioactive isotopes of the same element because chemical properties are based on electrons only, and isotopes of an element have the same electron structures (Fig. 11–4).

Radioisotopes are used in medicine in the diagnosis and treatment of disease. The main criterion in choosing radioisotopes is the dosage required. If the dose is small but large enough to be detected, the radioisotope can be used to diagnose abnormalities.

Iodine-131 is an isotope used in the diagnosis and treatment of thyroid disorders. The thyroid gland requires iodine to function normally, and, in turn, converts the available iodine into thyroxine, which acts to control the metabolic rate. If a patient is suspected of having a metabolic problem, he is given a drink of I^{131} in the form of sodium iodide. If the thyroid is functioning normally, the gland should take up about 12 per cent of the radioactive iodine within a few hours. The patient

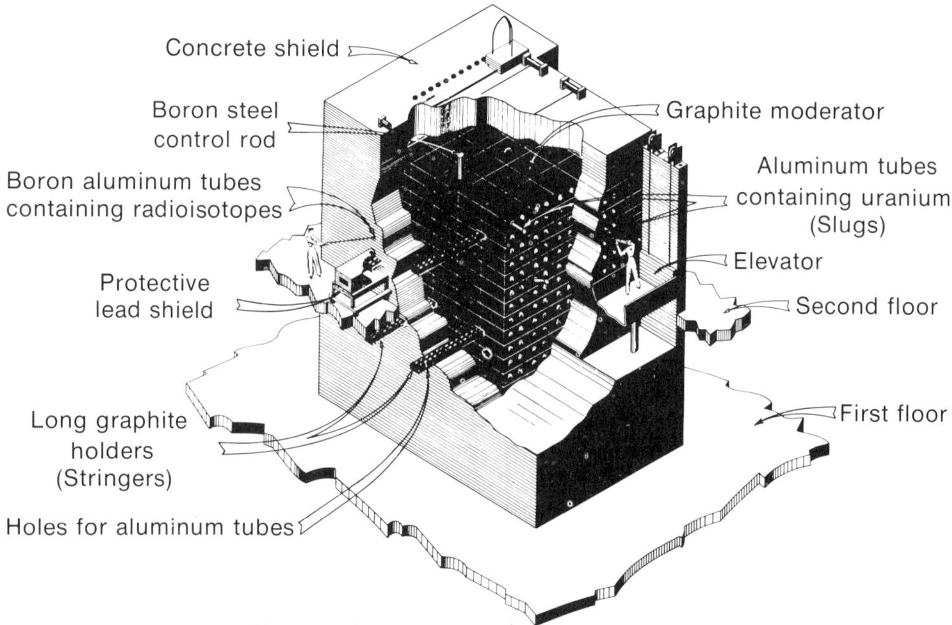

Figure 11–3 A nuclear reactor where isotopes are formed: an atomic "pile." (Courtesy of United States Atomic Energy Commission.)

is then placed in front of a specialized Geiger counter (Fig. 11–5), and the amount of radiation is measured. If less than the normal amount of I^{131} has been taken up, the patient may have a hypothyroid condition. Conversely, if the amount of I^{131} taken up by the thyroid is greater than normal, the patient may be hyperthyroid—may have an increased activity of the thyroid gland.

Other radiation counters involve the use of a scanner (Fig. 11–6) to detect the presence of localized radiation and to produce a picture, called a "scan," of the area involved. Proper interpretation of a scan can tell the physician, for example, whether the thyroid is normal or abnormal (Fig. 11–7). A patient is given a selected radioisotope that will accumulate in the body area suspected of having a problem. The scanner moves back and forth across the site and detects the radiation present.

Phosphorus-32 and technetium-99 are used in the location of brain tumors. Such

Text continued on page 221

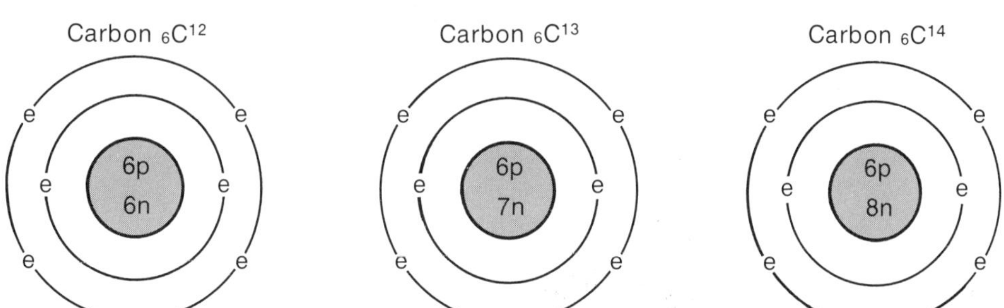

Figure 11–4 Isotopes of carbon. The nucleus contains six protons (p) and six, seven, or eight neutrons (n), whereas the outer energy rings or shells contain six electrons (e).

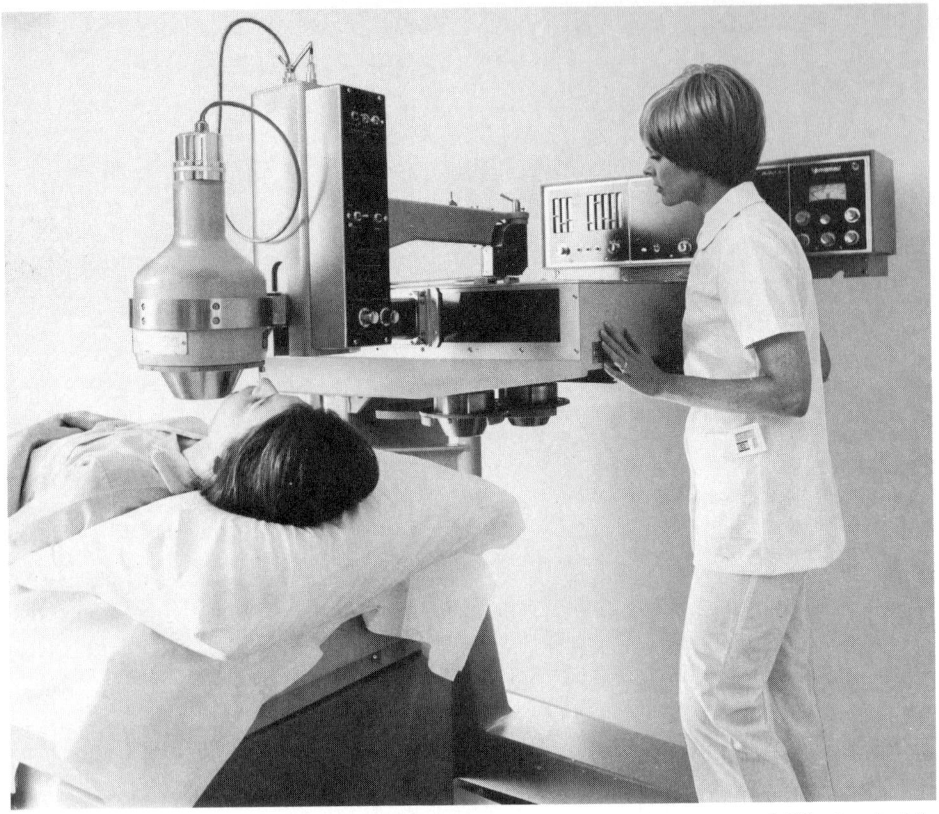

Figure 11–5 Thyroid uptake probe. A Geiger counter that measures the amount of I^{131} absorbed by the thyroid gland. (Courtesy of Searle Radiographics Inc., Des Plaines, Illinois.)

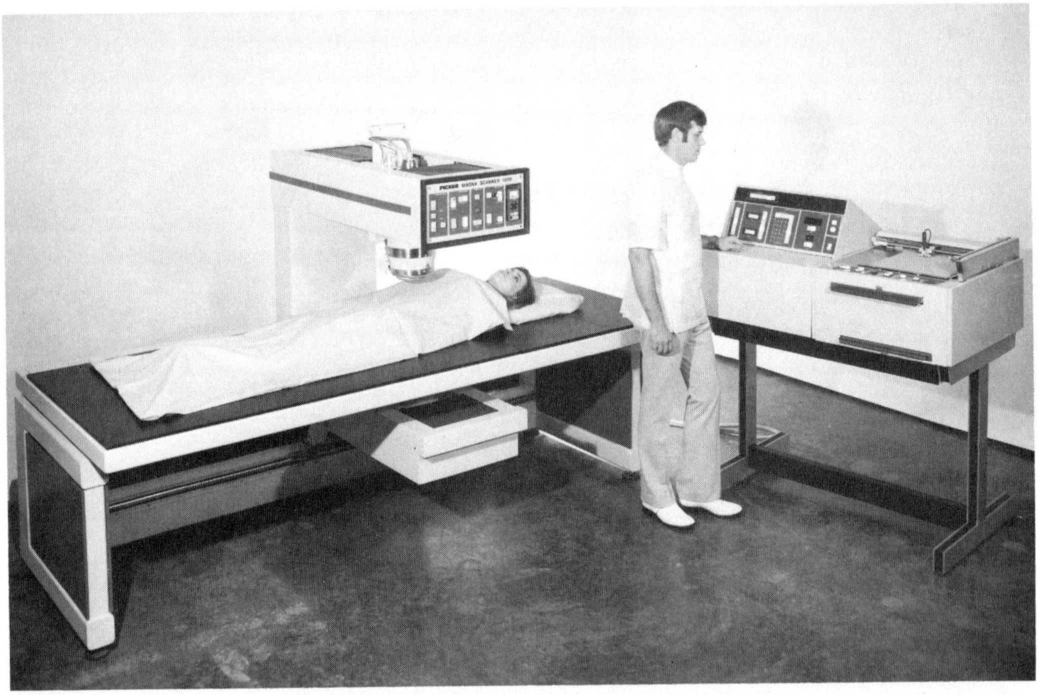

Figure 11–6 Magna Scanner 1000, an instrument used to detect and record the distribution of radioactive material in the human body. (Courtesy of Picker Corporation, Cleveland, Ohio.)

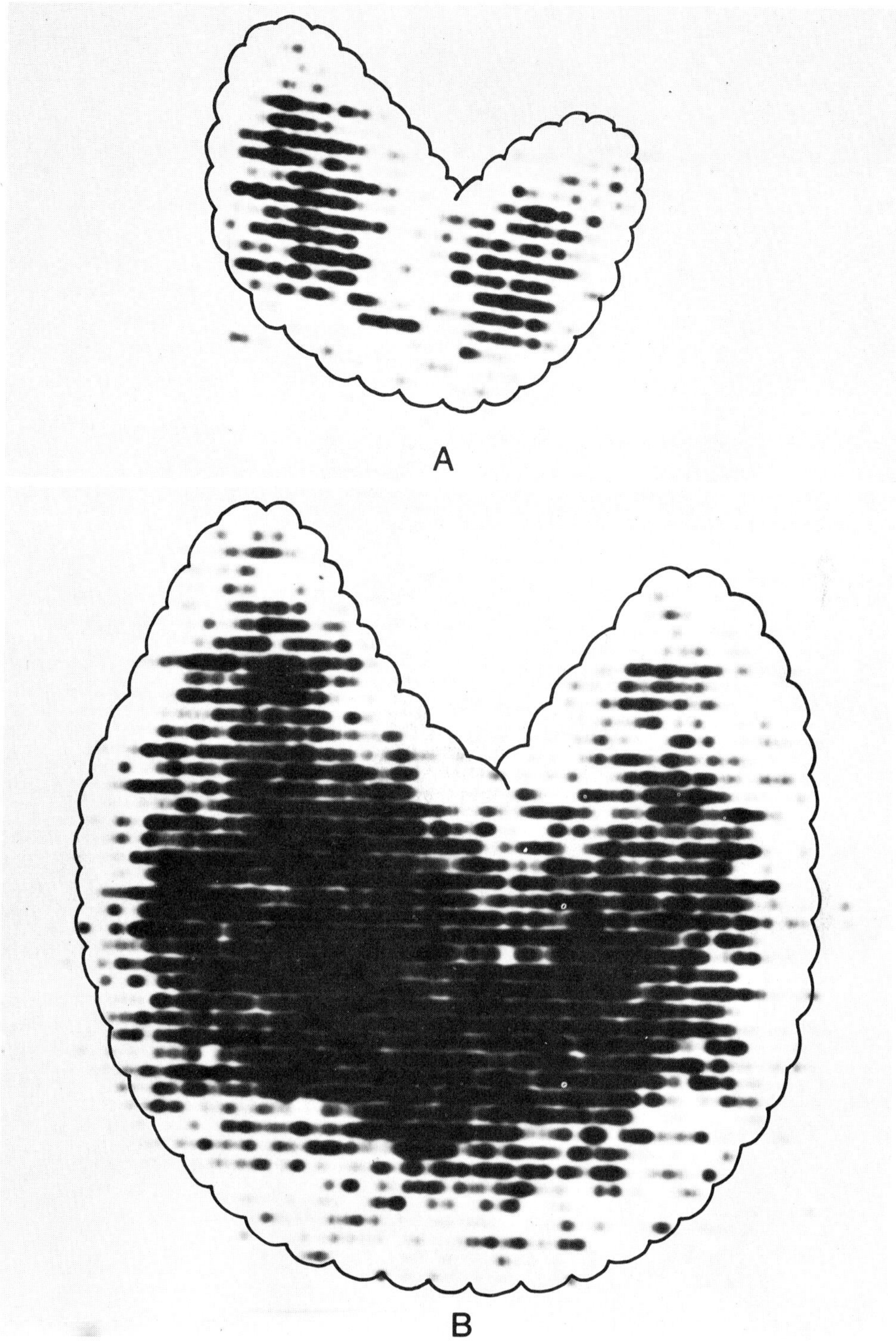

Figure 11–7 Thyroid scans, normal and abnormal. *A*, Normal scan. Note the even distribution of radioiodine in both lobes of the thyroid. *B*, Abnormal scan. Poor concentration of radioiodine indicates a malfunctioning thyroid.

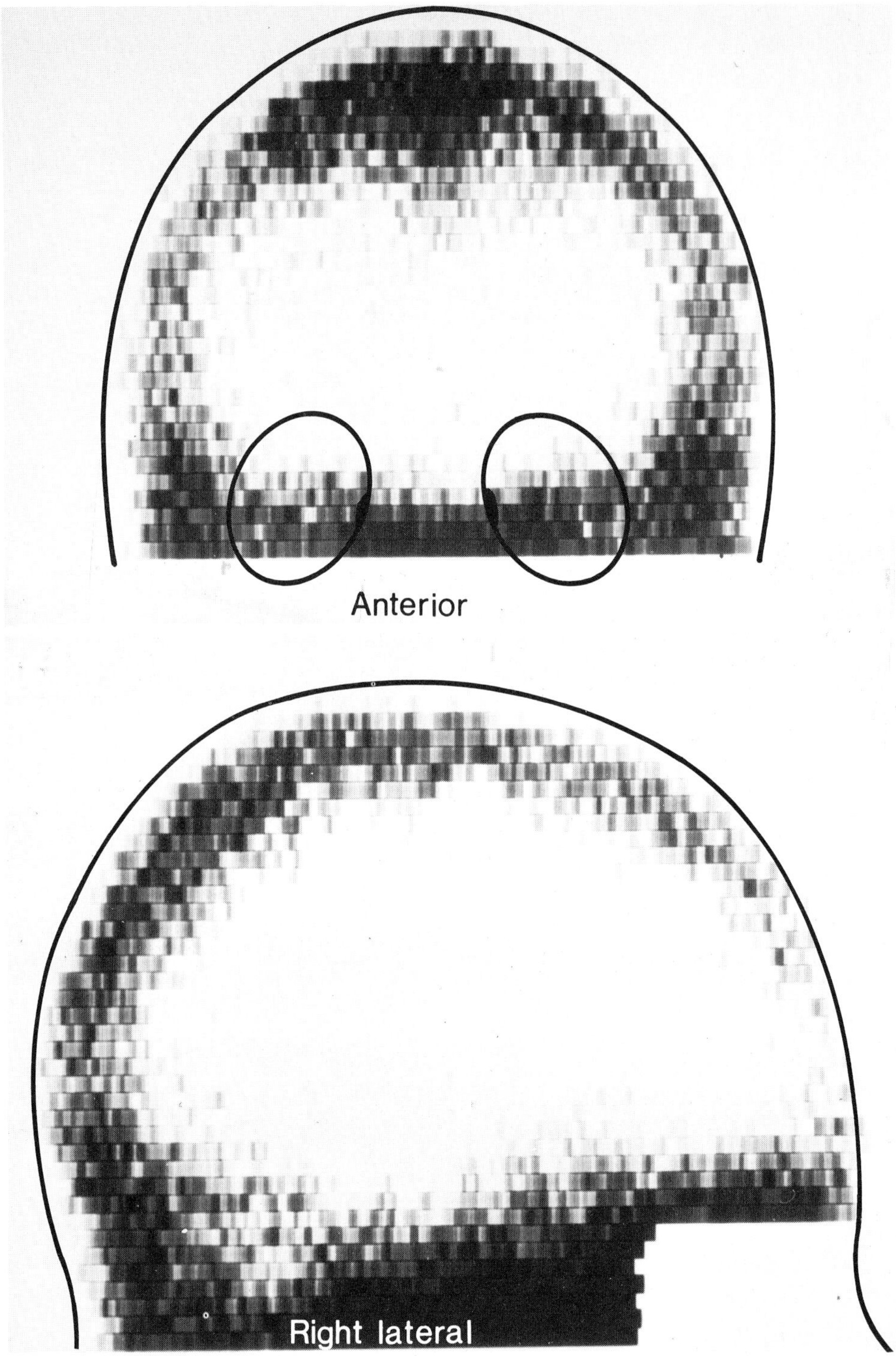

A

Figure 11–8 Brain scans, normal and abnormal. *A*, Normal scan, anterior and right lateral aspects. In this instance, the P^{32} or Tc^{99} radioisotope is not absorbed by healthy brain tissue. *B*, Abnormal scan, anterior and left lateral aspects. Note the radioisotope uptake, indicating a pathologic condition in the right parietal lobe area.

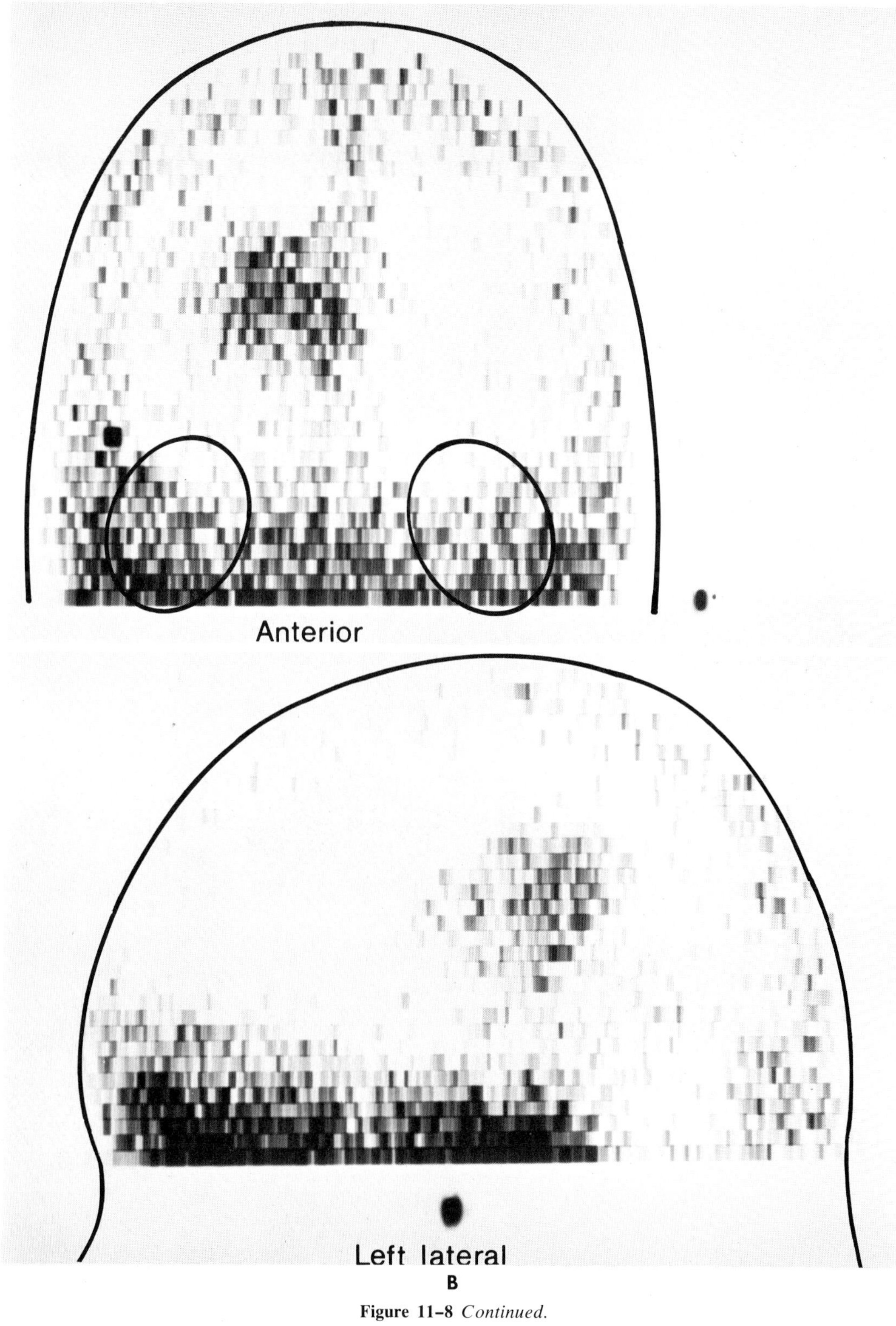

B

Figure 11–8 *Continued.*

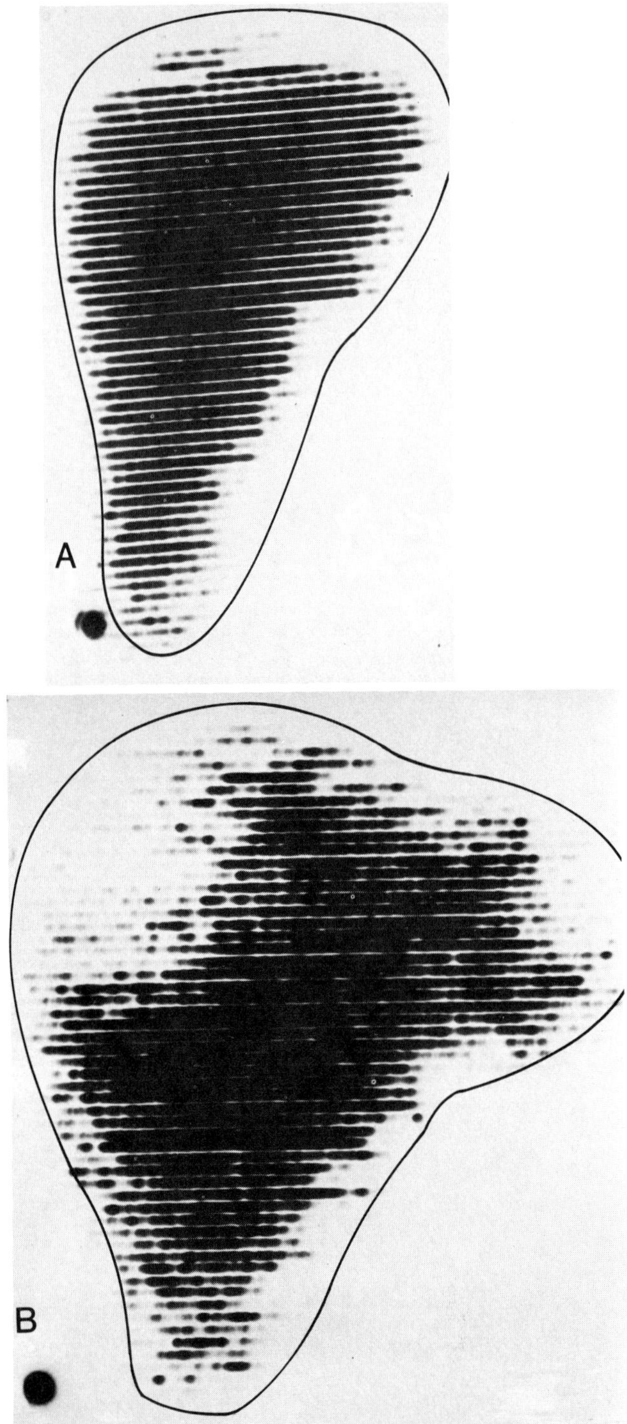

Figure 11-9 Liver scans, normal and abnormal. *A*, Normal scan using Hg^{198}, anterior aspect. The distribution of radioactivity indicates normal structures. *B*, Abnormal scan, anterior aspect, indicating an enlarged liver with a lesion in the upper right lobe.

tumors are known to accumulate these radioactive elements more readily than do healthy cells, and the tumor appears as a darkened area on the film (Fig. 11–8). Phosphorus-32 has also been used in the study of bones and teeth and in treating skin cancers. Because developing blood cells pick up phosphorus more rapidly than do mature red cells, P^{32} is used to treat polycythemia vera, a blood disorder in which an overproduction of red blood cells occurs.

In another diagnostic procedure, sodium-24 is injected into the bloodstream and its path followed carefully to determine whether any blood circulating through the arms or legs is obstructed. This examination is very helpful in patients considered for possible amputations and in detecting blood clots or circulatory constrictions.

Gold-198 is used extensively in the detection of liver problems. Normal, healthy liver tissue tends to absorb the radioisotope readily, whereas diseased tissue will reject it. Therefore, an even-textured scan indicates normal tissue, but a lightened area may indicate a problem (Fig. 11–9). Gold-198 is also used in treating chronic leukemia, and sometimes is injected directly into the chest cavity for the treatment of lung cancer.

Following an injection of mercury-197, a patient is placed in front of a probe (Fig. 11–10), which will measure the uptake of the isotope by his kidneys. Since Hg^{197} accumulates in the renal tubular cells more rapidly than it is excreted, a scan can

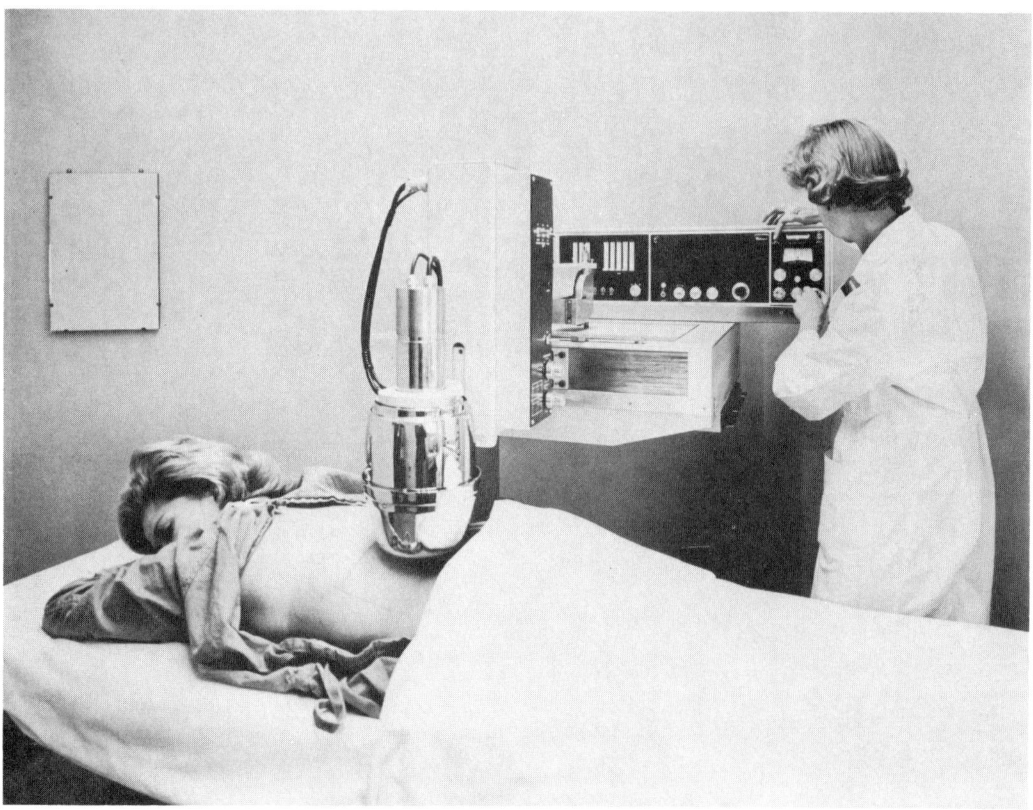

Figure 11–10 Renal uptake probe. (Courtesy of Searle Radiographics Inc., Des Plaines, Illinois.)

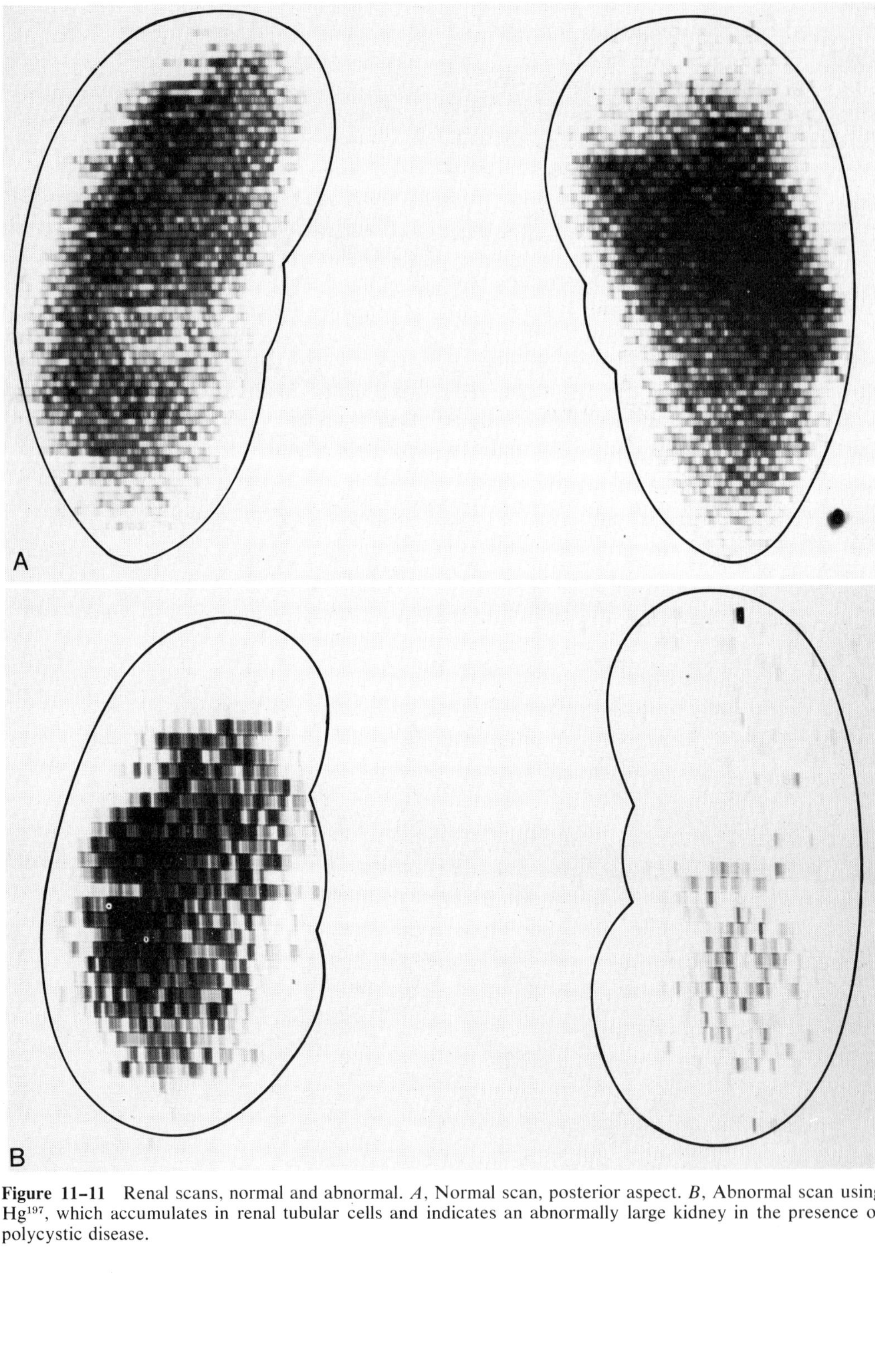

Figure 11–11 Renal scans, normal and abnormal. *A*, Normal scan, posterior aspect. *B*, Abnormal scan using Hg^{197}, which accumulates in renal tubular cells and indicates an abnormally large kidney in the presence of polycystic disease.

be made for the visualization of healthy or diseased renal tissue (Fig. 11–11).

Some other radioisotopes used in medicine and in biological research are iron-59, used to determine the rates of blood formation and destruction; carbon-14, used to study the paths of carbohydrates, fats, and proteins in the body; calcium-45 and strontium-89, used to detect and treat bone cancers; and hydrogen-3 (tritium), used to measure the water content of the body.

To this point we have been concerned with those radioisotopes used primarily in detecting and treating abnormal tissue. Cobalt-60, a radioisotope so strong that it is used exclusively for the treatment of many different types of cancer, is cheaper and easier to handle than radium, and emits powerful gamma rays. Both radium and cobalt are used by hospitals in the treatment of cervical cancer (Fig. 11–12), the latter either as a cobalt "stick" or in large machines containing a cobalt "bomb" to deliver strong radiation directly to the cancer site.

X-rays, a penetrating type of radiation, are like gamma rays, but less penetrating. They are used to detect internal abnormalities and to treat localized skin conditions. Deep-seated lesions require an x-ray machine of higher voltage. The area to be treated is clearly demarcated to prevent destruction of surrounding healthy tissue. X-ray therapy (Fig. 11–13) is now commonly used both before cancer surgery and after cobalt therapy.

DETECTION AND MEASUREMENTS OF RADIATION

How do we know whether something is radioactive? This is a very important question—particularly in the protection of medical personnel. One device used for this purpose is the Geiger counter (Fig.

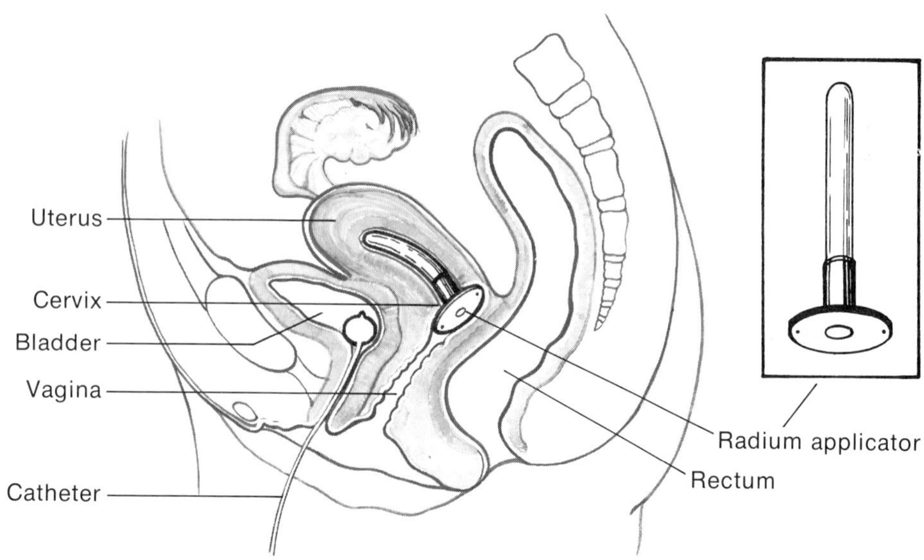

Figure 11–12 Radium implantation in treatment of cervical cancer. The radium applicator (inset) is placed in the uterus, and the energy rays from the radioactive radium are allowed to come in direct contact with the cancer site. Note the catheter inserted into and decompressing the bladder so that it will not come in contact with the uterus.

Figure 11–13 System Seventy x-ray therapy machine used in the treatment of deep malignancies. (Courtesy of Baird-Atomic, Inc., Bedford, Massachusetts.)

11–14, *B*), which consists of a glass tube containing a gas at low pressure. When the counter is exposed to a radioactive substance, the energy given off causes a current to flow through a small wire inside the glass tube. This wire carries the current to an amplifier, which gives off a "clicking" sound and records the intensity of energy emitted by the radioactive source. It does not tell you what the source is—just that it is there.

X-ray technicians and other workers continually exposed to radiation usually are required to wear a badge or a "pencil" (Fig. 11–14, *C*), devices that register the total amount of radiation to which they have been exposed. The badge can be developed like an x-ray film; the pencil can be held up to the light and read directly. In either case, if the device shows that a dangerous level of exposure has been reached, the person then must leave the area until his body has recovered from the accumulated radiation. Depending on the extent of exposure, this period may be up to several weeks, because the physiologic effects of radiation are cumulative.

RADIATION PROTECTION

If radiation is so powerful and so potentially dangerous, how do we protect ourselves from it? Shielding is one very practical method. Research has shown that the denser the material used, the greater the protection provided against radiation injury. Thus a very dense metal like lead is needed to stop most penetration by gamma or x-radiations. Less powerful emissions, such as alpha or beta particles, can be stopped by paper or wood (Fig. 11–15).

Because gamma and x-rays can be stopped by lead, aprons of this metal are made to be worn by technicians who must be in the room during x-ray procedures (Fig. 11–16). Normally, however, the technician either stands behind a lead-lined screen with a window in it or leaves the room.

A bombardment of *cosmic radiation* from outside the earth's atmosphere occurs constantly. These cosmic rays produce so much energy that no amount of shielding can protect us from them. Cosmic radiation is known to be the cause of

Figure 11–14 Radiation detection devices. *A*, Ionization type. *B*, Geiger, Mueller type. *C*, Ionizing self-reader "dosimeter" (pencil) worn by all personnel who are continually exposed to radiation of any type. *D*, Health physics radiation measuring system. (Courtesy of Victoreen Instrument Division, Cleveland, Ohio.)

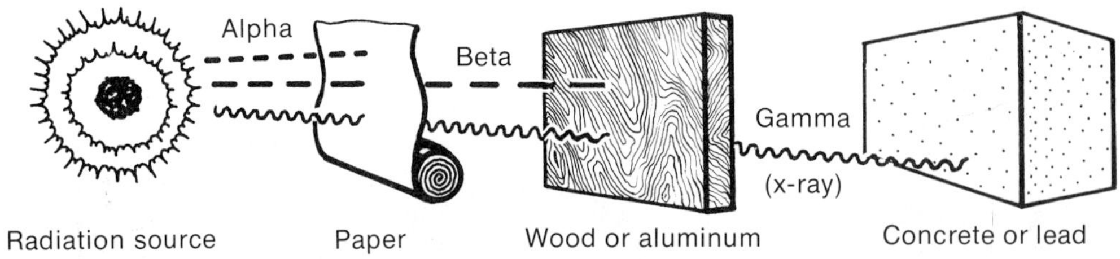

Figure 11–15 Radiation shielding. Note the increased density of the material necessary to protect the user against alpha, beta, and gamma (x-ray) radiation types.

many of the biological mutations that occur in nature.

BIOLOGICAL EFFECTS OF RADIATION

Alpha and beta rays are relatively harmless because of their slight penetrating power. Gamma rays, however, can cause ionization of the atoms inside the cells, leading to the production of substances harmful to the cell's metabolism. Such radiation also may affect the genes, causing a cell to grow abnormally, as in cancer, or to die.

The physiologic effects of radiation are cumulative, and they can shorten life. Radiation of the whole body can be lethal if enough radiation (measured in roentgen units) is absorbed. Such a dose is called LD_{50}—the amount of radiation necessary to kill 50 per cent of a population. Table 11–1 lists the levels of radiation necessary to kill 50 per cent of the population of certain types of organisms.

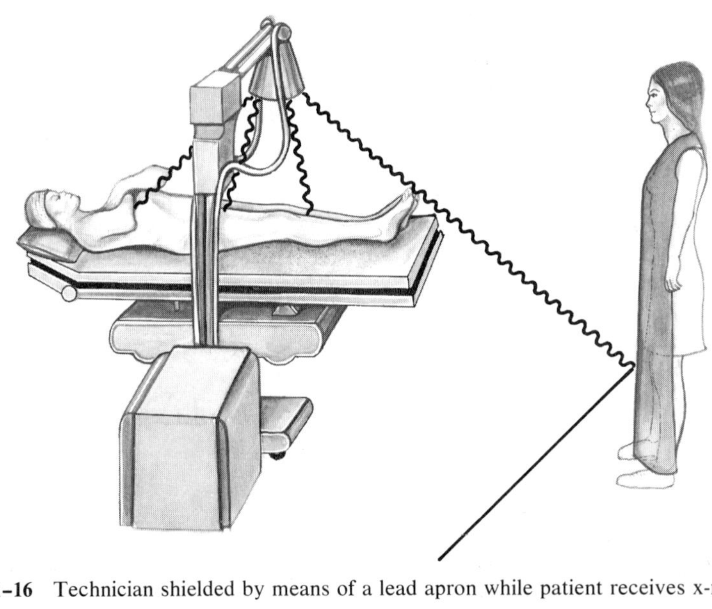

Figure 11–16 Technician shielded by means of a lead apron while patient receives x-ray therapy.

Table 11-1 LD_{50} OF VARIOUS ORGANISMS*

SUBJECT	*LD_{50} IN ROENTGENS*
Guinea pig	250
Dog	325
Pig	375
Man	450
Monkey	600
Rat	675
Bacteria	100,000
Viruses	1,000,000

*Notice that the microscopic organisms have a decidedly greater resistance to radiation than do the larger animals.

OUTLINE

I. Introduction
  A. Radiologic technology
    1. Practical science
II. Nature of Radiation
  A. Electromagnetic spectrum
    1. Photon
    2. Quantum theory
    3. Ionizing radiation if greater than 13.6 ev
    4. Hydrogen peroxide production
      a. Cellular poison
III. Types of Radiation
  A. Particulate
    1. Alpha—positive charge
      a. Relatively harmless
    2. Beta—negative charge
      a. Can penetrate the skin
    3. Gamma—powerful neutral charge
      a. Can penetrate deep tissues
    4. Neutron—emitted during nuclear reactions
  B. Electromagnetic
    1. Photon
IV. Natural and Artificial Radioactivity
  A. Natural radioactivity
    1. Heavy elements
  B. Artificial radioactivity
    1. Lighter elements
    2. Produced by nuclear bombardment
    3. Radioisotopes
  C. Radioisotopes used in medicine
    1. Iodine-131—thyroid
    2. Phosphorus-32—brain, blood, bones, skin
    3. Technetium-99—brain tumors
    4. Sodium-24—blood
    5. Gold-198—liver diseases, leukemia
    6. Mercury-197—renal uptake
    7. Iron-59—blood formation
    8. Calcium-45 } bone tumors
    9. Strontium-89 }
    10. Hydrogen-3—water content
    11. Cobalt-60—cancer treatment
    12. Carbon-14—paths of carbohydrates, fats, and proteins
  D. X-radiation
    1. Penetrating photon
    2. Diagnosis and treatment
V. Detection and Measurement of Radiation
  A. Geiger counter
  B. X-ray badges and pencils
VI. Radiation Protection
  A. Shielding
    1. Alpha—stopped by paper
    2. Beta—stopped by wood
    3. Gamma } stopped by lead,
    4. X-ray } concrete
  B. Aprons lined with lead to protect technicians
VII. Biological Effects of Radiation
  A. Alpha and beta rays—little effect externally
  B. Gamma rays—penetrating
    1. Ionizing
    2. Cellular poisons
    3. Genetic effect
  C. LD_{50}
    1. Dosage needed to kill 50 per cent of a population
    2. LD_{50} for man much lower than for microscopic organisms

STUDY QUESTIONS AND PROBLEMS

1. What is the quantum theory?
2. Differentiate between particulate and electromagnetic radiation.
3. What is artificial radioactivity? How is it produced?
4. Define radioisotope and describe how a particular radioisotope can be used in medicine.
5. How is radiation detected?
6. For what are x-rays used?
7. What factor determines whether an isotope is used for diagnosis or for therapy?
8. What may be done to minimize the effects of radiation on the body?
9. What are some of the biological effects of radiation?
10. If the earth were covered with lethal radiation, which animals might persist?

Chapter 12

The Special Senses

SENSORY MECHANISMS

A sense organ or receptor is a specialized nervous tissue situated at the peripheral endings of the dendrites of afferent neurons. The receptor's primary function is to provide the body with information, both conscious and unconscious, about degrees of change in the organism's external and internal environments.

Three important features of sense organs should be emphasized:

1. Specific receptors are particularly sensitive to specific stimuli. However, they can also respond to other stimuli of sufficient intensity. For example, pressure on the eyeball causes a sensation of light.

2. Specific sensitivity to certain stimuli is due to the structure and composition of the receptor (e.g., light-absorbing pigments in the photoreceptors of the eye).

3. The type of sensation elicited by a receptor depends on *which* nerve pathways are activated, not on *how* they are activated. All nerve impulses are essentially alike, regardless of the stimulus that initiates them. The impulse is transmitted to the central nervous system, where it is interpreted. For example, the region of the cerebrum receiving impulses from an olfactory receptor decodes and interprets them as a specific odor or aroma. If an olfactory receptor is artificially stimulated, the same nerve pathways to the brain will be activated, and the brain will interpret the arriving impulses as a specific odor.

Classification of Sense Organs

Numerous attempts have been made to classify the senses into groups, but none of the suggested systems has been entirely successful. Traditionally, the *special senses* are smell, vision, hearing, equilibrium, and taste; the *cutaneous senses* are those associated with receptors in the skin; and the *visceral senses* are concerned with perception of the body's internal environment.

The different sensations are evoked by stimulation of various specialized sensory receptors. Some of these may be grouped according to the nature of the stimulus:

1. *Chemoreceptors*—receptors sensitive to chemical stimuli (e.g., taste and smell).

2. *Pressoreceptors*—receptors sensitive to mechanical stimuli (e.g., touch, pain, temperature, pressure, sound, and motion).

3. *Photoreceptors*—receptors sensitive to visible light (sight).

Another classification of the various receptors categorizes them according to their somatic or visceral locations (Fig. 12–1):

1. *Teleceptors*—nerve endings that detect environmental changes occurring at some distance from the body (e.g., light rays, sound waves, smell—receptors of eye, ear, and nose).

2. *Exteroceptors*—nerve endings that

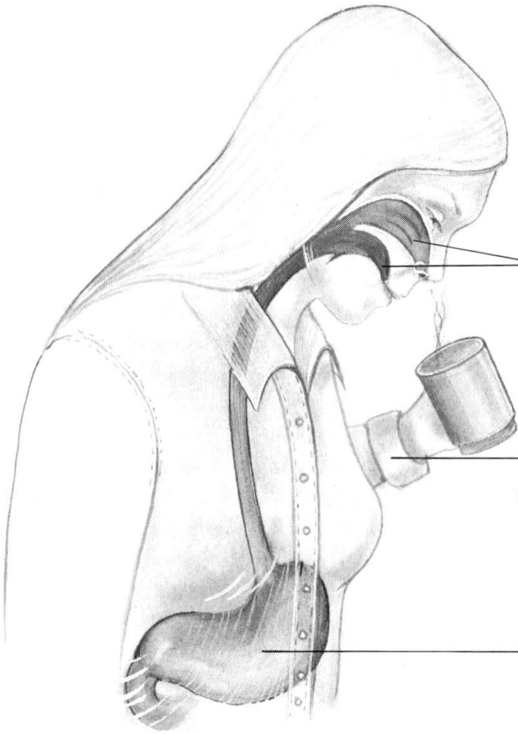

Figure 12–1 Somatic receptors. Exteroceptors and teleceptors recognize the smell, taste, and temperature of the coffee. Proprioceptors in the muscles of the arm provide movement, whereas interoceptors in the visceral organs recognize the sensations of hunger and thirst.

detect environmental changes that directly affect the skin (temperature, touch, pressure, or pain).

3. *Proprioceptors*—receptors that provide the body with information about its position in space (muscles, tendons, and joints).

4. *Interoceptors*—receptors located within organs equipped with motor innervation from the autonomic system. These visceral receptors are concerned with maintenance of the internal environment (smooth muscle contraction, peristalsis, visceral contractions).

Projection

No matter where a particular sensory pathway is stimulated (Fig. 12–2, *A*) along its course to the cerebral cortex, the conscious sensation produced by such stimulation is referred back to the location of the receptor. This principle is called the *law of projection.* For example, when the cortical receiving area for impulses from the left hand is stimulated, the patient reports sensations in his left hand, not in his head. Should you hit yourself on the left hand with a hammer, causing considerable pain, the cerebral cortex will interpret the sensation of pain but project the feeling back to the hand. It is not your brain that experiences the pain—it is your hand.

A particularly dramatic instance of projection occurs in amputees, who may complain of pain and proprioceptive sensations in the absent limb. For example, a patient has experienced a midthigh amputation of one leg, and is now conscious after the operation. The nurse has placed bed covers over the patient to keep him warm and to restrict his movements. The pressure of the covers on the stump of his leg stimulates the nerve tract that once had come from the receptor in his foot; consequently, the pressure is interpreted as coming from the foot. The patient complains of discomfort caused by the "too tight" blanket on his foot—even though the foot is no longer there. This phenomenon is called "phantom limb" pain (Fig. 12–2). A reorientation of the cerebrum to new nerve stimuli is essential in all amputees following surgery. The patient must, in a sense, retrain his brain to realize that impulses coming from the amputated area are now originating along the tract of the remaining sensory mechanism.

CUTANEOUS SENSES

Touch, pressure, heat, cold, and pain are the cutaneous senses.

Touch (Tactile) and Pressure Sensations

At least four kinds of receptors—free nerve endings, Meissner's tactile corpuscles, pacinian corpuscles, and nerve fibers at the base of each hair (Fig. 12–3)—give rise to touch sensations. The hair nerve ending is stimulated by a change of position of the hair; the Meissner's and pacinian corpuscles sense a quick touch or de-

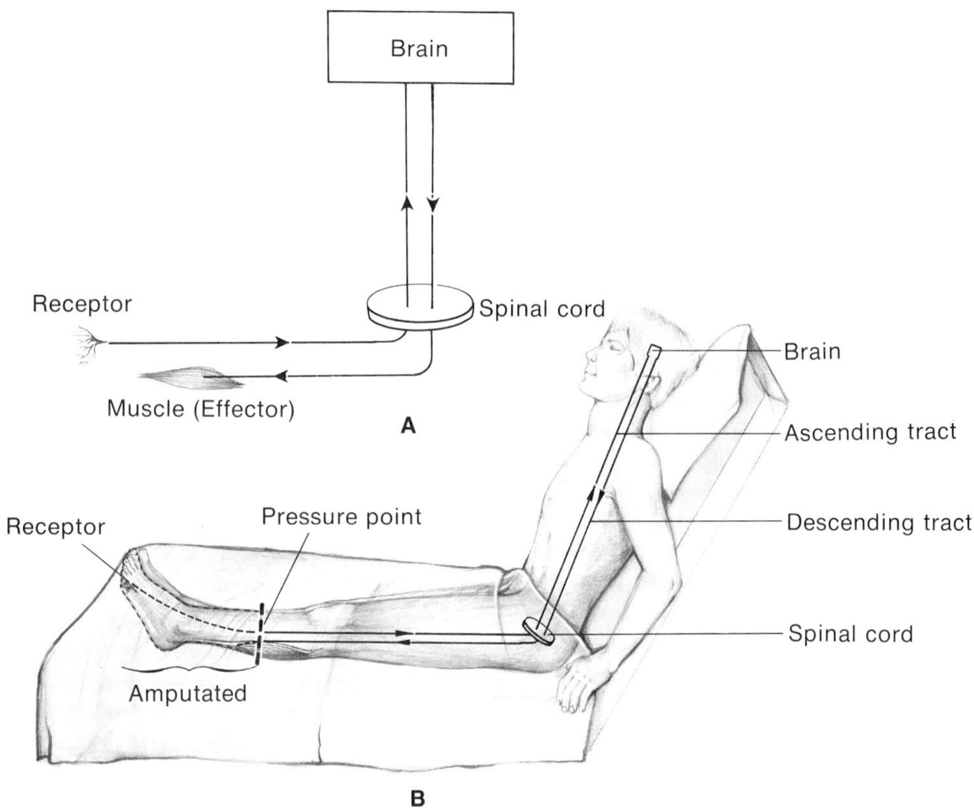

Figure 12–2 "Phantom limb" pain. *A*, The sensory mechanism, beginning with the receptor and ending at the muscle. *B*, Pressure exerted on the lower leg by the bed clothes, after surgical removal of the foot and the end of the lower leg, stimulates the nerve tract that once originated in the foot. The patient falsely interprets this impulse as coming from the foot that is no longer there—"phantom limb" pain.

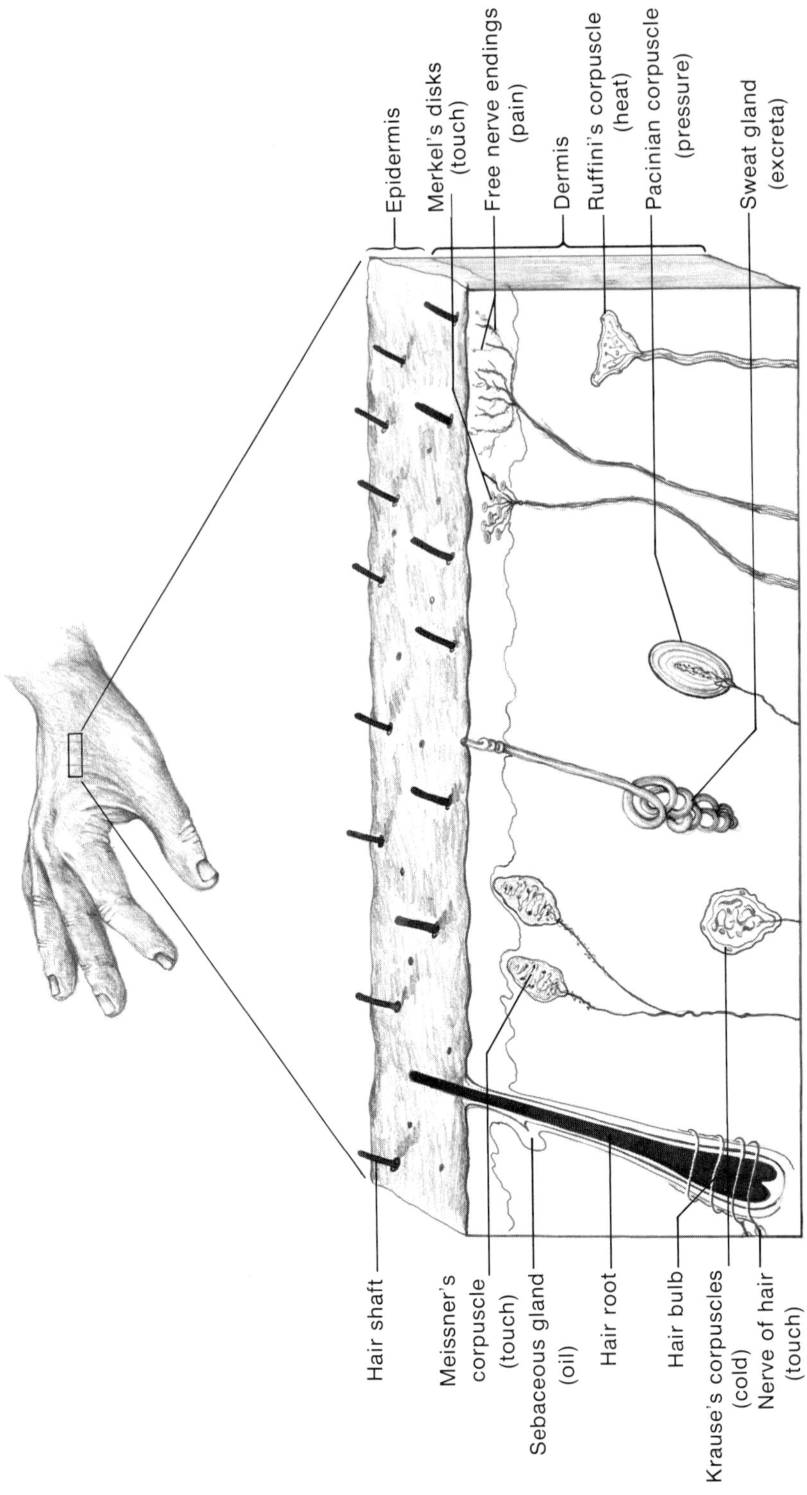

Figure 12–3 Cutaneous senses. Various types of exteroceptors and their sensations found in the skin.

formation of tissue, but not a sustained touch. Free nerve endings adapt slowly to continuous stimulation. For example, when we remove a particle from the eye, the free nerve endings in the cornea continue to "feel" it for a short time thereafter. Merkel's corpuscles or disks, located deep in the epidermal layer of the skin, are another example of free nerve endings concerned with touch sensation.

Pressure sensations differ from touch sensations in that they last longer. Light pressure can be discriminated from hard pressure by the depth of the receptor recognizing the stimulus. Free nerve endings in the superficial epidermis and Meissner's corpuscles are receptors of light pressure, whereas pacinian corpuscles, found deep in the dermis, recognize deep or firm pressure.

The brain's sensory area for touch is situated in the general sensory area of the parietal lobe of the cerebral cortex, just posterior to the central fissure (see Fig. 8–16). The adaptation to touch is very rapid and exceedingly important in acquiring information from the external environment. New techniques in learning still use the age-old principle of "take it in your hands and feel it—you'll never forget it."

Temperature Sense

Humans perceive temperatures ranging from 12° to 50° C. Some regions of the body, such as the fingertips and face, are equipped with specific structures for particular temperature ranges (Fig. 12–3). The bulboid Krause's corpuscles, which lie in the dermis of the skin, are receptors for cold (12° to 35° C). The receptors for heat (25° to 50° C), the Ruffini's end-organs or corpuscles, are also found in the dermis. In skin areas lacking specific receptors, temperature sensation is less acute and is subserved by free nerve endings. Hot and burning sensations result from the combined sensory inputs from "cold," "warm," and free nerve endings.

The sensory area for temperature, like that for touch, is in the general sensory area of the parietal lobe of the cerebral cortex, just posterior to the central fissure. Crude sensations of temperature are experienced in the thalamus. After-images can persist for a long time in the sensation of temperature. A cold object placed on the skin and quickly removed can still be "felt" for a few minutes. The skin of the face and hands is less sensitive to temperature changes than are the parts of the body usually covered by clothing, having adapted to frequent temperature changes over the years.

Somatic Pain

The receptors for pain are free nerve endings found in the epidermis (Fig. 12–3), muscles, joints, and tendons. Another stimulus for pain is excessive stimulation of any of the tactile senses. For example, an ice cube placed in the hand feels cold at first, but then begins to hurt. Overstimulation of the cold receptors resulted in pain.

Impulses for pain are transmitted to the thalamus by way of the lateral spinothalamic tract (Fig. 12–4, *A*). Awareness of pain takes place in the thalamus, but localization and recognition of the kind and intensity of pain occur in the parietal lobe of the cortex, just posterior to the central fissure (Fig. 12–4, *A*; see also Fig. 8–15).

Adaptation to pain does not exist. This is an extremely important defense—a warning signal. If we became accustomed to pain, damage to the body would follow our ignoring this signal. Everyone has about the same threshold of pain, but reactions to it vary widely among individuals, depending on such factors as ethnic background, childhood experiences, and emotional status.

The term *referred pain* is used when pain that seems to be in one part of the body, particularly in an outer part such as the skin, actually originates in a different body organ or structure. It has been found,

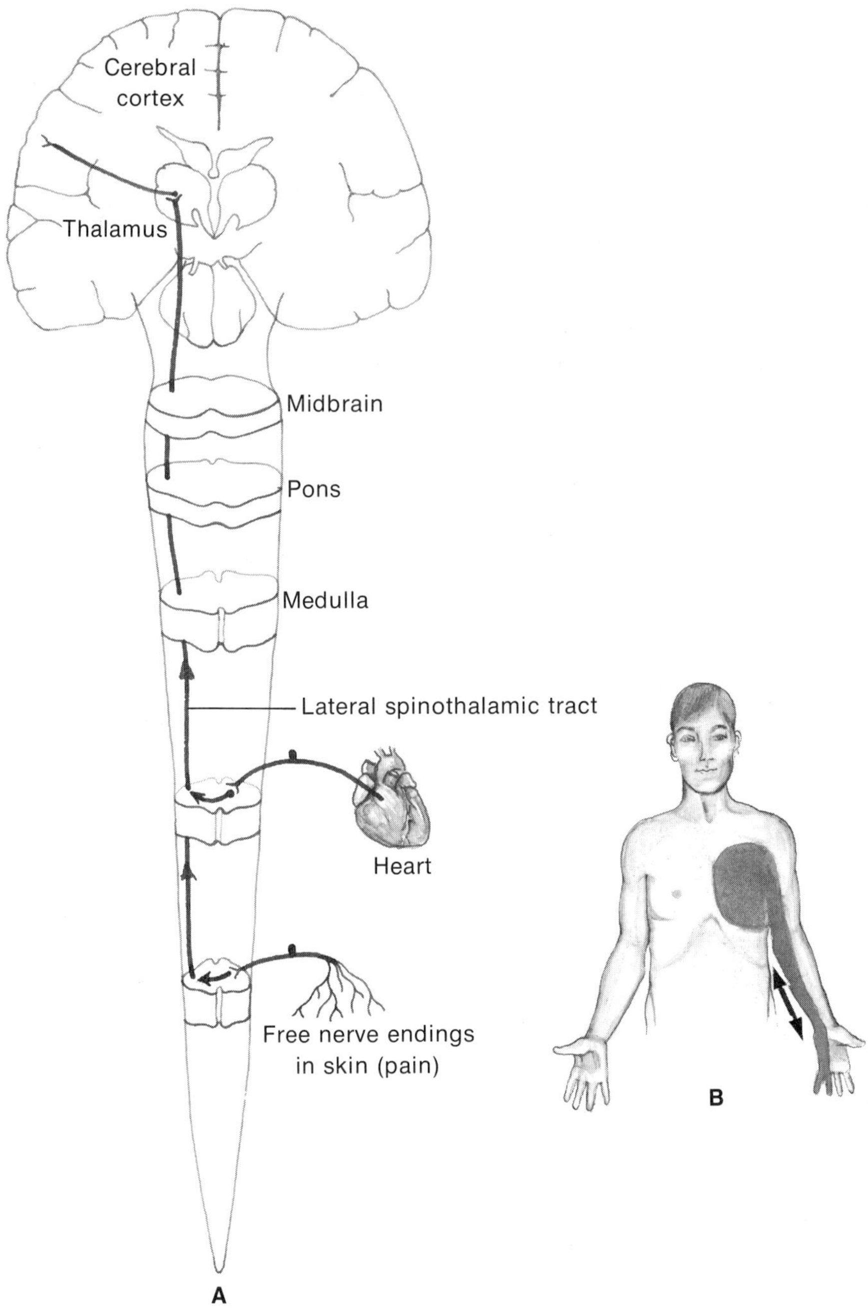

Figure 12–4 Referred pain. *A*, Nerve network that transmits pain sensations from the skin and visceral organs (heart). *B*, Pain generating from the heart is commonly interpreted as coming from the left arm.

for example, that liver and gallbladder disease often cause referred pain in the skin over the right shoulder. Spasm of the coronary arteries that supply the heart may cause pain in the left shoulder and the left arm (Fig. 12–4, *B*). One reason for this phenomenon is that certain neurons have the twofold duty of conducting impulses both from visceral (organ) pain receptors and from pain receptors in neighboring areas of the skin. The brain cannot differentiate between these two possible sources; however, since most pain sensations originate in the skin, the brain automatically assigns the pain to this more likely site of origin.

Pain sense differs from other senses in that continued stimulation does not result in adaptation, as has been mentioned. Sometimes the cause of pain cannot be remedied quickly, and it is necessary to use the following methods to help the patient:

1. Application of cold to dull the senses.
2. Compression of the area to prevent the nerve from carrying the impulse to the brain.
3. Analgesics—mild pain-relieving drugs (e.g., aspirin).
4. Narcotics—strong drugs that produce stupor and sleep (e.g., morphine).
5. Anesthetics (local or general)—chemical agents that produce lack of sensation in a specific area or total unconsciousness.

POSITION SENSE

Receptors located in muscles, tendons, and joints relay impulses that help us judge the position and changes in location of body parts in relation to one another (Fig. 12–5). They also inform the brain of the extent of muscle contraction and tendon tension. These widely spread end-organs—proprioceptors—are aided in their function by the semicircular canals of the inner ear (p. 249). This sense of position, also called the kinesthetic sense, enables us to know the positions of various body parts without actually seeing them. Information from the proprioceptors is needed for the coordination of muscles, and is important in everyday activities, such as walking and running, and in more complex skilled movements, such as playing a musical instrument. These muscle sense end-organs (Ruffini's end-organ and Golgi's tendon organ) also play an important part in maintaining muscle tone and posture.

The pathway for muscle sense impulses is provided by large myelinated

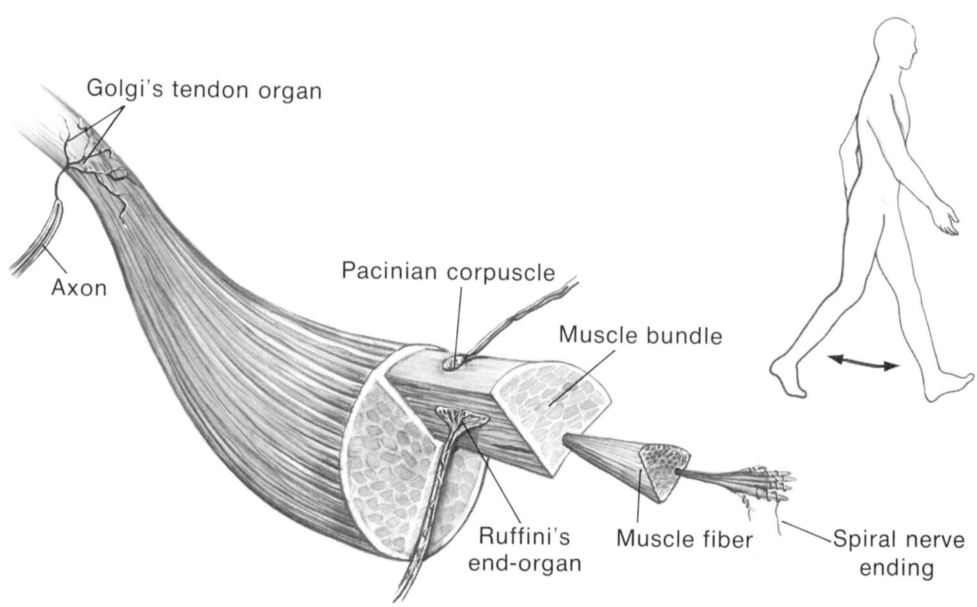

Figure 12–5 Proprioceptors in skeletal muscle.

fibers in the cranial and spinal nerves. Impulses dealing with conscious activities travel to the medulla by way of posterior columns, where they are then relayed to the thalamus and cerebral cortex. Impulses concerned with reflex activities travel to the cerebellum by way of the spinocerebellar tracts. The sensory area for conscious muscle sense lies just posterior to the central fissure in the sensory area of the parietal lobe of the cortex.

Syphilis and other diseases may involve the posterior spinal cord, causing its degeneration and a loss of position sense—a disorder called *tabes dorsalis* (p. 168). The lower part of the body is affected first, resulting in a loss of both position sense and muscular coordination. As the disease progresses, the patient may lose his ability to walk. Once the nerve tissue has been damaged, no cure is possible; therefore, early treatment is important to prevent permanent tissue destruction.

HUNGER AND THIRST

Hunger and thirst are organic sensations that follow the law of projection. The feeling of hunger is projected to the stomach; the feeling of thirst is projected primarily to the throat (pharynx).

Hunger

The feeling commonly called "hunger" occurs normally at a certain time before meals, and presumably is due to contractions of the empty stomach that stimulate receptors distributed throughout the organ's mucous membrane. If food is not taken, hunger increases in intensity for a time, and is likely to cause headaches, nausea, and overall weakness. The cliché, "never make a decision on an empty stomach," is based on the fact that a hungry individual can become quite excitable. More than one "big deal" has been made after the banquet—not before.

Appetite differs from hunger. Although it is basically a desire for food, it often has no relationship to a *need* for food. Appetite is also a pleasant sensation; conversely, hunger is disagreeable. A loss of appetite is known as *anorexia*, and may be due to a great variety of physical and mental disorders.

Thirst

Although thirst may reflect a generalized lack of water in the tissues, the sense of thirst seems to be localized largely in the mouth, tongue, and pharynx. It is a very unpleasant sensation, and is continuous until relieved or until death ensues. An excessive excretion of water, as in diabetes, may provoke excessive thirst, called *polydipsia*. The physical condition in which the tissues lack water is *dehydration*.

Visceral afferent impulses concerned with the organic sensations of hunger and thirst are conducted to the hypothalamus, where appetite is regulated. The location of the nerve receptors that transmit hunger and thirst impulses is still unknown. They are probably located in the muscles lining each respective organ.

TASTE AND SMELL

Taste and smell are considered to be chemical sensations because the taste buds and olfactory receptors respond to substances dissolved in fluids secreted by the oral and nasal epithelial tissues.

Taste

The sense of taste (gustation) involves receptors in the tongue and three different nerves that carry taste impulses to the brain. The taste receptors are called *taste buds*, and are located along the edges of small depressed fissures separating the pa-

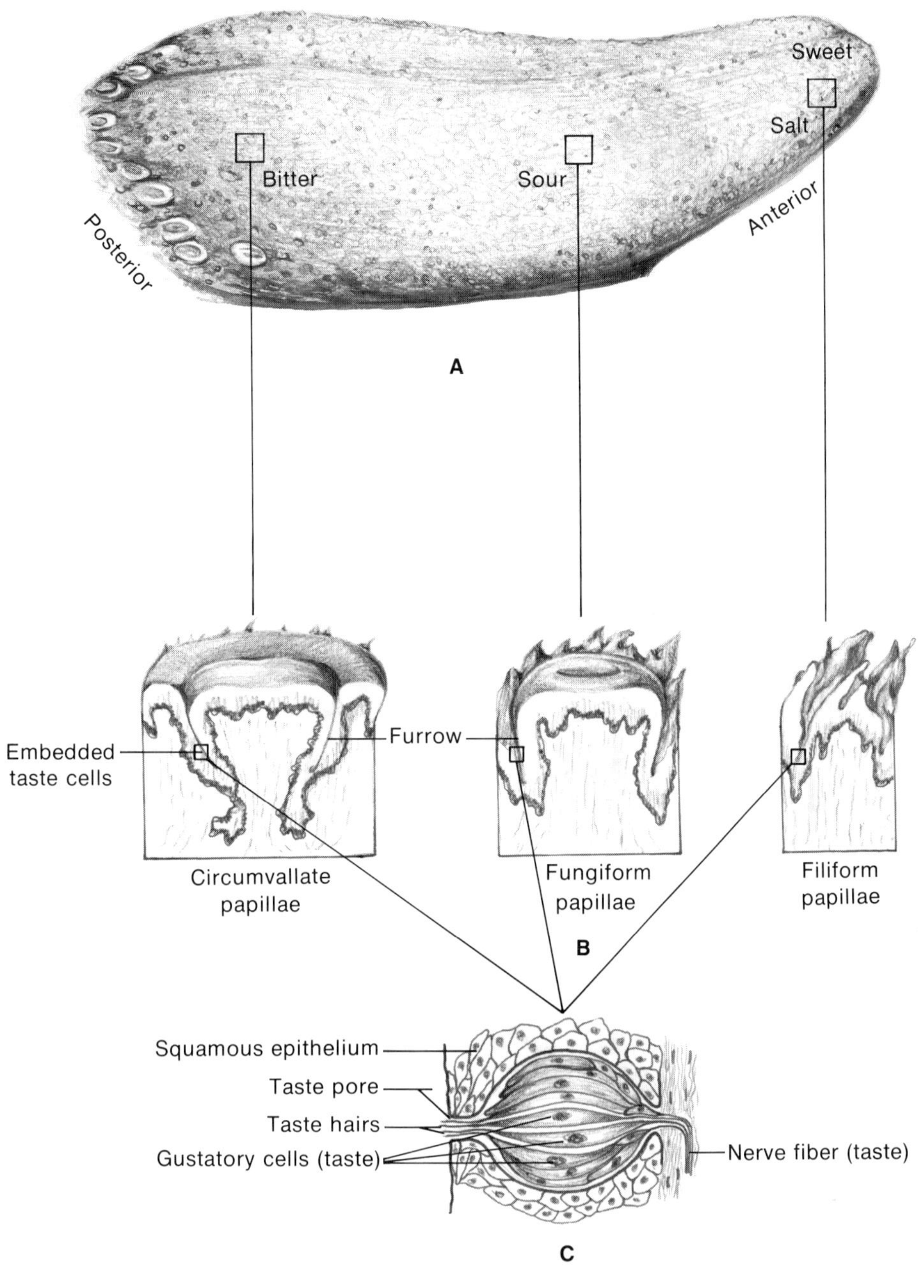

Figure 12–6 Taste. *A*, Regions of the tongue that are most sensitive to the various tastes. *B* and *C*, The tongue appears rough owing to the three types of papillae *(B)* that contain the taste (gustatory) cells or buds *(C)* embedded deep within each furrow.

pillae of the tongue (Fig. 12–6, *B*). These papillae vary in shape—pointed at the front of the tongue and fat at its base. The papillae give the tongue its characteristic rough appearance. All papillae contain taste buds.

There are essentially four kinds of taste (Fig. 12–6, *A*):

1. Sweet—acutely experienced at the tip of the tongue.

2. Sour—detected at the sides of the tongue.

3. Salt—felt at the tip of the tongue and along its sides.

4. Bitter—detected at the back of the tongue.

To be tasted, substances must be in solution. At the top of each bud an opening called the taste pore (Fig. 12–6, *C*) admits solutions (materials dissolved in saliva) which stimulate the receptors.

The sensory area for taste is located in the lateral fissure of the cerebral cortex near the postcentral gyrus. The sensory nerve fibers from the anterior two-thirds of the tongue travel to the brain via the facial (seventh cranial) nerve and from the posterior third of the tongue by the glossopharyngeal (ninth cranial) nerve. The vagus (tenth cranial) nerve conducts taste sensations from the pharyngeal region and from the base of the tongue.

Although the taste buds are structurally alike, they are physiologically different in sensitivity. Bitterness can be detected in a dilution of 1:2,000,000; sweetness, 1:250; sourness, 1:135,000; and salt, 1:500. Sourness is a response to acids, and is said to be the most specific of the four taste sensations. Taste sensations other than the basic four are a result of a blending of the fundamental sensations with one another and with the sense of smell.

Smell

The olfactory receptors are located in the mucous membrane lining the upper part of the nasal cavity. Their afferent pathway is the olfactory (first cranial) nerve, which leads to the olfactory center in the temporal lobe. A layer of mucus secreted by the surrounding supporting cells (Fig. 12–7) covers the epithelium and dissolves and absorbs gaseous particles in inspired air. Only volatile substances in solution can stimulate the receptors of smell. The amount of air reaching the olfactory region is greatly increased by sniffing, a semireflex response that usually occurs when a new odor attracts attention.

Naked endings of the trigeminal (fifth cranial) nerve are found in the olfactory mucous membrane. These are stimulated by irritating substances (chlorine, peppermint, menthol, etc.), producing sneezing, choking, and short breaths. This response can be advantageous at times, such as in using ammonia to arouse an individual from a fainting spell.

Man's sense of taste depends a great deal on his sense of smell. If you block your nose or have a severe cold, your food is tasteless. This phenomenon is due to the gustatory nerve, which blends sensations of smell and taste. It also creates the strange illusion that you feel you are able to taste something you actually are smelling. For example, if you enter a room that has been perfumed, you can taste sweetness from breathing in the odor. The smell of cooked cabbage can almost be tasted. A principal characteristic of smell is that only a minute quantity of the stimulating agent in the air is required to evoke the sensation.

VISION

Vision is usually considered to be the most important of all the senses. Blind people can and do learn to depend on their other senses to a remarkable degree, but nothing approaches complete compensation for loss of vision. Fundamentally, the sense of vision is similar to all the other senses in that there are photoreceptors (sensitive to light), an afferent pathway (optic nerve) which conveys the impulses to the cerebral cortex, and an area (occipital lobe) of the cerebrum necessary for interpretation of these impulses. But vision is far more complex than such senses as hunger and thirst, and therefore it is necessary to consider the physics, anatomy, physiology, and biochemistry of sight.

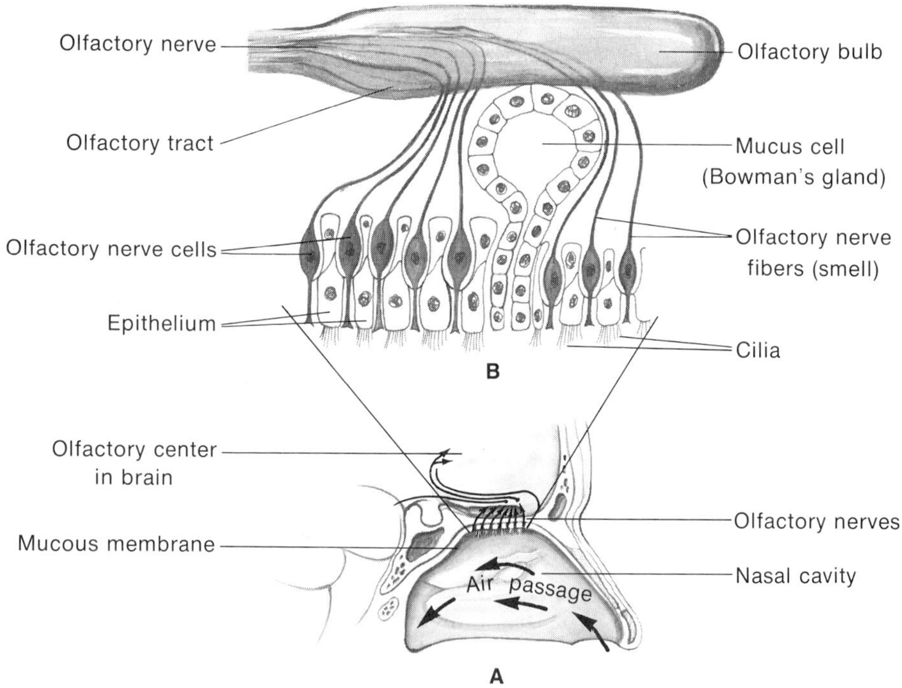

Figure 12–7 Smell. *A*, Olfactory receptive area in the roof of the nasal cavity, showing the gross relationship of the olfactory nerves to the mucous membrane. *B*, a microscopic sketch of a supporting mucus cell surrounded by olfactory cells that collectively form the olfactory nerve.

Accessory Organs of the Eye

These include the eyebrows, eyelids, conjunctiva, lacrimal apparatus, and muscles of the eyeball.

The *eyebrows* are thickened ridges of skin, covered with short hairs. They are positioned on the upper and lower borders of the orbit of the eye, and protect the eye from strong light, foreign materials, perspiration, and so forth.

The *eyelids* are two movable folds of skin that are placed in front of the eyeball and cover it. They are lined internally with a mucous membrane, the conjunctiva, which begins at the lower edge of the upper lid, progresses down over the eyeball, and ends at the upper edge of the lower lid (Fig. 12–10). A small muscle is attached to the upper lid and elevates it. The orbicularis oculi muscle (Fig. 7–16) is arranged as a sphincter around both lids and opens and closes them. Eyelids protect the eye and spread its lubricating secretions (tears) over the surface of the eyeball.

The *conjunctiva,* which lines the eyelids and is reflected onto the surface of the eyeball, is kept moist by tears flowing across the front of the eye. Inflammation of this membrane is called *conjunctivitis* and may be acute or chronic. A variety of irritants and pathogens can produce the condition. "Pink eye" is a highly contagious acute conjunctivitis caused by a bacterium. *Trachoma,* or granular conjunctivitis, is caused by a virus and is contagious in its early stages. This disease is characterized by granulation of the eyelids, which may result in an irritation of the cornea leading to blindness.

An eye infection of the newborn, *ophthalmia neonatorum,* is caused by venereal bacteria. Prevention of blindness in this instance, and in other possible congenital disorders of the eye due to bacterial invasion from the vaginal tract, is carried out by instilling a mild antiseptic (silver nitrate) into the infant's eyes at the time of delivery.

The eyelids bear rows of short, thick hairs, the *eyelashes*. The follicles of these hairs receive a lubricating fluid from sebaceous glands that open into them. The glandular secretions lubricate the edges of the eyelid and prevent adhesion of the two lids. If the glands become infected, a *stye* results. Distention of a sebaceous gland is called a *chalazion*.

The *lacrimal apparatus* (Fig. 12–8) consists of the lacrimal gland, the lacrimal sac, and the nasolacrimal duct. The lacrimal gland produces tears and is located above the eye toward one side; that is, superior and lateral to the eyeball. Tiny ducts carry the tears to the front surface of the eyeball, where they constantly wash the conjunctiva that separates the front part of the eyeball from the larger back portion (see Fig. 12–10). Six to 12 minute ducts lead from the gland to the surface of the conjunctiva of the upper lid. Tears pass from the gland, through the ducts, over the surface of the eyeball, and into two tiny lacrimal ducts at the inner angle of the eye. The lacrimal sac is the expanded upper end of the nasolacrimal duct, a small canal that opens into the nose.

Tears are secretions made up of salts, water, and mucin. These secretions normally are carried away by the nasolacrimal duct as quickly as they are formed, but under certain circumstances (conjunctivitis, intense emotion, hay fever, etc.), the secretion exceeds the drainage capacity, and the fluid overflows onto the cheeks. Because an infant's lacrimal glands do not develop sufficiently until he is about four months old, a baby's eyes must be protected from bright light and dust. Occasionally, a newborn's lacrimal ducts become blocked and cause a great deal of "caking" of the eyelids in the early morning. If soaking the eyelids in warm water does not remedy the situation, the infant should be seen by a physician, who will use a blunt metal probe to unblock the ducts.

The eye is moved within the orbit by six ocular muscles (Table 12–1) innervated by the oculomotor (III), trochlear (IV), and abducent (VI) cranial nerves. These muscles and the directions in which they move the eyeball are shown in Figure 12–9. The six muscles connected with each eye are ribbon-like, and extend forward from the apex of the orbit behind the eyeball. One end of each muscle is attached to a bone of the skull; the other end is attached to the white (sclera) of the eye. These muscles pull on the eyeball in a coordinated fashion that causes the two eyes to move together in order to center on one visual field.

Anatomy of the Eyeball

The eyeball has three coats or layers of tissue (Fig. 12–10):

1. An outer layer of tough, fibrous tissue, the opaque, white *sclera* (meaning "hard"), gives shape to the eyeball and protects its delicate inner layers. The most

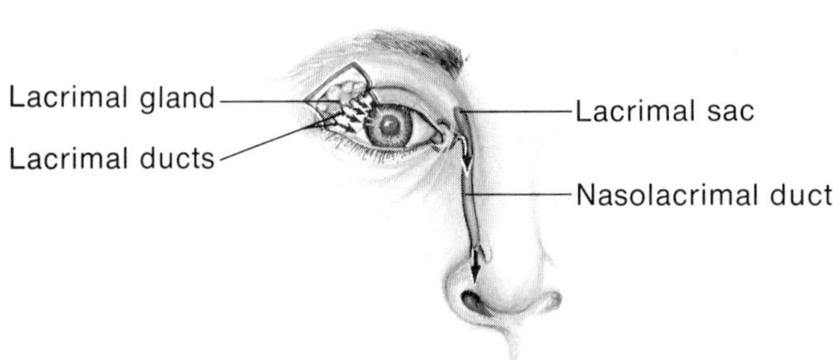

Figure 12–8 Lacrimal apparatus in relation to the eye and nose. Note that the direction of flow is from the outside in, to flush the eye continually and drain into the nose.

Table 12–1 MUSCLES OF THE EYE

NAME	ORIGIN	INSERTION	FUNCTION
Superior rectus	Apex of orbital cavity	Upper side of sclera	Rolls eyeball upward and medially
Inferior rectus	" "	Under side of sclera	Rolls eyeball downward and medially
Lateral rectus	" "	Lateral side of sclera	Rolls eyeball outward
Medial rectus	" "	Medial side of sclera	Rolls eyeball inward
Superior oblique	Lesser wing of sphenoid above orbital cavity	Sclera	Rolls eyeball on its axis, downward and laterally
Inferior oblique	Orbital plate of maxilla	Sclera	Rolls eyeball on its axis, upward and laterally

anterior or forward continuation of the sclera is the transparent cornea. Both the sclera and cornea are fibrous connective tissue and lack blood vessels and nerves—a fact essential to know when transplants are undertaken.

The transparent and colorless *cornea* is referred to as the "window" of the eye. It bulges forward slightly and is the eye's most important refracting structure. Injuries caused by foreign objects or by infection may result in scar formation and a slight opacity through which light rays cannot pass. Because the cornea is free of blood and nerves, it is one of the parts of the eyeball that can be transplanted (keratoplasty). Eye banks arrange for corneal transplants from donors to recipients who are blind because of a corneal defect. The entire transplant procedure should be performed within 72 hours after the death of the donor. The operation has been remarkably successful.

2. The middle layer of the eye comprises primarily the *choroid*, which is highly vascular and resembles skin. Its principal function is to supply blood to the other layers of the eyeball, especially the innermost retina. The forward extension of the choroid is a thin muscular layer, the *iris* (Fig. 12–10). The iris is the colored part of the eyeball; an opening at its center allows light to pass. This circular opening is the *pupil*.

The function of the iris is to regulate the amount of light entering the eye, and thus to assist in obtaining clear images. This regulation is accomplished by its smooth muscles, which contract or dilate, causing the pupil to become larger or smaller. The larger the pupil, the greater the amount of light that reaches the inner eye. This mechanism is an involuntary reflex that allows the eye to adjust to changes in light intensity, a property called *adaptation* (light reflex). When you are in a dimly lit room, your pupils are dilated to permit as much light as possible to reach the retina. When a light is turned on in the room, the pupils automatically contract to protect the retina against damage from too much light.

Homatropine (a drug used in eye examinations), epinephrine, atropine, and cocaine all make the pupil larger. Such drugs are called *mydriatics*. Pilocarpine and morphine cause the pupil to contract; such agents are *miotics*.

The *ciliary muscles*, which support and modify the shape of the lens, are another anterior modification of the choroid layer. The *lens*, a flexible, transparent, colorless body, consists of a more or less semisolid substance enveloped by a thin capsule. It is biconvex (thicker in the middle), and this property serves to converge light rays and to focus them on the retina. During youth the lens is elastic and plays

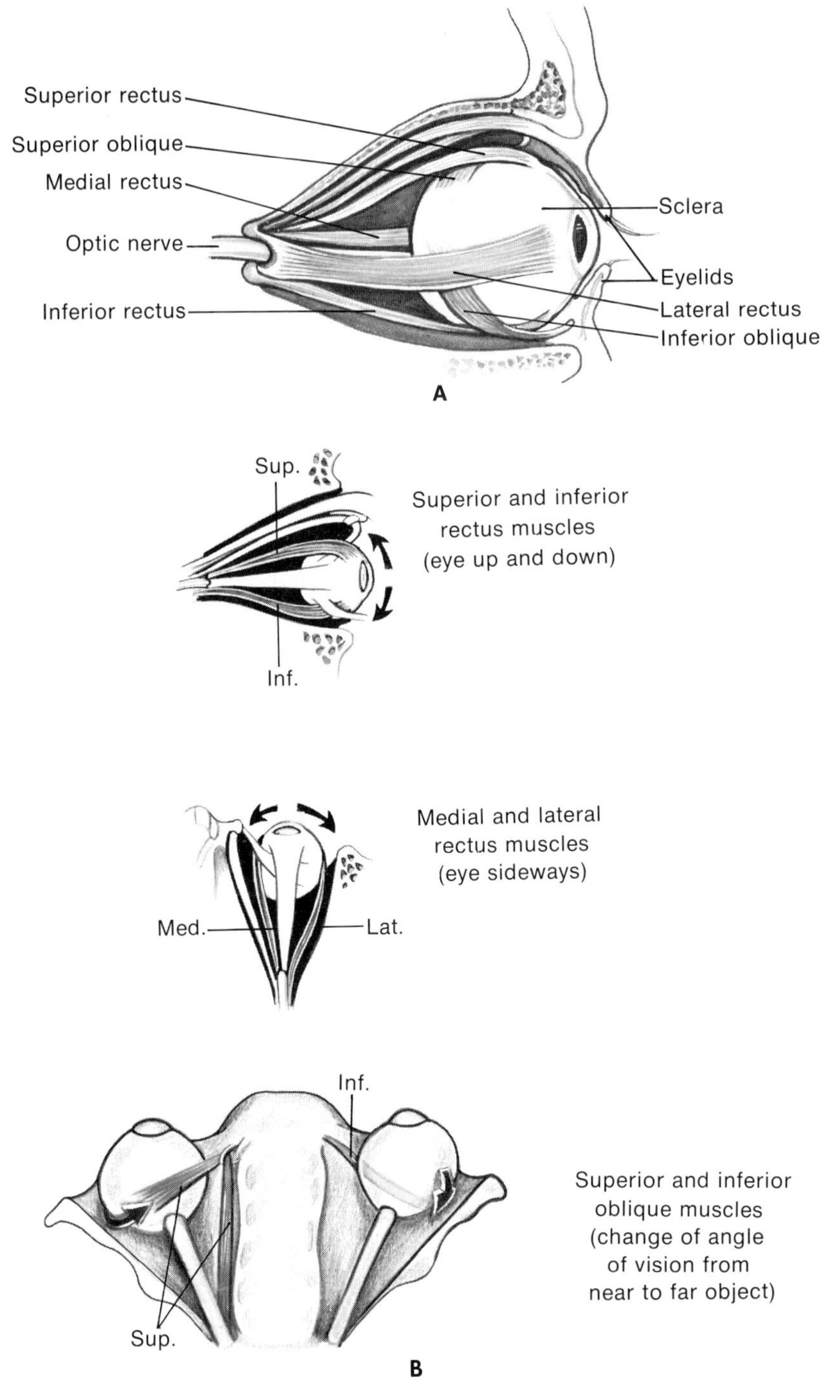

Figure 12–9 Muscles of the right eye. *A*, Extrinsic eye muscles, lateral aspect. *B*, Eye movements.

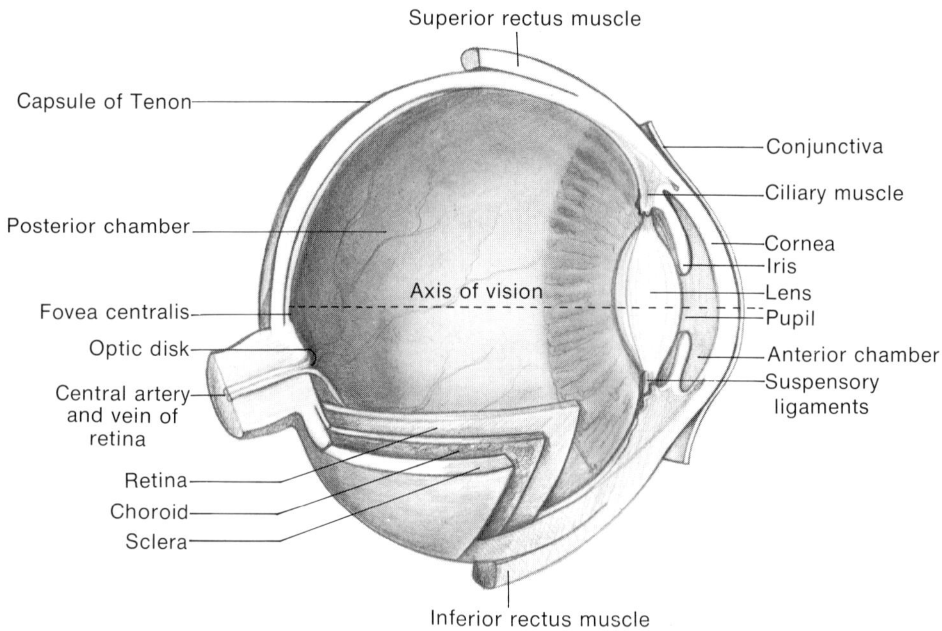

Figure 12–10 The structure of the eye, transverse section.

an important role in the eye's ability to focus on near objects. With aging, the lens loses its elasticity, and therefore its ability to adjust by thickening, resulting in a condition known as *presbyopia.*

3. The third and innermost layer of the eyeball is the all-important *retina.* The word retina means "a net," and that accurately describes the network of highly specialized nerve cells (rods and cones) and their processes. There are approximately 110,000,000 such photoreceptors in each eye. The image is focused on this layer, the receptors are activated, and the impulses are transmitted via the optic (second cranial) nerve to the occipital lobe of the cortex. More will be said about the physiology of sight later in this chapter.

Figure 12–10 shows two spaces, one between the cornea and lens and another between the lens and retina. The more anterior space is the *anterior chamber,* which contains a watery fluid, the *aqueous humor.* Aqueous humor fills much of the eyeball in front of the lens, and is continually produced by the ciliary body and absorbed by the choroid blood capillaries through the canals of Schlemm. This is a matter of physiologic importance in that blockage of these canals can result in a build-up of fluid in the anterior chamber, producing an increased pressure on both the cornea and lens (Fig. 12–11). Continued high pressure on the lens can cause premature hardening and opacity of the lens, a disorder called *glaucoma.* It is said that this is the most painful of all diseases, but it can be detected in its early stages by measuring the pressure on the anterior eyeball with a drumlike instrument—a procedure called tonometry. Iridectomy, the surgical removal of a portion of the iris, or drugs affecting the eye blood vessels (vasodilators) have been successful in relieving glaucoma.

The eye's *posterior chamber* contains a proteinaceous fluid similar to that of the anterior chamber. However, the aqueous humor is watery, whereas the *vitreous humor* of the posterior chamber is jelly-

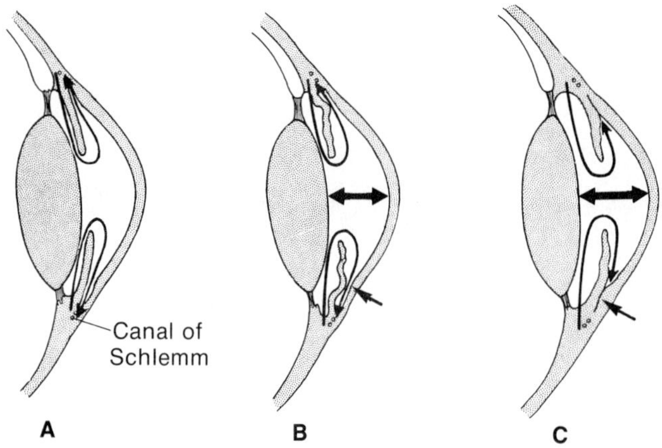

Figure 12–11 Glaucoma. *A,* Normal aqueous humor drainage through the canal of Schlemm. *B,* Open-angle glaucoma, showing a chronic condition treated by drugs. *C,* Closed-angle glaucoma requiring surgery (iridectomy) for correction.

like and fills the entire space behind the lens. This substance is necessary to keep the eyeball in its spherical shape.

Visual Pathways

The cell bodies that give rise to the primary fibers of the visual pathways reside in the retina. Processes from these cell bodies sweep across the internal surface of the retina to be gathered at a depressed point on the retina, the *optic disk* (Fig. 12–10). It can be seen that fibers emerge from the eyeball at the optic nerve; neurons that innervate the medial half of each retina cross at the *optic chiasma* (Fig. 12–12), whereas those that supply the lateral half do not cross but continue back on the same side.

Having two eyes rather than one makes possible true binocular or stereo-

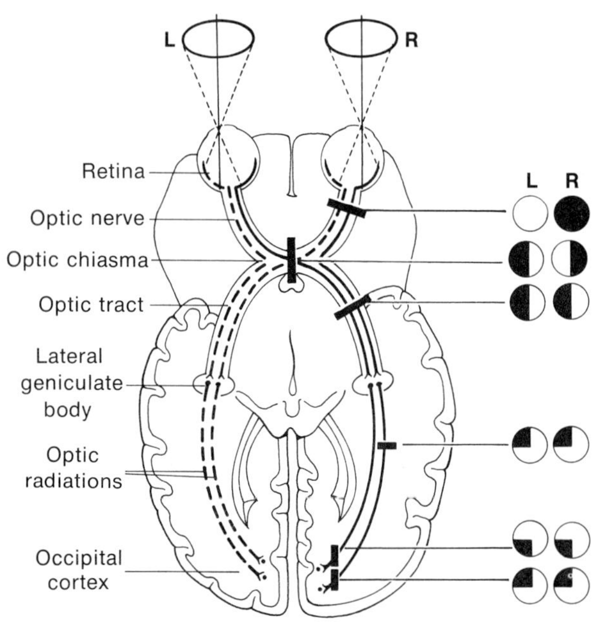

Figure 12–12 The visual pathway. On the right are diagrams of the visual fields, with areas of blindness darkened to show the effects of injuries in various locations.

scopic vision. Two optical images are received from slightly different angles, achieving the impression of distance and depth, thereby adding a third dimension to the visual field. In binocular vision it is necessary to turn the eyes inward so that two images of an object will lie on what are called corresponding points of the two retinas. Excitation of two corresponding points causes only one sensation, a perfect superimposition, in the occipital lobe of the cortex.

Convergence of the eyes is brought about by activation of the extrinsic eye muscles. If one or both eyes are deviated from their proper direction, convergence will be defective, and a form of *strabismus* will result. A strabismus in which one sees two views or pictures, resulting in *diplopia* or double vision, is called *exotropia. Esotropia* (crossed eyes), another form of strabismus, may be due to a muscle defect or to a weakness in coordination, and results in an inability of the two eyes to work together. The affected eyeball is pulled inward (medially). Esotropia is fairly common early in life, and "lazy eye" clinics in elementary schools have played an extremely important role in detecting the condition and treating it immediately. In some patients, glasses and exercise may correct the defect; in others, surgery may be required. If correction is not accomplished early, the affected eye may become blind through disuse.

It is far beyond the scope of this section to describe the many alterations in visual fields resulting from disease states; suffice it to say that the exact disturbance depends almost exclusively on the site of interruption of the conducting pathway (Fig. 12–12) and on the extent of that interruption. Characteristic types of defects are associated with lesions of these pathways.

Physiology of Vision

Light rays, while passing through a series of transparent, colorless eye parts (cornea, aqueous humor, lens, etc.), undergo a process of bending known as *refraction*. Refraction of light rays make it possible for light from a large area to be focused (pinpointed) on a very small surface, the retina, where the photoreceptors are located (Fig. 12–12; Fig. 12–13, *C*). Light rays bend as they pass from one medium into another of a different density. Parallel light rays striking a *convex* lens (Fig. 12–13, *A*) are refracted *toward* a focal point behind the lens. This focal point is on a line or axis of light passing through the center of the curvature of the lens. Conversely, a *concave* lens (Fig. 12–13, *B*) causes light rays to diverge or spread *away* from the axis of light.

The greater the curvature of a lens, the greater its refractive power. When the ciliary muscle is relaxed, parallel light rays striking the optically normal (*emmetropic*) eye are brought into focus on the retina. The problem of bringing diverging rays from objects closer than 20 feet to a focus on the retina can be solved by increasing the curvature or refractive power of the lens, a process called *accommodation*. At rest, the lens is held under tension by the suspensory ligaments (Fig. 12–10). When the gaze is directed at a distant object, the ciliary muscle contracts, producing a narrow, thin lens. This decrease in curvature facilitates focus on a distant object. Relaxation of the ciliary muscles produces a greater curvature in the lens for refraction of light rays from a near object (Fig. 12–13, *C*).

In some individuals, the eyeball may be shorter than normal or the refractive index of the lens too weak, bringing the light rays to a focus *behind* the retina (Fig. 12–13, *E*). This abnormality of accommodation is called *hyperopia*, or farsightedness. Sustained accommodation may compensate for this defect, but the prolonged muscular effort is tiring, and may cause headaches and blurred vision. A biconvex lens will augment the refractive power of the eye in these patients by shortening the focal distance. In *myopia*,

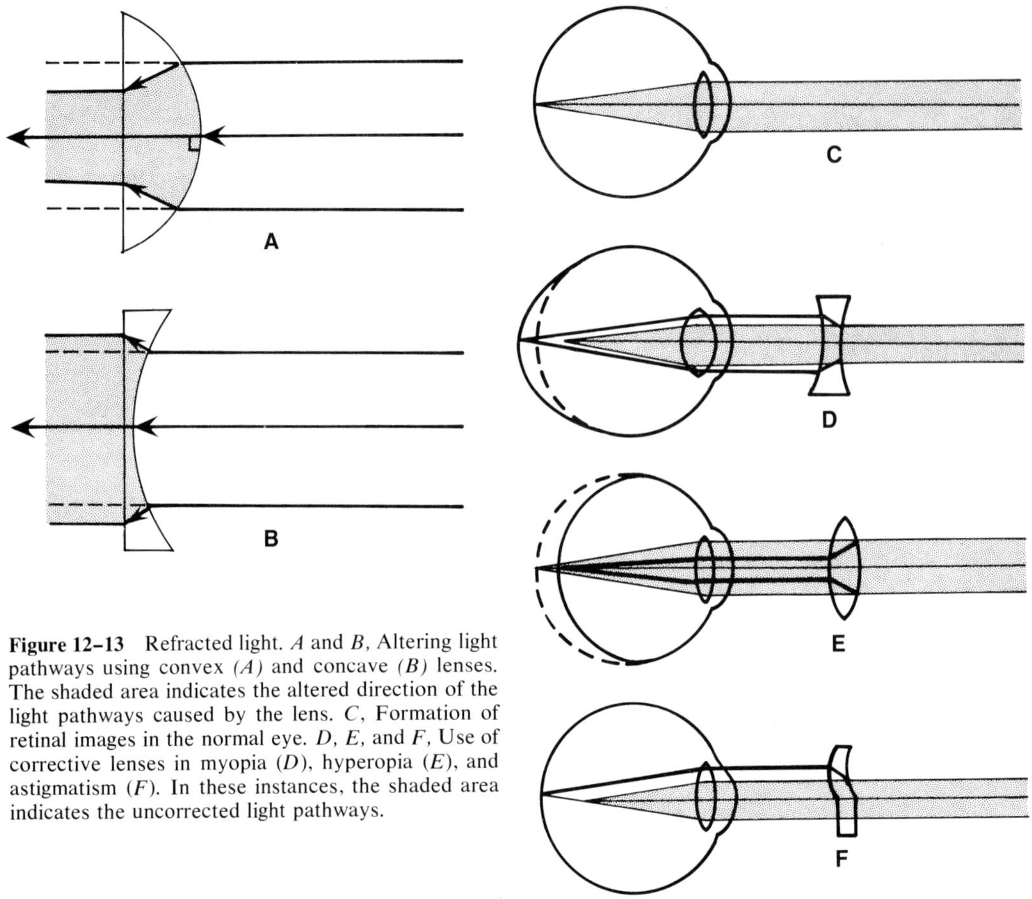

Figure 12–13 Refracted light. *A* and *B*, Altering light pathways using convex *(A)* and concave *(B)* lenses. The shaded area indicates the altered direction of the light pathways caused by the lens. *C*, Formation of retinal images in the normal eye. *D*, *E*, and *F*, Use of corrective lenses in myopia (*D*), hyperopia (*E*), and astigmatism (*F*). In these instances, the shaded area indicates the uncorrected light pathways.

or nearsightedness, the eyeball is too long or the refractive power of the lens is too strong. This defect can be corrected by using glasses with biconcave lenses so that the light rays are made to diverge before they strike the eye (Fig. 12–13, *D*).

Astigmatism is a common condition in which the curvature of the refractive surfaces is not uniform (Fig. 12–13, *F*). Parallel light rays are refracted in such a way that they fail to focus sharply and part of the retinal image is blurred. Astigmatism may be caused by trauma or by a metabolic discrepancy, and can be corrected if cylindric lenses are placed so that refraction is equalized.

Presbyopia, as mentioned before (p. 243), is a form of hyperopia, especially common among the elderly, in which distant objects are seen distinctly, but near objects are indistinct. During the aging process a loss of elasticity of the lens and a developing lack of tone in the ciliary muscle necessitate increasing the refractive power of the eye by using convex lenses for near vision. Elderly people often wear bifocals or trifocals for close, near, and distant vision.

The most common form of blindness in the elderly is cataract formation. A *cataract* involves the lens or its capsule, resulting in a loss of transparency. At times, areas of opacity may be seen through a pupil that greatly enlarges to compensate for the decreased quantity of light entering the eyeball. In other patients the loss of vision is gradual, and frequent changes of glasses may sustain adequate vision for some time. Surgical removal of the lens allows the cornea to carry on its chief refracting responsibilities without interference from a nonfunctional lens. Glasses are then a supplement to sight.

Perception of Light and Color

Light Vision

Light is a form of electromagnetic radiation made visible to us by a highly specialized receptor organ, the eye. The visible spectrum encompasses vibrations from 400 to 800 micrometers; these are the light and color waves. Vibrations shorter than this (ultraviolet, x-ray, and cosmic) and longer than this (infrared, radio, and electrical) are not ordinarily detected by the human eye.

The nervous structure of the retina is made up of *rods* and *cones* (Fig. 12–14), which are the photoreceptors for the sense of sight. There are about 7 million cones and over 100 million rods in each human retina.

The rods are very sensitive to movements of objects in the field of vision. They contain a unique photoreceptor chemical, a pigment called *rhodopsin* or visual purple, which is made up of a protein (opsin) and vitamin A. The rods, plus the rhodopsin, are used in twilight vision *(scotopia)*, when we see outlines of objects but no color or detail. Vitamin A is essential for the synthesis of rhodopsin in the rods; a deficiency may result in poor vision (night blindness) in dim light.

Far less is known about the physiologic or biochemical mechanisms of action of the cones, which are the receptors responsible for detailed vision and color. Near the center of the retina, as one looks into the eyeball through an ophthalmoscope, a small, oval, yellowish spot, the *macula lutea*, may be seen. The center of the macula lutea is depressed, and this region, called the *fovea centralis* (Fig. 12–10), is the area of highest visual acuity in which only cones are present. Cones are used in daylight vision (photopia), when we see the details and colors of objects. The cones are less sensitive to light than are the rods, and require 100 times more light intensity to be stimulated.

Visual Acuity

Clinically, visual acuity is usually determined by use of Snellen letter charts

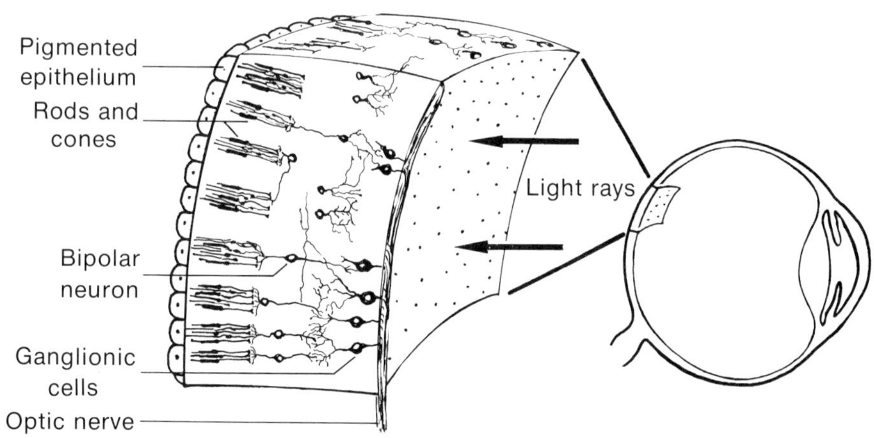

Figure 12–14 Microscopic section of the retina, showing detail of the three layers of neurons: the rods and cones on the inner surface, the bipolar connecting cells, and the ganglionic cells that form the outer optic nerve fibers.

viewed at a distance of 20 feet. The individual being tested reads aloud the smallest line he can distinguish, and the results of the test are expressed as a fraction. The numerator of this fraction is 20, the distance at which the subject reads the chart; the denominator indicates the greatest distance at which the smallest line can be read. Normal visual acuity is 20/20; a subject with 20/15 has better than normal vision, and one with 20/100 has subnormal vision. Ratings of 20/40 or 20/80, for example, mean that the eye must be at 20 feet to see clearly letters that can be seen by the normal eye at 40 or 80 feet, representing a visual acuity of 50 per cent or 25 per cent, respectively.

Color Vision

Although there are numerous theories and hypotheses about color vision, the Young-Helmholtz theory is probably the most widely accepted. According to this theory, there are three fundamental color cones (receptors)—red, green, and blue (violet)—each containing a different photochemical substance. Simultaneous stimulation of all three produces a sensation of white; total lack of stimulation produces a sensation of black. Varied colors are due to stimulation of various combinations of the three types of cones.

Numerous tests can be used to detect *color blindness*. The Ishihara color plates, which utilize printed numbers made up of colored spots on a background of similar colored spots (green on orange, for example), are the test most commonly used. The numbers or figures are intentionally made up of colors that are likely to look the same as the background color to an individual who is color blind. Individuals with normal color vision are called *trichromatic;* in *dichromatic* vision only two functional color cones exist. A nonfunctional dark receptor will allow an individual to distinguish the light colors but not the dark; conversely, light colors cannot be distinguished if the light receptor is nonfunctional. In *monochromatic* vision, only one color cone is operable. Apparently, only black and white and shades of gray may then be seen.

Abnormal color vision is found in about 8 per cent of the male population and 1 per cent of the female population. Some occupations require recognition of colors, and a job candidate must pass a color test before securing employment.

The Ophthalmoscope

In order to see the retina through the pupil, reflected light is essential. The ophthalmoscope is an instrument that uses reflected light and a mirror with a hole in it. The examiner looks through the hole into the retina, and the mirror enables him to throw more light into the eye. A short distance medially from the fovea centralis is the optic disk. This region is called the "blind spot" since it bears no receptors; it is the point where the optic nerve leaves the eyeball. If intraocular pressure rises, the disk becomes cup-shaped (cupped disk); a rise in intercranial pressure distorts the disk, giving an appearance called "choked disk." The central artery, the main blood supply to the eye, enters the eyeball at the center of the disk and divides into many branches. Its appearance through the ophthalmoscope contributes information useful in the diagnosis of high blood pressure (hypertension) and diabetes mellitus.

HEARING AND EQUILIBRIUM

The ear contains receptors for two sensations, hearing and equilibrium. The external ear, middle ear, and cochlea of the inner ear are concerned with hearing (Fig. 12–15); the semicircular canals, utricle, and saccule of the inner ear are concerned with equilibrium (Fig. 12–16).

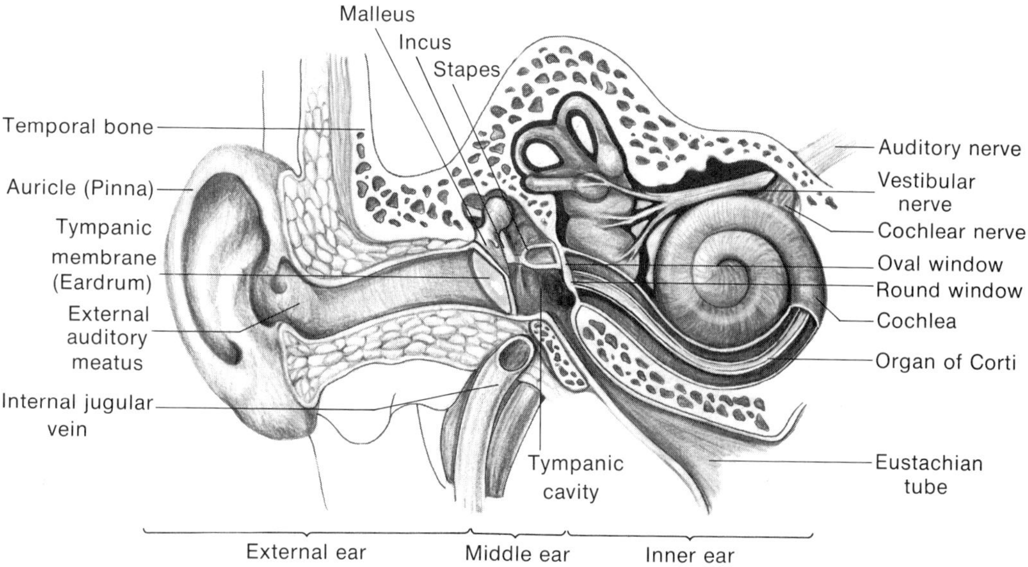

Figure 12–15 Frontal diagram of the outer ear, middle ear, and internal ear. A section of the cochlear duct has been cut away to show the position of the organ of Corti. For the structures involved in position sense, see Figure 12–16.

The External Ear

The external ear, which receives sound waves, consists of an expanded portion, the *pinna,* and the external auditory canal. The pinna (or *auricle*) projects from the side of the head as a framework of cartilage covered with skin and joined to the surrounding parts by ligaments and muscles (Fig. 12–15). It is irregular in shape and serves to collect sound waves and direct them toward the external audi-

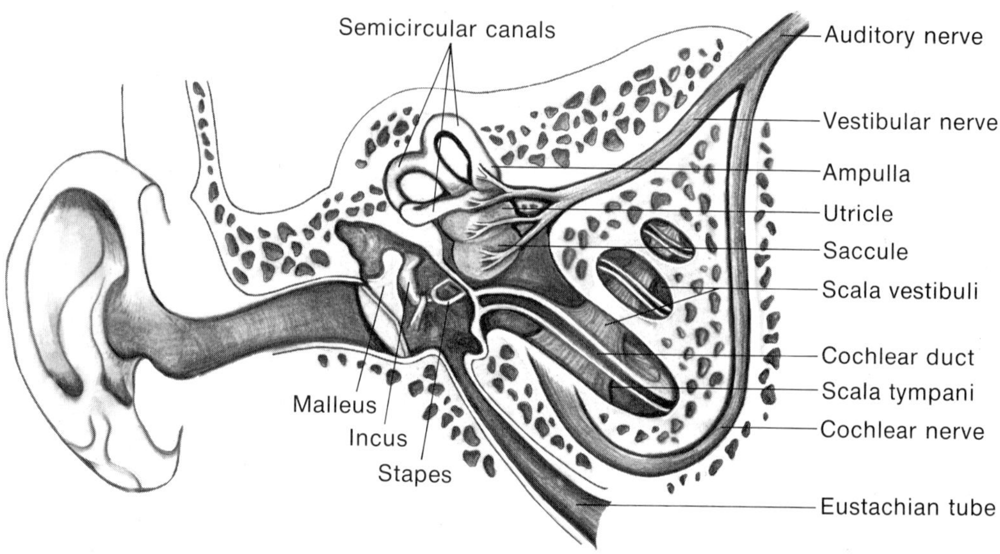

Figure 12–16 Cross-section through the cochlea, showing the relationship between the membranous and bony labyrinths, and the structures concerned with balance.

tory meatus. Although considered cosmetically unattractive, a projected or "flapping" ear is actually much more effective in trapping sound waves than is an ear set very close to the head.

The *external auditory meatus* (Fig. 12–15) is the opening of the *auditory canal,* a tubular passage about one and one-half inches long. It leads from the pinna to the inner eardrum, forming an S-shaped curve as it progresses inward. Lifting the pinna upward and backward tends to straighten the canal during an examination with an otoscope. The external portion of the auditory canal consists of cartilage; the inner portion is hollowed out of the temporal bone. The canal is lined with skin bearing hairs and with ceruminous glands that secrete the yellow, pasty cerumen commonly called earwax. Cerumen acts as a protective agent, but excessive amounts may block the ear canal and exert undue pressure on the eardrum, causing earaches. The hairs of the canal protect the ear from foreign substances.

The Middle Ear

The *tympanic membrane* (eardrum) separates the auditory canal from the tympanic cavity of the middle ear (Fig. 12–15). It is a thin layer of fibrous tissue covered externally with skin and internally with mucous membrane. The eardrum is attached so that it can vibrate freely when sound waves enter the auditory canal.

The *tympanic cavity* is a small, irregular bony cavity in the temporal bone (Fig. 12–15). It is separated from the auditory canal by the eardrum and from the inner ear by two small openings, the oval window and the round window. In the round window there is an opening into the mastoid process; through this, an infection of the middle ear may extend into the mastoid cells, causing mastoiditis.

The middle ear cavity is filled with air that reaches it from the upper part of the throat (nasopharynx) by way of the *eustachian tube* (Fig. 12–15). The walls of the cavity are lined with a mucous membrane continuous with that of the nasopharynx. Thus it is possible for infection to spread from the nose or throat to the middle ear (otitis media). Accumulations of pus in the middle ear can cause increased pressure on the eardrum; therefore, the membrane should be lanced (myringotomy) to alleviate this situation and to prevent further problems.

The eustachian tube acts to equalize the air pressure of the middle ear with atmospheric pressure. In inflammatory conditions the tube may become blocked, thus preventing this equalization; hearing is impaired, and permanent damage may result if the inflammation is not corrected. The pharyngeal opening of the tube is closed except during swallowing, yawning, or sneezing. It tends to remain closed when the external pressure is greater. Sudden changes in pressure exerted on the tympanic membrane by loud explosions or rapid falling, as in an airplane's descent, may cause the membrane to rupture. Such an occurrence often may be prevented by swallowing or yawning, which will open the eustachian tube, thereby balancing the pressure. Divers must be examined regularly to prevent ear damage that may result from prolonged intervals spent deep under the water; a diver with a blocked eustachian tube due to a cold may easily rupture his eardrums.

Stretching across the cavity of the middle ear from the tympanic membrane to the oval window are three tiny, movable bones (ossicles) named according to their shapes—the *malleus* (hammer); the *incus* (anvil); and the *stapes* (stirrup) (Fig. 12–15). The handle of the malleus is attached to the tympanic membrane; its head is attached to the incus. The long process of the incus is attached to the stapes, and the footpiece of the stapes fits into the oval window. These bones transmit sound waves from the outer ear to the inner ear. They are attached freely to one another by ligaments, and are set in motion by any movement of the tympanic membrane. The bones of the middle ear form a series of levers, the effect of which magnifies the force of sound vibrations to about 10 times that at the oval window.

Inflammation of the joints between the ossicles may prevent normal vibration and amplification of sound waves. *Otosclerosis* is a type of deafness found in adults, and occurs more frequently in women. Bone changes (fusion of the joints) prevent normal vibration of the third ossicle (stapes). An operation to release the stapes so that it will move again has had considerable success.

The Inner Ear

The inner ear contains receptors for hearing and for position sense. It consists of a bony labyrinth enclosing a membranous labyrinth. The bony labyrinth is a series of channels in the temporal bone filled with a fluid called *perilymph*. The inner membranous labyrinth more or less duplicates the shape of the bony channels (Fig. 12–16) and is filled with a fluid called *endolymph*. There is no communication between the spaces filled with perilymph and endolymph.

The *cochlea*, a portion of the labyrinth about one and one-half inches long, is a coiled tube resembling a snail shell. Internally, the cochlea is divided into three spiral membranous channels that run the full length of the organ. The organization and arrangement of these channels can be visualized by hypothetically unwinding the spirals of the cochlea, as shown in Figures 12–17 and 12–18. The uppermost channel, the *vestibular channel* (scala vestibuli), is attached to the oval window. At the tip (apex) of the cochlea, the vestibular channel communicates with the lowermost channel, the *tympanic channel* (scala tympani). The base end of this channel terminates at the round window, leading back to the middle ear. Both the tympanic and vestibular channels are filled with perilymph. The third and smallest channel, the *cochlear duct*, rests between the other two

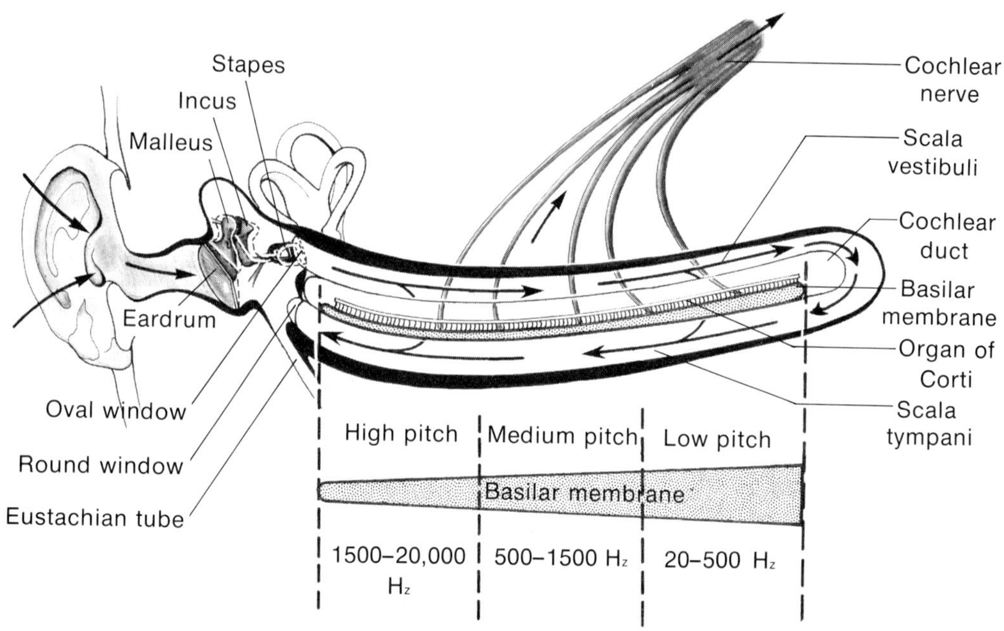

Figure 12–17 Diagram of an "unrolled" cochlea, showing the transmission of sound through the canals.

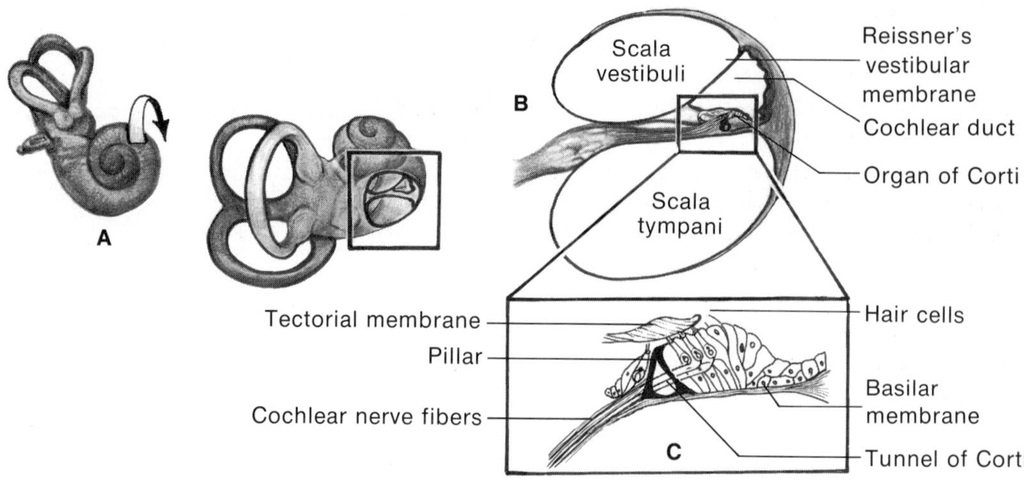

Figure 12–18 Diagrams of the inner ear. *A*, External appearance of the cochlea and semicircular canals. *B*, Internal structure of the cochlea. *C*, A microscopic sketch of the organ of Corti, showing the hair cells.

and is filled with endolymph. It is separated from the overlying vestibular channel by the vestibular (Reissner's) membrane, and from the underlying tympanic channel by a ledge-like projection of the bony cochlear wall and by the basilar membrane.

The actual receptors for hearing are found in the cochlear duct as several rows of specialized "hair" cells, approximately 24,000 in number (Fig. 12–18). These bear cilia that project into the endolymph from the free end of each cell. The hair cells, the supporting cells, and the nerves constitute the *organ of Corti,* which rests on the basilar membrane within the cochlear duct.

Physiology of Hearing

Hearing depends on a succession of events. The first is entry of sound waves into the external auditory canal, which causes the tympanic membrane to vibrate. These vibrations are transmitted mechanically by the malleus, incus, and stapes of the middle ear to the oval window at the entrance to the inner ear. The vibrations of the oval window set up pressure waves in the perilymph of the vestibular and tympanic channels; these pressure waves, in turn, are transmitted via the basilar membrane to the endolymph of the cochlear duct, where the hair cells of the organ of Corti are stimulated. The resulting stimulation of the hair cells arouses nerve impulses that are transmitted by the cochlear nerve to the hearing center in the temporal lobe of the cerebral cortex. This process suggests that a certain portion of the basilar membrane and certain hair cells are affected by sound waves of one frequency. Waves of a different frequency set up vibrations in another portion of the basilar membrane.

According to the just described place theory of hearing, age and other factors determine the range of hearing among individuals. The human ear can detect sounds ranging in frequency from 20 to 20,000 cycles per second; sounds below or above this range are not heard. The squeaking of a mouse is higher than 20,000 cycles per second. We do not hear it, but the cat does. A dog whistle can be made to produce sounds higher than our hearing capacity. The dog's hearing range is said to go as high as 24,000 cycles per second; therefore, when we blow the whistle, the dog hears it and we don't.

Deafness

Any impairment or defect in the ability to hear sounds is known as deafness. There are three common types:

1. Conductive. Defects in the sound-transmitting mechanism may result from an obstruction in the external auditory canal, damage to the eardrum, or damage to or stiffness of the middle ear bones. For example, excessive wax may block the external canal; scar tissue or injury may affect the tympanic membrane; fusion of the middle ear bones (ankylosis) or stiffness of the ear bones may occur. Conduction through the air is prevented in each situation, but hearing aids devised for conduction through bone from the outside to the inner ear fluids may help these patients.

2. Perceptive. This kind of deafness is due to cochlear damage, usually to the hair cells in the organ of Corti and to the cochlear nerve itself. Pilots and artillery men often are deaf to particular sounds following destruction of a specific group of hair cells in the organ of Corti. The cochlear nerve is most often impaired by growth of tumors.

3. Central. Central deafness is usually due to a physical or psychological disorder in the auditory sensory region on the cerebral cortex or in the tracts leading to it. This type of deafness is relatively rare.

Hearing Acuity

The instrument used to detect or measure a person's ability to hear is the *audiometer*. School children must rely extensively on hearing, and new methods employing sound vibrations are being used to help children born with hearing defects. Currently, all children of a certain school age are tested to detect hearing problems so that such weaknesses may be identified and corrected immediately.

Hearing tests not only help to determine the type of hearing defect present, but also establish the potential of the patient's hearing. In the Weber test an inexpensive tuning fork is used to differentiate between conductive and perceptive deafness. The fork is placed on the patient's forehead in order to compare hearing in the two ears. In normal hearing or deafness equal in both ears, the patient hears vibrations in the middle of his head. Variations suggest hearing inequality. In conductive hearing loss, bone-conducted sounds shift to the poorer ear; in perceptive hearing loss, sounds are heard louder in the better ear.

Equilibrium

In addition to the cochlea, the inner ear contains three organs—the utricle, saccule, and semicircular canals—that play a major role in the sense of equilibrium or balance (Figs. 12–16, 12–19, and 12–20).

The *vestibule* is a chamber that contains the first two membranous sacs, the *utricle* and the *saccule,* each of which is filled with endolymph. Within the walls of these two sacs are special thickenings or *maculae,* containing receptor "hair" cells and nerve endings that project into the overlying endolymph. Enmeshed in the hairs are tiny particles of calcium carbonate, called *otoliths* or ear stones (Fig. 12–19). Any changes in position of the head or body that cause the otoliths to bend in the direction of the pull of gravity initiate nerve impulses that travel to the brain via the vestibular nerve (Figs. 12–15 and 12–19), which joins the acoustic nerve to become collectively the eighth cranial nerve.

The three *semicircular canals* are hollow tubes containing endolymph. They are arranged at right angles to one another so that all three planes of space are represented (Figs. 12–16 and 12–19). One end of each canal is enlarged into a swelling called the *ampulla.* Within each

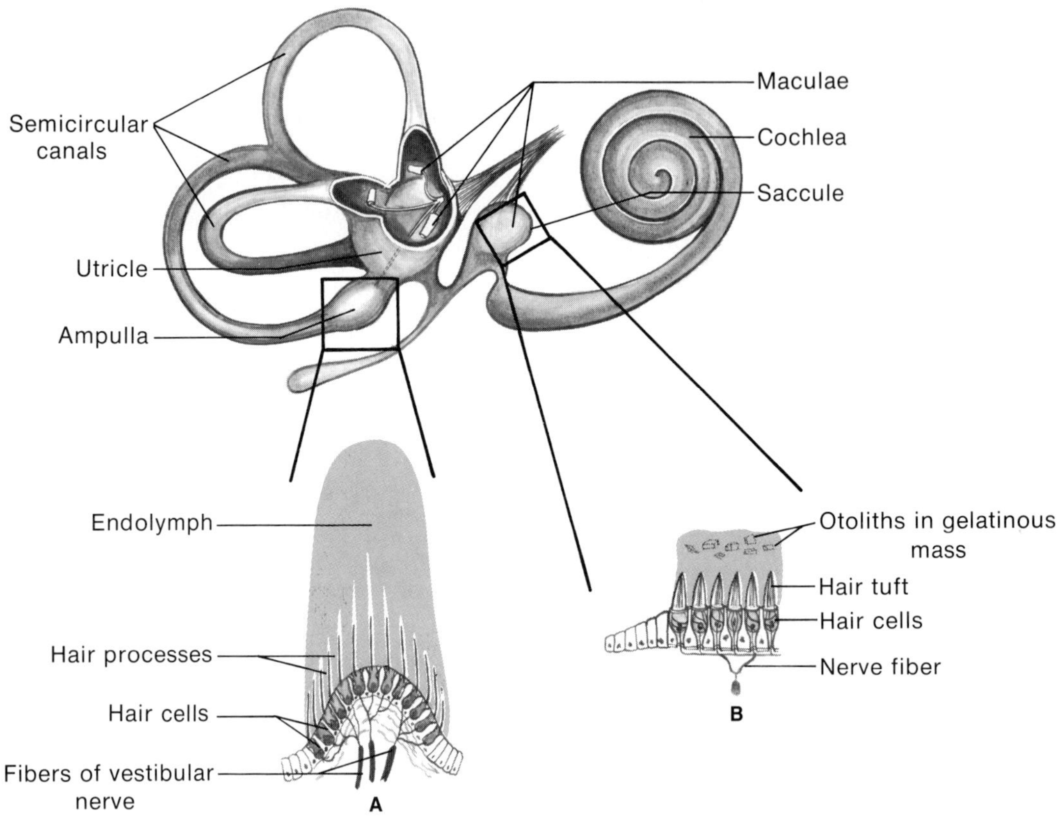

Figure 12–19 Equilibrium. Diagram of the organization of the maculae and cupula of the inner ear. Both contain receptors covered by a gelatinous mass. *A*, Otoliths move in the maculae, mechanically stimulating the hair processes of the hair cells. *B*, Endolymph movement stimulates the hair cells in the cupula. Each stimulation is carried to the brain, which interprets the movement.

ampulla is a small cluster of hair cells and nerve endings that project into the endolymph. Movement of the hair cells sets up impulses in the vestibular nerve; whenever movement begins or ends, goes faster or slower, or changes direction, impulses are initiated.

A change in position of the head causes impulses from receptors in the semicircular canals and vestibule to initiate the *righting reflex*. Balance against gravity is thus maintained by the coordinated effective response of the antigravity muscles. The cerebellum links the impulses that arise from stimulation of the sensory nerves of the canals, joints, and so forth, and sends nerve impulses to the motor centers of the cerebrum and spinal cord.

Disorders of Equilibrium

Tinnitus is a noise in the ear, usually a buzzing or ringing sound. It may be caused by an accumulation of wax in the outer canal, by otitis media, and possibly by a degeneration of the acoustic nerve itself.

Vertigo is a sense of movement, either of self or surroundings, due to stimulation of the hair cells in the semicircular canals by movements of the endolymph. *Nystagmus,* an oscillating eye movement, is closely associated with vertigo. Both produce a sensation of nausea and are collectively called *motion sickness*. The inability of the equilibrating mechanism to keep up with the constant motion of the body brings on this condition.

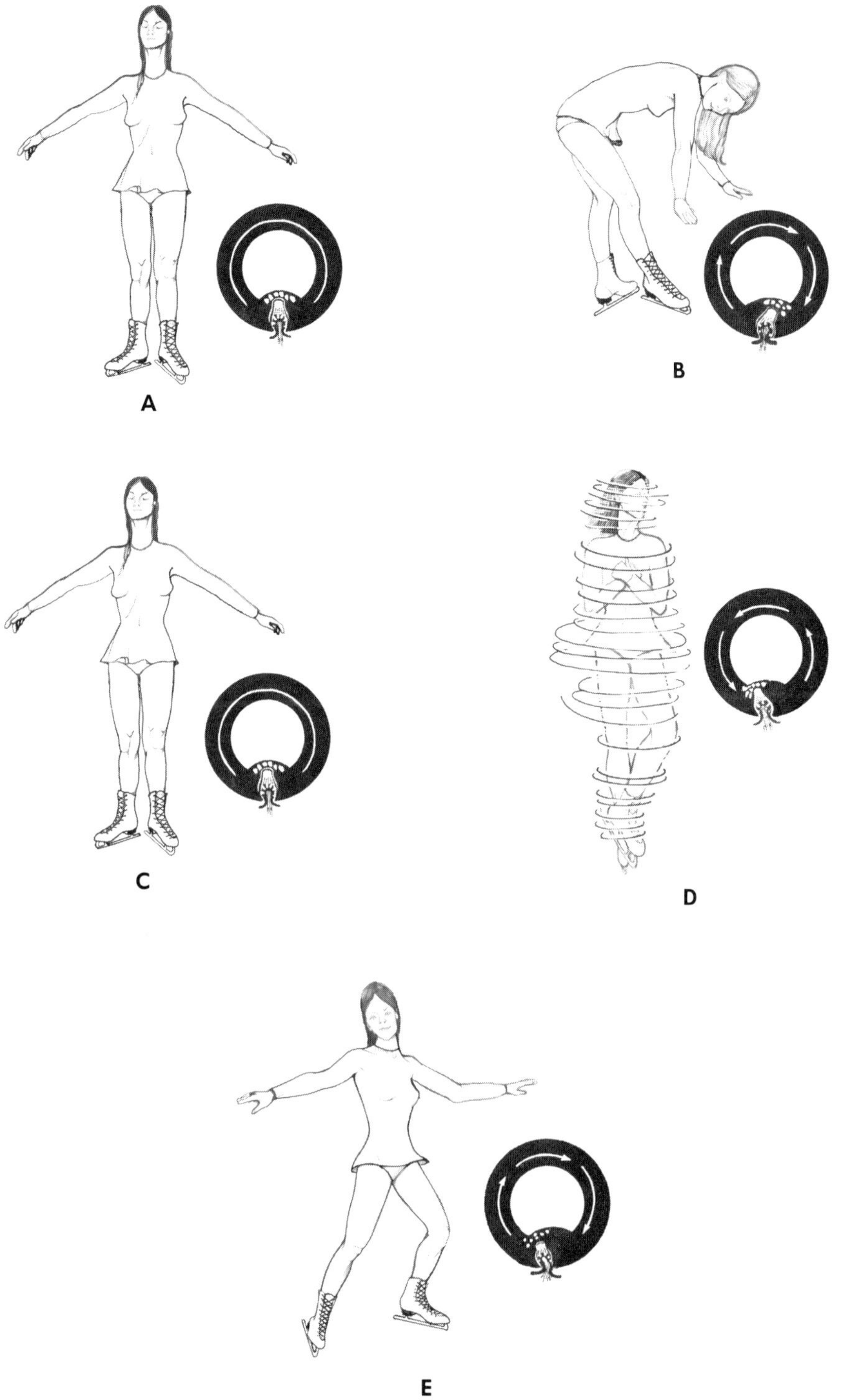

Figure 12–20 Sense of equilibrium. In the utricle and saccule, the maculae contain receptors that are stimulated by a change in position of the head (from *A* to *B*)—static equilibrium. Each ampulla is stimulated by the change of the endolymph fluid and therefore detects such movements as starting and stopping (*C, D, and E*)—dynamic equilibrium.

OUTLINE

I. Sensory Mechanisms
  A. Awareness of environment
    1. Receptor, neuron, brain
  B. Important features of sense organs
  C. Classification of sense organs
    1. Special, cutaneous, and visceral
    2. Nature of stimulus
      a. Chemoreceptors, pressoreceptors, and photoreceptors
    3. Location
      a. Teleceptors, exteroceptors, proprioceptors, and interoceptors
  D. Law of projection
    1. Phantom limb pain
II. Cutaneous Senses
  A. Touch and pressure sensations
    1. Free nerve endings, Meissner's and pacinian corpuscles, hair nerve fibers
    2. Merkel's disks
    3. Sensory area in parietal lobe of cerebral cortex
  B. Temperature sense
    1. Cold – Krause's corpuscles
    2. Warm – Ruffini's end-organs
    3. Sensory area in parietal lobe of cerebral cortex and in thalamus
  C. Somatic pain
    1. Free nerve endings in skin, muscles, joints, tendons
    2. Overstimulation of any tactile senses
    3. Sensory area in parietal lobe and thalamus
    4. Referred pain
    5. Pain remedies
III. Position Sense (Kinesthesia)
  A. Muscles, tendons, and joints
    1. Ruffini's end-organ
    2. Golgi's tendon organ
    3. Proprioceptors
    4. Sensory areas in thalamus and cerebral cortex (conscious activities), and in cerebellum (reflex activities)
IV. Hunger and Thirst
  A. Organic sensations
    1. Hunger – projected from body cells to stomach
      a. Appetite – pleasant sensation
    2. Thirst – projected from body cells primarily to pharynx
      a. Dehydration – tissues lack water
      b. Polydipsia – excessive thirst
    3. Sensory area for organic sensation in the hypothalamus
V. Taste and Smell
  A. Chemical sensations
    1. Taste – gustatory taste buds found on tongue and oropharynx
      a. Sweet, sour, bitter, and salt
      b. Sensory area – lateral fissure of cerebral cortex
    2. Smell – receptors in nasal mucosa
      a. Volatile substance detection
      b. Sensory area in olfactory center of temporal lobe
    3. Gustatory nerve
      a. Blends sensations of taste and smell
VI. Vision
  A. Accessory organs
    1. Eyebrows, eyelids, eyelashes
    2. Conjunctiva – disease barrier
    3. Lacrimal apparatus
      a. Tears
    4. Six muscles (extrinsic)
  B. Anatomy of eyeball
    1. Layers
      a. Sclera – outer white of eye
        1) Cornea
      b. Choroid – middle vascular layer
        1) Iris, pupil, ciliary muscles
        2) Lens
        3) Adaptation

c. Retina—inner layer of nerve cells
   1) Rods and cones—photoreceptors
2. Anterior chamber
   a. Aqueous humor
   b. Canal of Schlemm
      1) Glaucoma
3. Posterior chamber
   a. Vitreous humor
   b. Shape of eyeball

C. Visual pathways
   1. Optic disk—convergence of fibers of optic nerve
   2. Binocular vision
      a. Strabismus
         1) Exotropia
         2) Esotropia

D. Physiology of vision
   1. Refraction
      a. Bending of light rays
      b. Two media of differing densities
   2. Convex and concave lenses
   3. Accommodation
   4. Hyperopia and myopia
   5. Astigmatism
   6. Presbyopia—old age
   7. Cataracts

E. Perception of light and color
   1. Light—electromagnetic radiation
      a. Visible light spectrum colors
      b. Rods—night vision
         1) Rhodopsin—visual purple
      c. Cones—light vision, color vision
         1) Macula lutea—fovea centralis
   2. Visual acuity
      a. 20/20 normal vision
   3. Color
      a. Young-Helmholtz theory—trichromatic vision
         1) Red, green, and blue cones
         2) Dichromatic and monochromatic color blindness
      b. Ishihara color plate test
      c. Hereditary color blindness
   4. Ophthalmoscope
      a. Eye examination
      b. Diagnostic tool

VII. Hearing and Equilibrium
   A. Anatomy of the ear
      1. External ear
         a. Pinna (auricle)
         b. Auditory canal
            1) Cerumen glands
      2. Middle ear
         a. Tympanic membrane (eardrum)
         b. Tympanic cavity
            1) Eustachian tube—maintains equal air pressure
            2) Otitis media, myringotomy
            3) Ossicles (ear bones)—malleus, incus, and stapes
            4) Conduction of sound vibration
            5) Otosclerosis, ankylosis
      3. Inner ear—labyrinths
         a. Cochlea—coiled, snail-like
            1) Channels filled with perilymph and endolymph
            2) Organ of Corti—sensory hair cells for hearing
            3) Cochlear nerve
   B. Physiology of hearing
      1. Sound waves
      2. Place theory
         a. Vibrations conducted from external to inner ear
         b. Hair cells of organ of Corti stimulated
         c. Impulses carried to temporal lobe of cerebral cortex by cochlear nerve
         d. Range of hearing—20 to 20,000 cycles per second
      3. Deafness

- a. Conductive, perceptive, and central defects
- 4. Audiometer
  - a. Measures person's ability to hear
- C. Equilibrium
  - 1. Anatomy
    - a. Vestibule (saccule and utricle) and semicircular canals
    - b. Hair cells in utricle, saccule, and ampulla of semicircular canals
    - c. Otoliths enmeshed in hair cells
  - 2. Righting reflex
  - 3. Disorders
    - a. Tinnitus, nystagmus, vertigo, and motion sickness

STUDY QUESTIONS AND PROBLEMS

1. An individual receives a third degree burn on the arm. The deep structures of the skin are destroyed.
   a. What sensations are lost?
   b. Which receptors are involved?
   c. Will there be much pain?
   Explain.
2. Explain how receptors function as a protective mechanism.
3. John Jones has received an injury to his right occipital lobe. Will he experience partial loss of sight in his right, left, or both eyes? Explain.
4. Colds are hunger antagonists. Explain.
5. If you have a cold, why is it not a good practice to blow your nose hard? Describe the resultant middle ear involvement.
6. What does the term "referred pain" mean? Give an example.
7. If my father cannot distinguish blue from black or purple, what do I tell him to explain this?
8. Describe the phenomenon of accommodation, using as examples your eyes, the windshield of your car, and a pedestrian walking across the street 50 feet in front of your moving car.
9. How does one know what is happening outside one's body? Inside one's body?
10. What is the law of refraction? Using it as a basis, explain the following:
    a. Myopia and hyperopia
    b. Astigmatism
    c. Presbyopia
11. What does "20/20" mean? If you have 20/10 vision, what is the condition of your eyes?
12. What are mydriatics and miotics? Describe their function and give an example of each.
13. How is equilibrium related to the ear? How is vertigo related to dizziness?
14. Differentiate among the types of deafness. Describe bone versus air conduction of sound.
15. What is the difference between hunger and appetite?

Chapter 13

The Blood

The body's internal environment is continually renewed and restored by the circulation of blood. Water and crystalloid components of blood plasma enter the intercellular spaces from the capillaries by ultrafiltration (filtration under pressure through filters with minute pores). At the distal ends of the capillaries, a similar volume of interstitial fluid returns to the plasma by osmosis. Meanwhile, the composition of the blood is regulated by certain of the internal organs. The process of circulation and exchange in the capillaries is so effective that the regulation of their relatively small volume is sufficient to maintain the entire internal environment at a near constant composition.

FUNCTIONS OF BLOOD

Everything required by the tissues for energy or synthesis is carried in the blood, and all waste products derived from the tissues are removed by the blood. As a result, the functions of blood are extremely complex. The following are some of the more evident ones:

1. It supplies the cells and tissues with oxygen and nutrients for growth, repair, and other life activities.
2. It transports the waste products of cellular metabolism, including carbon dioxide, to the excretory organs for elimination.
3. Its white blood cells, antibodies, and other protective substances provide an important defense against damage from injury and disease.
4. It helps regulate and equalize body temperature by distributing heat to the body parts.
5. It transports hormones, the secretions of the endocrine glands, to the cells and tissues.
6. It plays an important role in maintaining the acid-base balance of the tissues.
7. It can help to maintain its own volume and composition by forming a clot, thereby reducing bleeding and preventing the escape of body fluids.
8. It bathes the various body cells and tissues in fluid, thus regulating the water and salt content of all tissues. At the same time it maintains the osmotic concentration throughout the body.

CHARACTERISTICS OF BLOOD

Blood is bright red (well oxygenated) in the arteries, and dark red (less oxygenated) in the veins. It flows four to five times more slowly than water because it is four to five times thicker (a property called *viscosity*). Its specific gravity (weight compared with water) varies between 1.045 to 1.065 (water is 1.000), and its temperature and pH values are 38° C (100.4° F) and 7.38, respectively. The volume of blood in the body has been measured in many ways, but obviously can be expected to vary with the size of the body. A useful estimate is that its weight is 9 per cent of the body's total weight. The blood volume of a man of average size is approximately five quarts (10 pints).

Seen with the naked eye, blood appears to be opaque and homogeneous; but, on microscopic examination, it is seen to consist of cells, cell fragments, and an intercellular fluid, the plasma (Fig. 13–1).

When blood is obtained from a person's vein (by venipuncture) and transferred to a test tube, and clotting is prevented by adding certain chemicals called anticoagulants, it separates into two distinct layers. The upper layer, which is yellow and transparent, contains most of the chemical components of the blood in solution; this is the *plasma* (Fig. 13–2). Plasma constitutes 55 per cent of the blood's volume. The lower portion of the blood sample consists of the *formed elements*: the red blood cells (*erythrocytes*), the white blood cells (*leukocytes*), and the platelets (*thrombocytes*). The percentage of total blood volume contributed by these formed elements is called the *hematocrit.* One of the two common tests for a basic anemia is determination of the hematocrit—normally about 45, that is, 45 per cent of the blood volume.

THE FORMED ELEMENTS

The formed elements of the blood—the erythrocytes, leukocytes, and platelets—often are called "corpuscles," a word meaning "little body." Each of these three types of corpuscle will be discussed here separately.

Erythrocytes

Erythrocytes, the red blood cells, are tiny, biconcave, disk-shaped cells. They are reddish-yellow, and, in their mature form, do not possess a nucleus (Fig. 13–3). These cells are manufactured in the bone marrow and are by far the most prevalent of the formed elements. There are approximately 5,500,000 red cells per cubic millimeter of blood in the male, and 4,800,000 per cubic millimeter in the female.

The primary role of erythrocytes is to carry molecular oxygen to all cells and tissues of the body. This is accomplished

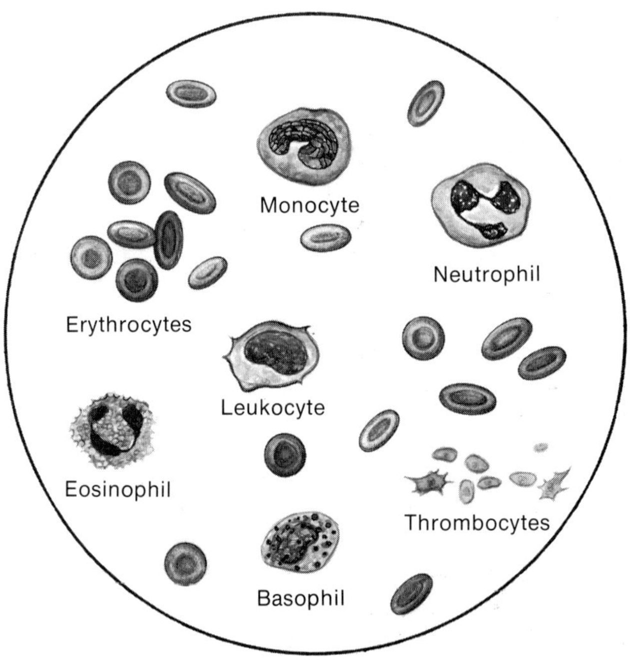

Figure 13–1 Microscopic blood smear.

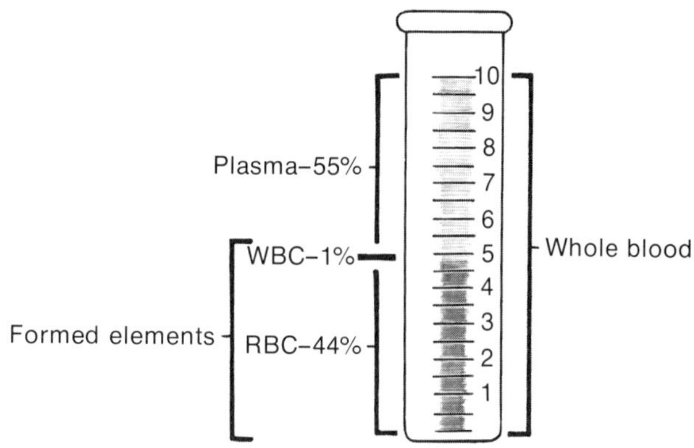

Figure 13–2 Centrifuged whole blood, showing the relationship of the fluid portion (plasma) to the formed elements. Abbreviations: WBC, white blood cells; RBC, red blood cells.

through the main ingredient of the red cells, a blood protein called *hemoglobin.*

Hemoglobin is a complex protein molecule formed of *globin* (blood protein) and a nonprotein portion called *heme* (hematin), which contains iron. One erythrocyte contains several million molecules of hemoglobin. The entire function of hemoglobin depends on its capacity to combine with inhaled oxygen in the lungs and then to release this oxygen readily in the capillaries of the tissues. In the tissues, hemoglobin picks up carbon dioxide and helps maintain the acid-base balance.

In adults, 100 ml of normal blood contains between 12 and 19 gm of hemoglobin. The normal average is 15 gm in adults and 12 gm in children four to twelve years old.

The manufacture of red blood cells is called *erythropoiesis.* In the unborn child this process is at first carried on in the mesenchymal cells in the yolk sac; later, in the tissues of the liver and spleen. Erythropoiesis is stimulated by a circulating hormone, *erythropoietin,* thought to be secreted by the kidney. This hormone promotes the differentiation of mesenchymal (stem) cells into mature erythrocytes. After the fifth month of fetal development, erythropoiesis decreases in the liver and takes place in bone marrow. After birth, red bone marrow is the only tissue concerned with red cell formation.

The stages in the development of a mature erythrocyte are shown in Figure 13–3. The cells are originally nucleated, but lose their nuclei while still in the bone marrow, before their release into the bloodstream. In certain anemias the release of cells is speeded up as if to make good the defect, and immature, nucleated cells (normoblasts and reticulocytes) appear in the circulating blood.

The life span of red blood cells is about 120 days. Towards the end of this time, erythrocytes become more and more fragile, and eventually rupture. In this process, *hemolysis* or "laking," the erythrocyte membrane is destroyed and the cell's hemoglobin content goes into solution in the plasma. When and how the erythrocytes are destroyed is not completely known. One probable method of destruction is by phagocytic cells of the reticuloendothelial system of the liver and spleen. Most of the iron freed by destruction of erythrocytes is reused.

For complete maturation of red blood cells, vitamin B_{12} is necessary. The inner mucosal lining of the stomach secretes a substance called the "intrinsic factor," which promotes absorption of vitamin B_{12} from the foods we eat. Absorbed vitamin B_{12} is then carried to the liver, where it acts as the principal antianemic factor. It is stored in the liver and taken by the circulation to the bone marrow, where it functions enzymatically to complete the maturation of red blood cells (Fig. 13–4).

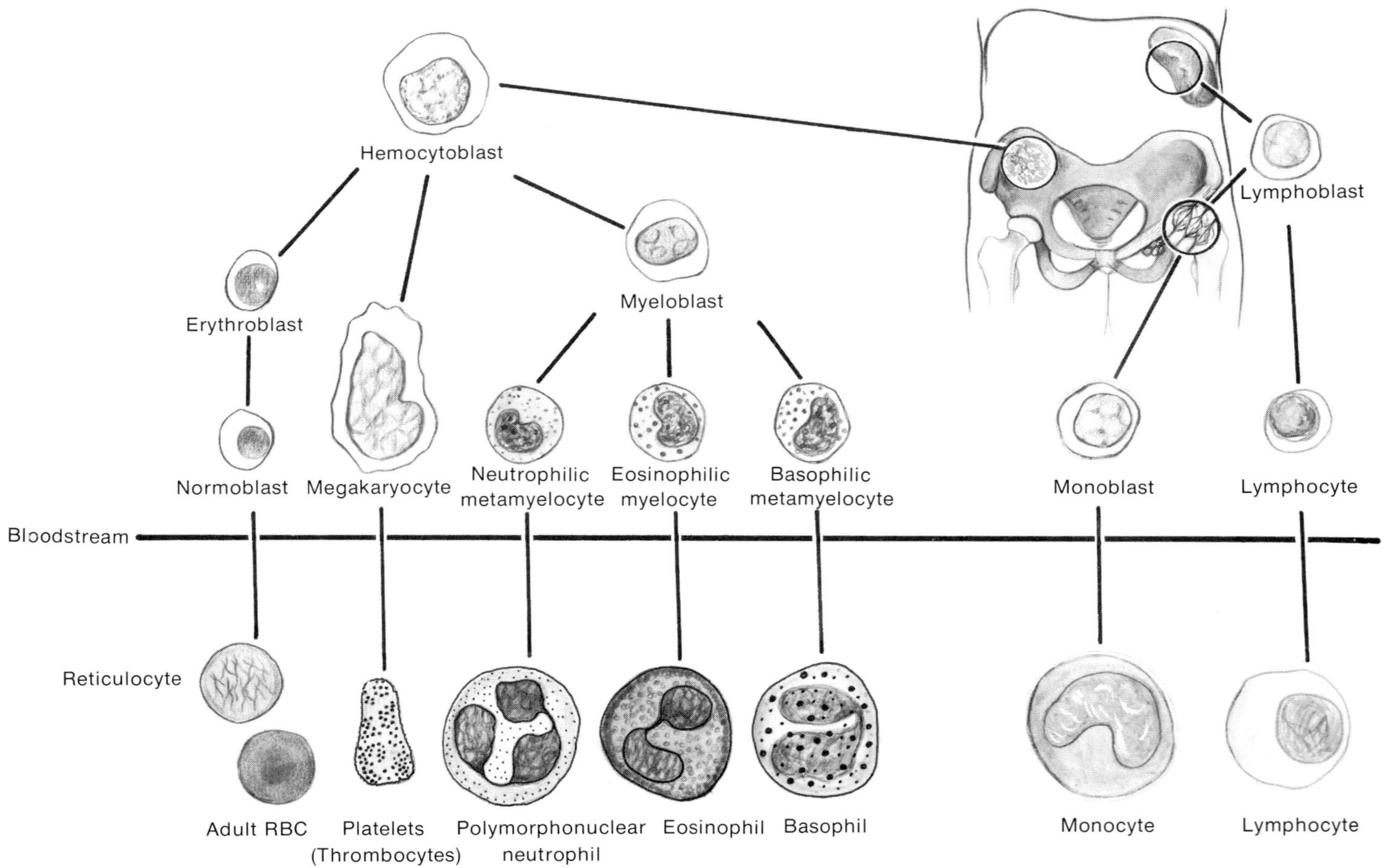

Figure 13–3 Development of the blood cells. Those cell forms above the black line are found in the marrow or lymphatic organ and are considered immature. Mature blood cells are found in the bloodstream. RBC, red blood cell.

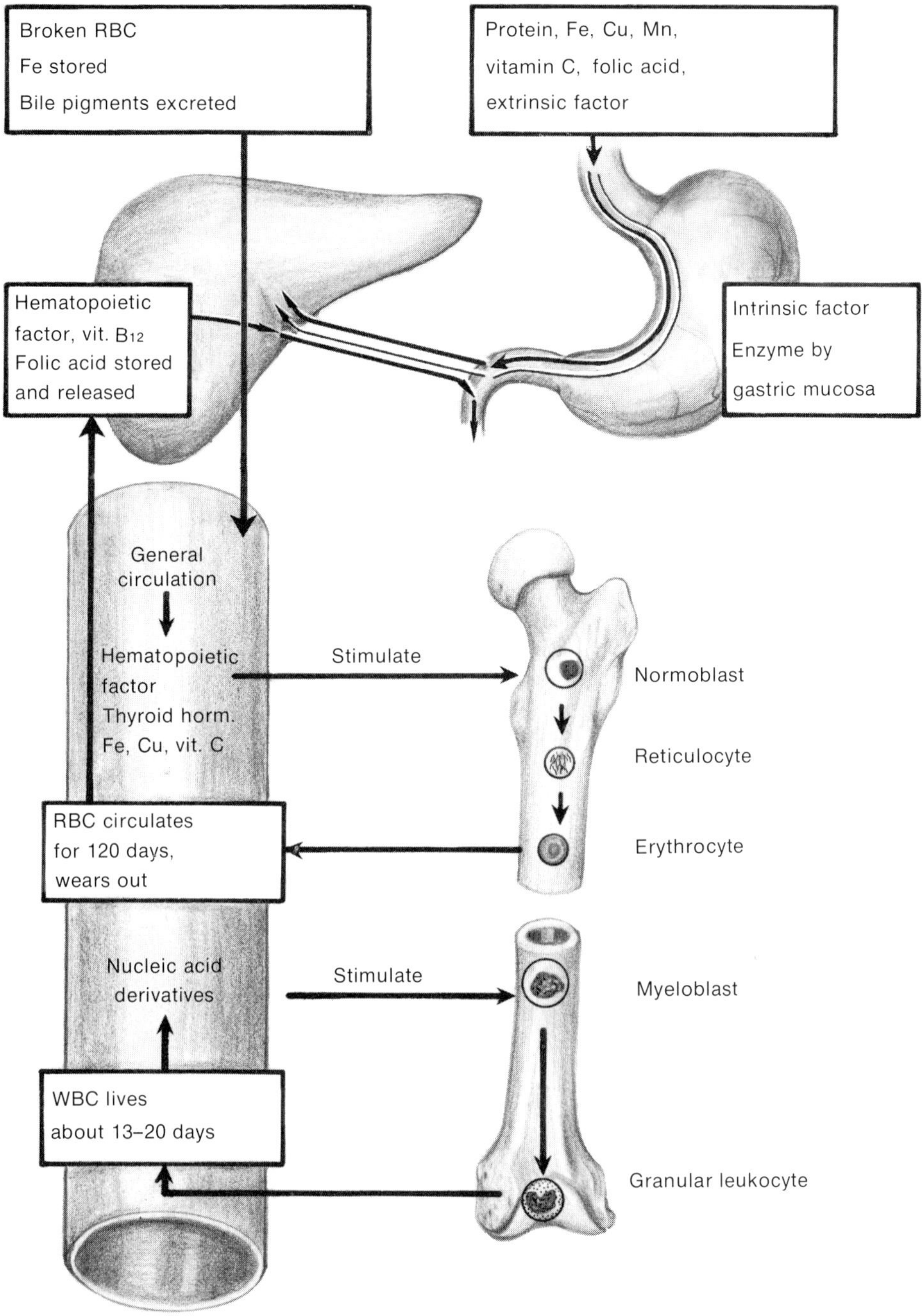

Figure 13–4 Sequence of events in, and elements essential for, normal blood formation. Abbreviations: RBC, red blood cells; WBC, white blood cell; vit., vitamin; horm., hormone.

Several proteins, folic acid, basic metals (cobalt, copper, and iron), and other vitamins are also essential for red blood cell formation.

Anemias

Anemia is an abnormal condition in which the blood lacks either its normal number of red cells or its normal concentration of hemoglobin.

Primary anemias are characterized by a disturbance in the red cell-forming mechanism, leading to a deficiency of erythrocytes. The most important primary type is *pernicious anemia.* Its initial cause is a permanent deficiency of the intrinsic factor in the stomach responsible for absorption of vitamin B_{12}, resulting in the accumulation of immature red blood cells in the bone marrow. In this disease the red cell count may go as low as 20 per cent of normal; a blood examination also reveals the presence of immature red cells in the bloodstream (Fig. 13–5). When untreated, pernicious anemia can bring about deterioration of the nervous system, causing difficulty in walking, generalized weakness, stiffness of the extremities, brain damage, and permanent damage to the spinal cord. Early treatment by injection of vitamin B_{12} and proper diet may assure a favorable outcome.

Aplastic anemia is another type of primary anemia. Here, marrow failure is frequently caused by poisonous agents. Chemical poisons (lead, benzene, arsenic, nitrogen mustard, and gold) discourage production of red cells, whereas physical agents (x-rays, atomic radiation, radium, and radioactive phosphorus) tend to injure the bone marrow itself. The marrow fails to produce either red or white blood cells, causing a drop in the total cell count (Fig. 13–5).

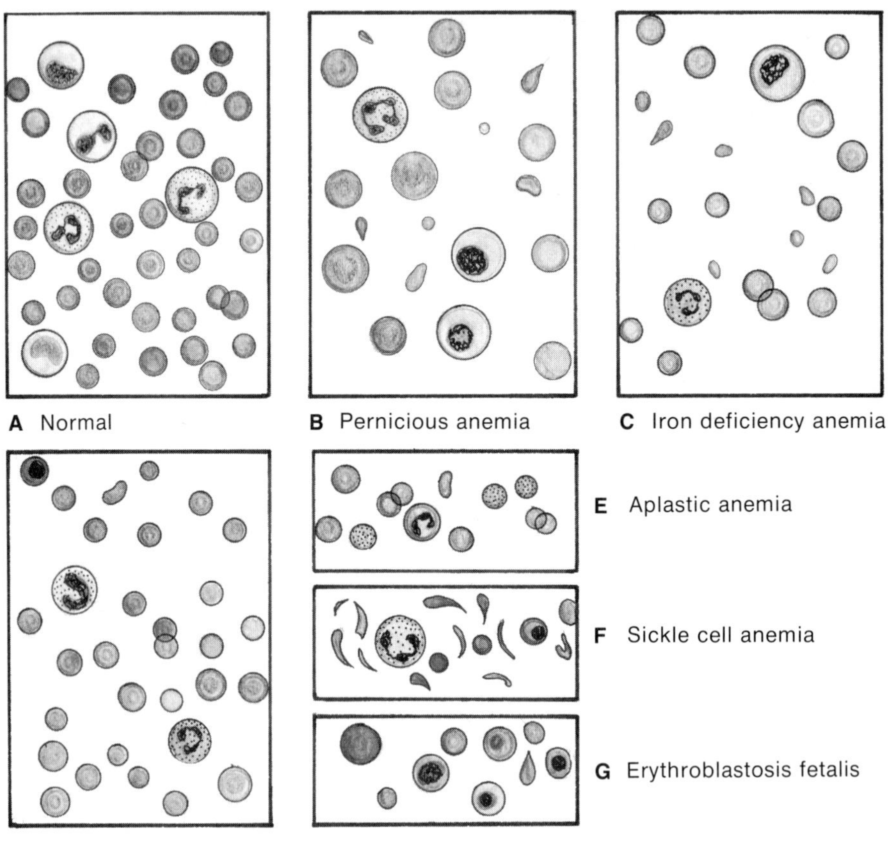

Figure 13–5 Diseases of red blood cells. Using the normal blood smear (*A*) as a reference, note the decrease in number of cells in all of the anemias, the larger-sized cell (macrocyte) in pernicious anemia (*B*), and the odd-shaped cells in the other forms of anemia (*C* to *G*).

Some anemias (*hemolytic* anemias) are brought about by excessive destruction of erythrocytes. Occasionally, the phagocytic system proceeds too rapidly. More commonly, infections and infestations (malaria and bacteria) are the cause of blood cell loss. As each blood cell is entered, the infecting organism multiplies until the cell bursts (hemolysis), resulting in anemia. Less common types of anemia resulting in hemolysis include *sickle cell anemia* (occurring almost exclusively in the colored races) and a hemolysis that occurs in infants as a result of antibodies from an Rh-negative mother. This latter condition, *erythroblastosis fetalis*, will be discussed later (p. 277).

Secondary anemias are due to simple loss of blood or to a defective formation of hemoglobin caused by other diseases. The average adult has five quarts of blood. If he or she loses two quarts suddenly, death will result. However, a gradual loss of the same quantity over a period of weeks can be handled and withstood by the body. In chronic bleeding it is important to locate and remedy the cause, which may be hemorrhoids, excessive menstrual flow, or ulcerations of the stomach or bowel. Malignant tumors, infections, liver cirrhosis, thyroid deficiency, and chronic kidney disease (nephritis) and other causes of secondary anemia. In one of the most common disorders, *macrocytic hyperchromic anemia*, the erythrocyte is larger than normal and abnormally high amounts of hemoglobin are present; in *microcytic hypochromic anemia* the reverse is true—smaller than normal red cells with decreased hemoglobin. Both of these are iron-deficiency anemias, and often occur in women of child-bearing age, during chronic hemorrhage, and in infants. In microcytic anemia, prolonged red cell production, due to loss of red cells in the circulation, imposes a continuous demand on the iron stores of the body, leading to iron deficiency. Consequently, the hemoglobin content of newly formed red cells is lower than normal (hypochromic) and the cells are smaller than normal (microcytic). Macrocytic anemia, on the other hand, is characterized by abnormally large red cells owing to the absence of vitamin B_{12} or the intrinsic stomach factor, or both. These cells contain a greater than normal amount of hemoglobin, and therefore are considered hyperchromic.

A person with a secondary anemia needs frequent blood checks (hemoglobin and hematocrit) and continued advice and treatment—not only to treat the cause of the disease, but also to restore the composition of the blood to normal.

Polycythemia

The condition known as polycythemia is characterized by the production of a greater than normal number of red cells. The erythrocyte-forming areas become overactive. Erythropoiesis is normally subject to a feedback control. The process is inhibited by a rise in the circulating red cell level to super-normal values and is stimulated by anemia. It is also stimulated by reduced oxygen (hypoxia) situations, such as residence at high altitudes. It is thought that the low atmospheric pressure existing at high altitudes decreases the ability of hemoglobin to combine with oxygen, and that this reduction tends to stimulate the formation of new red cells. As a result, there is an increased viscosity of the blood and a rise in blood pressure. The heart may become overloaded, and there is an increased danger of clotting within the circulation.

Reactions of Hemoglobin

On exposure to atmospheric conditions, hemoglobin forms a covalent bond with oxygen to produce oxyhemoglobin (HbO_2), oxygen attaching to the iron (Fe^{++}) in the heme. The details of the oxidation and reduction of hemoglobin will be described later in the chapter on respiration (ch. 17). Previously it was mentioned that the normal average hemoglobin content of blood is 15 gm per 100 ml, all of it in red blood cells. When old red cells are destroyed in the liver and spleen, the globin portion of the hemoglobin molecule is split off, and the heme is converted to bilirubin, which is excreted by the liver in the form of bile (Fig. 13–4). The iron is reused for hemoglobin synthesis, and if blood is lost from the body and the iron deficiency is not corrected by dietary intake, iron deficiency anemia results.

When excessive amounts of hemoglobin are broken down, or when biliary excretion of bilirubin is decreased by liver disease or bile duct obstruction, the plasma bilirubin level rises, imparting a yellow tinge to the skin (jaundice). A jaundiced condition usually accompanies an infected liver (infectious hepatitis).

Leukocytes

The white cells of the blood, the leukocytes, are concerned primarily with the body's defense against infection (Table 13–1). Leukocytes vary in size and shape (Fig. 13–3). They are less numerous than erythrocytes, numbering 5000 to 9000 per cubic millimeter. They lack hemoglobin, are colorless, contain a nucleus, and are larger than the red cells. White cells are extremely active, and move under their own power by ameboid motion, often against the flow of the bloodstream. Unlike erythrocytes, white cells pass out of the bloodstream and into the intracellular spaces to phagocytize foreign materials (e.g., bacteria) found between the cells (Fig. 13–6; see also Fig. 13–10).

Leukopoiesis is white blood cell formation. Except that it is stimulated by the presence of bacteria, little is known about it.

Neutrophils, eosinophils, and basophils are called *granulocytic* leukocytes (granulocytes) because of the presence of granules in their cytoplasm. Granulocytes have horseshoe-shaped nuclei. They are formed from stem cells in the red bone marrow (Fig. 13–3) and survive in the blood for no more than two weeks. When the body is invaded by bacteria, the number of circulating neutrophils increases dramatically. They squeeze through the walls of the capillaries and ingest and destroy the bacteria (phagocytosis). This may be seen in the formation of *pus*, a fluid containing bacteria and millions of phagocytes (neutrophils).

All granulocytes contain histamine. The granules of the neutrophils and eosinophils appear to be lysosomes that function in the digestion of material taken into the cells. The basophils contain heparin, but their role in the clotting and anticlotting systems is uncertain. The eosinophils apparently phagocytize antigen-antibody

Table 13–1 CLASSIFICATION OF WHITE CELLS (LEUKOCYTES)

LEUKOCYTE	*PER CENT OF TOTAL WHITE CELLS*	*SPECIALIZED FUNCTION*
Granular (polymorphonuclear)		
Neutrophil	65–70	Defense against bacteria
Eosinophil	2–4	Allergy
Basophil	0–1	Heparin formation
Agranular (mononuclear)		
Lymphocyte	20–30	Immunity—antibody formation
Monocyte	2–8	Defense against bacteria

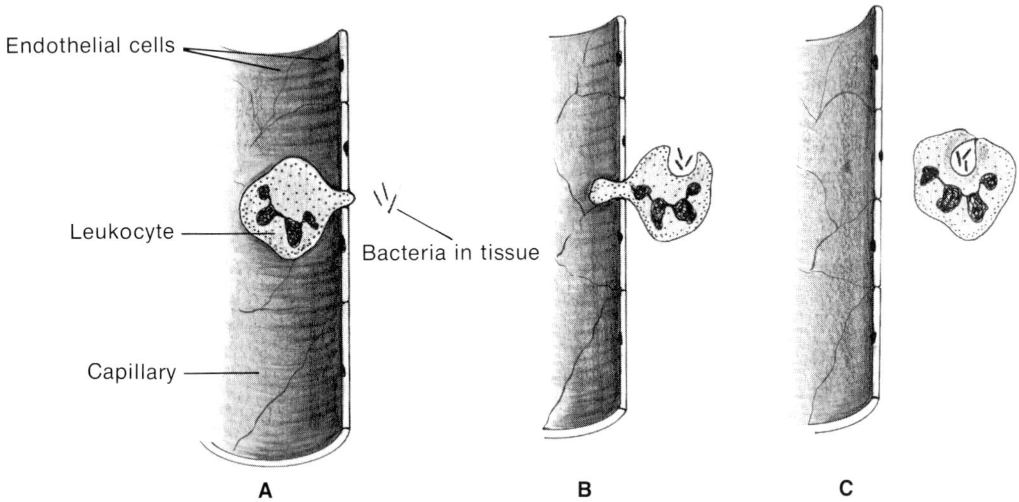

Figure 13–6 Phagocytosis. *A* to *C*, A white blood cell squeezes through a capillary wall and engulfs the bacteria in the tissue.

complexes; therefore, their number increases in patients with allergic diseases.

Old granulocytes are destroyed in the spleen and liver.

Agranulocytes (non-granulocytes) are white cells with a scanty, clear cytoplasm and large, round nuclei.

Lymphocytes are white cells formed in the lymphatic tissue of the lymph nodes, thymus, and spleen (Fig. 13–3). They enter the bloodstream by way of the lymphatic circulation. Lymphocytes contain the antibodies responsible for delayed hypersensitivity reactions. They produce the beta- and gamma-globulins of the plasma, and therefore also produce the antibodies described later in this chapter.

Monocytes are large white cells, each with an indented central nucleus (Figs. 13–1 and 13–3). They function effectively as phagocytes. There are comparatively few monocytes in the blood, and, like the other phagocytic white cells, these are actively motile. They are produced in bone marrow.

Children have more lymphocytes than do adults, as well as a higher total white cell count. In bacterial infections the numbers increase (*leukocytosis*) to as high as 30,000/per cubic millimeter, with most of the increase in the neutrophils. In more chronic infections, the number of lymphocytes increases. A decrease in the total white cell count (*leukopenia*) to 2000 to 3000 per cubic millimeter is said to be the result of a viral infection. A differential count of the white cells is made when infections are suspected in order to determine the composition of the leukocyte population.

Inflammation

The protective action of white cells can be illustrated by so common an incident as cutting a finger. The laceration (cut) injures the surrounding tissues, and bacteria in the air or on the skin enter the wound, releasing toxic products that can destroy these tissues. Irritation of the tissues brings an increased blood supply to the damaged area, which in turn produces a reddening of the skin and increases the local temperature.

Leukocytes become active and migrate in large numbers through the blood vessel walls (*diapedesis*) into the infected tissues. These cells are actively phagocytic, and rapidly ingest both the bacteria and the damaged cells. Blood plasma exudes, causing a puffing of the area (tissue edema). Some of the phagocytes are destroyed by the bacterial toxins; most, however, manage to control the bacterial invasion. All these events constitute *inflammation*: the redness, heat, swelling, pain, and irritation due to toxins released by the invading bacteria.

The festering that may occur around the base of a laceration is called *suppuration* (formation of pus). In this instance, the bacteria have destroyed large quantities of phagocytes and tissue cells. In the usual situation, however, the white cells clean up the infected area (resolution), and regeneration of the skin and tissue cells restores the area to normal.

We have seen that when any pathogen (e.g., bacteria) enters the tissues, leukocytes are attracted to the site. At the same time, the leukocyte-forming tissue increases its production, and the number of white cells in the blood is enormously increased. Therefore, if, in the course of a blood examination, abnormally large numbers of white cells are observed, this sometimes is indicative of a deep-seated infection (e.g., appendicitis). A rise in the white cell count affords excellent evidence to a physician that there is infection within the body. The severity may be indicated by the extent of leukocytosis in the blood. Often both a white cell count and a white cell differential examination are prescribed to determine not only the total white cell count, but also the relative numbers of each type present—in particular, the number of neutrophils. Most bacterial infections cause an increase in neutrophils. Some chronic infections, on the other hand, lead to an increase in lymphocytes;

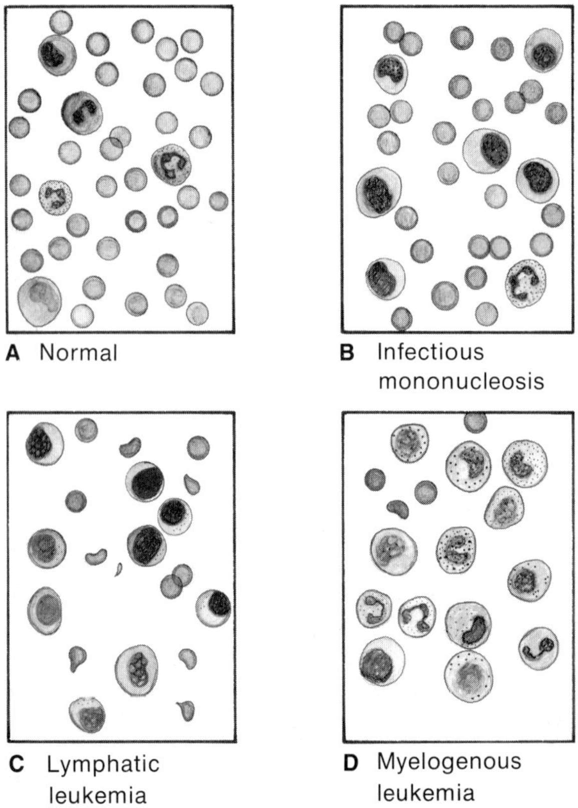

Figure 13–7 Disorders of white blood cells. *A*, Normal distribution and appearance of white blood cells (nucleated). *B*, An increase in the number of white blood cells; mononucleosis. *C*, If the white blood cell increase involves the monocytes alone, the disease may be a lymphatic leukemia. *D*, If the granular white cells are primarily involved, the condition typifies a myelogenous or marrow leukemia.

others (e.g., asthma, skin diseases, and parasitic infestations) cause an increase in the number of eosinophils. Basophils increase in chickenpox, smallpox, and Hodgkin's disease.

Neoplastic Blood Diseases

This group of diseases is characterized by an enormous increase in the number of white cells, owing to cancer of the leukocyte-producing tissues. As previously noted, white cells have two main sources: the red bone marrow (granulocytes) and lymphatic tissue (lymphocytes). When an increase in white cells stems from a tumor of the bone marrow, the condition is given the name *myelogenous leukemia* and the increase is noted in the number of granulocytes. If the leukocytosis results from overactivity of lymphoid tissue, the condition is called *lymphatic leukemia* (Fig. 13–7), and the white cells showing increased numbers will be the lymphocytes.

A special needle is usually employed to obtain a sample of red marrow from the spongy bone of the upper sternum (breast bone) in a procedure called "sternal puncture." A smear on a glass slide is prepared in much the same manner as the blood slide for a differential white cell count, and examination aids the physician in diagnosing leukemia and other blood diseases.

At the present time, the suggested causes of leukemia are many. There may be hereditary implications or inherent chemical factors, or various environmental agents may aggravate existing conditions, causing the disease. The patient with leukemia exhibits general symptoms of anemia and tends to bleed easily. The spleen may be enlarged owing to the increased white cell count. X-ray treatments, along with drugs (chemotherapy), have proved beneficial in some types of leukemia, but the disease is usually fatal.

Mononucleosis

Infectious mononucleosis most often affects young people between 15 and 25 years of age. It is usually considered a viral disease, but to this day no cause has been definitively established. It is characterized by fever, enlargement of the lymph nodes in the neck region, and increased lymphocyte production. The blood counts of both red and white cells are not altered, but the differential shows abnormal (atypical) lymphocyte production, both in number and in shape (Fig. 13–7). Surprisingly, it is not a contagious disease, such as measles or mumps. Treatment usually consists of complete rest for a period of four to six weeks.

Thrombocytes

Blood platelets (thrombocytes) are small, oval fragments of specialized bone marrow cells. They lack a nucleus (Figs. 13–1 and 13–3), average around 400,000 per cubic millimeter of blood, and are formed from giant cells (megakaryocytes) in the bone marrow by fragmentation of the cytoplasm.

If it were not for platelets, man would not last long; we would all bleed to death. When blood is shed, the platelets disintegrate and release an enzyme (thromboplastinogenase) which reacts with a plasma protein called fibrinogen to initiate a chain of reactions that finally forms the blood clot.

Blood platelets cause blood clots to shrink, drawing together the edges of the vessel defect, reducing its size, and further stemming the leakage. Platelet deficiency (thrombocytopenia) leads to spontaneous rupture of vessels in the skin (thrombocytopenic purpura), a basic hemorrhagic disorder. An increased number of platelets (thrombocytosis) is prevalent in granulocytic leukemia and polycythemia vera (elevated red cell count).

Thrombocytes that escape disintegration at the site of injured tissues probably are destroyed by the liver and spleen. Their life span is approximately eight days.

BLOOD PLASMA

Over half (55 per cent) of the total volume of blood is plasma, and plasma itself is approximately 92 per cent water. The remaining 8 per cent is accounted for by numerous different substances dissolved or suspended in this water (Fig. 13–8). The content of plasma varies, since blood carries substances to and from organs, where some are utilized and others are introduced into the plasma.

Plasma Proteins

Hemoglobin of the red blood cells accounts for two-thirds of the blood proteins; the remaining one-third is made up of the plasma proteins—serum albumin, globulin, fibrinogen, and prothrombin.

Serum albumin, which constitutes about 53 per cent of the total plasma proteins, is synthesized in the liver. All of the plasma proteins, but especially serum albumin, are concerned with the regulation of blood volume. Solutions containing solid materials are called colloidal solutions, and such protein solutions as plasma are in this category. The difference of osmotic (water) pressure across the capillary membrane is due to the impermeability of

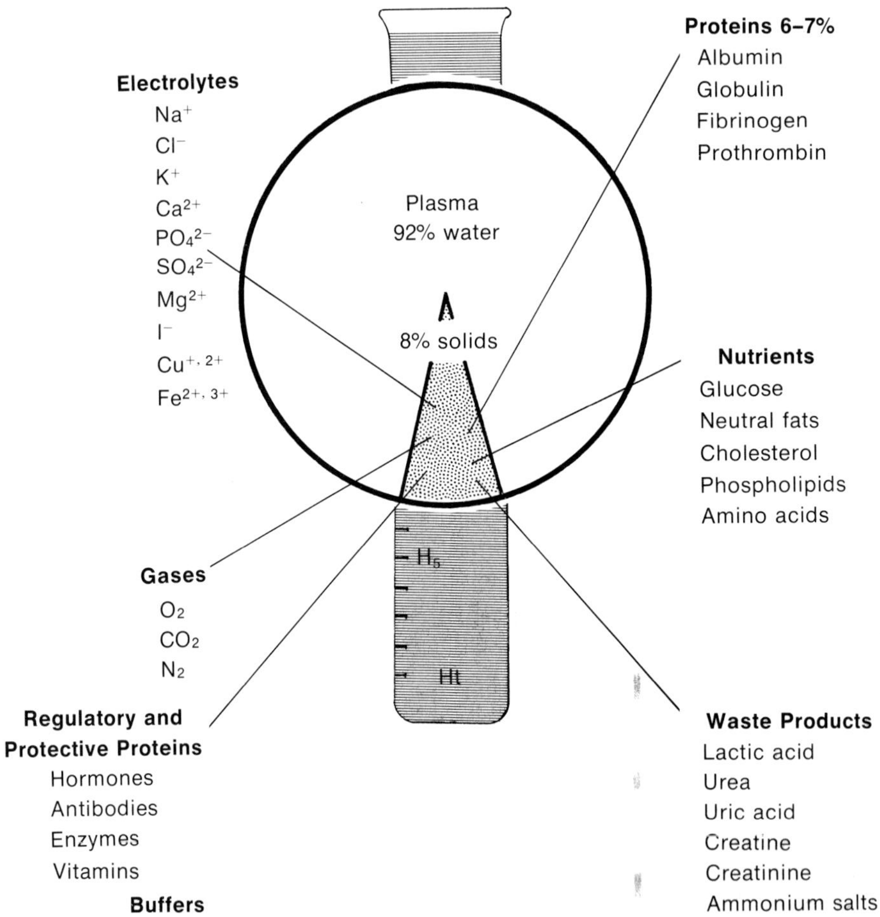

Figure 13–8 Composition of plasma.

the membrane to the large colloidal protein molecules. This difference, called the colloidal osmotic pressure, provides the "pulling" pressure of the plasma essential for holding and pulling water from the tissue fluid into the blood vessels. Serum albumin, therefore, serves a vital role in the exchange and restoration of the internal environment.

Globulin components of the plasma can be separated by chemical methods, by ultracentrifugation, and, most effectively of all, by electrophoresis, into alpha, beta, and gamma fractions. The gamma fraction consists of specific protein molecules that act as antibodies in the immune reactions that constitute the body's defenses against infection. Plasma cells (lymphocytes) are the main source of globulins active in the immune response.

Fibrinogen is concerned with the coagulation of the blood.

Prothrombin, a plasma protein formed in the liver, also participates in coagulation of blood. Vitamin K is essential for prothrombin synthesis.

Plasma Electrolytes

The major ions of the plasma are sodium, chloride, potassium, calcium, phosphate, sulfate, and magnesium. There are others, such as iodide, copper (as cuprous and cupric) and iron (as ferrous and ferric) in trace quantities. The essential role of the electrolytes is best described by indicating their part in the respective systems covered separately in this text. We have already seen the importance of some of them: for example, phosphate in ATP, sodium and potassium in permeability, iron in hematopoiesis, and calcium in bone growth. The remaining electrolytes will be introduced in their specific area of concern.

Nutrients

Carbohydrate is present in the blood, chiefly in the form of glucose, a simple sugar. Glucose is dissolved in blood and delivered to all parts of the body to provide a source of energy; it is also stored in the liver in the form of glycogen until it is needed.

Lipids (neutral fats, cholesterol, and phospholipids) are found in all tissues and body fluids. Their chief source is saturated fats formed in large quantities by liver cells. Lipids are an essential component of nerve cells and steroid hormones; they also control water evaporation through the skin.

Waste Products

Lactic acid and such nonprotein nitrogen compounds as urea, uric acid, creatine, creatinine, and the ammonium salts are the most important metabolic waste products carried by the plasma.

Gases

Oxygen (about 2 per cent), carbon dioxide (about 60 per cent), and nitrogen are the gases found in solution in the plasma.

Regulatory and Protective Proteins

As we have seen, the hormones secreted by the endocrine glands are needed for control of cellular activity. Antibodies (lysins, opsonins, agglutinins, and antitoxins) are formed by the reticuloendothelial system (spleen, liver, lymph nodes, and bone marrow); their site of formation is related to the portal of entry of the invading organism.

Enzymes and vitamins are found in variable concentrations directly associated with the organs of digestion and metabolism.

Buffer Systems

Several buffer systems protect the organism from being overwhelmed by major shifts in acid-base balance. The one that reacts most readily is the blood buffer system.

Buffers are chemical compounds that prevent a major change in pH. In the body, the reaction of the blood is slightly alkaline—pH of 7.35 to 7.45, under normal conditions. Most buffers occur in pairs, consisting of weak acids or bases and their respective salt substances. These pairs react with strong acids or bases, and replace them with less harmful substances that are easily eliminated from the body. The major purpose of the blood buffer system is to maintain a proper ratio of acids (H^+) and bases (OH^-) in the body—the acid-base balance.

Perhaps the best way to explain buffering action is to discuss an actual simple buffer system. A buffer of great importance is the combination of sodium bicarbonate ($NaHCO_3$) and its associated weak acid, carbonic acid (H_2CO_3). If a strong acid (HCl) is added to the solution, the salt, sodium bicarbonate, will neutralize it by producing sodium chloride (table salt) and a weak acid:

$$NaHCO_3 + HCl \longrightarrow NaCl + H_2CO_3$$

Although the bicarbonate buffer system has been illustrated in the above reactions as a mixture of carbonic acid and sodium bicarbonate, other salts of bicarbonate, not only sodium, will perform identically the same function.

The two other important buffer systems of the body are the phosphate buffer system (NaH_2PO_4 and Na_2HPO_4) and the protein buffer system ($-COOH$, $-NH_3^+$, and OH^-). These can react in both acidic and basic situations.

THE CLOTTING MECHANISM

The properties of blood that enable it to form a clot are essential to life. Without this self-sealing process, fatal hemorrhages would quickly occur. Small capillaries are constantly being ruptured in the course of normal daily activity; at these sites of rupture, the blood coagulates, sealing off the opening.

The formation of a blood clot, or *thrombus*, involves a sequence of steps, each dependent on the one before it. This sequence leads to the conversion of the soluble plasma protein, fibrinogen, into insoluble fibrin. Fibrin forms as a network of microscopic fibers that enmesh the blood's formed elements to produce a solid clot.

The factors essential for blood coagulation are:

1. *Thromboplastinogen*—a globulin component of the plasma and the precursor of thromboplastin.
2. *Antihemophilic factor (AHF)*—plasma globulins that speed the breakdown of blood platelets.
3. *Thromboplastinogenase*—an enzyme found in blood platelets and injured tissues.
4. *Calcium ions*—in conjunction with thromboplastin, these catalyze the conversion of prothrombin to thrombin.
5. *Prothrombin* and *fibrinogen*—two essential plasma proteins.

In summary, the enzyme thromboplastinogenase (liberated from disintegrated blood platelets) and the antihemophilic factors convert thromboplastinogen to thromboplastin. The latter, in conjunction with calcium ions, activates prothrombin to produce the enzyme thrombin, which catalyzes the conversion of soluble fibrinogen to insoluble fibrin, or a blood clot (Fig. 13–9).

The clotting process is actually much more complicated than the preceding description would indicate; several proposed factors are not mentioned. The detailed interactions of all the chemicals are not yet well understood.

Blood samples taken from the body may be prevented from clotting by removal of the fibrinogen or the calcium ions. The body itself has other mechanisms to

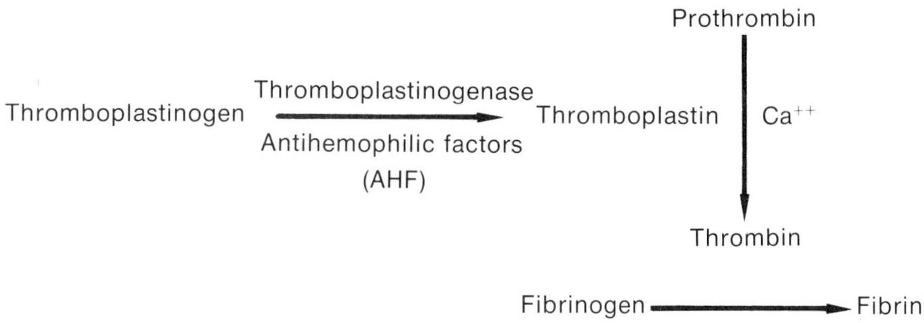

Figure 13–9 The blood clotting mechanism.

inhibit clotting—not only to prevent unwanted intravascular clotting but also to prevent a desirable and useful clot from extending indefinitely. Plasma contains neutralizing substances, or *antithrombins,* which counteract the action of the small amount of thrombin that is always being produced. An example is *heparin,* a potent anticoagulant polysaccharide (complex sugar) that acts by blocking the action of thrombin. It is produced in the liver.

When an artery becomes obstructed by a clot (thrombus), the supply of oxygen and nutrients to the dependent tissues is cut off. An obstructing clot is one kind of *embolus* (a plug in a vessel). *Embolism* is the blocking of a vessel by an object (embolus) that has been carried by the bloodstream from another site. More deaths occur in the United States from the formation of clots in blood vessels than from any other single disorder. Blood clots that obstruct the arteries that nourish the heart are responsible for the most common form of heart attack, a *coronary thrombosis.* This disease will be discussed in the chapter on the heart (see p. 296).

One of the most effective anticoagulants in use today is dicumarol. When taken by mouth, it requires 24 hours to produce its effects, a reduction in the clotting ability of blood caused by an inhibition of the synthesis of prothrombin in the liver. The drug competes with vitamin K, which is required for prothrombin formation.

Hemorrhagic Disorders

In these diseases a disruption of the coagulation process brings about abnormal bleeding. A familiar example of the group is *hemophilia.*

Hemophilia is a hereditary disease transmitted as a sex-linked recessive trait; it occurs in males but is passed through the female to subsequent generations. The hypothesis is generally accepted that the antihemophilic factor (AHF) is absent from the blood of hemophiliacs, and that this abnormality causes the blood to clot very slowly, resulting in extensive internal bleeding. When normal human plasma is added to hemophiliac blood, clotting proceeds normally; therefore, transfusion of normal blood into a hemophiliac can prevent a bleeding crisis. There is, however, no known cure. Several kinds of hemophilia attributable to absence of different factors from the blood are now recognized.

Another type of "bleeding" disease is *purpura,* in which hemorrhages occur in the skin and mucous membranes. This disorder is frequently caused by a reduction in the number of platelets in the blood (thrombocytopenia), and may be due to a cancer that destroys the bone marrow where platelets originate.

Excessive menstrual bleeding is sometimes caused by a deficiency or lack of thrombin or fibrinogen (afibrinogenemia). In this situation the blood has already formed clots, thus tying up the thrombin, and the fibrin usually has been dissolved by enzymes in the lining of the uterus. The result is a prolonged menstrual flow (period).

Diagnosis of Coagulation Abnormalities

Coagulation (Clotting) Time

Nature prevents excessive loss of blood by the formation of a clot. It is sometimes desirable to find out whether a patient's blood clots in the normal length of time. The most common means for measuring the clotting time is the Lee-White method, which uses a regular blood tube. Other methods require a small capillary tube. By the Lee-White method the normal clotting time is from 5 to 15 minutes.

Bleeding Time

In this test the blood is blotted at half-minute intervals after a pin prick is made in the ear lobe or finger. A piece of absorbent paper is touched to the blood drop repeatedly; after one to three minutes, no more drops should form. Prolonged bleeding generally indicates a low blood platelet count, pernicious anemia, or one of the acute leukemias.

Prothrombin Time

Prothrombin is the precursor of thrombin, and therefore is an important factor in the sequence of clot formation. A patient's prothrombin activity can be determined in the laboratory by mixing a sample of his plasma with standardized amounts of thromboplastin and calcium ions, and recording the time needed for coagulation to occur. A normal prothrombin time ranges from 15 to 20 seconds.

IMMUNITY

The body is continually invaded by microorganisms that enter through the respiratory passages and through small wounds in the skin. In most cases, these invasions are harmless, since they are dealt with immediately by the phagocytic leukocytes and by the immune reactions.

Invading organisms and the toxins they produce are composed of proteins. The danger of a foreign protein lies in the reactive chemical groups on its surface, for these can combine with groups in the body and interfere with normal processes. Such effects can be avoided if the invading protein is either destroyed or made inactive by combination with another protein (Fig. 13–10). Such a combination or action is, in essence, the *immune reaction.* A foreign protein that will waken the immune reac-

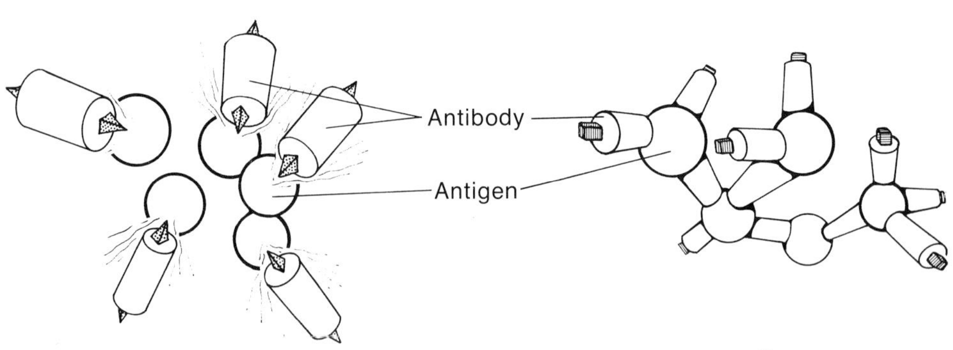

Figure 13–10 The immune reaction. A foreign protein (antibody) enters the bloodstream and is confronted by its respective antigen (*A*), which "ties up" the antibody (*B*) and inactivates it.

tion is called an *antigen*, and the protein in the individual's system that is produced to react with it is called an *antibody*.

Antibodies are part of the globulin fraction of the blood, and are produced by the lymphoid tissues and lymphocytes. Each antigen has a specific antibody that alone can react with it. The particular configuration of the combining sites is seemingly complementary to the determining portions of the antigens that induce their formation—an arrangement comparable to a lock and key (Fig. 13–10). The antigen molecule illustrated has many active chemical sites or determining groups protruding from its surface as specific configurations. These determine antigenic specificity, and fit into the pocket-like combining sites of the antibody. The number of antibody molecules produced is usually many times greater than the number of antigen molecules present.

Two broad groups of antibodies are recognized: natural antibodies, which are formed without apparent antigenic stimulation; and acquired antibodies, which appear only after exposure to a known antigen. The anti-A and anti-B blood group antibodies involved in blood reactions are examples of the natural type. Antibodies that are developed during the course of immunization (as for polio, measles, typhus, etc.) are examples of the acquired type.

The combination or binding of antibodies to antigens is manifested in several different ways:

1. *Agglutination*—when antigens such as large particles or whole cells (red blood cells or bacteria) are clumped together in a fluid.

2. *Precipitation*—when soluble antigen molecules become insoluble, as when bacteria have combined with an antibody and thus become more readily phagocytized by a white cell or macrophage.

3. *Lysis*—when red cells or bacteria rupture their cell membranes, releasing hemoglobin and cellular contents into solution.

4. *Opsonization*—when antibodies (opsonins) combine with bacteria or viruses and make them susceptible to phagocytosis.

The thymus gland, located in the chest cavity near the lower end of the trachea in children, is a prominent mass of lymphoid tissue. During early childhood the gland produces lymphoid cells responsible for antibody production during most of the life of the individual. Removal of the thymus gland in newborn mice has proved fatal, for the mice die from infection within a few weeks. Although the thymus is naturally atrophied and inactive in an adult, early removal of the gland in children can almost entirely prevent antibody production and, therefore, blocks the formation of an important line of defense against disease.

Allergy

Normally, the antigen-antibody reaction takes place in the bloodstream, and the debris produced is removed by the reticuloendothelial system. If an acquired immunity is weak, a foreign protein may penetrate the interstitial tissue before it comes into contact with an antibody. When the immune reaction occurs in the tissues themselves, the surrounding cells are irritated or destroyed by its side effects.

Allergies result from weak immunities, and the reactions in the tissues give rise to the allergic symptoms. In infancy, there may be allergic reaction to food proteins because the child has not yet developed the very strong immunity to them possessed by adults. More commonly, however, allergic reactions are due to materials that are protein-like and induce some antibody formation, but not enough to produce full immunity. These materials are know as *allergens* (e.g., dust and flower pollens).

The deleterious effects of the immune reaction in the tissue spaces release *histamine*, which is a potent vasodilator that causes redness and swelling of the skin (e.g., hives, urticaria) and in the irritated membranes of the eyes and sinuses (as in hay fever). If the release of histamine is excessive, a sudden vasodilation can occur throughout the body, lowering the blood pressure and causing shock. A massive response to an allergen is known as *anaphylactic shock.* This can occur when a drug or vaccine, or any allergen to which a person is sensitive, has entered the circulatory system. In other allergies, the immune reaction causes constriction of the smooth muscle fiber of the respiratory tract, producing asthma—a difficulty in getting enough air into and out of the lungs.

The symptoms of a hay fever type of allergy may be relieved by the antihistamine drugs, and those of asthma by epinephrine. More positive treatment can be carried out following intensive skin testing with small amounts of various allergens. Serum injections are now commonly used over an extended period, either to alleviate the temporary discomfort or to eliminate the problem entirely.

BLOOD TYPING AND TRANSFUSIONS

If, for some reason, the amount of blood in the body is severely reduced, whether through hemorrhage, disease, or surgical procedures, the body cells suffer from lack of oxygen and nutrients. The usual routine in such cases is to replenish the lost blood by providing an injection of blood from another person. This is called a *blood transfusion.*

Such procedures may be complicated, for the blood of some persons is incompatible with that of others. The plasma proteins and red cells of each individual are slightly different from those of everyone else. Therefore, a transfusion of blood from one person to another can cause an immune reaction. From birth, an infant's plasma develops antibody-like proteins called *agglutinins.* These will react specifically with proteins in the red cells from some other blood, and the resulting immune reaction causes clumping (agglutination) and lysis of the red cells (hemolysis). The red cell proteins (antigens) involved are called *agglutinogens.*

ABO Blood Grouping

In 1900, Karl Landsteiner, a Viennese pathologist, discovered that human blood is not the same in all individuals. He found that human blood can be classified into four well-defined groups, based on the presence or absence of two specific antigens located on the surface of the red blood cells. These agglutinogens, labeled A and B, may be present on the erythrocytes of an individual in one of three possible combinations, or they may be completely absent. When an individual has only one of the antigens, the blood group is called A or B, depending on which is present. When both antigens are present, the blood is known as AB; when both are absent, the blood group is known as O (Fig. 13–11).

Fortunately, when an agglutinogen is present in the blood, the opposite agglutinin is absent. Thus the serum of blood type A always contains the specific antibody designated as anti-B. Type B blood has an anti-A antibody; type AB blood has both antigens, and therefore has neither antibody. Type O blood has neither antigen, so it contains both anti-A and anti-B antibodies.

If a group A individual should through error receive a transfusion of type B blood, his anti-B plasma antibodies will react with the incoming red cell B antigen, resulting in the clumping of the red cells of the transfused donor blood (Fig. 13–10). Similarly, the A antigen of the donor's blood would react with the anti-A of the patient's B type blood, causing clumping of the former's red cells. Individuals who belong to the AB group are "universal recipients" since they do not have any antibod-

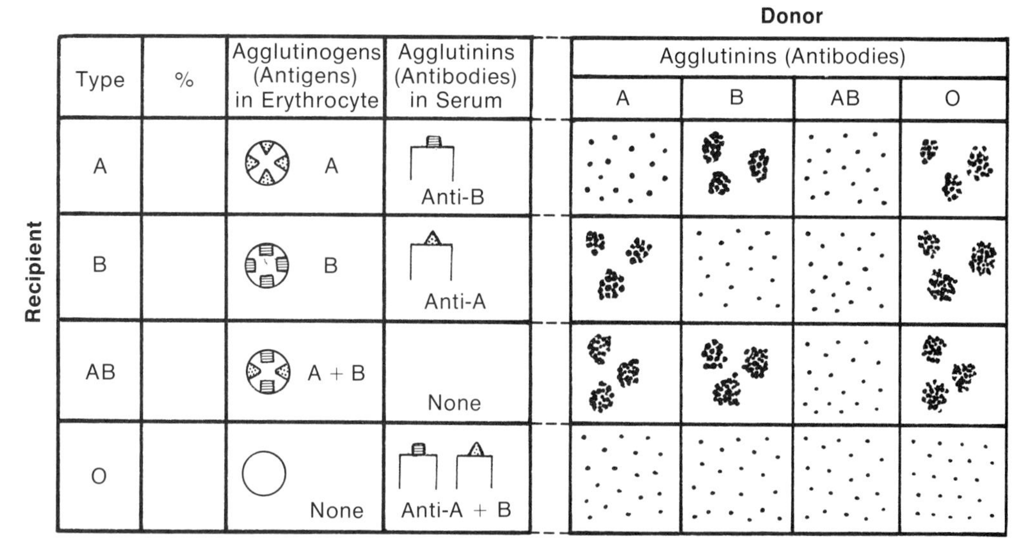

Figure 13–11 ABO blood group system. If a cross-match of two persons is necessary, the two blood samples are mixed and observed under the microscope. If the recipient and donor are both type A, the blood cells will be evenly distributed. If, however, the recipient is type A and the donor is type B, the antibodies present in the recipient's blood will cause the donor's blood cells to clump or agglutinate, proving the two blood specimens to be incompatible.

ies in their plasma and are able to receive any type of blood. Individuals of the type O group are called "universal donors" since they do not have either of the antigens, and therefore cannot cause any clumping in the receiver's blood. Whatever a person's blood type, however, he or she can usually give blood safely to another with the same blood type. In all cases, before a transfusion is given, determination of compatibility is made (cross-matching of bloods).

Rh Factor

About 85 per cent of the populace have another red cell protein called the Rh factor. Such individuals are said to be Rh+ (positive); the remaining 15 per cent lack the protein and are said to be Rh− (negative). If Rh-positive blood is given by transfusion to an Rh-negative person who has previously been sensitized to the Rh factor, the blood of the recipient will produce counteracting antibodies (anti-D) that destroy the red cells containing the "foreign" Rh-positive blood.

This phenomenon may have serious consequences during pregnancy, for it can result in a blood condition fatal to the fetus or newborn child of an Rh-negative mother and Rh-positive father. If the fetus inherits the Rh-positive blood of the father (Fig. 13–12), the mother may form antibodies to Rh-positive erythrocytes during or after childbirth, since there is a slight mixing of the fetal and maternal bloods (Fig. 13–13). During subsequent pregnancies, such antibodies may pass from her blood into the blood of the child, thereby causing destruction of red cells in the baby (erythroblastosis fetalis). The child may be born dead (stillborn); if it lives, replacement transfusions must be begun immediately (Table 13–2). Rh-negative blood seems to give the best results.

Erythroblastosis fetalis does not, however, invariably occur in these situations, for not all Rh-negative individuals develop anti-D antibodies in the presence of Rh-positive antigens. Furthermore, only a small percentage of Rh-negative women react in this way when pregnant with an Rh-positive fetus. Even if the mother's

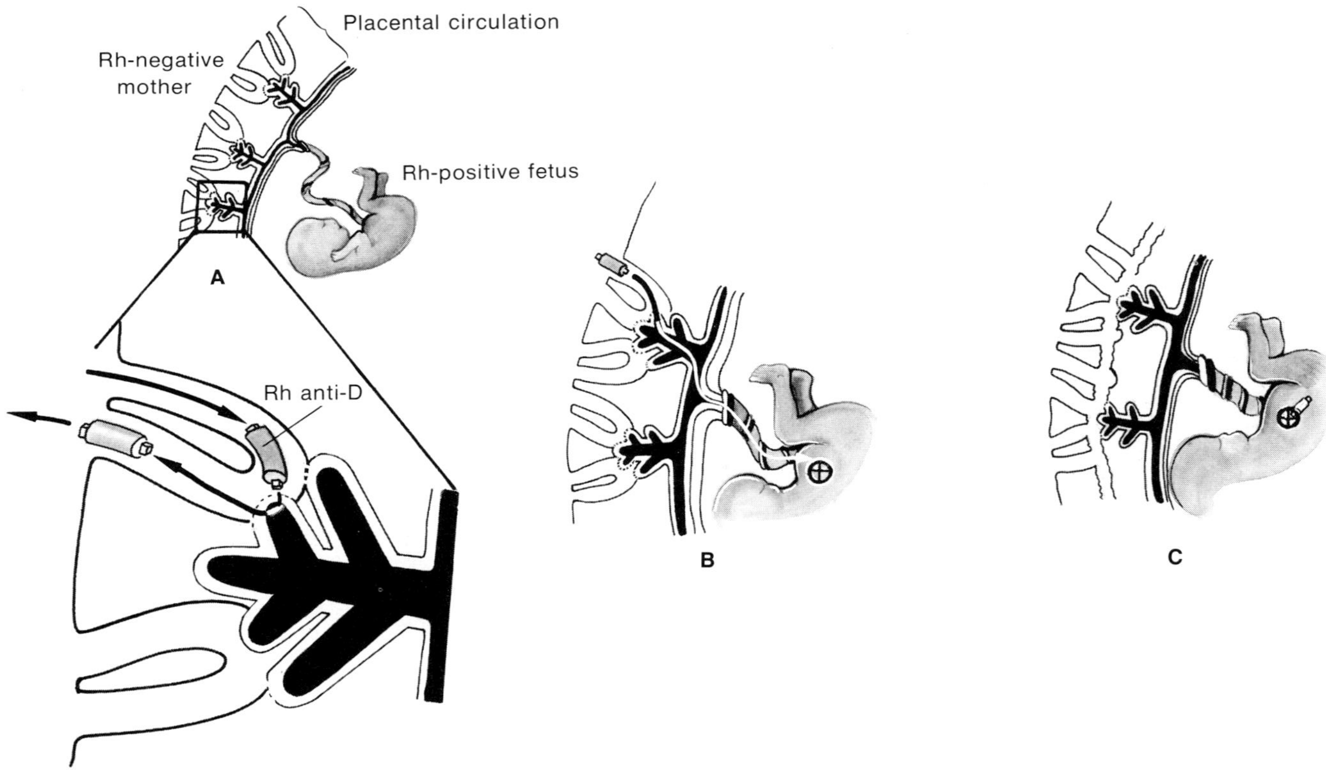

Figure 13–12 Rh factor. An Rh-negative mother may bear an Rh-positive fetus if the father is Rh-positive. If the mother passes her Rh antibodies to the fetus (*A* and *B*), the blood cells of the fetus will agglutinate and eventually be destroyed (*C*).

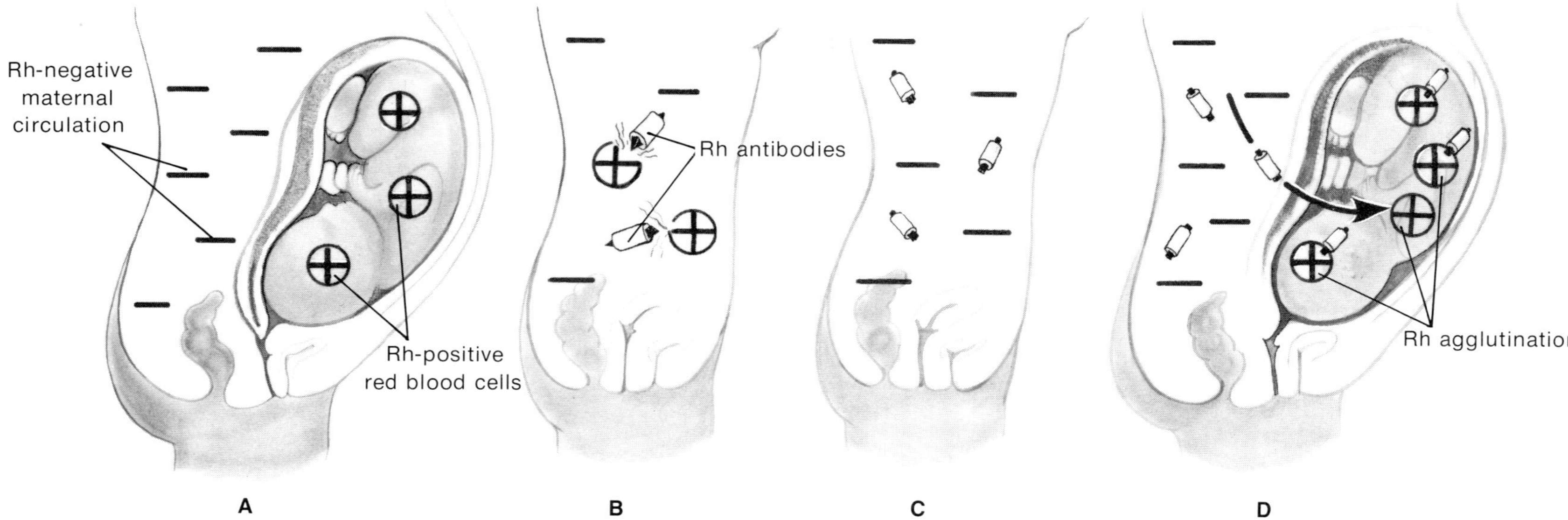

Figure 13–13 Erythroblastosis fetalis. *A*, Carriage of an Rh-positive fetus by an Rh-negative mother may not cause complications during the first pregnancy. *B*, Following delivery, however, mixing of the maternal and fetal bloods in the uterus will result in Rh antibody production (filter). *C* and *D*, In ensuing pregnancies, fetal blood cells will be destroyed.

Table 13–2 SELECTION OF BLOOD FOR EXCHANGE TRANSFUSIONS*

MOTHER	*BABY*	*BLOOD FOR EXCHANGE*
Group O	O A B	O
Group A	O B	O
	A AB	A or O
Group B	O A	O
	B AB	B or O
Group AB	A	A or O
	B	B or O
	AB	A, B, AB or O

*When the groups of the mother and baby differ, the baby's blood may not be compatible with the maternal serum because of the expected agglutinins anti-A or anti-B.

plasma does develop anti-D antibodies, in most cases the child will be quite normal. Trouble usually develops not during the first pregnancy but rather during subsequent ones.

RhoGAM (immune globulin-human) is the commercially prepared chemical designed to suppress maternal antibody production as a result of exposure to Rh-(D) positive fetal red cells. The principle is that passive antibody administration of RhoGAM in the proper dose to the Rh-negative mother at or soon after delivery (within 72 hours) prevents her from responding actively to the antigenic stimulus of incompatible fetal cells that entered her circulation at delivery. The protection given at the delivery of the first baby does not, however, prevent the mother from generating anti-D antibodies in subsequent pregnancies. Therefore, RhoGAM must be given immediately following each delivery.

ROUTINE BLOOD STUDIES

Many different studies may be made of the blood. A few of these have already been mentioned (clotting and bleeding time, prothrombin activity, etc.); the following are done routinely as part of a physical examination.

Hemoglobin (Hb) and Hematocrit (HCT)

Cellular metabolism requires an adequate supply of oxygen to assure the body sufficient energy yield and use. The hemoglobinometer (hemometer) is a device that measures the hemoglobin by comparing the blood with a standard color scale. The normal average is 15 gm of hemoglobin per 100 ml of blood; as has been mentioned, a low content of iron-rich Hb results in basic anemia.

The hematocrit, a measurement of the erythrocytes of the blood, is determined by filling a capillary tube with citrated blood (blood that has been treated with sodium citrate to prevent coagulation), packing the cells by centrifugation, and then reading the percentage of packed cells per total volume of the blood sample. A hematocrit greater than 45 per cent can indicate too many red blood cells (polycythemia); a hematocrit less than 45 per cent generally indicates anemia.

Blood Cell Count

A hemocytometer (Fig. 13–14) is a device used for counting the blood cells in a sample. Two blood-drawing tubes are used, one for red cells and the other for white cells, together with an accurately ruled glass slide for viewing the blood samples under the microscope. The purpose of the gridiron rulings is to provide a fixed area in which the cells may be counted. A normal red cell count varies

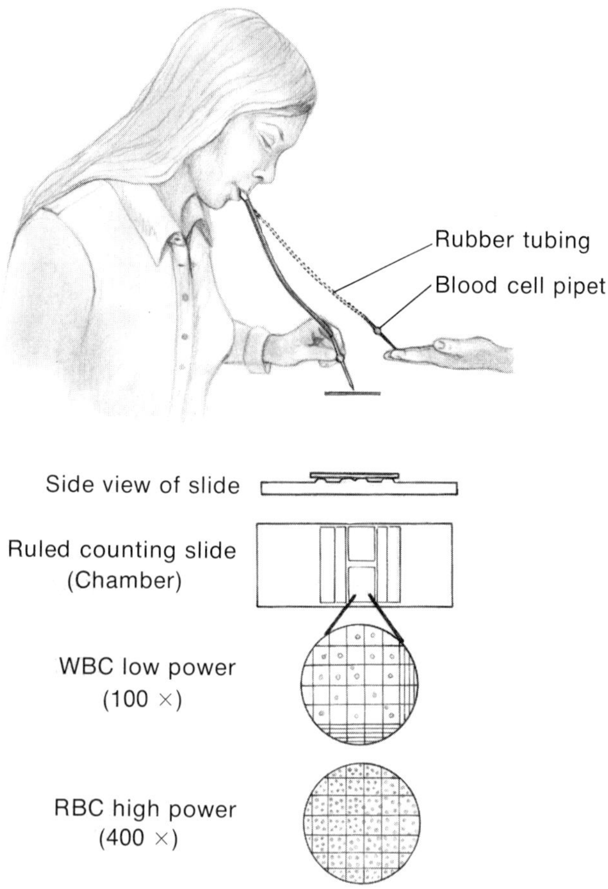

Figure 13–14 Parts of a hemocytometer, an apparatus commonly used to describe the number of red and white blood cells in a sample of blood. The technician draws the sample from the end of a finger using a calibrated pipet, dilutes the sample, places a drop of the mixture on the slide chamber, and counts the number of cells per square under a microscope. Abbreviations: WBC, white blood cells; RBC, red blood cells.

from 4.5 to 5.5 million cells per cubic millimeter; the white cell count varies from 5000 to 9000 per cubic millimeter. Erythrocytosis and leukocytosis are increases in either cell count, resulting, respectively, from overproduction of red cells and infection. Erythropenia and leukopenia reflect a decreased cell count due to disease or hemorrhage.

White Blood Cell Differential

A drop of blood is spread very thinly and evenly over a glass slide. A special stain (Wright's) is applied to differentiate the colorless white cells, and the number and shape of the different white blood cells are counted and noted under the microscope. Abnormal red cells, parasitic infective stages, and abnormal shifts in the white cells indicate different types of anemia, infection, infestation, and leukemia. Such a count is an important aid to the physician in making a diagnosis.

Figure 13–14 demonstrates the manual technique of performing a complete blood count (CBC). Today, there are new instruments on the market (Counter Counters, Autoanalyzers, etc.) that use a small quantity of blood and automatically perform a number of tests simultaneously, saving the technician valuable time and providing a more rapid means of obtaining the clinical results so necessary for the physician in treating his patient properly.

OUTLINE

- I. Functions of Blood
  - A. Carries oxygen and nutrients to tissues
  - B. Carries waste products away from tissues
  - C. Defends against injury and disease
  - D. Regulates body temperature
  - E. Carries hormones to cells
  - F. Maintains acid-base balance
  - G. Prevents loss of blood through its clotting mechanism
  - H. Maintains osmotic concentration in tissues
- II. Characteristics of Blood
  - A. Arterial blood—scarlet red
  - B. Venous blood—dark red
  - C. Thicker than water (average sp. gr. = 1.050)
  - D. pH = 7.38
  - E. Temperature = 38° C (100.4° F)
  - F. Blood volume = 5 quarts (9% body weight)
  - G. Made up of plasma (55%) and formed elements (45%)
- III. Formed Elements
  - A. Erythrocytes—red blood cells
    1. Biconcave, disk-shaped
    2. Produced in bone marrow
    3. Lack a nucleus
    4. Average 4.5 to 5.5 million per cubic millimeter
    5. Carry oxygen to tissues using hemoglobin
       - a. Hb = average 15 gm per 100 ml
    6. Life span = 120 days
    7. Phagocytized by reticuloendothelial system—hemolysis
    8. Intrinsic factor, extrinsic factor (vitamin B_{12}), and iron necessary for erythropoiesis
    9. Anemias
       - a. Primary—disturbance of red cell mechanism: pernicious anemia, aplastic anemia, hemolytic anemia, sickle cell anemia
       - b. Secondary—loss of blood, and defective hemoglobin formation (macrocytic hyperchromic and microcytic hypochromic anemias)
    10. Polycythemia—greater than normal number of red cells
    11. Hemoglobin—carries blood gases; destruction indicated by bilirubin level
  - B. Leukocytes—white blood cells
    1. Nucleated and polymorphic
    2. Granulated or nongranulated
    3. Primary function—defense against infection (phagocytosis)
    4. Secondary function—antibody production
    5. Life span = about two weeks
    6. Formed in bone marrow and in lymphatic tissue
    7. Destroyed in the liver
    8. Inflammation
       - a. Injury to tissue permits bacteria to enter the skin
       - b. Bacteria release toxic substances that damage the tissue
       - c. Increased blood supply to the damaged area produces reddening of the skin and increased temperature
       - d. Migrating white cells ingest the bacteria and the damaged tissue cells
       - e. Exuding blood plasma causes tissue edema
       - f. Suppuration—formation of pus around a laceration
    9. Neoplastic blood diseases
       - a. Myelogenous leukemia—increased granulocytes
       - b. Lymphatic leukemia—increased lymphocytes
       - c. Sternal puncture—means of detecting bone marrow abnormality
    10. Infectious mononucleosis—possible viral etiology
  - C. Thrombocytes—blood platelets
    1. Fragments of specialized bone marrow cells

2. Formed from megakaryocytes in the bone marrow
3. Lack a nucleus
4. Active in blood clotting mechanism

IV. Blood Plasma
- A. 55 per cent of total blood volume—the liquid portion approximately 92 per cent water
- B. Proteins
  1. Serum albumin—53 per cent of total plasma protein
     a. Regulates blood volume
     b. Synthesized in liver
  2. Globulin
     a. Alpha, beta, and gamma fractions
     b. Gamma fraction participates in immune reactions
  3. Fibrinogen—coagulation mechanism
  4. Prothrombin—synthesized with aid of vitamin K—coagulation mechanism
- C. Plasma electrolytes
  1. Sodium, chloride, potassium, calcium, phosphate, sulfate, magnesium—major plasma ions
  2. Iodide, copper, iron—trace quantities
- D. Nutrients
  1. Glucose—primary carbohydrate fraction
  2. Lipids (neutral fats, cholesterol, and phospholipids)—formed in liver cells
- E. Waste products
  1. Lactic acid, uric acid, urea, creatine, creatinine, ammonium salts
- F. Gases
  1. Oxygen, carbon dioxide, nitrogen—found in solution
- G. Regulatory and protective proteins
  1. Hormones, antibodies, enzymes, vitamins—found in variable concentrations
- H. Buffer systems
  1. Prevent major shifts in acid-base balance
  2. A weak acid (e.g., carbonic acid) and the salt of that acid (e.g., sodium bicarbonate), or a weak base in its salt
  3. Other important body buffer systems
     a. Phosphate buffer system
     b. Protein buffer system
- I. Clotting mechanism
  1. Prevents fatal loss of blood
  2. Clotting sequence
     a. The enzyme thromboplastinogenase and antihemophilic factors convert thromboplastinogen to thromboplastin
     b. Thromboplastin, with calcium ions, activates prothrombin to produce the enzyme thrombin
     c. Thrombin catalyzes conversion of soluble fibrinogen to insoluble fibrin—a blood clot
  3. Thrombus—stationary clot; embolus—clot brought by bloodstream from another site
  4. Hemorrhagic disorders
     a. Hemophilia—absence of antihemophilic factor (AHF)
     b. Purpura
     c. Excessive menstrual bleeding
  5. Anticoagulants that prevent excess clotting of blood—heparin, dicumarol
  6. Diagnosis of coagulation abnormalities
     a. Coagulation (clotting) time—5 to 15 minutes (Lee-White method)
     b. Bleeding time—1 to 3 minutes
     c. Prothrombin time—15 to 20 seconds

V. Immunity
- A. Body's defense against infection and disease
- B. Antigen—foreign protein that wakens immune reaction

C. Antibody—protein produced to react with a specific antigen
D. Antigen-antibody reaction comparable to lock and key
E. Types of antigen-antibody reactions
   1. Agglutination
   2. Precipitation
   3. Lysis
   4. Opsonization
F. Allergy—weak immunity due to protein-like substances in air, food, etc. that fail to induce sufficient antibody formation
   1. Redness and swelling (hives) —manifestations of allergy
   2. Antihistamine drugs and epinephrine—used to relieve hay fever, asthma, and anaphylactic shock

VI. Blood Typing and Transfusions
A. ABO blood grouping—based on presence or absence of two specific antigens on surface of erythrocytes
   1. Group A—only A antigen is present
   2. Group B—only B antigen is present
   3. Group AB—both antigens are present ("universal recipient")
   4. Group O—both antigens are absent ("universal donor")
B. Blood type compatibility—necessary to prevent red cell hemolysis (clumping) during transfusions
   1. Cross-matching
C. Rh factor
   1. Additional red cell protein possessed by 85 per cent of individuals (Rh-positive)
   2. Individuals lacking Rh factor—Rh-negative
   3. Serious complications may arise if Rh-negative mother gives birth to Rh-positive child
      a. If Rh-positive and Rh-negative bloods are mixed during birth of first child, antibodies from mother may hemolyze baby's blood during subsequent pregnancies —erythroblastosis fetalis
      b. RhoGAM—commercially prepared globulin used to prevent maternal antibody production

VII. Routine Blood Studies
A. Hemoglobin and hematocrit—used to detect anemia and polycythemia
B. Blood cell count—number of red and white blood cells; important in diagnosing hemorrhage, infections, etc.
C. White blood cell differential—shape and structure of white and red cells can indicate leukemias, infections, and infestations

STUDY QUESTIONS AND PROBLEMS

1. What are the functions of:
   a. Erythrocytes
   b. Leukocytes
   c. Thrombocytes
2. What is the advantage of transfusing whole blood rather than simply plasma?
3. When might a doctor order a transfusion of "packed cells"—plasma-free—rather than of whole blood?
4. What is cross-matching?
5. What is an Rh baby? Describe the probable blood type situation leading up to the birth of such a baby.
6. What is an "H & H"? Why would doctors order it?
7. Describe agglutination. Differentiate between coagulation and hemolysis.
8. What are the functions of blood?
9. Differentiate between plasma and serum.
10. Mrs. Jones had a baby and hemorrhaged. The diagnosis was afibrinogenemia. Is this a cause for bleeding? Describe it.
11. What are the different types of leukocytes? Describe their functions and indicate which of them has a migratory capability.
12. Why don't we form clots in our blood system?
13. What is meant by the ABO blood system? List the four blood types and their frequency of occurrence.
14. Describe the "lock-and-key" theory. What does it have to do with immunity?
15. What is the reticuloendothelial system? Why do we have one?

Chapter 14
The Heart

The heart is a hollow, muscular organ that can be thought of as two separate pumps operating in parallel. The right heart receives blood from the tissues and pumps it, via the pulmonary arteries, to the lungs. There it passes through capillaries, where it is oxygenated and where its carbon dioxide is removed. The pulmonary veins return the blood to the left heart, which pumps it out again through the aorta to supply the tissues of the body.

STRUCTURE OF THE HEART

Heart muscle, the *myocardium,* resembles voluntary skeletal muscle (Fig. 7–4) in that its fibers are striated—but there the likeness ends. In voluntary muscle, the fibers are separate; in heart muscle, which is involuntary, they are joined to one another by side branches, and each fiber has one central nucleus. This joining of muscle fibers, which results in a mass known as a *syncytium,* has an important bearing on the function of the heart, for it permits very rapid contractions of the whole organ. The characteristic appearance of myocardial fibers is created by short bands, the *intercalated disks,* which cross the fiber in a steplike formation.

The myocardium is made up of spirally wound sheets of muscle that start at a central point—the tip (apex) of the heart—wind around the heart, and return to their point of origin. In the process, these sheets surround the chambers of the heart and, upon contraction, propel the blood through them.

The heart and the bases of the great blood vessels are enclosed in a two-layered membrane, the *pericardium* (Fig. 14–1). Its outer layer is composed of fibrous tissue; the more delicate inner layer is a serous membrane. The serous membrane is closely applied to the heart (epicardium), whereas the fibrous portion forms a closed sac, the pericardial cavity. A small amount of pericardial fluid is produced by the inner serous membrane to prevent friction as the heart beats.

The *endocardium,* the inner lining of the heart, is also a serous membrane (Fig. 14–2). It covers all the heart's projections and irregularities, and is modified to form the cardiac valves. An inflammation of this lining is called *endocarditis,* a secondary infection that commonly accompanies such diseases as scarlet fever and rheumatic fever.

Cavities of the Heart

The right and left sides of the heart are separated by a muscular partition, the *septum.* Each half is divided into an upper compartment, the *atrium* or auricle, and a lower compartment, the *ventricle* (Fig. 14–2). Before birth there is an opening in the partition between the atria, but soon after birth this normally closes, so that there is no longer any communication between the right and left sides of the heart. The partition between the atria is called the interatrial septum; that between the ventricles is the interventricular septum. Each atrium communicates with the ventricle below it

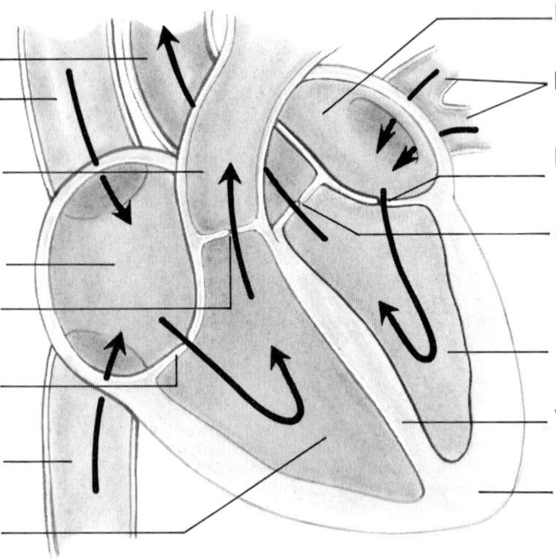

Figure 14–1 Diagram of the heart. Arrows indicate direction of blood flow.

on the same side by the *atrioventricular orifice,* an opening on the atrial floor. The left ventricle ejects blood into the extensive systemic circulatory system under higher pressure than is required of the right ventricle for ejecting blood into the relatively short pulmonary (lungs) circulation. The structural arrangement of muscle fibers provides for the thicker muscle to be on the left side, where it is needed in this stronger pumping action. The right and left sides of the heart contract and relax almost simultaneously.

Valves of the Heart

The atrioventricular orifice between each atrium and ventricle is protected by valves. On the right, the *atrioventricular (AV)* valve is composed of three irregularly shaped flaps or cusps, and is there-

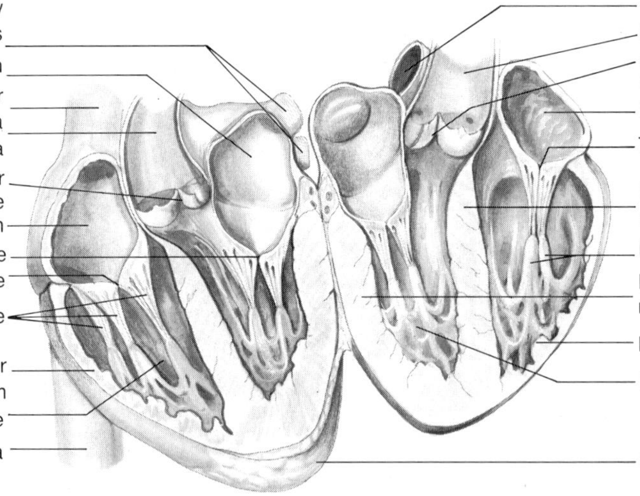

Figure 14–2 Internal anatomy of the heart. The heart has been dissected and the anterior portion "hinged open" to its left.

fore called the *tricuspid* valve (Figs. 14–1 to 14–3). The flaps are formed mainly of fibrous connective tissue covered by endocardium; their pointed ends project down into the ventricle and are attached to small *papillary* muscles by fibrous cords, the *chordae tendineae* (Figs. 14–2 and 14–3). These muscles contract when the ventricular walls contract, thus preventing the downward positioned flaps from being pushed back into the atria by the force of blood ejected from the ventricle.

The left atrioventricular valve bears two flaps or cusps, and is named the *bicuspid* or *mitral* valve (Figs. 14–1 and 14–2). It is attached in the same manner as the tricuspid valve, which it resembles closely; however, it is much stronger and thicker in all parts. The object of both valves is to allow blood to flow freely from the atria to the ventricles, but to prevent its flow in the reverse direction.

The orifice between the right ventricle and the pulmonary artery is guarded by the *pulmonary* valve (Fig. 14–3), and the orifice between the left ventricle and the aorta is guarded by the *aortic* valve (Fig. 14–2). These two valves are called *semilunar* valves because they consist of three cusps, each being half-moon-shaped

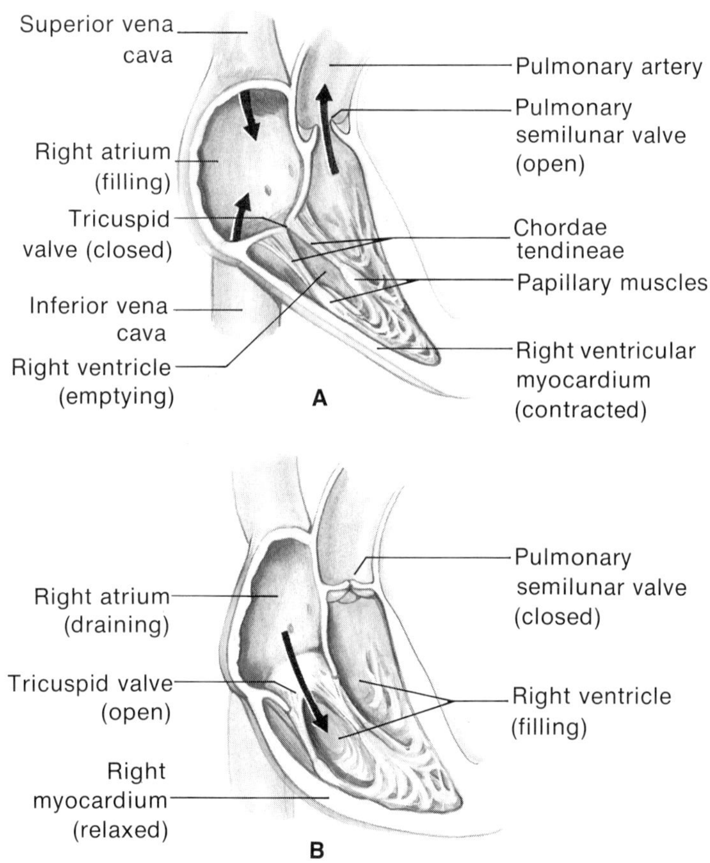

Figure 14–3 Valve action. Blood returning to the heart from the body fills the right atrium. *A*, At the same time, the tricuspid valve is closed to allow the ventricle to contract and to pump its contents out through the pulmonary semilunar valve into the pulmonary artery. Once the ventricle is empty, blood from the right atrium drains down through the tricuspid valve. *B*, Filling the relaxed right ventricle. The pulmonary semilunar valve closes, preventing any backflow of blood from the pulmonary artery. An identical cycle occurs simultaneously on the left side of the heart.

and attached by its outer curvature to the inside of the artery, where it joins the ventricle.

The semilunar valves offer no resistance to the passage of blood from the heart into the arteries, but they form an impenetrable barrier to the passage of arterial blood back into the heart. Each cusp fills with blood, and the borders distend so that they meet in the center of the vessel, preventing blood from flowing in a reverse direction. The aortic semilunar valve is larger and stronger so that it can withstand the greater pressure created by the action of the left ventricle.

THE CARDIAC CYCLE OR HEARTBEAT

The task of the heart is to propel blood to all parts of the body. In doing this, it acts as a pump working in definite strokes or beats. In the healthy adult these occur, on the average, about 72 times per minute.

Heartbeats may be felt by placing the hand flat over the region of the apex of the heart; that is, in the area just under the left breast about three and one-half inches from the midline. The sensation felt is one of bumping or thrusting, each thrust corresponding to a contraction of the heart.

The series of different actions that the heart performs regularly and in succession is spoken of as its *cycle*. Starting with any one of the separate actions or movements of the heart, the sequence of cardiac events that follows until the original action commences to repeat itself constitutes a complete cardiac cycle. The contraction of the heart is called *systole;* its relaxation is called *diastole* (Fig. 14–4).

For convenience, we can consider the events of the heartbeat starting from diastole, when the muscle is relaxed and relatively flaccid. Venous blood from the superior and inferior venae cavae pours into the right atrium; at the same time, oxygenated blood from the pulmonary veins pours into the left atrium. Meanwhile, the valves between the atria and the ventricles on each side are open so that the blood can flow through the atria and into the ventricles. Next, the atria contract and squeeze their blood content into the ventricles. It is a combination of the mild muscular contraction of the atria and the increased pressure of the full atrial chambers that ultimately drains the atrial blood into the ventricles.

Very shortly thereafter, the ventricles contract. The rising pressure closes the atrioventricular valves through which the blood has just been squeezed by atrial contraction, the valves being supported by the papillary muscles and the chordae tendineae (Fig. 14–2). As the pressure rises, it

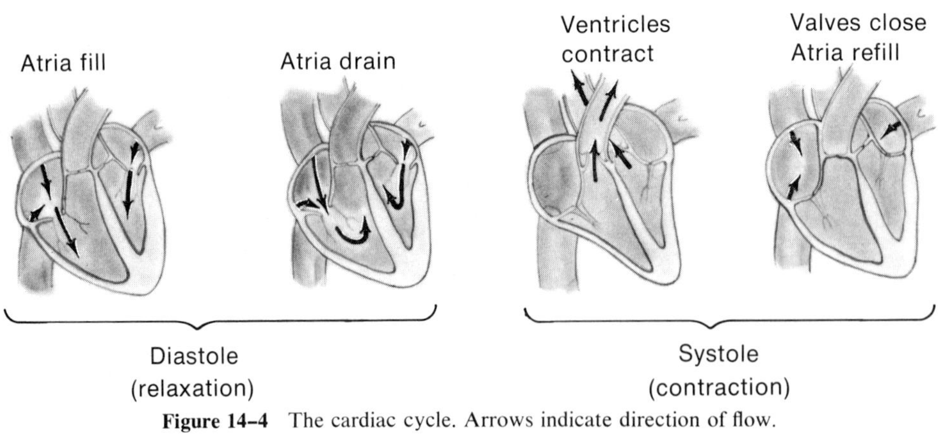

Figure 14–4 The cardiac cycle. Arrows indicate direction of flow.

forces open the semilunar valves on the pulmonary arteries and aorta, so that the blood is driven out into the circulation. When contraction is completed, the ventricles relax. The blood in the aorta is under high pressure, and the falling pressure in the ventricles causes this blood to attempt to reflux, or flow back into the ventricles. The semilunar valves of the aorta and the pulmonary artery prevent this reverse flow. During ventricular contraction, the atria once more relax and venous blood starts to fill them. When the ventricles are completely relaxed, the atrioventricular valves open again, blood flows into the ventricles, and the next cycle begins.

The Heart Sounds

Today's physician derives a great deal of information by carefully listening to the heart with a stethoscope. The first heart sound is associated with the initiation of ventricular systole, closing of the AV valves, and opening of the semilunar valves (Fig. 14–5). This first sound is lower in pitch and longer in duration than the second heart sound. Both sounds can be heard over the entire area of the heart, but the first sound, "lubb," is heard most clearly in the region of the apex of the heart.

The second sound, "dupp," is sharper and shorter, and has a higher pitch. It is heard best over the second right rib, for it is here that the aorta approaches nearest to the surface. The second sound is caused mainly by the closing of the semilunar valves during ventricular diastole. The momentum of the moving blood closes the valve cusps, preventing a backflow into the heart. Following the second sound there is a period of silence. The sequence sounds like:

—lubb, dupp, silence, lubb, dupp, silence, etc.—

Defects in the valves can cause excessive turbulence or regurgitation of the blood, and these are audible through the stethoscope as extra sounds. Abnormal sounds are called *murmurs,* and are due to faulty action of the valves. If, for example, the valves fail to close tightly and the blood leaks back, a "hissing" murmur is

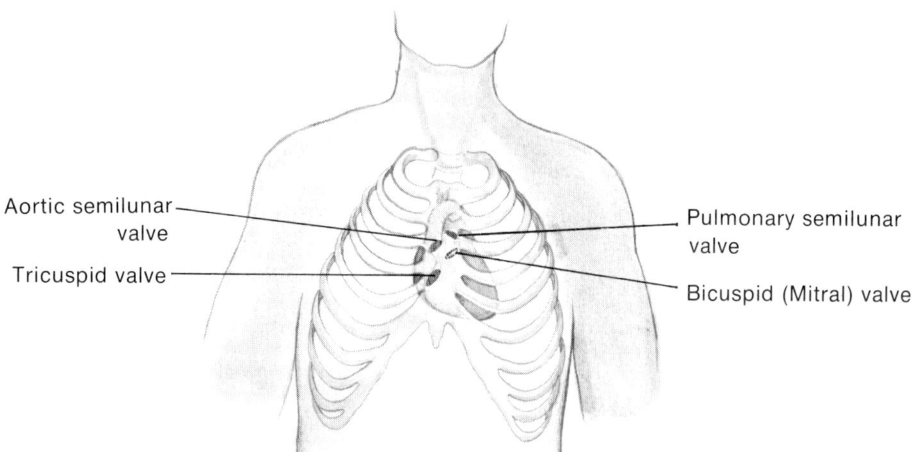

Figure 14–5 Heart sounds. The first heart sound is due to closure of the tricuspid and bicuspid valves and is longer and lower in pitch (systole = "lubb"). The second sound is due to closure of the semilunar valves and is shorter in duration but much higher in pitch (diastole = "dupp").

heard. Determination of which set of valves is faulty may be made by the sequence of sounds. If the AV valves are affected, the hissing occurs between the first and second heart sounds: —lubb, ssss, dupp, silence, etc.—

Another condition giving rise to abnormal sound is the narrowing (*stenosis*) or stiffening of a valve orifice. Mitral stenosis prevents an adequate flow of blood from the left atrium to the left ventricle, so that there is congestion in the pulmonary circulation. In addition, the left ventricle is forced to work much harder than usual to compensate for this condition, and may eventually fail.

The Heartbeat

Cardiac muscle, unlike skeletal muscle, is able to contract rhythmically. The heart is liberally innervated by the autonomic nervous system, which can modify the beat of the heart; but the fundamental rhythm of the normal heart muscle is completely independent of any nerve supply. The heart may be removed from an animal and continue to beat.

The contraction of the heart is due to a wave of depolarization spreading from a small area of specialized tissue in the right atrium, near the entry of the superior vena cava. This is the sinoatrial (SA) node, the principal pacemaker of the heart. It depolarizes spontaneously at regular intervals, the impulse spreading out over both atria in more or less concentric rings. The fibers of cardiac muscle are branching, and, because of this syncytial arrangement, the wave of excitation passes through the atrial muscle to the junction of the ventricles, where the atrioventricular (AV) node, the secondary pacemaker, is located (Fig. 14–6). Here there is a brief pause; then the AV node fires, and the depolarization spreads through the ventricles. There is a special conducting tissue (bundle of His) leading to the interventricular septum, where it divides and spreads through each ventricle (Purkinje fibers). The pause at the AV junction is important in that it allows time for the contracting atria to empty their contents into the ventricles.

Cardiac Rhythms

Cardiac muscle, as already mentioned, has an inherent rhythm; but this rhythm varies in different parts of the heart. The

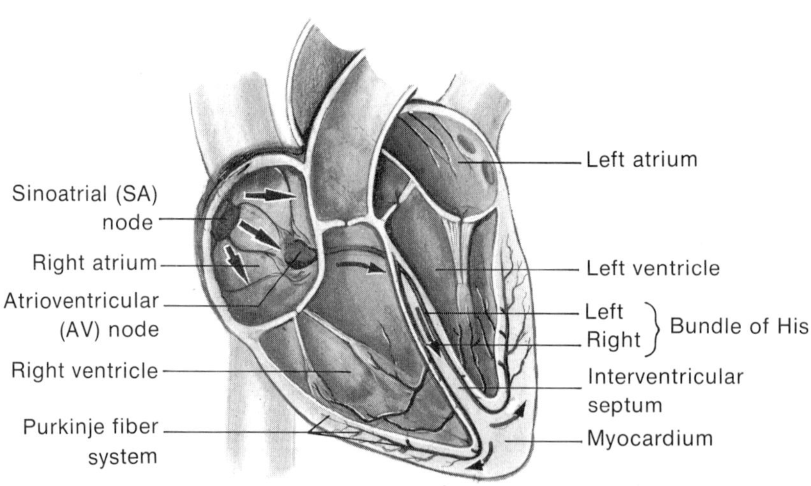

Figure 14–6 Excitation-conduction system. Diagram showing how an impulse is conducted from its point of origin, the right atrium, to its destination, the myocardium.

SA node has the fastest rhythm (70 to 74 beats per minute); therefore, the impulse spreads from the SA node to depolarize the rest of the heart. The rhythm of other areas is thus obscured. But the SA node may be destroyed or inhibited, thereby permitting the inherent rhythm of the AV node (45 to 65 beats per minute) to dominate. If the SA node's influence is removed by destruction or other means, the AV node takes over the initiation of the beat.

Abnormality in propagation of the wave of depolarization results in *heart block.* If the damage is total, and no impulses reach the ventricles, the ventricles take on a regular rhythm of their own, completely independent of the atria and bearing no relation to them. This is known as *complete* heart block. If, however, the damage to the bundle of His is not complete, every second or third impulse may pass down to the ventricles so that the ventricles contract once to every two or three atrial contractions—a condition known as *partial* heart block.

If the heart rate exceeds its normal maximum of about 74 beats per minute, the condition is termed *tachycardia.* Cases have been reported involving rates higher than 300 beats per minute. This is termed *flutter,* and generally involves the atria, while the ventricles contract at a rate of 175 to 200 beats per minute.

A complete breakdown in the rhythm and propagation of the impulse results in *fibrillation.* In this instance, the impulse follows no regular pathway, and many impulses seem to be traversing various parts of the myocardium at the same time. The result is a lack of rhythm or coordination of the beat and ultimate loss of the blood supply to the myocardium.

An electrical current or shock can stop fibrillation. Defibrillation is accomplished by throwing all parts of the myocardium into simultaneous activity. The heart will stop for an instant; then, following a rest period, the normal beat may take over.

Electrocardiogram (ECG)

On stimulation of a muscle (or nerve), the cell membranes are depolarized, and on recovery they are repolarized. These electrical events are spread throughout the body, and can be detected with suitable instruments applied to the skin surface at considerable distances from the sites of origin. The study of the heart's electrical activity is termed *electrocardiography.* The instrument used to record the electrocardiogram is called the electrocardiograph (ECG or EKG).

The electrodes, or leads, are usually made of metal and are slightly concave so as to make good contact with the skin in the regions of the wrists and ankles. A jelly containing an electrolyte is rubbed over the surface of the skin, and the electrode attached to this area facilitates conduction of the impulse from the skin to the machine.

Standard Leads

Lead I—right arm and left arm
Lead II—right arm and left leg
Lead III—left arm and left leg

For diagnostic purposes it has been found valuable to use leads other than these standard ones. In cardiac diagnosis, six standard positions on the chest are recognized (Fig. 14–7). These chest leads are referred to as V_1, V_2, and so forth. By comparing the records obtained from all six positions, one achieves a composite picture of the electrical activity of the heart that will disclose an abnormality more readily than will the three standard leads.

A normal electrocardiogram is shown in Figure 14–8.

1. *P wave*—The impulse is initiated by the SA node, and spreads over the atria. The P wave represents the electrical activity associated with the spread of the impulse over the atria; that is, the wave of depolarization or activity of the atria.

2. *QRS complex*—The wave of depolarization spreads from the SA node over the bundle branches (His) and the Purkinje

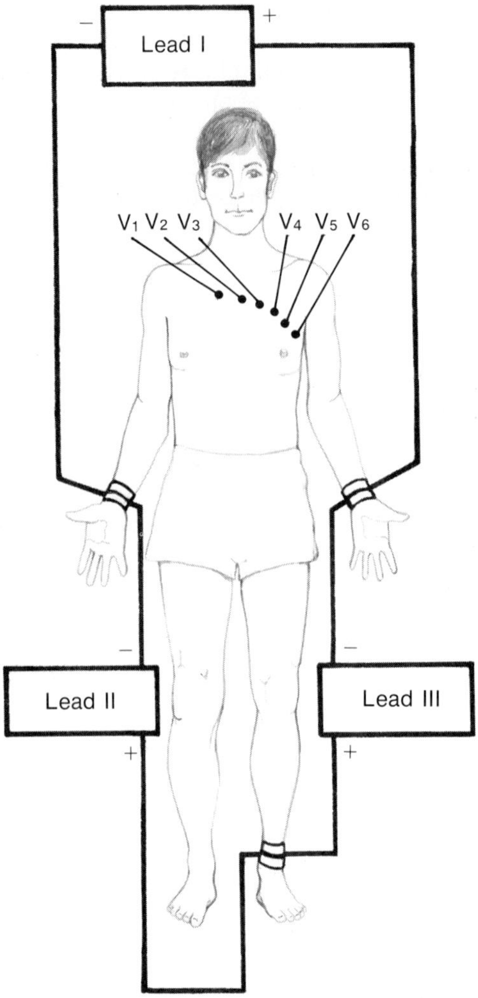

Figure 14–7 The standard EKG leads, including the chest leads V_1 to V_6.

system to activate both ventricles simultaneously. The QRS complex is the result of all electrical activity occurring in the ventricles.

3. *P-R interval*—This is measured from the beginning of the P wave to the beginning of the QRS complex, and indicates the time that elapses between the activation of the SA node and the activation of the AV node.

4. T wave—This represents ventricular repolarization.

Only a few of the more obvious abnormalities that may be detected by the electrocardiogram will be outlined here (Fig. 14–9):

1. *Abnormal rates and rhythms*—The normal rate is about 70 beats per minute. A very slow rate, *bradycardia,* is not considered pathological. An increased rate, tachycardia, is more frequently associated with disease, and is demonstrated by rapid beats (175 to 200 per minute).

2. *Heart block*—A P-R interval in excess of 0.2 second is indicative of this condition, together with an occasional missing QRS complex.

3. *Atrial fibrillation*—An atrial rate in excess of 180 to 200 is termed an atrial fibrillation or flutter.

No matter how refined the recording methods used, it should be emphasized that the ECG is simply a record of the fluctuations originating from the heart and spreading through the body. Some largely empirical rules for interpreting ECG's have been worked out over the years, and the intelligent application of this accumulated experience makes the ECG a valuable diagnostic and research tool.

CORONARY CIRCULATION

The blood vessels that supply the heart muscle are the coronary arteries and coronary veins.

Coronary Arteries

The left coronary artery originates in the left aortic sinus, gives off an anterior descending branch, runs into the left atrium, and, as the circumflex branch, joins with the right coronary artery in the left part of the coronary groove. The anterior branch supplies the ventricles; the circumflex branch supplies the left atrium and ventricle (Fig. 14–10). The right coronary artery originates in the right aortic sinus, and supplies the right atrium and ventricle.

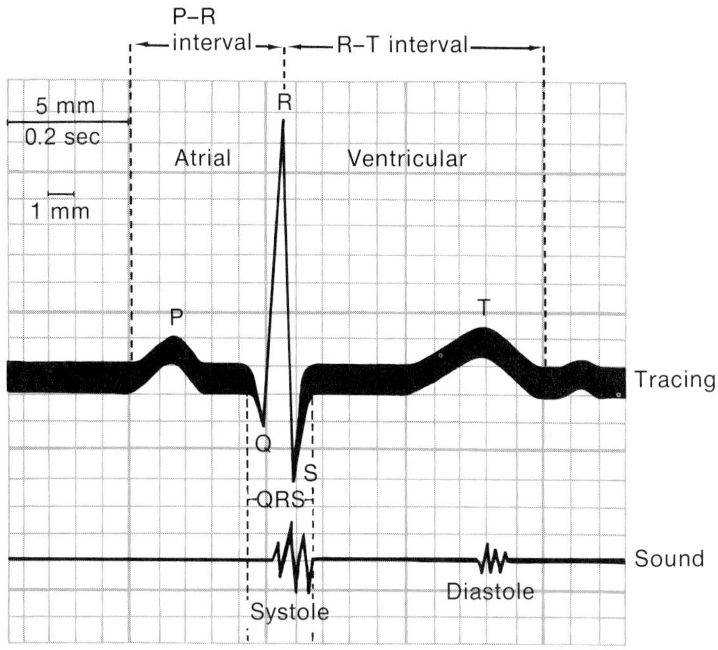

A

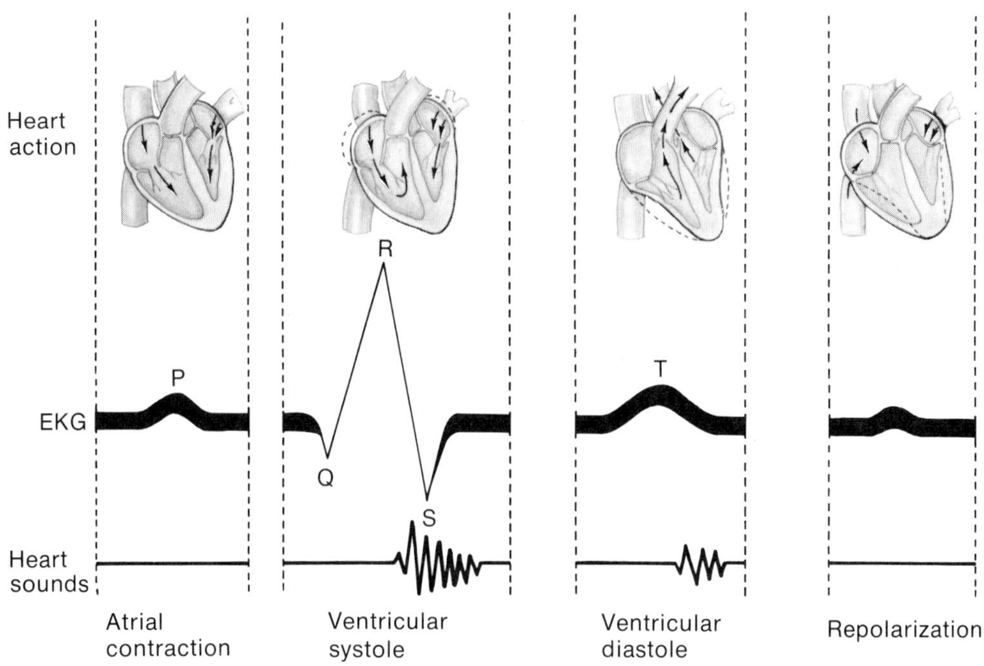

B

Figure 14–8 Electrocardiogram. *A*, Typical tracing or record of an EKG. *B*, Events during the cycle of heart contraction. A comparison of how the heart works, what it looks like while it works (EKG), and how it sounds during its activity.

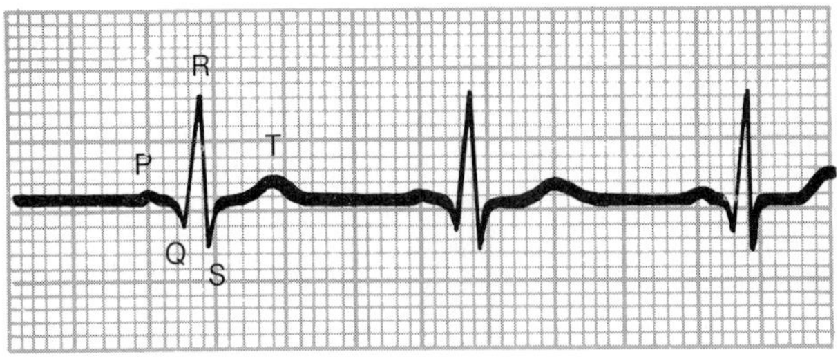

Normal EKG

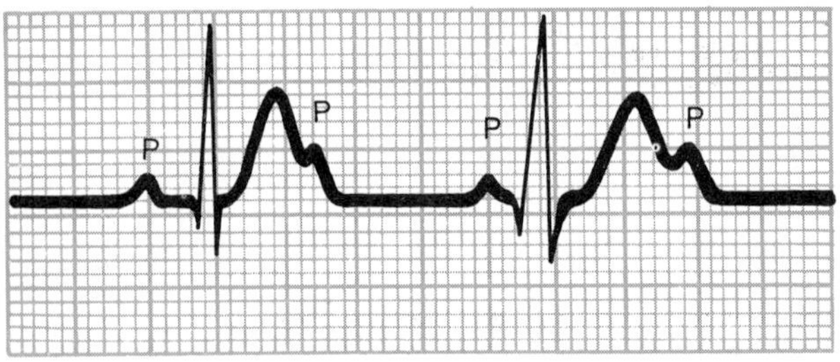

Heart block: P–R interval > 0.2 sec; missing QRS

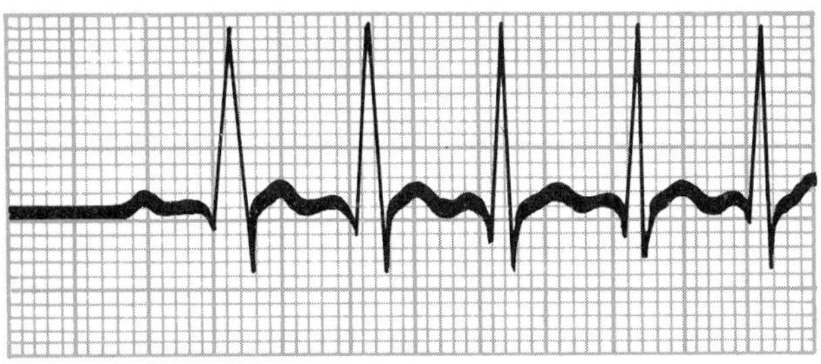

Atrial fibrillation: Heart rate 180+

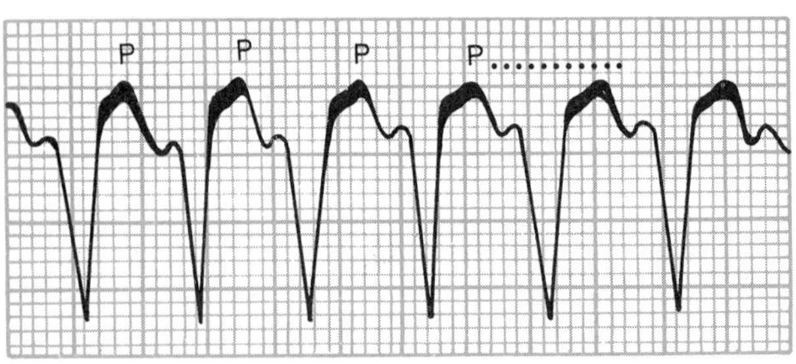

Atrial tachycardia: Heart rate 170+

Figure 14–9 EKG tracings, normal and abnormal.

These coronary arteries supply the heart muscle with blood. Obviously, any blockage within them is serious, because the heart muscle cannot continue its work without adequate oxygen and nutrition.

Coronary Veins

The coronary sinus receives most of the veins of the heart, and terminates in the right atrium. The great and small car-

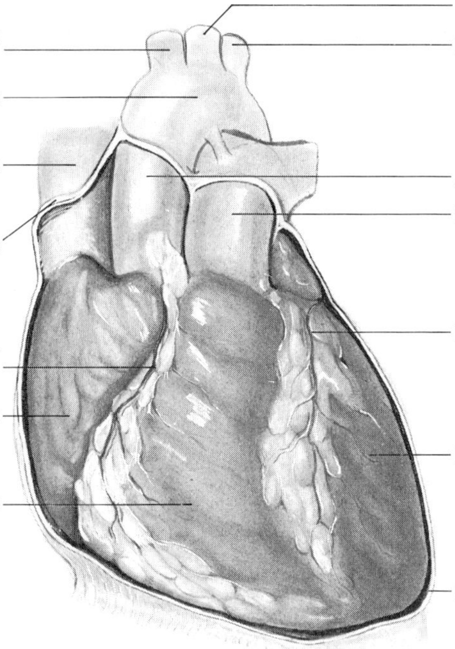

Anterior

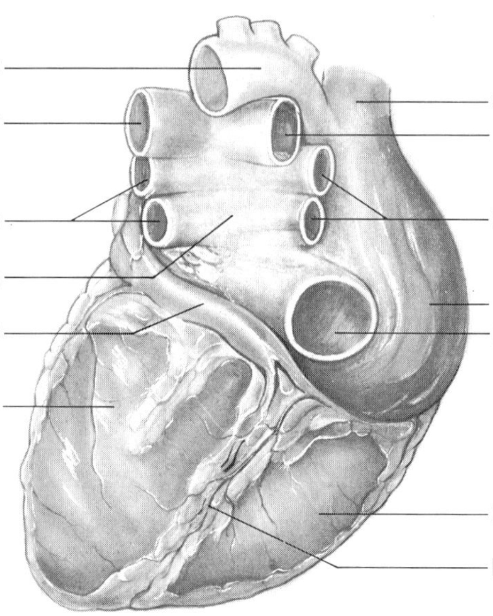

Posterior

Figure 14–10 External structure of the heart.

diac veins begin at the apex of the heart and ascend to enter the coronary sinus in the right atrium.

Coronary Artery Disease

A common cause of sudden death from heart disease is *coronary occlusion*—blockage of the heart's arterial supply by obstruction of a coronary artery. This interference with cardiac blood supply damages the heart muscle.

When an artery wall undergoes degenerative changes, its inside surface often becomes roughened. Such roughening is highly conducive to the formation or lodging of a thrombus—a blood clot that may cause closure of the vessel. When this occurs in a coronary artery, the condition is known as *coronary thrombosis*. If a smaller branch is blocked, the heart may continue to function, but severe obstruction of the major coronary arteries will cause sudden death.

The commonest cause of coronary artery disease is *arteriosclerosis*, a group of diseases characterized by degenerative hardening of the arteries. Hypertension, diabetes mellitus, hypothyroidism, excessive blood levels of cholesterol, and obesity are among the factors thought to contribute to the development of these conditions. Symptoms arise only when heart muscle damage becomes severe enough to cause a serious blood deficit. The chief clinical consequences are angina pectoris and myocardial infarction.

Angina Pectoris. Angina pectoris is chest pain so severe that the person must stop whatever he is doing. The pain is due to a temporary inability of the coronary arteries to supply enough oxygenated blood to the heart muscle, and is believed to be caused by stimulation of afferent nerve endings in the myocardium.

Myocardial Infarction. Myocardial infarction results when sudden occlusion of a coronary artery blocks the blood supply to the heart. A part of the myocardium dies because its oxygen supply is cut off. If the attack is acute and a great deal of the heart muscle is affected, the patient will die suddenly, and no necrosis (tissue breakdown) or infarction will be evident. If the person survives the initial attack, on the other hand, the affected part of the myocardium will become necrotic and die. In many instances, the collateral circulation—new branches of arteries that form to bring blood to the damaged area—is sufficient to maintain a reasonably adequate supply to the damaged area, at least for a time.

Direct myocardial revascularization by surgical implantation of double internal mammary arteries (Fig. 14–11), or by grafting of segments of vessels from elsewhere in the body, has brightened the hopes of people with coronary artery disease. The most important therapy, however, has been the rehabilitation of the patient so that he can return to a productive life. Drugs, proper diet, and rest have proved invaluable.

The myocardium is relatively rich in enzymes. Necrosis of muscle fibers following a myocardial infarction results in an increase in their levels in the blood (Fig. 14–12). Elevated levels are observed in practically all patients within a few hours after the occurrence of infarction. The degree of elevation appears to be roughly proportional to the extent of the myocardial necrosis.

NERVOUS CONTROL OF THE HEART

The heart, as we have seen, is innervated by the autonomic nervous system (Fig. 14–13). The vagus nerve, also known as the tenth cranial nerve, contributes fibers to the parasympathetic portion of the autonomic system. The vagus is a mixed nerve containing both motor and sensory components. The motor parasympathetic neurons innervate the SA node and the AV node. Although the heartbeat is started by the SA node pacemaker, the

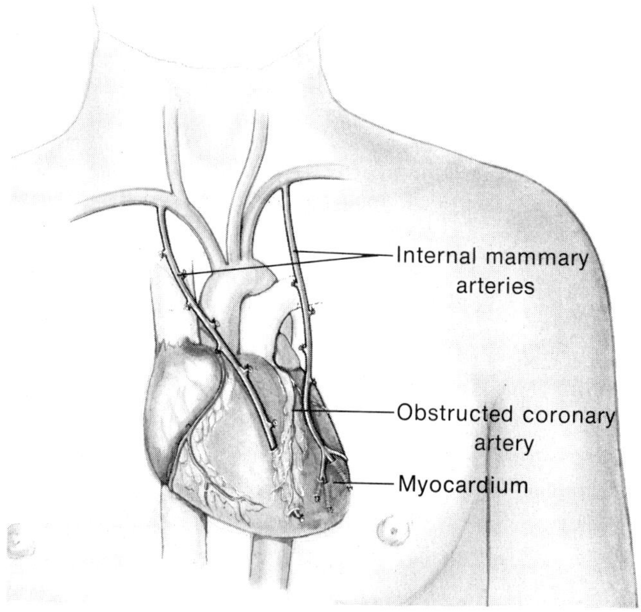

Figure 14–11 Mammary artery implantation. The internal mammary arteries are dissected from their points of attachment in the inner thoracic cavity and implanted in the myocardium below the obstructed coronary arteries.

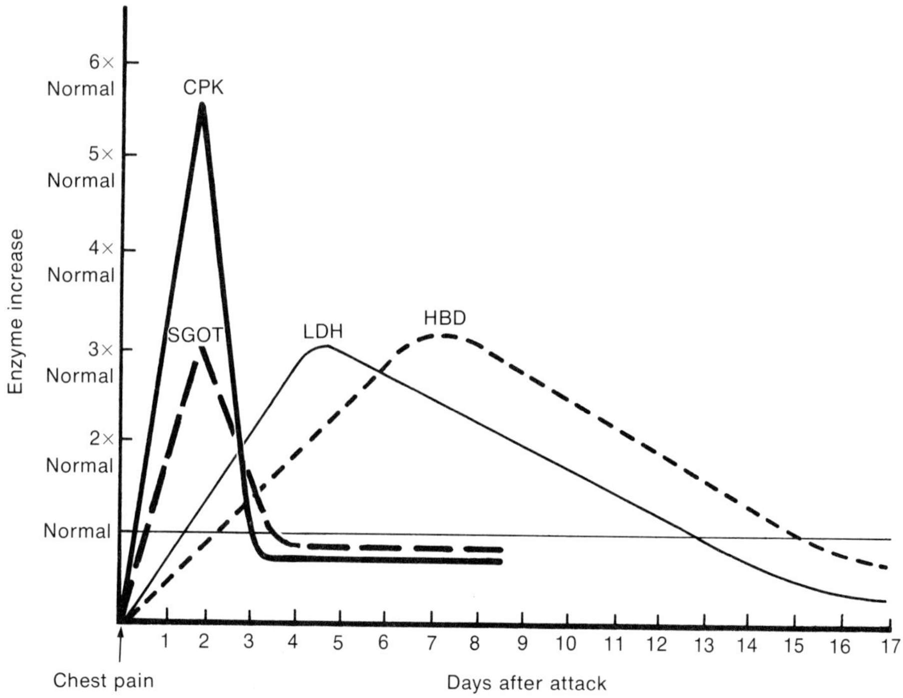

Figure 14–12 Typical changes in serum enzyme levels following myocardial infarction (MI). Note that the CPK level rises sharply but briefly, whereas the HBD level rises slowly but persists for a longer period of time. Abbreviations: CPK, creatine phosphokinase; SGOT, serum glutamic oxalacetic transaminase; LDH, lactic dehydrogenase; HBD, hydroxybutyric dehydrogenase.

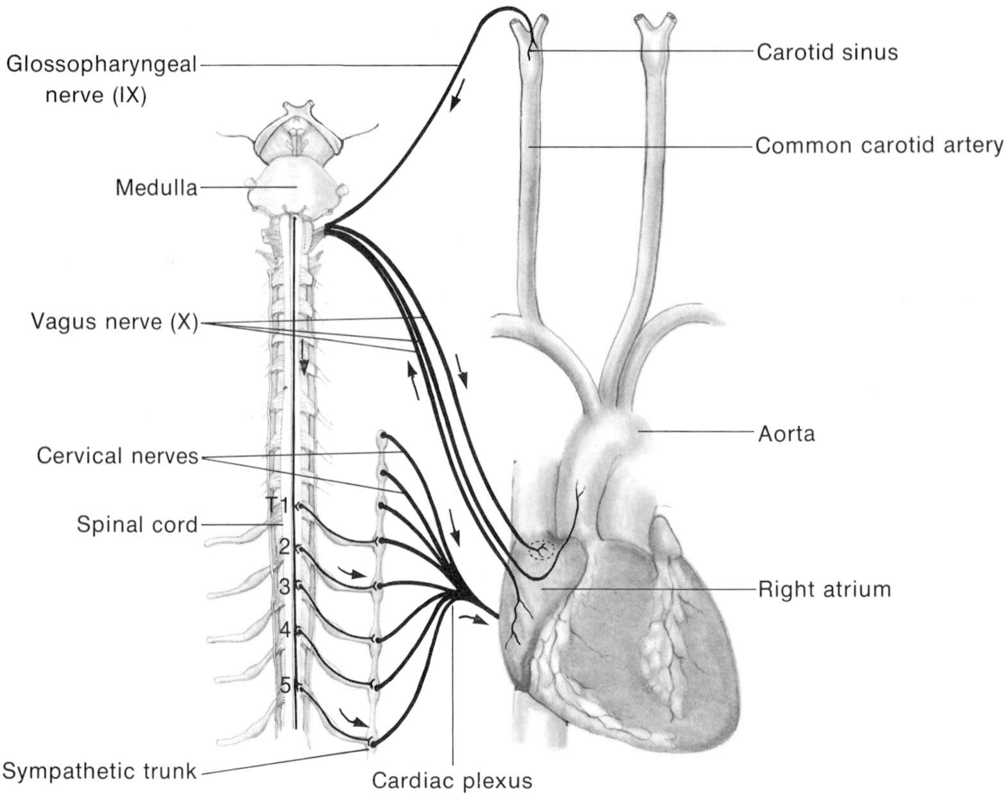

Figure 14–13 Nervous regulation of heart action. Arrows indicate the direction of nerve impulse travel.

frequency of the beat may vary considerably.

Stimulation of the vagal neurons results in a marked slowing of the heart and a decreased force of contraction. The vagus action is said to be inhibitory, in that it secretes acetylcholine at the nerve endings; this decreases the rate and rhythm at the node. It also decreases the excitability of the AV junctional fibers between the muscles of the atria and the Purkinje system, slowing transmission of impulses.

Sympathetic neurons arise from the first five thoracic segments of the spinal cord and then pass to the cervical ganglia. They innervate the SA and AV nodes and also the myocardium of the atria and the ventricles.

Stimulation of the sympathetic (accelerator) nerves speeds the heart and increases the force of contraction. In general, stimulation of the sympathetic nerves increases the activity of the heart through secretion of epinephrine, thereby increasing the effectiveness of the heart as a pump.

Cardiac Reflexes

The dual control of the heart just described is coordinated by centers located in the medulla oblongata (Fig. 14–14). The cardioinhibitor center gives rise to the parasympathetic neurons that travel along the vagus nerve to the heart. Activation of this center decreases cardiac action. The cardioaccelerator center sends neurons down the spinal cord to activate the accelerator (sympathetic) neurons, producing an increase in cardiac activity. The two centers also send neurons to each

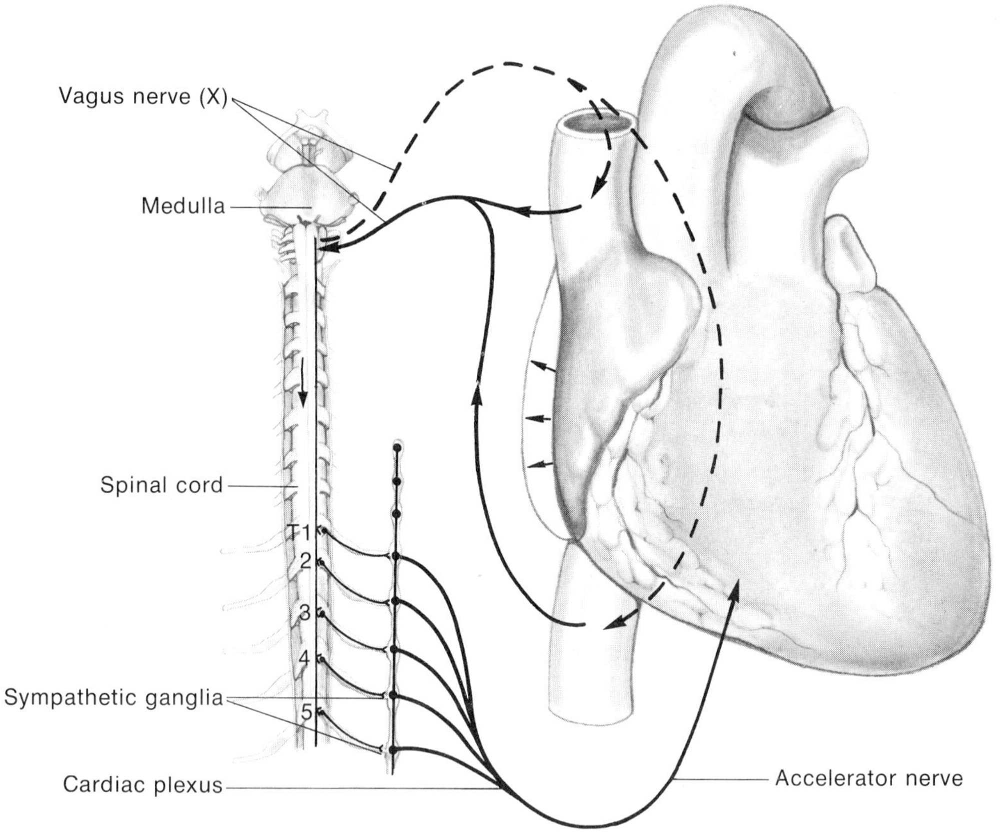

Figure 14–14 Cardiac reflexes. A stimulus from the accelerator nerves of the sympathetic nervous division increases the arterial blood pressure by an increased ventricular contraction. As the aorta stretches to accommodate the increased blood volume, stretch receptors in the aortic arch send an impulse to the brain by way of the vagus nerve, which acts to inhibit the impulses coming from the brain, resulting in slowing of the heart's activity and lowering of the blood pressure.

other, and these have an inhibitory effect. Therefore, activity of the cardioinhibitor center depresses the accelerator center, and vice versa. These centers, then, coordinate vagal and accelerator function, thus regulating cardiac activity.

Reflexes adjusting heart rate may be classified according to whether they are initiated by pressure receptors or by chemoreceptors.

1. *Pressure receptors,* also called *baroreceptors,* are sensitive to changes in blood pressure. They are found at the division of each common carotid artery, at the carotid sinus, and at the arch of the aorta (Fig. 14–13). The carotid and aortic sinuses are innervated by the glossopharyngeal (Cr. IX) and vagus (Cr. X) nerves, respectively. An elevation in blood pressure causes the pressure receptors to fire more frequently, which activates the cardioinhibitor center. As a result, the cardioaccelerator center is depressed, resulting in a slowing of the heart and a weakening of the force of contraction. Cardiac output falls, and therefore so does the arterial blood pressure.

2. *Chemoreceptors* are sensitive to oxygen and carbon dioxide tension of the blood. These receptors are found in the carotid and aortic bodies and are sensitive to lack of oxygen (hypoxemia). Impulses from these receptors are conveyed to the cardiac center, and the heart rate is accel-

erated, increasing the cardiac output, so that more blood is moved on to the tissue cells. Chemoreceptors are also stimulated by an increase in carbon dioxide (hypercapnia).

CLINICAL IMPLICATIONS

Bacterial Endocarditis

Acute and subacute bacterial endocarditis are characterized by an invasion of the endocardium by pathogenic bacteria that leads to progressive destruction of the heart valves. As a rule, death is due to toxemia and anemia rather than to heart failure.

Anticoagulant Therapy

Anticoagulant drugs are commonly used in the treatment of coronary thrombosis with myocardial infarction. In all cases, the use of anticoagulant drugs must be controlled by correctly chosen and carefully performed laboratory tests.

Two types of anticoagulant agents are in general use:

1. *Heparin*—which prevents extension of existing blood clots and formation of new ones.
2. Sodium warfarin (*Coumadin*)—which acts by depressing the production of prothrombin and certain other blood factors.

Chemotherapy

One of the oldest chemotherapeutic agents, and still a very valuable drug for many heart patients, is *digitalis,* which acts as a regulator in certain types of heart disease. It stimulates the myocardium directly, causing an increase in contractility and excitability. Other members of the digitalis family, *digoxin* and *digitoxin,* are also heart stimulants. They provide action of short and prolonged duration, respectively.

Nitroglycerin is used to dilate the coronary arteries, thus reducing blood pressure and cardiac work.

Heart Catheterization

Straightforward radiography can reveal the existence of a cardiac defect, but it can only hint at its severity. To answer questions concerning abnormal heart action, a measuring instrument, a *catheter,* is fed into the working heart itself. It is inserted through a vein in the arm or neck, fed (steered) down the vein, and then, with the aid of a fluoroscopic monitor, rotated until it enters the heart. Its bent tip will lead it in different forward directions so that the desired route may be selected without the catheter being pushed into other veins. The instrument can be fed into the right atrium, down through the tricuspid valve, and into the ventricle below, where blood samples can be withdrawn for analysis. Manometers can then be used to record instantaneous pressure changes. Harmless dyes that are opaque to x-rays provide a record of the catheter's route, revealing septal defects, faulty valves, and abnormalities in the great veins and arteries (Fig. 14–15).

Pacemakers

Electric battery-operated pacemakers that supply impulses to regulate the heartbeat have been implanted under the skin in the left abdominal area below the beltline. Electrode catheters attached to the artificial pacemakers are then passed beneath the skin to the myocardium, where they are returned to the surface through a previously made incision in the chest wall (Fig. 14–16). The batteries last from three to five years, and are easily replaced and maintained. Many people whose hearts cannot beat effectively because their biological pacemaker does not function properly have been saved by this simple device.

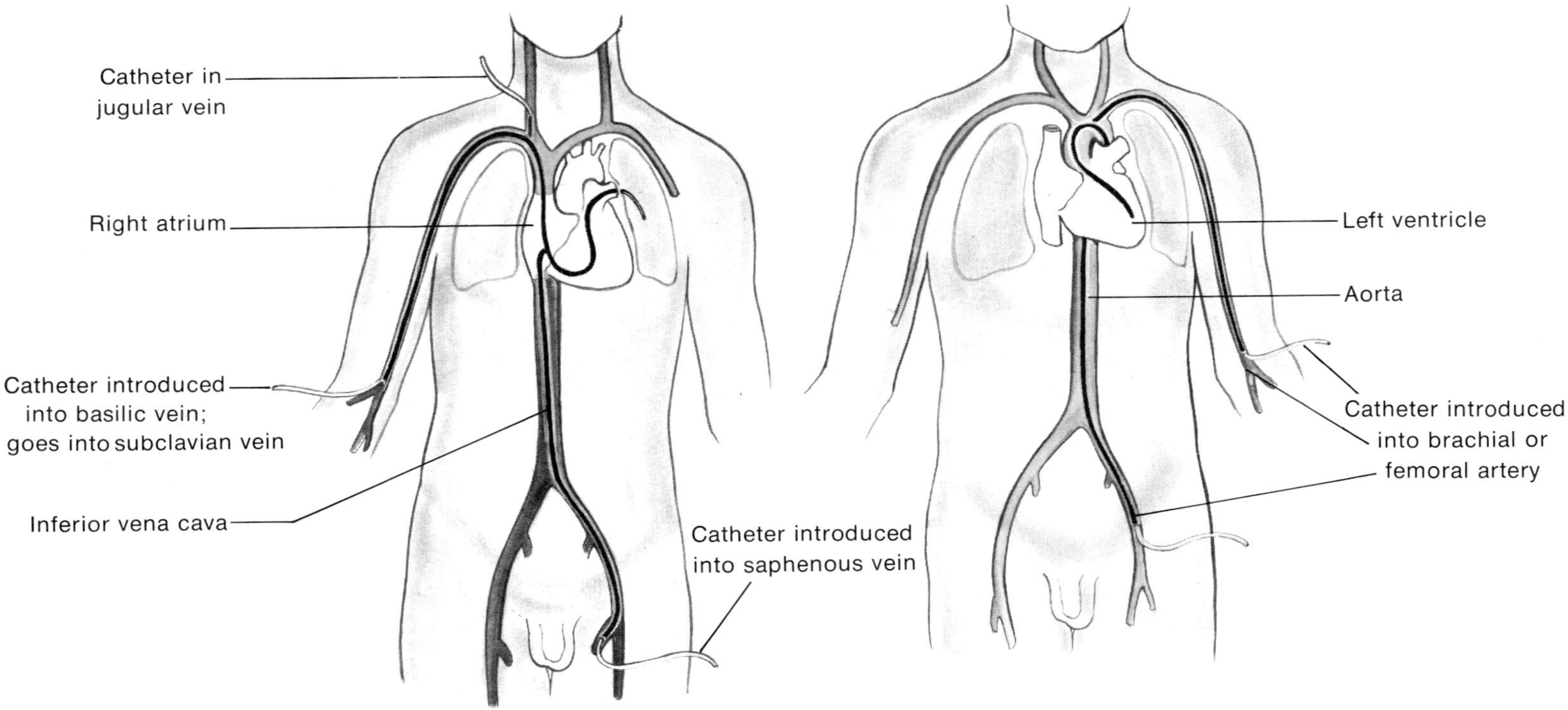

Figure 14–15 Cardiac catheterization. Insertion of a catheter into the right heart via the subclavian or saphenous veins or into the left heart via the brachial or femoral arteries to detect structural and other defects.

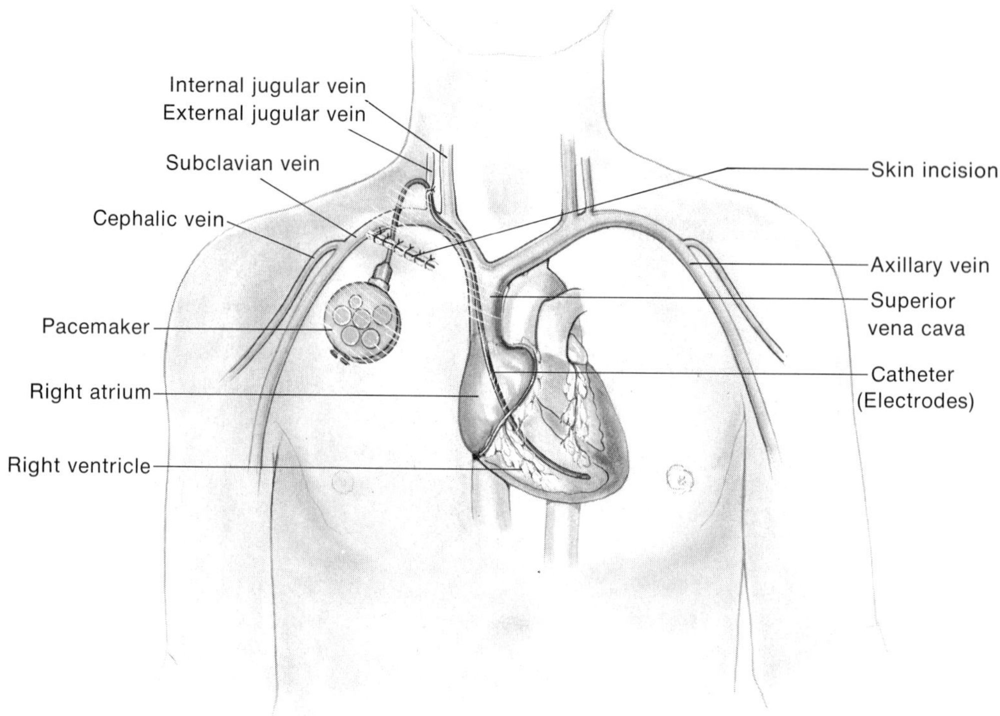

Figure 14–16 Pacemaker implantation, a current treatment for complete heart block. The electronic pacemaker replaces a nonfunctioning SA node.

The Artificial Heart

The heart-lung machine has made it possible to perform many surgical operations on the heart that previously could not be undertaken. The machine siphons off blood from the large vessels entering the heart on the right side so that no blood passes through the heart and lungs. While passing through the machine, the blood is oxygenated and carbon dioxide is removed. It is then returned to the general circulation through one of the large arteries. The blood is also "defoamed" to ensure that all air bubbles are removed, since this air could obstruct the blood vessels and be fatal to the patient.

A diseased valve may be repaired or replaced by an artificial valve. More spectacular has been the transplantation of a heart from the body of a person who has recently died to substitute for the diseased heart of a cardiac patient. Artificial hearts or parts of hearts designed to assist the ventricles in their pumping function have not proved as successful as hoped; however, research continues, and the outlook is optimistic.

OUTLINE

I. Introduction
II. Structure of the Heart
  A. Muscle and membranes
    1. Myocardium
      a. Heart muscle
      b. Fibers joined by side branches
    2. Pericardium
      a. Two-layered membrane enclosing heart
      b. Outer layer—fibrous tissue
      c. Inner layer—serous membrane (epicardium)

3. Endocardium
   a. Inner lining of heart
   b. Serous membrane
   c. Forms valves of heart

B. Cavities
1. Two upper atria and two lower ventricles
   a. More muscular on left side owing to greater work load

C. Valves
1. A right three-cusped AV valve and a left two-cusped AV valve
2. Two semilunar valves at the origins of the aortic and pulmonary arteries

III. Cardiac Cycle

A. Action of heart
1. Contraction of myocardium—systole
2. Relaxation—diastole
3. Sequence of events
   a. Blood from venae cavae and pulmonary veins enters both atria simultaneously
   b. Increased atrial pressure forces AV valves open, and blood enters ventricles by gravity flow
   c. Ventricular contraction forces blood upward, closing AV valves, opening semilunar valves, and discharging blood to arteries
   d. Relaxation of ventricles creates a backflow from arteries, closing the semilunar valves
   e. AV valves reopen, and next cycle begins

B. Heart sounds
1. Stethoscope
   a. Ventricular systole, closing of AV valves—"lubb"
   b. Ventricular diastole, closing of semilunar valves—"dupp"
   c. Murmurs—leakage through valves
   d. Mitral stenosis—congestion

C. Heartbeat
1. Depolarization spreads from specialized neuromuscular tissue
   a. Pacemaker in right atrium—SA node
   b. Impulses carried by atrial muscle to AV node at ventricular junction
   c. Bundle of His conducts impulses down ventricular septum to apex and up into muscle
   d. Purkinje fibers attach to heart muscles directly and innervate them
2. Cardiac rhythms
   a. SA node—70–74 beats per minute
   b. AV node—45–65 beats per minute
3. Abnormal conduction
   a. Heart block
   b. Tachycardia—up to 300 beats per minute
   c. Bradycardia—lower than 70 beats per minute
   d. Fibrillation—complete breakdown of rhythm and impulse propagation
4. Electrocardiogram (ECG)
   a. Records electrical events of heart
   b. Various leads from different positions on chest
   c. Normal ECG
      1) P wave—atrial activity
      2) QRS complex—ventricular activity
      3) T wave—ventricular repolarization

IV. Coronary Circulation

A. Arteries
1. Right and left coronary arteries supply myocardium with blood

B. Veins
1. Coronary sinus collects blood received from cardiac veins and returns it to right atrium

C. Coronary artery disease
  1. Coronary occlusion
    a. Coronary thrombosis
  2. Arteriosclerosis
    a. Angina pectoris
      1) Inadequate oxygenated blood supply to heart muscle
    b. Myocardial infarction
      1) Sudden occlusion of coronary artery damages myocardium
  3. Surgical repair
    a. Mammary artery implantation
  4. Clinical diagnosis
    a. Blood serum enzyme levels —used to detect myocardial necrosis

V. Nervous Control of the Heart
  A. Heartbeat starts itself (pacemaker) but is further controlled by the ANS
    1. Vagus nerve—inhibits
      a. Sympathetic nerves—accelerate
  B. Cardiac reflexes
    1. Medulla oblongata
      a. Cardioinhibitor center—parasympathetic (inhibits)
      b. Cardioaccelerator center—sympathetic (stimulates)
    2. Localized centers
      a. Pressure receptors – sensitive to changes in blood pressure
      b. Chemoreceptors – sensitive to oxygen and carbon dioxide tension of blood
      c. Counteractivity acts to regulate heart activity

VI. Clinical Implications
  A. Bacterial endocarditis
    1. Toxemia and anemia
  B. Anticoagulant therapy
    1. Heparin—antithrombin
    2. Coumadin—antiprothrombin
  C. Chemotherapy
    1. Digitalis family—stimulants
    2. Nitroglycerin—dilator
  D. Heart catheterization
    1. Hollow, narrow catheter
    2. Accompanying fluoroscopy
    3. Venous entrance to heart
    4. Blood samples analyzed
    5. Pressure gradients measured
    6. Radiographs detect abnormalities
  E. Pacemakers
    1. Electronic devices
    2. Battery inserted near hip
      a. Replaced every 3–5 years
  F. Artificial heart
    1. Heart-lung machine
    2. Exchange of gases and removal of air bubbles
    3. Effective in heart transplants and other forms of cardiac surgery

STUDY QUESTIONS AND PROBLEMS

1. Describe the valves of the heart and explain their relation to the heart sounds.
2. Why does edema occur when an individual has a congestive heart attack?
3. Where are the pressure receptors found?
4. What are chemoreceptors and what are they sensitive to?
5. Describe the ANS antagonistic control of the heart muscle.
6. What does the pacemaker have to do with the excitation-conduction pathway?
7. Discuss the principle of the electrocardiogram and its use in diagnosis.
8. Why are so many leads used in taking an ECG?
9. What does a thrombus have to do with coronary occlusion?
10. How is blood pressure modified?
11. Name and describe the four chambers of the heart and discuss their roles in the cardiac cycle.
12. Describe the two sets of valves of the heart and justify their structure and position.
13. Explain systole and diastole and relate them to the cardiac cycle.
14. What part does infection play in heart disease? Give examples.
15. Of what value is heart catheterization, and how is this procedure carried out?
16. Of what value to heart patients are anticoagulants and antibiotics?

Chapter 15
Circulation

Stores of oxygen and nutrients in the internal environment would be rapidly used up if they were not replenished by the turnover of the interstitial fluid accomplished, as we have seen, by the circulation of blood. This circulation permits the exchange of fluid and dissolved materials in the capillaries. The blood pressure provides the driving force for the ultrafiltration of fluid into the interstitial spaces, and the pressure of the plasma proteins causes a similar return of fluid at the distal ends of the capillaries. For this exchange to maintain the composition of the internal environment effectively, not only must the plasma composition be well regulated by the internal organs, but the plasma must be delivered to the tissues in sufficient volume to meet their oxygen and nutrient requirements—and at sufficient pressure to ensure the all-important ultrafiltration.

GENERAL CIRCULATION

It is appropriate to begin with a short review of the general plan of the circulation, which is shown diagrammatically in Figure 15–1.

As described in the preceding chapter, the heart is a muscular organ that can be thought of as two separate pumps in parallel. The right heart receives blood from the veins draining the tissues, and pumps it via the pulmonary arteries to the lungs. Here it passes through capillaries in which it is oxygenated and its carbon dioxide removed. The pulmonary veins return the blood to the left heart, which pumps it out again through the aorta, from which arise the various systemic arteries that distribute blood to the other tissues.

We see, therefore, that there are *two* parallel circulations: the pulmonary circulation for gas exchange in the lungs (see also p. 320, and the systemic circulation for maintaining a constant internal environment in the other tissues. Note that all blood goes through both systems.

THE BLOOD VESSELS

Arteries

Arteries carry blood away from the heart to the tissues; veins return blood from the tissues to the heart. The large arteries nearest the heart divide into smaller vessels. These, in turn, divide into still smaller ones, and so on. The smallest arteries are called *arterioles*. These lead into vessels of narrow caliber, the *capillaries*. The capillaries branch and form a network where the exchange of materials between blood and tissue fluid takes place. After passing through the capillary network, the blood is collected into the smallest veins, the *venules*. The venules unite to form small veins; these combine to make ever larger vessels until eventually the great veins that enter the heart are reached.

A typical artery in cross section (Fig. 15–2) consists of three layers:

1. An inner coat *(tunica intima)*, which itself consists of three layers: a layer of endothelial cells lining the arterial

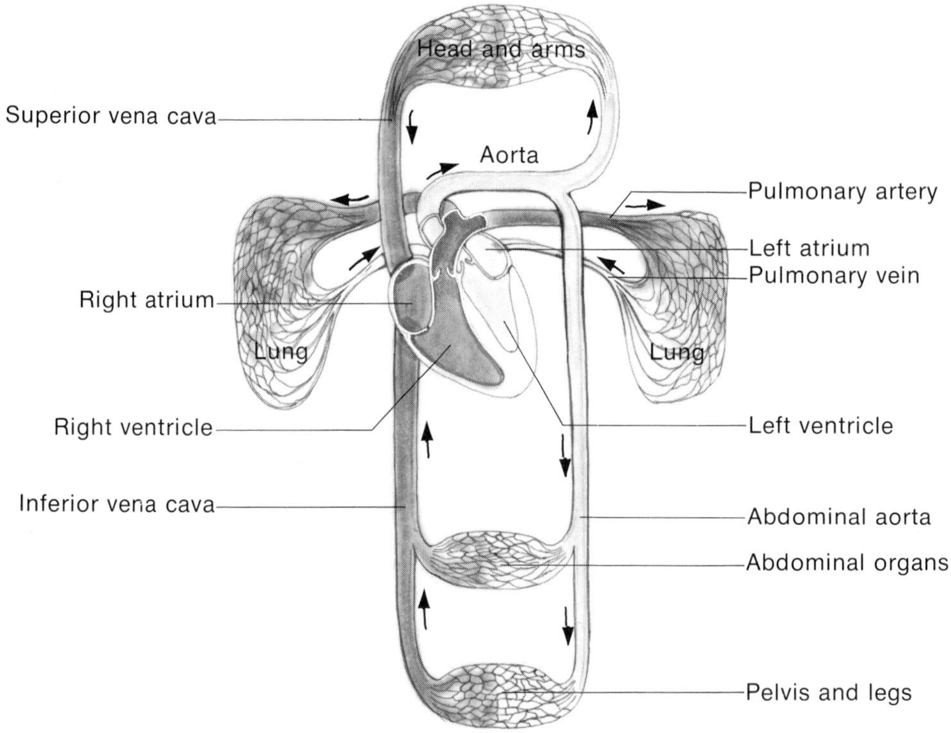

Figure 15–1 Diagram of general circulation.

passage (lumen), a layer of delicate connective tissue, and an elastic layer made up of a network of elastic fibers.

2. A middle coat *(tunica media),* which consists mainly of smooth muscle fibers with elastic and collagenous tissue.

3. An external coat *(tunica externa* or *adventitia),* which is composed of loose

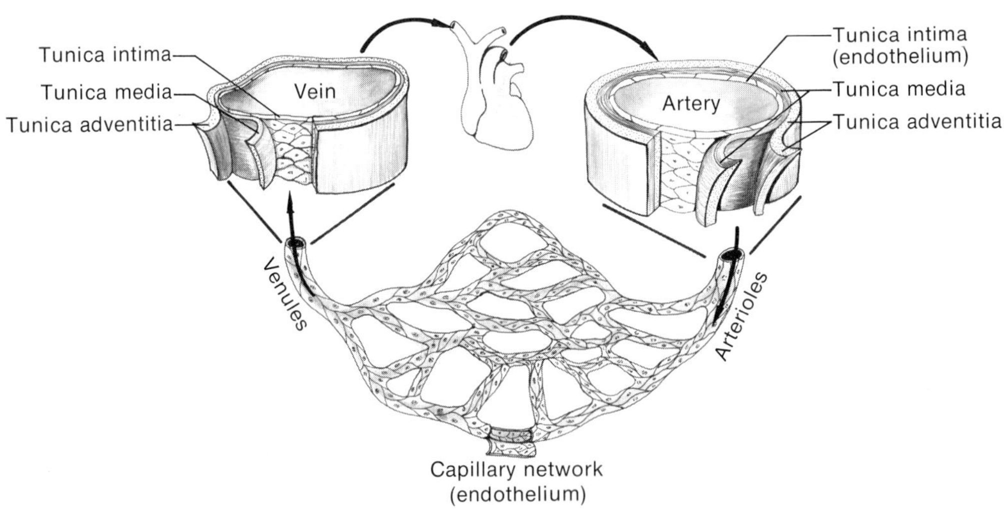

Figure 15–2 Blood vessels. Diagrammatic sketch showing the single-cell endothelium of all the vessels and the layered muscular coats of arteries and veins.

connective tissue with bundles of smooth muscle fibers and elastic tissue.

In the smaller arteries there is proportionately less elastic tissue and more smooth muscle; this is especially true of the arterioles. The elasticity of the large arteries is of great importance in maintaining a steadier blood flow than that initiated by the pulsating pumping of the heart. As the blood is forced into the aorta at systole, the great pressure stretches the arteries, which temporarily absorb some of the pumping energy. Then, when the pressure falls during diastole, the elastic rebound of the arterial walls compresses the blood and continues to drive it into the tissues, even though the heart is relaxed (Fig. 15–3).

The distention of the arteries during systole extends to the periphery, and can be detected as a throbbing sensation, the *pulse,* in several arteries near the surface. One of the most useful areas for measuring the pulse is the radial artery at the wrist, but the femoral pulse also may be palpated with ease. The latter is more convenient in examining infants. The pulse has the same frequency as the heart, but its propagation along an artery is determined by the structure of the vessel. The pulse wave travels in a manner similar to the wave of vibrations resulting from plucking a taut violin string.

Veins

The veins have the same three layers as the arteries, but differ in having a much thinner tunica media (Fig. 15–2). They have little elastic tissue or muscle, and therefore, when dissected, appear "collapsed."

The unique feature of the veins is the presence of valves made of loose pockets of endothelium, which permit the blood to flow in one direction only—toward the heart (Fig. 15–4). Veins have a much larger total diameter than arteries, and the blood moves back to the heart rather sluggishly compared with the surging force of the arterial circulation. The return of venous blood is helped by the "milking" action of muscular contractions (Fig. 15–4). When a vein is compressed by movement

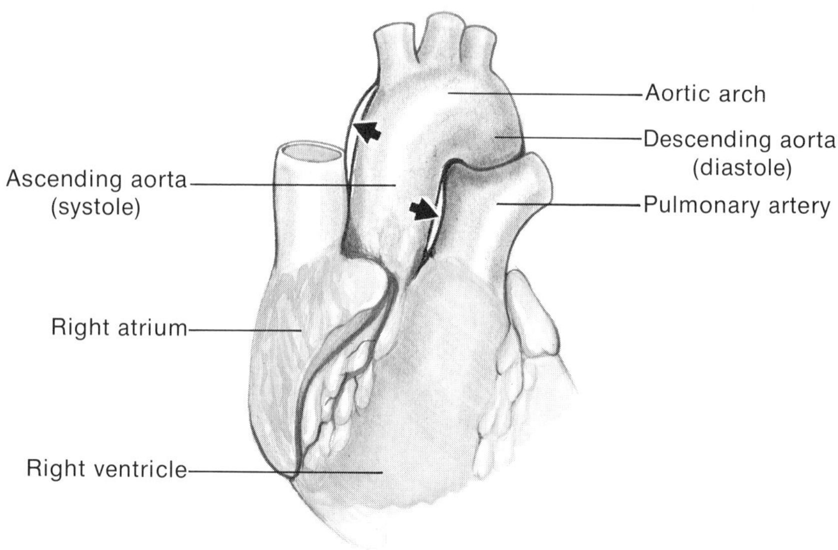

Figure 15–3 Elastic arteries stretch during ventricular systole owing to the increased pressure and volume of blood. When the ventricles relax (diastole), the pressure decreases and the arteries return to their normal shape.

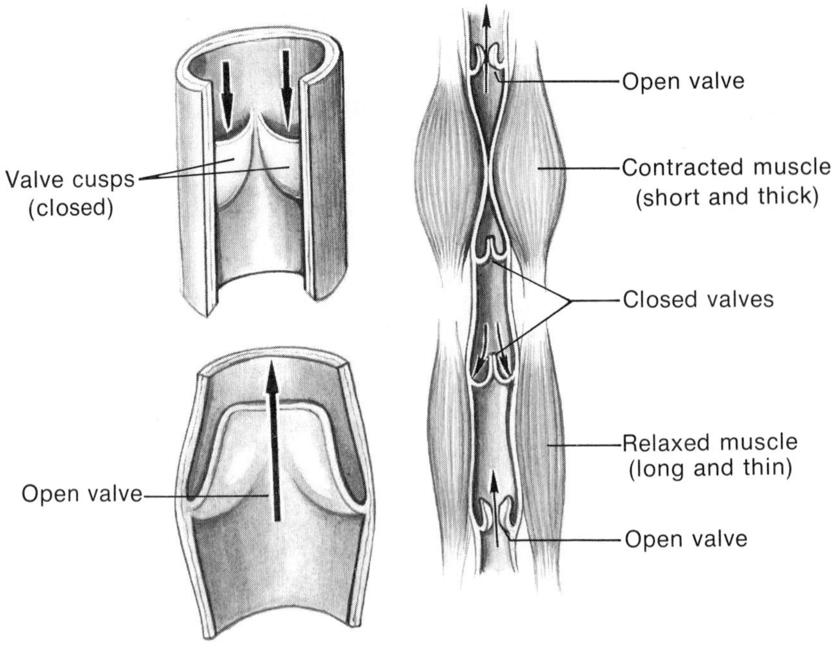

Figure 15–4 Valve action in the veins. *A*, Section of a vein, showing an open and a closed valve. *B*, Action of a contracted muscle forces the blood in a vein up toward the heart. This squeezing or "milking" action moves blood along the veins back toward the heart.

of a limb, its valves allow the blood to be squeezed only in the proper direction.

Capillaries

The walls of the capillaries consist of one layer of endothelial cells continuous with the layer that lines the arteries, veins, and heart (Fig. 15–2). The capillaries form an interlacing network of variable size.

As noted, it is in the capillary beds that the main work of blood is done—the exchange of materials between cells and blood. A molecule diffuses through the capillary wall into the tissue fluid and from the tissue fluid into the cell, or in the reverse direction as the case may be (Fig. 15–5). In addition to diffusion, there is some filtration of the fluid part of the blood through the capillary walls, especially at the arterioles, where the blood pressure is high, so that the blood continually contributes to the tissue fluid.

Capillaries in glands supply the substances needed for secretion; in the ductless glands they also take up the products of secretion. In the alimentary canal, capillaries collect the end-products of digested food; in the lungs they absorb oxygen and give up carbon dioxide; in the kidneys they discharge waste products accumulated from other organs. All these functions collectively maintain homeostasis.

PRINCIPAL SYSTEMIC ARTERIES

The systemic circulation leaves the left ventricle of the heart by way of the great artery called the aorta.

Aorta

The aorta is the largest artery in the body. It arises from the left ventricle and ascends behind the pulmonary artery. It

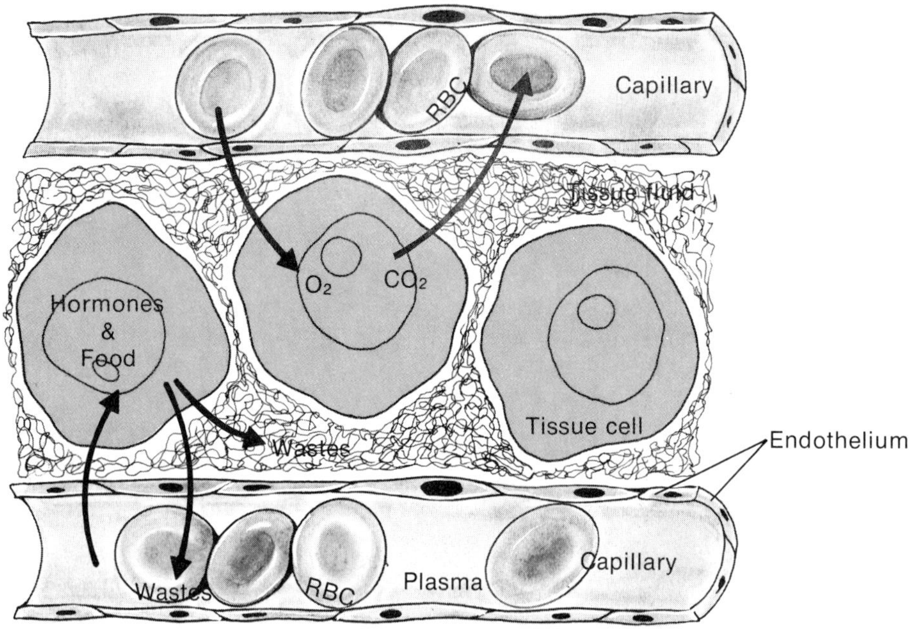

Figure 15–5 Diagram illustrating some of the physiologic exchanges between capillaries and body cells. RBC, red blood cell.

then arches to the left and passes backward, where it descends along the posterior thoracic and abdominal walls to the fourth lumbar vertebra (Fig. 15–6) and then divides into the right and left common iliac arteries. The aorta is divided into three sections: the ascending aorta, the aortic arch, and the descending aorta. The descending aorta is further divided into thoracic and abdominal portions (Fig. 15–7).

1. The *ascending aorta* is short—about two inches long—and is contained within the pericardium. The only branches are the right and left coronary arteries, which arise just above the aortic valve. They divide into a number of branches that supply the heart itself.

2. The *aortic arch* extends from the ascending aorta upward, backward, and to the left of the trachea; it then continues downward on the left side of the body, where it becomes the descending aorta. The brachiocephalic, left common carotid, and left subclavian arteries supply the head and the upper extremities.

3. The *descending aorta* extends from the aortic arch downward to the fourth lumbar vertebra. The *thoracic aorta* is that portion above the diaphragm; its branches supply the body wall of the chest cavity and the organs contained within it. The *abdominal aorta* begins at the opening of the diaphragm and terminates by dividing into the two common iliac arteries in the lumbar region. Branches from the abdominal aorta supply the body wall of the abdominal cavity and the organs contained within it.

The principal arteries stemming from the aorta are as follows:

- Ascending aorta and aortic arch
  - Coronary, right and left
  - Brachiocephalic (Fig. 15–7)
  - Right common carotid (Fig. 15–7)
  - Right subclavian (Fig. 15–7)
  - Left common carotid (Fig. 15–7)
  - Left subclavian (Fig. 15–7)
- Descending aorta
  - Thoracic aorta (Fig. 15–7)
    - Pericardial
    - Bronchial

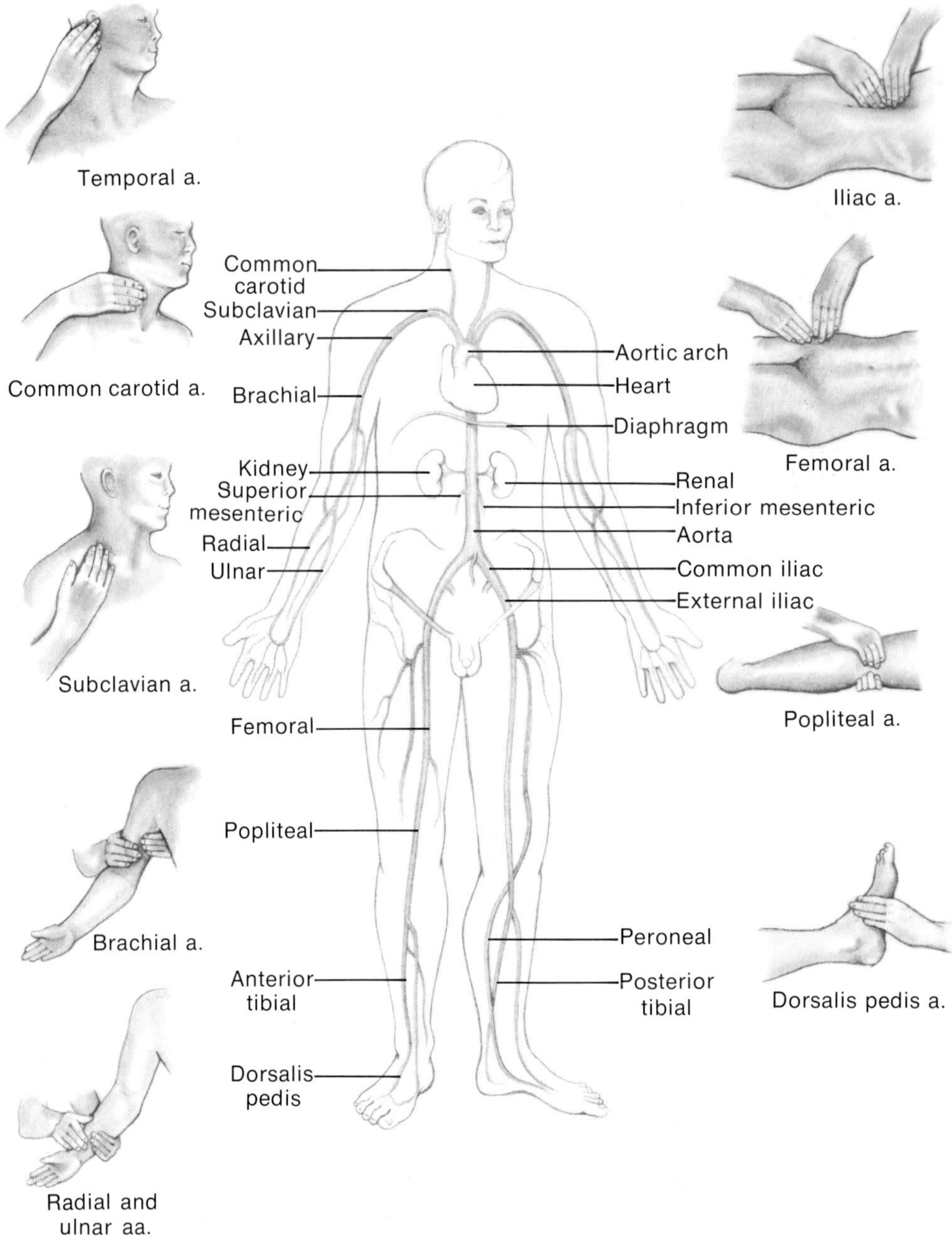

Figure 15–6 Major arteries and pressure points of the body.

Esophageal
Mediastinal
Intercostal
Superior phrenic
Abdominal aorta
Inferior phrenic
Celiac (Fig. 15–7)
Superior mesenteric (Figs. 15–6, 15–7)
Renal (Fig. 15–6)
Spermatic or ovarian (Fig. 15–7)
Inferior mesenteric (Figs. 15–6, 15–7)

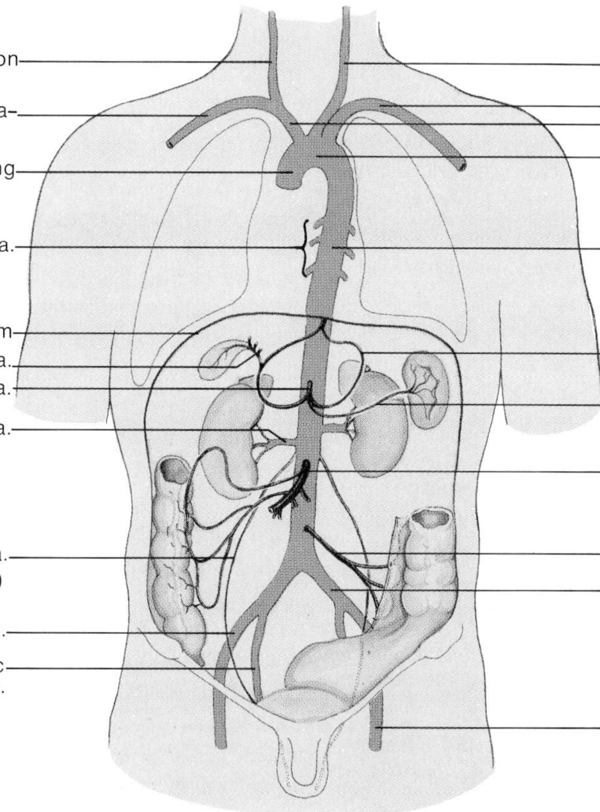

Figure 15–7 The aorta and its branches.

Arteries of the Head and Neck

The right and left *common carotid* arteries (Figs. 15–7 and 15–8) pass upward in the neck, the right arising at the bifurcation of the brachiocephalic trunk, the left as a branch of the aortic arch. Each ends by dividing into external and internal ca-

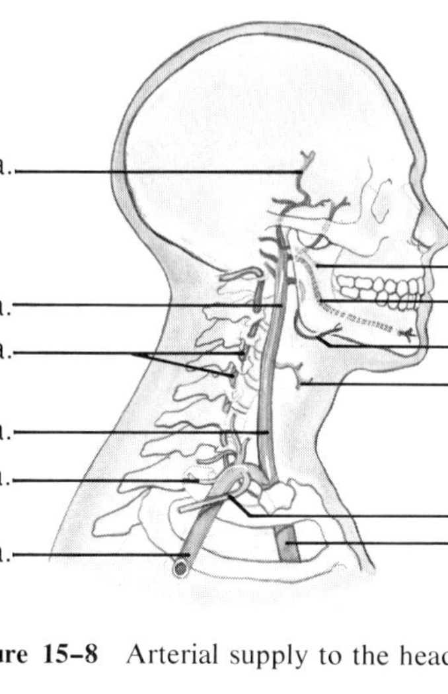

Figure 15–8 Arterial supply to the head and neck.

rotid arteries. The *external carotid* gives off many small branches and rapidly decreases in size; the *internal carotid* passes upward, turns inward, and enters the cranium. It supplies the anterior portion of the brain, the forehead, the nose, and the orbit of the eye. Branches of the internal carotid, with the aid of other arteries supplying the brain, form an arterial network at the base of the brain—the arterial *circle of Willis* (Fig. 15–9). This represents the union of the anterior cerebral and the posterior cerebral arteries, which are branches of the internal carotid and basilar arteries, respectively. Such an arrangement, a complete circle, equalizes the circulation of the blood in the brain, and in cases of destruction (hemorrhage) of one of the arteries, allows blood to reach the brain through other vessels. Massive hemorrhage in this area is commonly referred to as a *stroke* (apoplexy) or a cerebrovascular accident (CVA).

Carotid Sinus and Carotid Body. There is an enlargement at the base of the internal carotid artery called the *carotid sinus* (Fig. 15–10). In the region between the bases of the external and internal carotid arteries, there lies a small structure composed of epithelioid tissue, the *carotid body*. Similar structures in the arch of the aorta are called the aortic bodies. Sensory receptors (chemoreceptors and pressoreceptors) are located here, and play an important role in the reflex regulation of circulation and respiration.

Arteries of the Upper Extremity

The *axillary* artery begins as a continuation of the subclavian artery. It travels through the axilla until it reaches the side of the upper arm, where it becomes the brachial artery (Fig. 15–11).

The *brachial, radial,* and *ulnar* arteries terminate to form an anastomosis of the superficial and deep volar arches (Fig. 15–11). The branches of these arches supply the fingers and the hand.

The pulse is commonly taken from the radial artery at the wrist (Fig. 15–11). Bleeding from the palm of the hand may be difficult to control because of the many anastomosing vessels. If one artery is tied (ligated), blood may continue to pour into the wound from another direction.

Arteries of the Thorax and Abdomen

The arteries of the thoracic region are small but numerous. The *pericardial* arte-

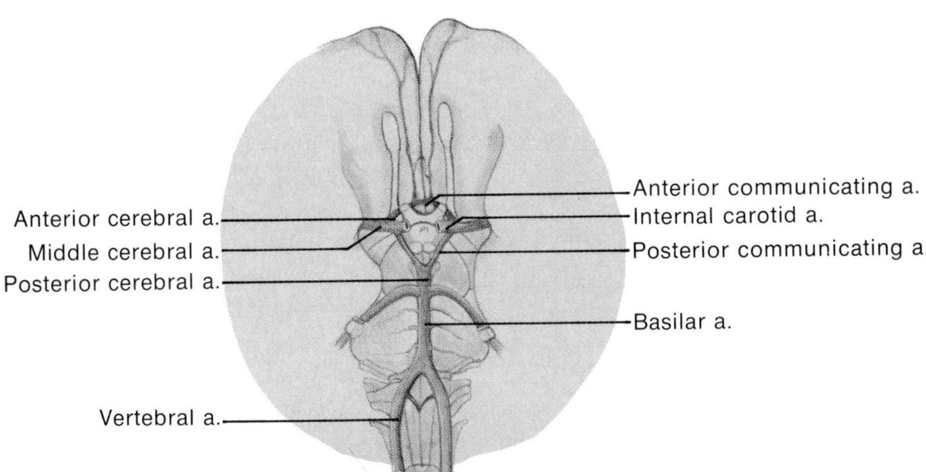

Figure 15–9 Circle of Willis, showing communicating branches from the basilar and internal carotid arteries.

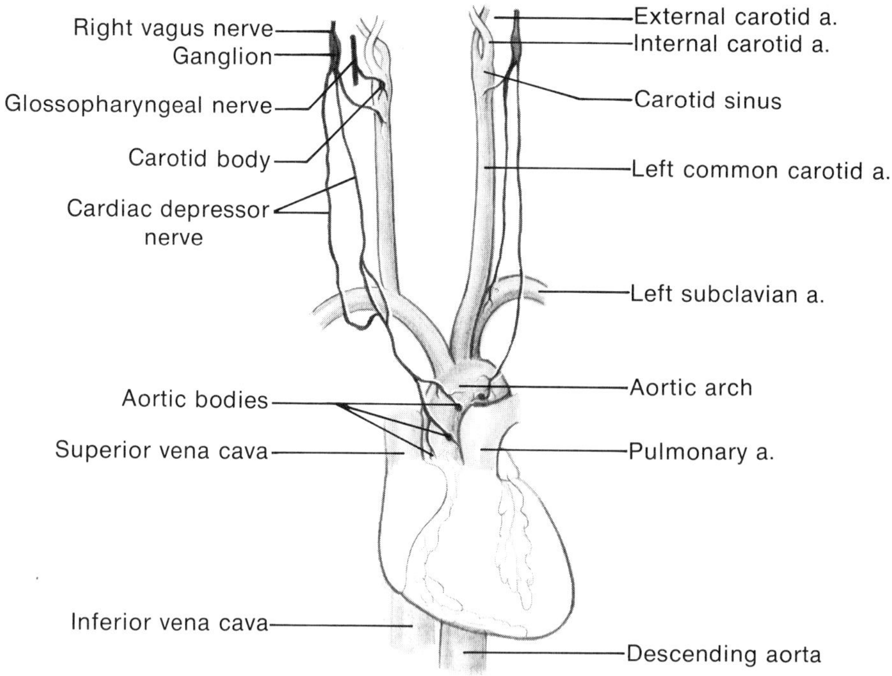

Figure 15–10 Diagram illustrating the location of the carotid and aortic bodies in relation to the aortic arch and its branches.

ries, as their name suggests, supply the pericardium. Similarly, branches of the *bronchial* arteries supply the bronchial tubes and the tissues of the lung. The *esophageal* arteries form a network supplying the esophagus. The *mediastinal* arteries supply the mediastinum, and the *intercostal* arteries supply the muscles and

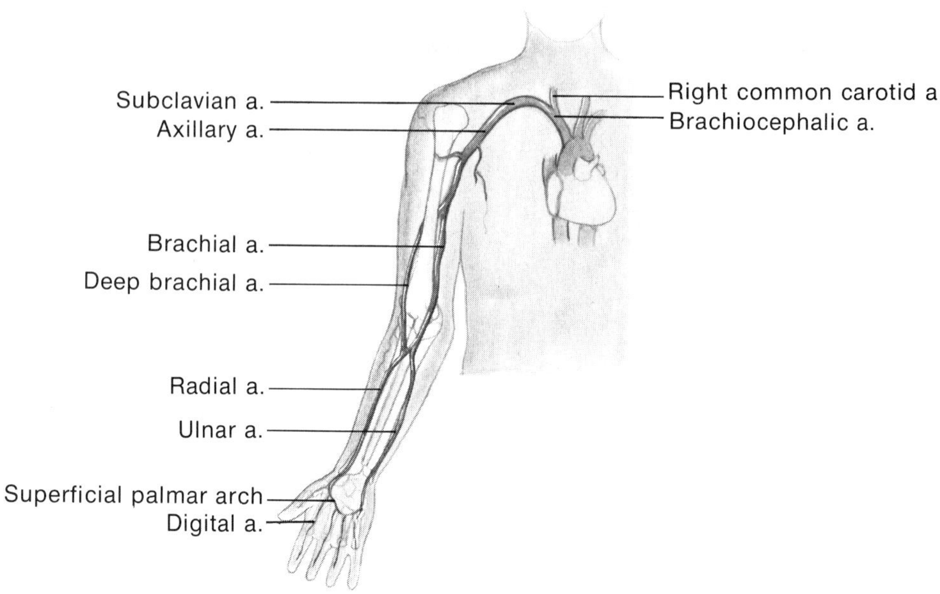

Figure 15–11 Major arteries of the right upper extremity.

the skin of the thoracic wall. The superior *phrenic* arteries arise from the lower part of the thoracic aorta, and supply the upper surface of the diaphragm.

The abdominal aorta diminishes in size as it progresses posteriorly. The *inferior phrenic* arteries arise just below the diaphragm and supply its abdominal surface.

The *celiac* artery arises from the abdominal aorta just below the diaphragm (Fig. 15–7), and divides into the *common hepatic* (liver), *left gastric* (stomach), *dorsal pancreatic* (not shown in Fig. 15–7), and *splenic* (spleen) arteries.

The *superior mesenteric* artery arises just below the origin of the celiac, and supplies the small intestine, the ascending colon, and part of the transverse colon. Its branches are supported by the mesenteric tissues as they approach the intestine.

The *renal* arteries are two short, wide vessels arising laterally from the aorta. They are distributed to the kidneys.

The *gonadal* (testicular or ovarian) arteries usually arise from the aorta just below the renal arteries. The *testicular* (spermatics) are long, slender vessels that pass downward and through the inguinal canals to the testes. In the female, the *ovarian* arteries descend into the pelvis, where each is distributed to an ovary.

The *inferior mesenteric* artery arises just above the branching of the aorta into the common iliac arteries. It supplies the descending and sigmoid colons and the rectum.

Arteries of the Pelvis

The aorta divides at the level of the fourth lumbar vertebra and gives rise to the *common iliac arteries*. They course downward and laterally to end opposite the lumbosacral junction by dividing into the hypogastric and external iliac arteries (Figs. 15–6 and 15–7).

The *internal iliac* (hypogastric) arteries descend into the pelvis. With their branches, they supply the bladder, rectum, prostate, and uterus. Branches extend also to the gluteal muscles, to the medial area of each thigh, and to the external genitalia.

Each of the *external iliac* arteries courses downward along the edge of the pelvis, passes behind the inguinal ligament, and enters the thigh area to become the femoral artery.

Arteries of the Lower Extremity

The *femoral* artery (Fig. 15–12, *A* and *B*) passes down the front of the thigh and becomes the *popliteal* artery behind the knee. As it passes downward, it lies along the depression on the inner aspect of the thigh. Here, pulsation of the artery may be felt, and circulation through the artery may be controlled. The femoral artery supplies the lower abdominal wall and the external genitalia, as well as the lower limb.

The popliteal artery (Fig. 15–12, *B*) gives rise to the posterior and anterior *tibial* arteries. Both arteries descend beneath the muscles of the calf of the leg. The posterior tibial arteries form the *plantar* arteries on the sole of the foot, whereas the anterior tibial arteries pass forward between the tibia and fibula and become the *dorsalis pedis* arteries on the anterior surface of each ankle. The dorsalis pedis arteries also branch over the top of the foot.

Clinical Considerations

Among arterial diseases, the most common are the group known as arterioscleroses, "hardening of the arteries," which produce many very serious effects. The vessels become hard and tortuous, and the loss of elasticity in the walls of the arteries increases resistance to outflow from the heart. As a result, the heart must work harder and may eventually fail if too severely overtaxed. The arterial pressure at each systole becomes high (hypertension), and the arteries, especially the smaller ones, become susceptible to rupture. A serious hemorrhage may result.

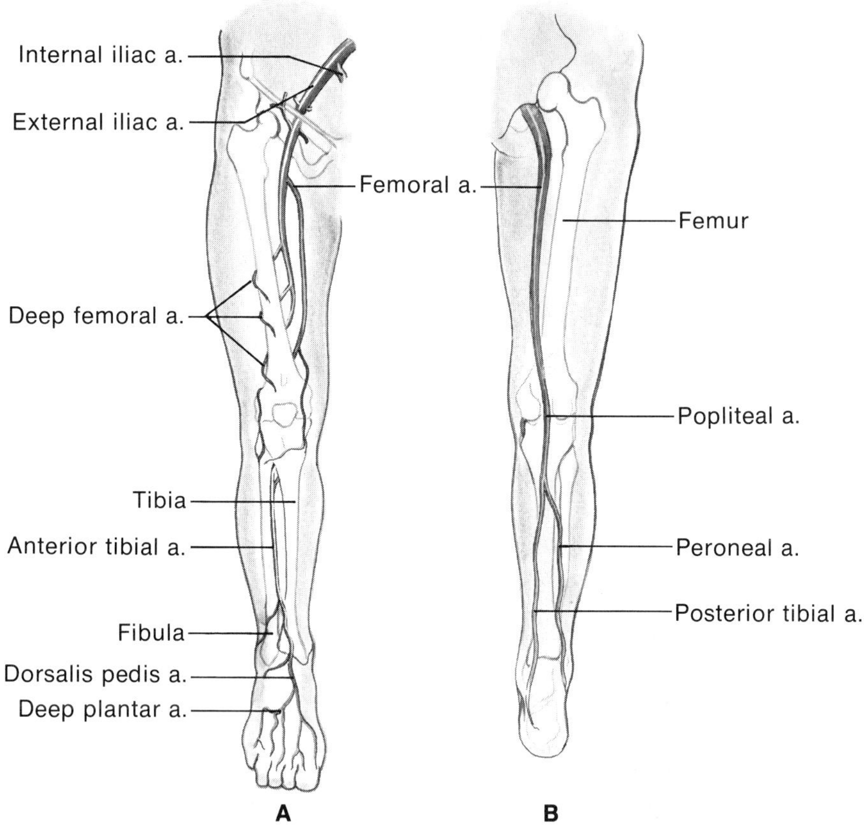

Figure 15-12 Major arteries of the right lower extremity. *A*, Anterior aspect. *B*, Posterior aspect.

Hemorrhage from a large blood vessel is obviously a serious matter. To control the bleeding from an artery, it must be tied off or compressed proximal to the wound—that is, on the side nearest the heart. As an emergency measure to control severe arterial bleeding, the main artery supply to the region should if possible be compressed against a bone. For example, the external maxillary artery can be compressed where it crosses the temple; the brachial artery can be pressed laterally against the humerus; the radial and ulnar arteries can be pressed against the bones behind them; the femoral artery is accessible at the groin; and the dorsalis pedis can be easily controlled at the ankle. These and other points, called *pressure points,* not only serve as locations for "reading" the pulse, but are extremely useful in first aid.

Replacement of damaged parts of arteries by synthetic tubes is proving very satisfactory. Such tubes have been substituted for parts of the aorta, the iliac, and other arteries. Various materials (polymers), such as nylon, Dacron, and plexiglass or medical silicones, have been used to make replacements for parts of the body.

PRINCIPAL SYSTEMIC VEINS

In general, the total diameter of the veins returning blood from any organ is about twice that of the arteries carrying blood to the organ.

Blood from the systemic group of vessels returns to the heart through three main channels (Fig. 15–13):

1. That from the wall of the heart returns through the coronary sinus.

2. That from the head, neck, upper extremity, and part of the trunk returns through the superior vena cava.

3. That from the remainder of the trunk and lower extremities returns through the inferior vena cava.

The coronary sinus receives most of the veins of the heart. It lies on the posterior aspect of the organ and terminates by emptying into the right atrium (Fig. 14–2).

Veins of the Neck and Head

The *external jugular* veins are small vessels that lie superficially in the lateral aspect of the neck, one on each side (Figs. 15–13 to 15–15). They descend over the sternocleidomastoid muscle, beneath the platysma, and empty into the subclavian vein.

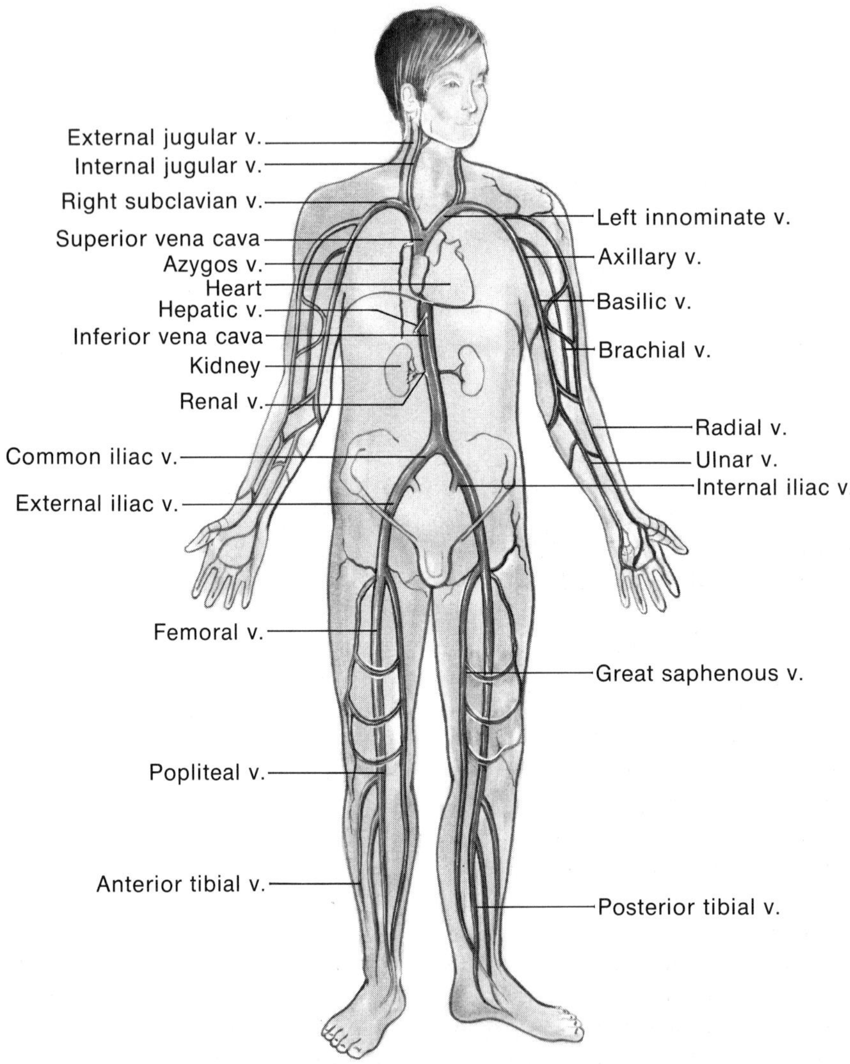

Figure 15–13 Major veins of the body.

An abundant blood supply reaches the brain through the internal carotid arteries, and there are numerous veins on the brain surface and within it concerned with gathering this blood and returning it by way of the internal jugular veins. Deep within the substance of the brain itself are many veins that drain into large sinuses (Fig. 15–14). These veins and sinuses have very thin walls and lack valves.

The *internal jugular* veins are the principal vessels draining the head and neck (Figs. 15–13 to 15–15). They arise from the transverse sinus and descend laterally beside the common carotid artery. At their base they join with the *subclavian* veins to form the *brachiocephalic* (innominate) veins (Fig. 15–15). These receive numerous tributaries from the face, neck, and cranial cavity. The right brachiocephalic is almost vertical and only about one inch long. The left is almost horizontal and about three inches long. It passes above the heart to join the right brachiocephalic and then form the superior vena cava.

The *superior vena cava* extends downward about three inches to drain into the right atrium of the heart (Figs. 15–13 and 15–15). The *azygos* vein is a long, slender, unpaired vessel that passes upward along the right posterior abdominal and thoracic walls and receives blood from the intercostal and lumbar veins. It terminates by emptying into the superior vena cava.

Veins of the Upper Extremity

Blood from the upper extremity is returned by deep and superficial sets of veins. The deep veins are called by the same name as their corresponding arteries. They frequently anastomose with one another and with the superficial veins. There are valves in both sets of veins, and both groups finally open into the axillary and subclavian veins.

The superficial veins lie just beneath the skin (Fig. 15–16), and may be easily seen on the back of the hand and the front of the elbow. They are larger than the deep veins and return more blood. They include (starting from the distal end of the limb) the *cephalic* vein (dorsal surface of the hand), the *accessory cephalic* (a lateral branch), the *basilic* (medial side of arm), the *median cubital* (at the bend of the elbow), the *axillary* (continuation of the basilic), and, finally, the *subclavian* vein that drains the axillary and cephalic veins.

The large superficial veins in front of

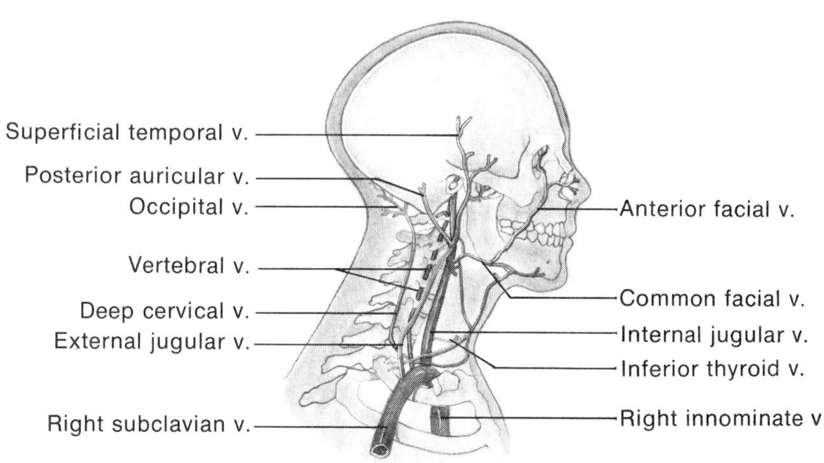

Figure 15–14 Veins of the head and neck.

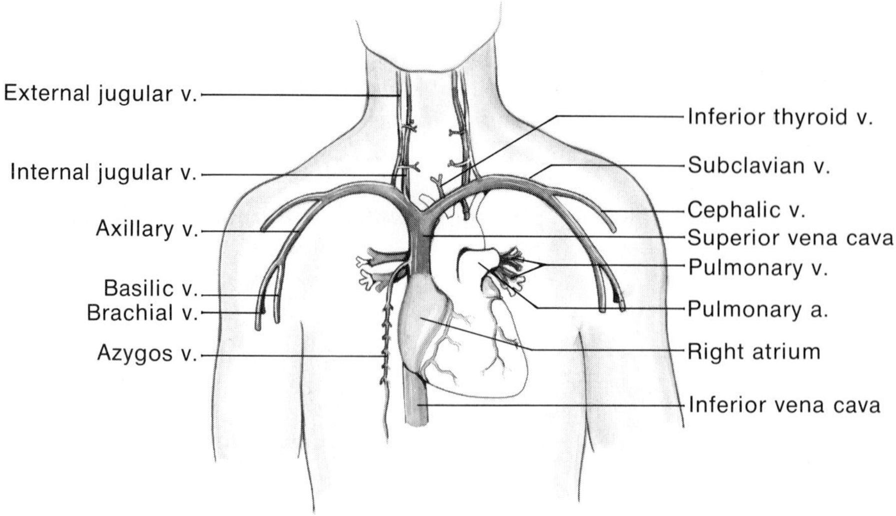

Figure 15-15 Veins of the upper extremity.

the elbow (basilic, median cubital, and cephalic) are commonly used by surgeons and medical teams when drawing blood from the body, or when injecting medicinal substances into the bloodstream (intravenous injection).

Veins of the Lower Extremity

Like those of the arm, the veins of the lower extremity are both superficial and deep. The superficial veins lie beneath the skin between the layers of fascia. The deep veins are covered by deep layers of muscle, and follow the large arteries, usually enclosed in the same sheath. Both sets are provided with valves, which are more numerous in the deep veins.

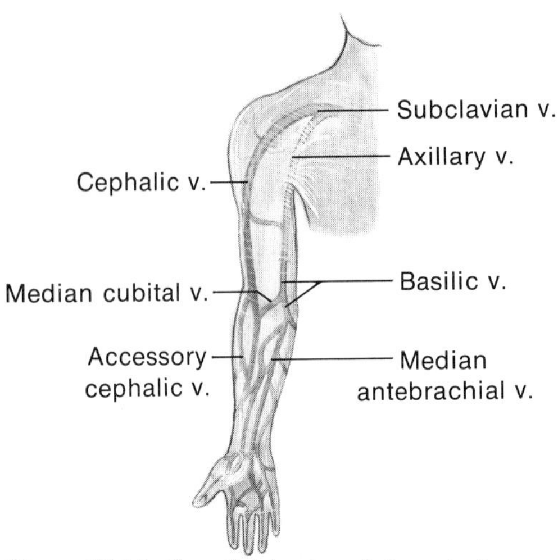

Figure 15-16 Superficial veins of the anterior surface of the right arm.

The superficial veins include the *great saphenous* vein, which begins in the marginal vein of the foot, extends up the medial side of the leg and thigh, and ends in the femoral vein just below the inguinal ligament. It receives blood from all aspects of the superficial layers of the leg and foot. A major tributary is the *small saphenous* vein, which passes up the back of the leg and frequently empties into the popliteal vein at the knee (Fig. 15-17).

Varicosities. The great saphenous vein, because of its dependent and superficial position, is particularly vulnerable to the development of varicosity ("varicose veins"). This condition is usually due to prolonged standing when the circulation is poor and the back pressure on the veins is high. Leg supports (stockings), heat treatments, and chemical injections are commonly used to relieve the varicosities. Varicose veins of the rectum are commonly referred to as "piles" or, more properly,

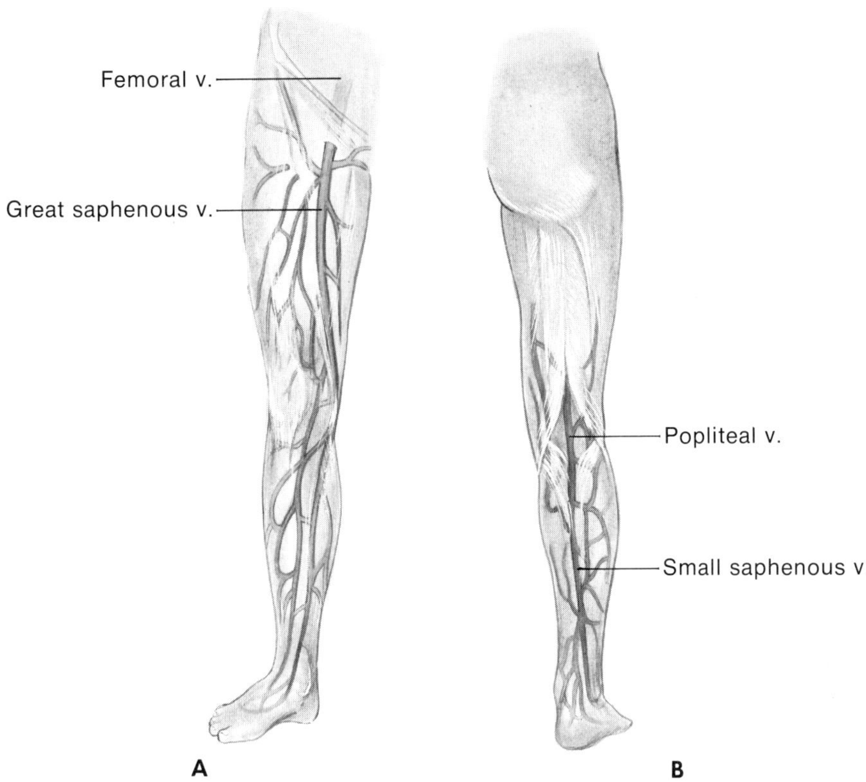

Figure 15–17 Superficial veins of the right lower extremity. *A*, Anterior aspect. *B*, Posterior aspect.

hemorrhoids; varicose veins of the esophagus are called esophageal varices. Surgical removal (vein ligation and stripping) of the great saphenous vein ultimately may be indicated if the problem recurs.

Deep veins accompany the arteries below the knee. They are paired and carry the same names as the corresponding arteries. Veins from the foot empty into the anterior and posterior *tibial* veins, which unite to form the *popliteal* vein. This is continued as the *femoral*, and becomes the *external iliac* vein (Fig. 15–13).

Veins of the Pelvis and Abdomen

The internal and external iliac veins unite to form the *common iliac* veins. The common iliacs progress from the sacral region to form the inferior vena cava in the lumbar region.

The *inferior vena cava* ascends along the right side of the aorta, passes through the diaphragm, and terminates by entering the right atrium of the heart (Fig. 15–13). It receives veins that have the same names as the peripheral branches of the aorta (*lumbar, renal, suprarenal, inferior phrenic, superior phrenic, spermatic or ovarian*).

The *hepatic* veins (Fig. 15–13) arise in the liver and flow into the inferior vena cava. These carry the return flow from the hepatic artery and the blood that comes to the liver through the portal system.

Hepatic Portal System. A portal system is a venous pathway that has capillaries at either end. Veins ordinarily transmit blood directly back to the heart. In certain pathways, however, blood passes through a capillary network before returning to the heart. This system occurs in the human liver and is called the hepatic portal system.

Blood returning from the spleen,

stomach, and intestines enters the portal vein and passes through the capillaries in the liver before leaving the liver via the hepatic veins to the heart (Fig. 15–18). The portal system is concerned with glucose metabolism, and therefore will be considered in detail when the digestive organs are discussed.

PULMONARY CIRCULATION

The pulmonary circulation (Figs. 14–1, 14–2, and 14–10) carries blood from the heart to the lungs and back to the heart again. It leaves the heart by way of the large pulmonary artery, and returns through the pulmonary veins.

Originating in the right ventricle, the *pulmonary artery* passes upward and backward for a short distance, and then divides into right and left branches that enter the lungs. Within each lung the artery branches widely, and each branch closely follows the subdivisions of the bronchi. The terminal arterioles open into the capillary networks that approximate the alveolar air sacs.

The *pulmonary veins* entering the left atrium have two branches from each lung. Their smallest tributaries arise as venules from the capillary networks of the alveolar air sacs.

Of paramount importance here is the *condition* of the blood in these vessels. As we have seen, blood vessels are named according to the direction of flow to or from the heart—not in terms of what they carry. Generally, a vessel leaving the heart is called an artery and carries oxygenated blood ("red"); a vessel returning to the heart is called a vein and carries carbon dioxide–laden blood ("blue"). However, in the instance of the pulmonary circulation only, the pulmonary artery carries carbon dioxide–laden blood *away* from the heart (to the lungs), and the pulmonary vein carries oxygenated blood *to* the heart (from the lungs).

Clinical Considerations

Inflammation of a vein is called *phlebitis*. There is marked pain, frequent swelling, and involvement of the entire

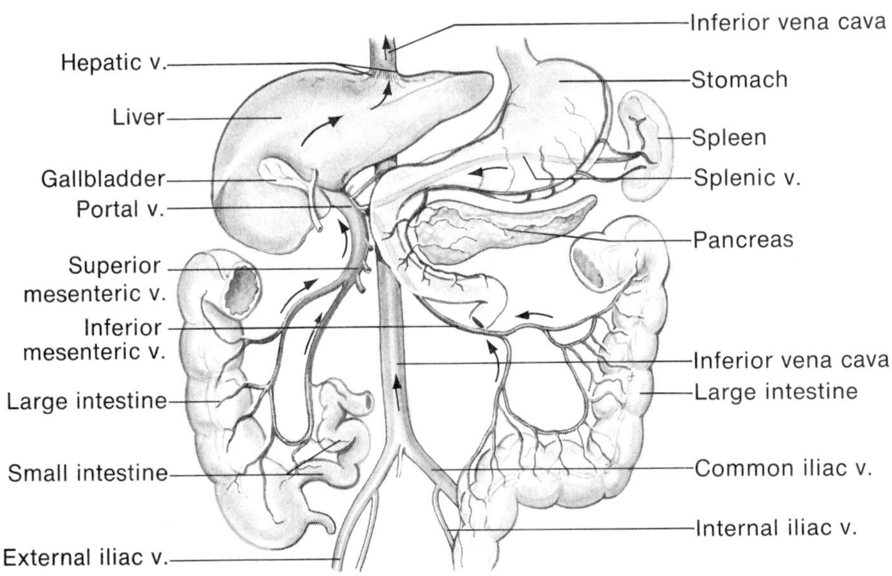

Figure 15–18 The hepatic portal system. Arrows indicate the direction of flow in all abdominal organs, through the liver, to the superior vena cava.

vein wall. A blood clot may form, causing the dangerous condition called *thrombophlebitis,* which entails the possibility that a piece of clot may loosen and float in the blood. If this clot reaches the lungs, as it so often does, sudden death from pulmonary embolism may result. Prevention of infection and the use of anticoagulant drugs have greatly reduced the threat of this disease.

FETAL CIRCULATION

The pathways through which blood flows are not the same before a child is born and after. The fetus receives oxygen and food from the placenta, not from its own pulmonary and digestive system. Following birth, changes in the blood flow occur, owing to the sudden need to use the lungs and gastrointestinal tract.

Before birth, venous blood from the aorta of the fetus is brought to the placenta via the umbilical arteries (Fig. 15–19) that branch from the iliac arteries of the fetus. After oxygenation at the placenta, the blood, now arterial, returns via the umbilical vein and the *ductus venosus* to the inferior vena cava of the fetus. It then enters the right atrium along with venous blood from the systemic capillaries of the fetus. It is at this point that the "pure" arterial blood from the umbilical vein becomes mixed before it enters the heart to be circulated. The umbilical vein, therefore, is

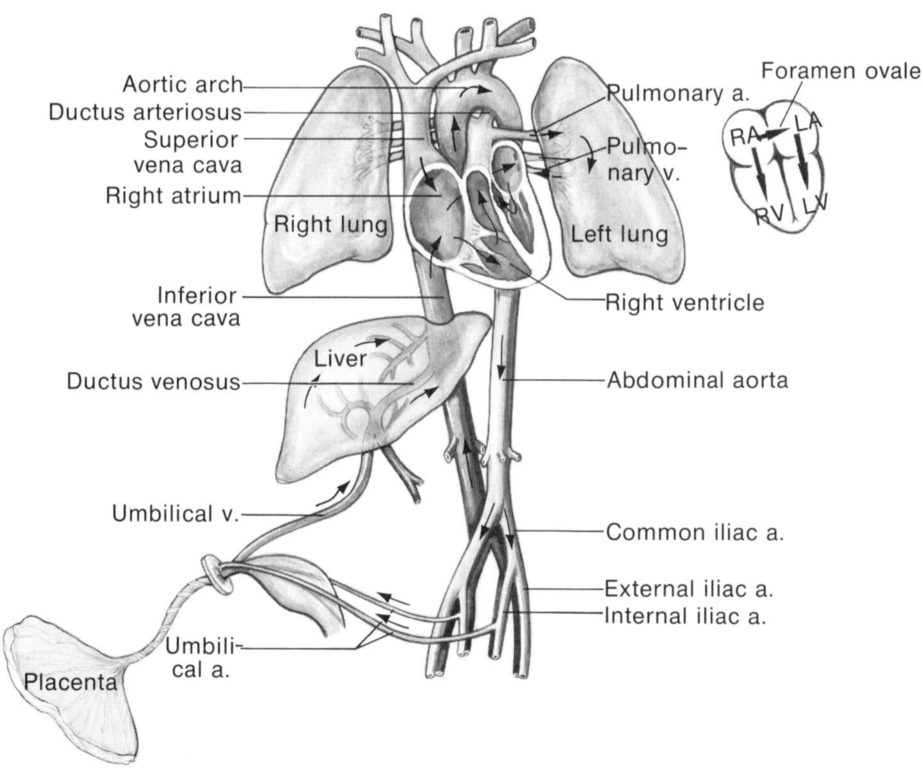

Figure 15–19 Plan of circulation in a mature fetus. Arrows indicate the direction of blood flow. The inset is a diagram of the blood flow in the fetal heart. Note the blood flow from the right atrium into both the right ventricle and left atrium through the foramen ovale. Abbreviations: RA, right atrium; LA, left atrium; RV, right ventricle; LV, left ventricle.

the only vein that carries almost pure oxgenated blood before birth.

The blood from the vena cava follows the route from the right atrium to the right ventricle. As the lungs of the fetus are collapsed and of small volume, only a fraction of the normal blood flow enters the pulmonary circuit. Two bypasses are used to prevent overtaking the pulmonary vessels:

1. The blood coming into the right atrium goes through a shortcut, the *foramen ovale*, into the left atrium, as well as down into the right ventricle (Fig. 15–19). This cuts the blood volume in half.

2. The *ductus arteriosus*, a short vessel that connects the pulmonary artery with the aortic arch (Fig. 15–19), also affords a reduction in blood volume by diverting most of the blood from the right ventricle back into the systemic circulation. Note that the arterial blood from the placenta and the venous blood from the fetus's systemic circulation have parts of their routes in common. Therefore, some mixing must occur.

In the fetus, the left and right ventricles make approximately equal contributions to the systemic blood supply. Following birth, the foramen ovale closes, and the blood entering the right atrium goes only to the right ventricle; the ductus arteriosus degenerates and remains as a suspensory ligament to support the positioning of the pulmonary artery and the aortic arch; the ductus venosus is obliterated once the placenta and its circulation have been discarded. The umbilical arteries no longer carry blood, and persist only as the umbilical ligaments.

Occasionally a congenital defect develops owing to failure of the foramen ovale or the ductus arteriosus to close properly. The pulmonary blood supply is inadequate; the blood is insufficiently oxygenated; and the "bluish" blood circulates through the system, creating a pallor (cyanosis) of the skin. Babies with such circulatory defects are commonly called "blue babies."

THE PHYSIOLOGY OF CIRCULATION

Clearly, the heart is of primary importance in moving the blood through the vascular system. Within limits and under ordinary conditions, the faster the beat, the more rapidly will blood be supplied to the tissues and the higher will be the pressure of blood as it is forced into the arteries. But the entry of blood into the organs is determined also by the degree to which the blood has free access to the capillaries. Arterioles control blood flow into the capillaries according to the degree to which they are dilated or constricted. The volume of blood moved per unit of time generally depends on the force, frequency, and volume of the heartbeat.

Factors Affecting Circulation

Cardiac Output. The action of the heart determines cardiac output, which, in turn, influences the circulation and blood pressure. Determination of the output of the ventricle per unit of time gives the volume of blood that is pumped into the arteries and the volume that determines blood pressure, as well as the volume that flows through the entire circulation in that time. An increase in rate increases the diastolic pressure, whereas an increase in force of contraction increases systolic pressure.

Blood Volume. The normal volume of blood is about five quarts, and this volume, contained within a closed tubular system, will exert a mean pressure on the walls of this sytem. An increase in blood volume above normal causes an increase in pressure; a decrease in volume, such as occurs in injury or shock, causes the blood pressure to drop. Blood transfusions readily rectify decreased volume.

Peripheral Resistance. Resistance depends on the viscosity of the fluid and the diameter of the tube. In relating this to the circulatory system, tubes of small diameter (capillaries) offer more resistance than do larger tubes. Thus, at a given time, the flow in a small tube is less than in a larger one. The velocity of the blood is greatest in the largest arteries—about 400 mm per second in man—and slows down as it approaches the capillaries (Fig. 15–20). In the capillaries, the rate is slowest of all, about 1 mm per second, and the flow then gradually increases as the venules pick up the blood until a rate of about 200 mm per second is reached in the large veins. Therefore, it can be seen that the velocity of flow is inversely proportional to the total cross-sectional area of the vessel; that is, the greater the area, in the capillaries, the slower the velocity of flow. In man, it takes approximately 25 to 30 seconds for a recognizable chemical to flow from jugular vein to jugular vein throughout the circulation.

The Pulse

As the heart beats, the ventricles force blood into the aorta. The contraction of the ventricles initiates a wave of increased pressure, which starts at the heart and travels along the arteries, like a ripple of water, until the arteries end. This wave of increased pressure, the pulse, already has been mentioned (p. 307). Its rate is the same as the heart rate (70 beats per minute), and, as the arteries lose their elasticity through age, the speed of the pulse wave increases.

A physician or a trained medical team member can detect the heart rate, blood pressure, arterial tension, and other physiologic indicators by taking the pulse. A soft or hard pulse can indicate a low or high blood pressure, or whether an artery has low or high tension. A slow pulse indicates a slow heart rate. A constant pulse indicates a constant blood volume, whereas a "thready" pulse may indicate internal hemorrhage or other loss of blood.

Blood Pressure

Even with the moderating effect of the elastic arterial walls, the arterial blood

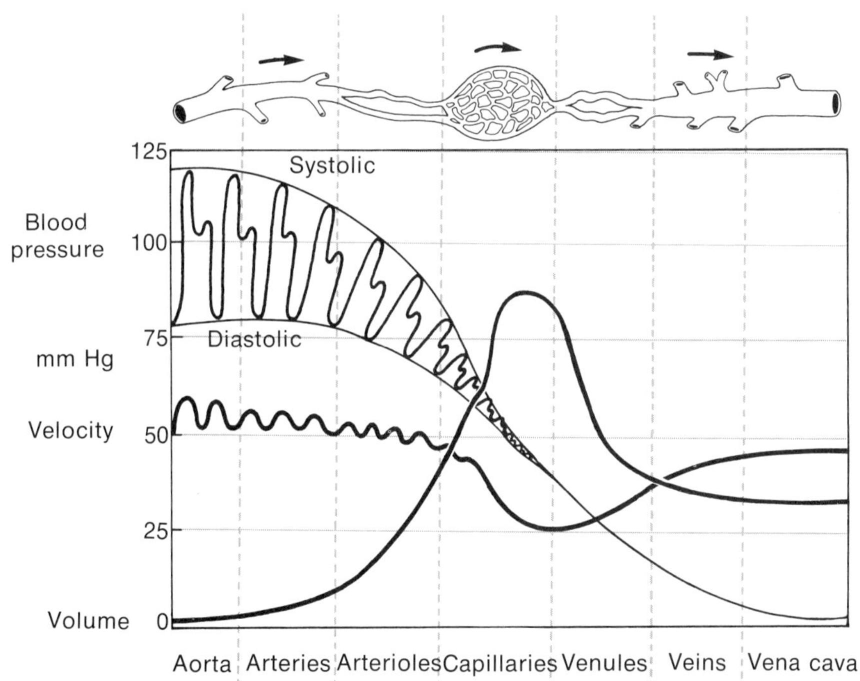

Figure 15–20 Diagram showing blood pressure, blood velocity, and blood volume in the various types of blood vessels.

pressure rises and falls with the heartbeat. The peak pressure is the *systolic pressure*; the second detected beat is the *diastolic pressure*. In normal young adults these are approximately 120 mm and 80 mm of mercury, respectively. The difference between the systolic and diastolic pressures is the *pulse pressure*.

For routine determination of blood pressure a *sphygmomanometer*, or blood pressure apparatus, is used (Fig. 15–21). A cuff is placed around the upper arm and inflated with air by squeezing a rubber bulb. The air pressure in the cuff is shown on the manometer. When the cuff is sufficiently inflated, the tissues of the arm compress the brachial artery against the humerus where it runs close to that bone, thereby completely closing the vessel. A stethoscope is placed over the brachial artery below the cuff. As the air is slowly released by the bulb, the pressure outside the artery diminishes until it is the same as that inside the artery. When it is exactly equal to the systolic pressure, blood will flow through the artery only at the peak of systolic pressure. This is heard in the stethoscope as a throbbing at the same rate as the heartbeat. Further release of the cuff pressure allows more blood to pass at lower and lower pressures, until it is low enough to allow flow during the diastolic pressure. From this point on, blood flow is continuous and is no longer audible in the stethoscope.

In practice, then, the physician or nurse gradually releases air from the cuff and notes the pressure shown by the manometer at the time the first sounds become audible (systolic), and again when the sounds disappear (diastolic).

Vasomotor Control of Blood Vessels

The tone of the smooth muscles in the arteriolar walls is the most controllable

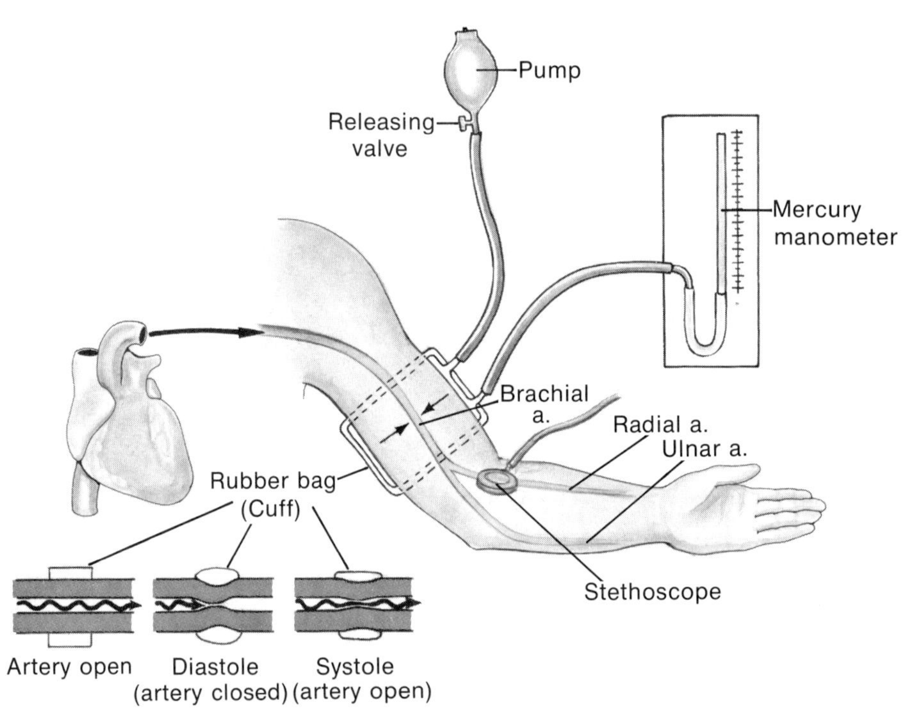

Figure 15–21 Measurement of arterial blood pressure. Application of an external pressure closes the brachial artery. As the external pressure is decreased, the manometer registers the first passage of blood through the constriction (systole). As the external pressure decreases further, the pulsation diminishes and disappears (diastole).

aspect of the peripheral resistance. These muscles are supplied with vasomotor nerves, and if the muscle fibers are stimulated to contract, the diameter of the blood vessel is reduced—a process known as *vasoconstriction*. If the muscle fibers relax, permitting the diameter of the blood vessels to increase, *vasodilation* occurs.

Impulses causing vasoconstriction are carried over sympathetic nerve fibers that arise from the same center in the medulla as the accelerator nerves supplying the heart (Fig. 15–22). The center is responsive to the pattern of afferent impulses from the baroreceptors in the aorta and the carotid sinuses, and sends the appropriate vasoconstrictor impulses. A fall in the arterial blood pressure causes a reflex vasoconstriction which, together with the cardioaccelerator reflex, helps to return the pressure to normal. Conversely, a rise in blood pressure causes peripheral vasodilation, which, along with the reflex slowing of the heart, helps to restore normal pressure.

In addition to the arterial pressure receptors, there are also pressure receptors on the venous side of the circulation.

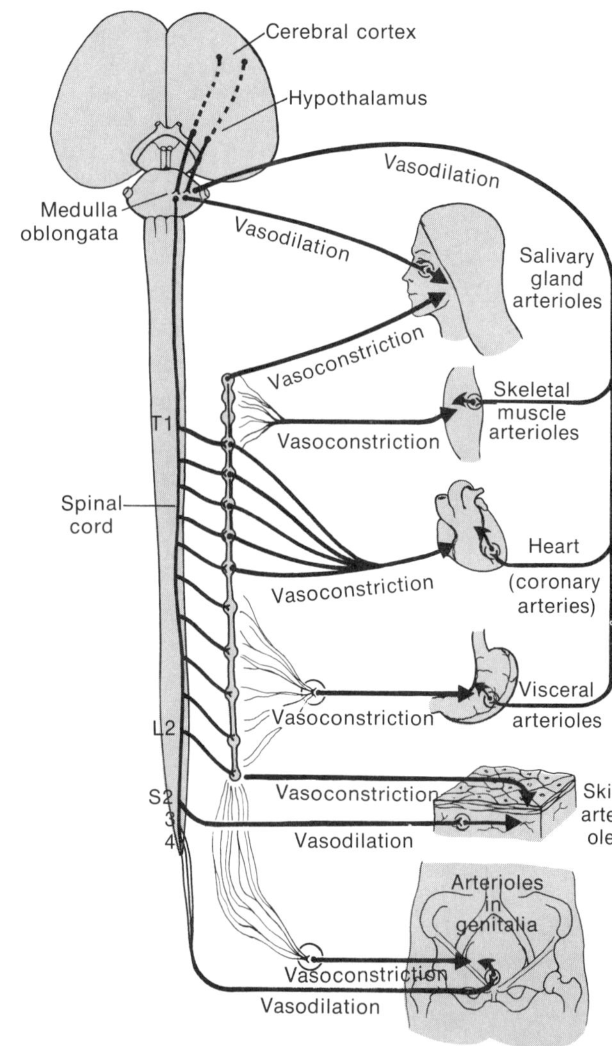

Figure 15–22 Nervous regulation of visceral and muscular arteries and arterioles. Stimulation of sympathetic nerves induces a vasoconstriction of the vessels, whereas stimulation of the parasympathetic nerves produces a vasodilation of the vessels.

For example, receptors in the right atrium signal pressure changes in that organ. An increase in pressure in this chamber implies that the blood is not moving rapidly enough out of the ventricles. Increased blood pressure in the right atrium is followed by stimulation of the cardioaccelerator center. Together with the cardioregulatory center, the vasomotor center is also affected in a manner appropriate to provide vasodilation if the blood pressure is too high, and vasoconstriction if the blood pressure is too low.

Chemoreceptors also help control blood distribution. A high CO_2 level in the blood implies that the blood is not being moved rapidly enough from the tissues to the lungs. It is therefore appropriate that CO_2 and pH directly stimulate the vasoconstrictor center. These factors also result in a more effective heartbeat.

Venous Pressure

Capillary pressure is about 30 mm Hg near the arterioles, and 15 mm Hg near the venules. A person at rest has virtually zero blood pressure in the large veins as they enter the right atrium. The "milking" action of muscles and the valves of the vein are the chief factors in maintaining venous circulation. There is no quick method of measuring venous pressure; it is too low to be measured accurately by a sphygmomanometer.

In brief, it can be seen that the arterioles, capillaries, and venules present an area of great resistance to the flow of blood from the arteriole to the venule. The effects exerted by each of the factors discussed previously can be summarized as follows:

1. *Elasticity of arteries*—the elastic layers in the arterial walls cause these vessels to constrict, increasing the pressure.

2. *Blood viscosity*—blood pressure varies directly with the viscosity of the blood.

3. *Cardiac output*—the greater the cardiac output, the higher the blood pressure.

4. *Blood volume*—the quantity of blood is constant throughout the circulatory system.

5. *Blood velocity*—this is directly proportional to the size of the vessel and its distance from the heart. Blood travels fast in large vessels (arteries), slows down in small vessels (capillaries), and speeds up again in the veins—although it does not flow as fast as it does in the arteries.

6. *Distance from the heart*—blood pressure is directly proportional to the distance it flows; the farther it flows, the lower the blood pressure.

Hemodynamics

Although the capillary wall, the barrier between the plasma and the interstitial fluid, permits free movement of water and small molecules, its pores are too small to permit plasma proteins and other colloids to pass through in significant quantities. This semipermeable membrane exerts an osmotic pressure of 25 mm Hg—the *colloidal osmotic pressure*, sometimes called the *oncotic pressure*. The hydrostatic (blood) pressure in the capillary and the plasma protein osmotic pressure represent two opposing forces that regulate the exchange of fluid between the blood and the tissues.

The rate of filtration at any point along a capillary depends on the filtration pressure (the hydrostatic-blood pressure in the capillary, minus the hydrostatic pressure of the interstitial fluid) at that point. The filtration pressure is opposed by the inwardly directed osmotic gradient due to the presence of plasma proteins in the capillary (oncotic pressure). The interstitial tissue hydrostatic pressure normally averages about 1 to 2 mm Hg. Therefore, fluid moves into the interstitial space at the arteriolar end of the capillary (Fig. 15–23) when the filtration pressure across the wall exceeds the oncotic pressure, and into the

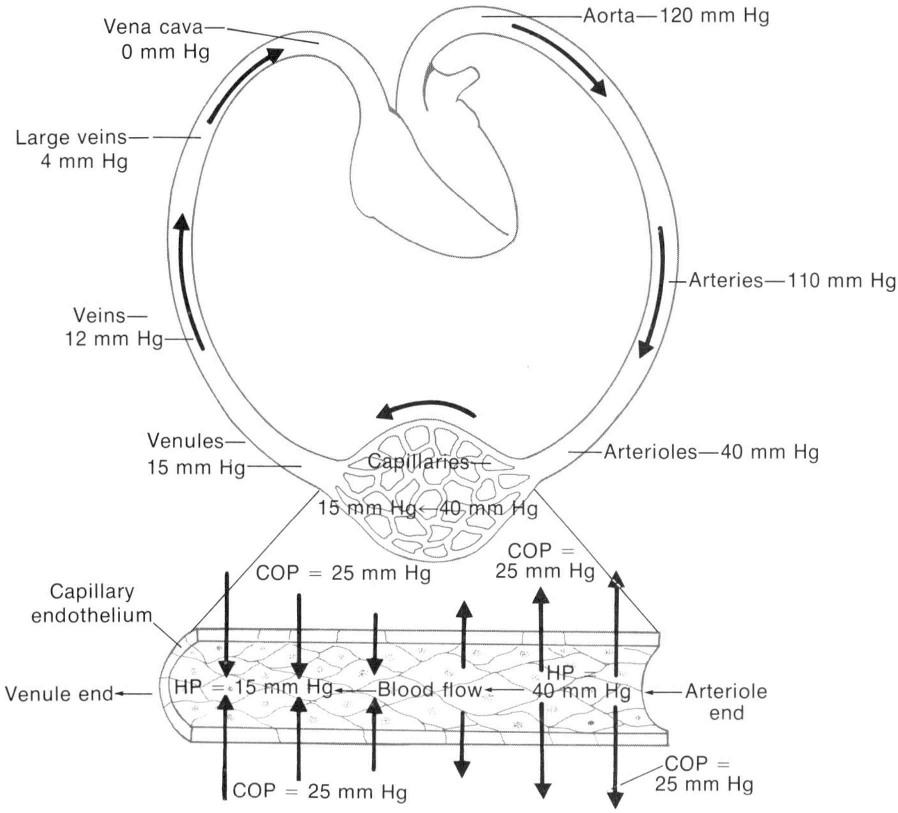

Figure 15–23 Hemodynamics. The two forces determining the exchange of materials between the blood and tissue fluids are the blood pressure (hydrostatic pressure) and the colloidal osmotic pressure (oncotic pressure). At the arteriole end of the capillary bed, the blood pressure is greater than the tissue pressure; therefore, the materials diffuse out of the blood vessel into the tissues. As the blood pressure decreases there is less diffusion, until at the venule end of the capillary bed the tissue pressure is greater than the blood pressure, reversing the flow so that fluid flows from the tissues into the blood vessel. The length and direction of the arrows indicate the extent and direction of flow of the fluid. Abbreviations: COP, colloidal osmotic pressure; HP, hydrostatic pressure.

capillary at the venule end when the oncotic pressure exceeds the filtration pressure. The amount of fluid that moves across the capillary walls is enormous. It is said that an amount equal to the entire plasma volume enters the tissues and an equal amount leaves them every minute.

CLINICAL CONSIDERATIONS

Exercise

The blood pressure rises considerably during physical exercise. Activities such as calisthenics, weight lifting, and so forth, that require a maximum exertion, raise the systolic blood pressure considerably. Systolic pressures of 200 to 225 mm Hg are not unusual for a normal man during exercise. The pulse remains relatively stable, and the blood pressure returns to normal immediately following the physical exertion.

Activities that require a prolonged expenditure of energy (long races, swimming, etc.) do not raise the blood pressure so markedly; however, they are followed by a long recovery time.

Shock

Shock is a highly dangerous condition caused by a sudden failure of the circulation. Severe hemorrhage, a sudden disturbance, burns, prolonged strain, drugs, insufficient oxygen in the blood, and factors creating cold and exhaustion are classic examples of situations evoking shock.

Primary shock is demonstrated directly after the causative incident. The person collapses, apparently in a fainting spell, but this unconscious state is short-lived, and the patient usually recovers with little damage. *Secondary* shock has its onset a longer time after the incident. In this case, however, a sequence of arteriolar vasoconstriction, tissue hypoxia, blood loss, and even death may ensue within a short period. The usual treatment is to administer blood plasma or whole blood intravenously. This increases the blood volume and helps restore the circulation to normal.

Aneurysm

An aneurysm is a sac or bulged-out section in the wall of an artery or a vein due to a localized weakness in that part of the vessel. The aorta and the circle of Willis (Fig. 15–24) are the vessels most commonly involved. The damage to the wall may be due to degenerative changes, such as those caused by hardening of the arteries or syphilis, or the aneurysm may be congenital. Whatever the cause, the

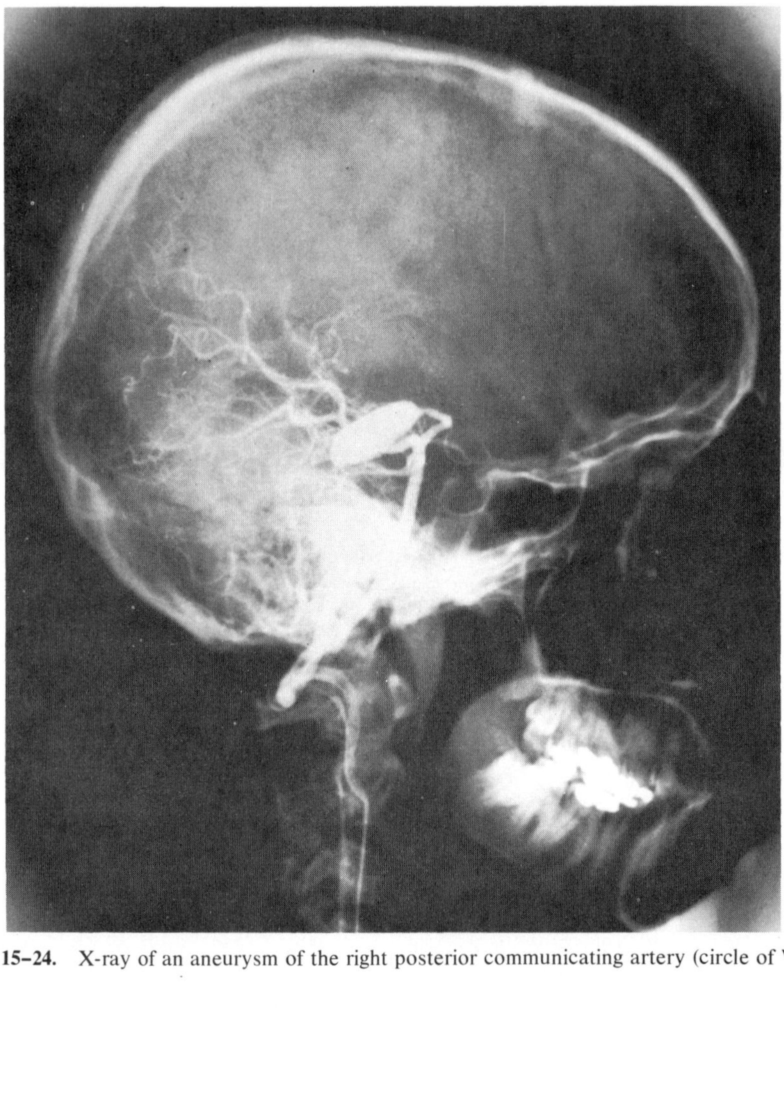

Figure 15–24. X-ray of an aneurysm of the right posterior communicating artery (circle of Willis).

weakness in the wall grows, deranging other structures in its immediate proximity. Definite symptoms may be manifested that do not necessarily indicate an aneurysm, but do show that something is wrong. Eventually, unless treated in time, the aneurysm will yield to pressure and burst (stroke), usually causing severe physical impairment or death.

Arterial Degeneration

Changes in the arterial walls frequently lead to loss of elasticity accompanied by an irregular hardening of the wall around the arterial lumen. Areas of fatlike material (cholesterol) may replace the muscle and connective tissue, leading to *atherosclerosis*. If the lining of the vessel is damaged, blood clots (thrombi) may form at this point. In other patients, calcium salts and other metabolic products may cause a hardening of the arteries known as *Mönckeberg's arteriosclerosis*.

OUTLINE

I. Introduction
   A. Blood maintains constant fluid environment
      1. Supplies tissues with nutrients
      2. Removes tissue waste products
II. General Circulation
   A. Heart
      1. Two pumps in parallel
         a. Pulmonary circulation
         b. Systemic circulation
III. The Blood Vessels
   A. Arteries
      1. Carry blood away from the heart
      2. Microscopic structure
         a. Tunica intima – inner layer of endothelial, connective, and elastic tissues
         b. Tunica media – middle layer of smooth muscle and collagenous fibers
         c. Tunica adventitia – outer layer of loose connective tissue, smooth muscle, and elastic fibers
      3. Elastic walls important in maintaining blood pressure
         a. Detected as pulse
   B. Veins
      1. Return blood to the heart
      2. Microscopic structure
         a. Same as in artery except for a diminished tunica media, giving it a "collapsed" look
         b. Valves aid blood transport – milking action
   C. Capillaries
      1. Connect the smaller arteries (arterioles) with the smaller veins (venules)
      2. Exchange materials to and from the tissues in a bedlike network
IV. Blood Circulation in Upper Extremities (See chart on following page.)
V. Blood Circulation in Lower Extremities (See chart on following page.)
VI. Blood Circulation in Trunk
   A. Arterial
      1. Thoracic aorta
         a. Pericardial, bronchial, esophageal, mediastinal, intercostal, and superior phrenic arteries supply chest wall, its organs, and diaphragm
      2. Abdominal aorta
         a. Inferior phrenic, celiac (including left gastric, common hepatic, splenic, and dorsal pancreatic), renal, superior and inferior mesenteric, and gonadal (testicular or ovarian) arteries supply the abdominal organs
      3. Pelvic branches
         a. The internal iliacs and their branches supply the

IV. Blood Circulation in Upper Extremities

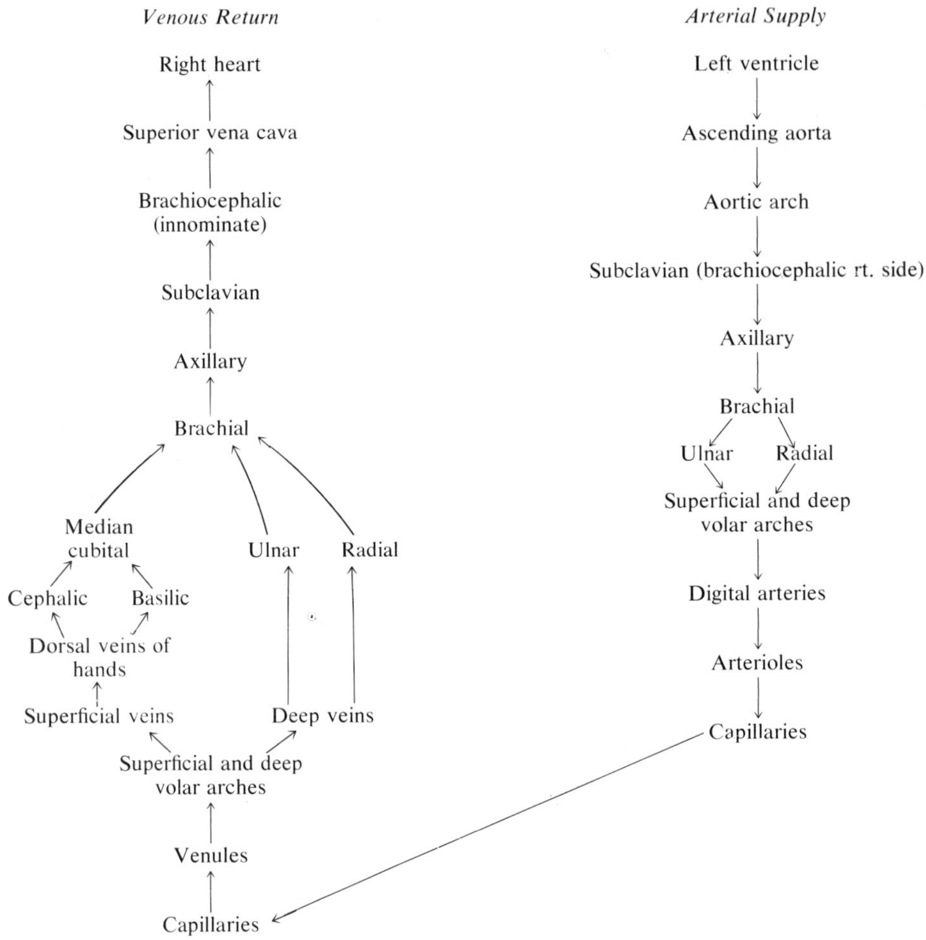

V. Blood Circulation in Lower Extremities

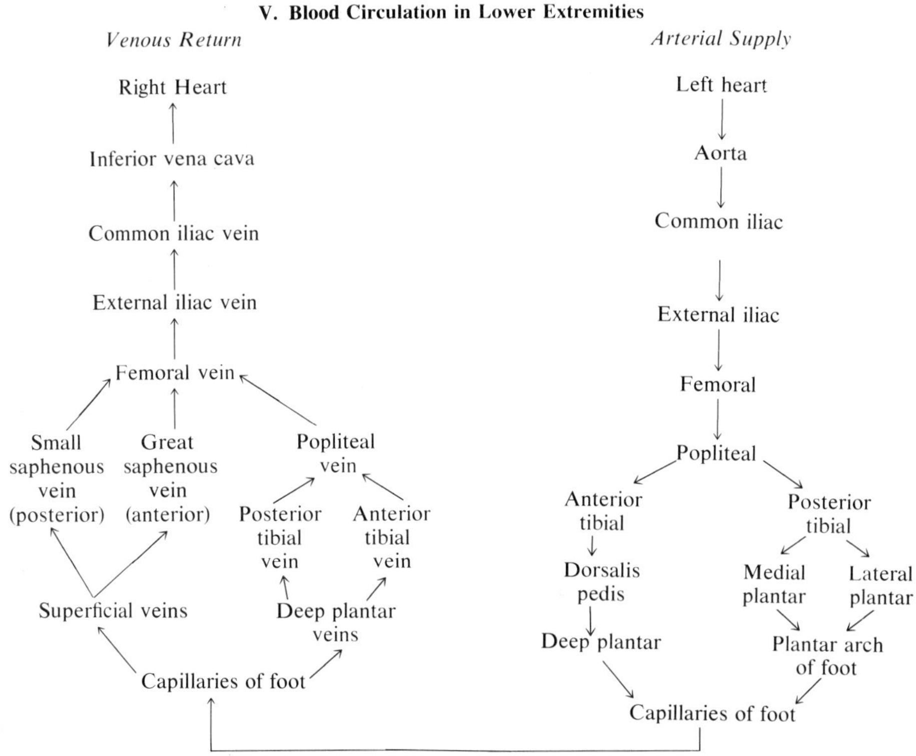

organs and tissues of the pelvis and perineum

B. Venous
1. Inferior vena cava drains the trunk by veins with the same names and locations as the peripheral branches of the aorta
2. Hepatic portal circulation
a. Returns blood to the inferior vena cava, through the capillaries in the liver, from the organs of digestion in the abdomen

VII. Pulmonary Circulation
A. Carries blood to and from the lungs
1. Pulmonary artery from right ventricle to lungs
2. Massive pulmonary capillary network
3. Pulmonary veins return blood to left atrium
4. Oxygenation and decarboxylation occur in lungs

VIII. Fetal Circulation
A. Exchange of gases and nutrients dependent on placenta
1. Umbilical vein supplies fetus with oxygenated blood and nutrients
2. Umbilical arteries remove waste materials from the fetus to the placenta
B. Structural modifications necessary to achieve homeostasis in the fetus
1. Ductus venosus—liver bypass from umbilical vein to inferior vena cava
2. Foramen ovale—valve connecting the atria
3. Ductus arteriosus—lung bypass from pulmonary artery to aortic arch
4. All modifications are corrected at birth; if not, a blue baby results

IX. Physiology of Circulation
A. Factors affecting circulation
1. Cardiac output
2. Blood volume
3. Peripheral resistance
a. Blood viscosity
b. Vessel diameter
c. Blood velocity
B. Pulse
1. Rhythmic contraction of arteries
C. Blood pressure
1. Pressure of blood in the arteries
2. Systole (120 mm Hg) vs. diastole (80 mm Hg)
3. Sphygmomanometer—blood pressure apparatus
4. Stethoscope—detects sounds
D. Vasomotor control of blood vessels
1. Sympathetic nerves—vasoconstriction
a. Aided by cardioaccelerator reflex—restores normal pressure
2. Vasodilation and cardioinhibitor reflex in medulla—lower blood pressure
3. Pressoreceptors and chemoreceptors—detect and regulate O_2–CO_2 tension and arterial pressures
E. Venous pressure
1. Diminishes as it approaches the heart
2. Aided by milking action of the valves and muscles
F. Hemodynamics
1. Semipermeable membrane of capillary wall
2. Tissue pressures involved
a. Hydrostatic pressure of tissues
b. Colloidal osmotic pressure of proteins and colloids in plasma
3. Filtration based on difference between opposing pressures in any given area
a. Arteriolar "feeding"—greater hydrostatic pressure than COP
b. Venular "draining"—great-

er COP than hydrostatic pressure

X. Clinical Considerations
  A. Arterial
    1. Aneurysm
    2. Thrombosis
    3. Arteriosclerosis
    4. Atherosclerosis
    5. Synthetic arterial transplants
  B. Venous
    1. Thrombophlebitis
    2. Varicosities (stripping)
    3. Hemorrhoids
  C. Systemic
    1. Circle of Willis—stroke
    2. Hemorrhage
    3. Shock
    4. Hypertension—hypotension
    5. Exercise
    6. Pulmonary embolism

STUDY QUESTIONS AND PROBLEMS

1. Where and how would you place pressure to stop bleeding from an artery?
2. Which veins are commonly used to obtain a sample of blood?
3. If a vessel were severed, how would you know whether it was a vein or an artery?
4. What is meant by the pulse? Where and how would you "take" a pulse?
5. Discuss the differences between arteries, veins, and capillaries according to their:
    a. Structure
    b. Function
    c. Location
6. Differentiate:
    a. Hydrostatic pressure of tissues
    b. Blood pressure
    c. Colloidal osmotic pressure
    d. Pulse pressure
7. Trace a drop of blood through the pulmonary circulation. Indicate the gaseous content along with the structural anatomy.
8. Where is the carotid sinus and what does it do?
9. How does the portal vein differ from other veins? Why?
10. What are some of the factors that may cause an increase in the pulse rate?
11. Explain the principle of the sphygmomanometer. Why is the stethoscope placed distal to the inflated cuff?
12. Correlate the two sounds heard through the stethoscope during the taking of the blood pressure with the cardiac cycle.
13. Define systolic and diastolic blood pressures in relation to the structural anatomy of the heart and arteries.
14. What are the symptoms of arteriosclerosis and how are they produced?
15. What is hemorrhage? If it occurred in the left region of the circle of Willis, what part of the body would be affected?
16. Define shock and indicate its seriousness.

Chapter 16

The Lymphatic System

Filtered fluid and cellular constituents that accumulate in the spaces between cells are drained and conducted back to the circulatory system by way of lymphatic networks.

GENERAL ANATOMY

Lymphatic Vessels

The lymphatic system is a one-way collecting system that gathers and drains the fluid collected between the cells. The larger lymphatic vessels drain into veins which return the lymph to the blood circulation.

Lymphatic capillaries resemble blood capillaries in structure. Both consist of a single layer of endothelial tissue. The major difference is that the lymphatic capillary has a closed terminal end; lymph fluid is absorbed from the tissue spaces through the endothelial membrane (Fig. 16–1). Most of the tissues of the body, with the notable exception of the central nervous system, are drained by lymphatics.

Larger lymphatic vessels drain the capillary network. The walls of these vessels resemble the walls of veins in structure. The muscle fibers in both the middle and outer layers are longitudinal and oblique, and therefore these larger vessels

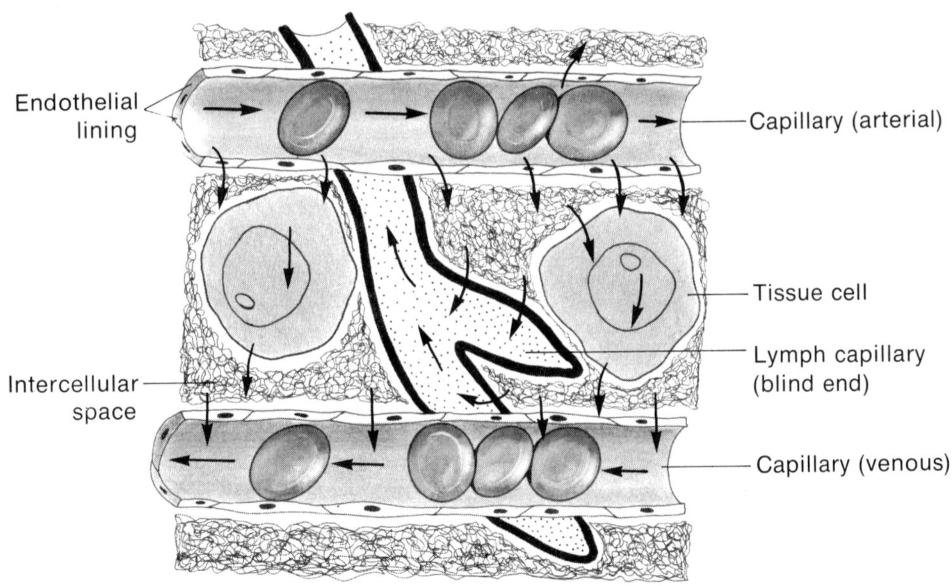

Figure 16–1 Diagram of a capillary bed, showing how materials diffuse between arterial capillaries and venous capillaries. Materials that are trapped in the intercellular tissue spaces are collected by lymphatic capillaries and returned to the blood system.

are contractile; lymphatic capillaries are not.

The larger lymphatics not only have the same basic three layers as veins; they also have valves on their inner surfaces to prevent backflow (Fig. 16–2). The valves are bicuspid or tricuspid, and occur at shorter intervals than do those of veins. Lymphatic capillaries do not contain valves.

Lymph Nodes

Along the course of the lymphatic vessels are small bodies of lymphoid tissue called lymph nodes (Fig. 16–3). These usually are oval, and are commonly referred to as "microkidneys."

Afferent lymphatic vessels enter the node at different points, but the efferent vessels leave only at a small depression, the *hilus* (Fig. 16–3). Blood vessels also enter and leave from the hilus. Internally, the lymph node resembles a sponge. Sinuses are spaced throughout the tissue and provide a filtering network for the lymph stream (Fig. 16–3). These sinuses consist of a maze of passageways lined with phagocytic cells that engulf bacteria or other foreign products from the lymph fluid.

Experimental evidence indicates that lymph nodes may afford some protection against infection or disease. Lymphocytes (leukocytes) of the lymph nodes and spleen actively destroy such foreign substances as bacteria and antigens that are introduced into the body. Lymph nodes commonly become swollen and inflamed (e.g., the tonsils) during severe bacterial infections.

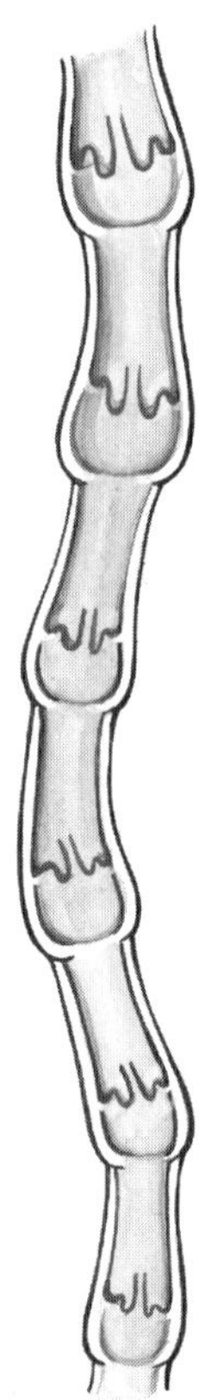

Figure 16–2 Diagram of valves in a lymph vessel.

Lymphatic Tissues

The following organs containing lymphoid tissue perform somewhat different functions, particularly with respect to the substances that they filter:

1. Tonsils. These masses are found just beneath the epithelium of the mouth and throat. There are several tonsils, including the *palatine* (at each side of the soft palate), the *pharyngeal* (behind the nose on the back wall of the upper pharynx, sometimes called the adenoids); and the *lingual* (back of the tongue). Any or all of these tonsils may become so infected with bacteria that their surgical removal may be indicated.

2. Thymus. This gland, usually comprising two lobes, plays a key role in the formation of antibodies in the first few weeks of life and in the development of immunity. It is located at the base of the neck and manufactures lymphocytes. Removal of the thymus causes a decrease in lymphocyte production throughout the body and also a decrease in size of the spleen and lymph nodes. The thymus is

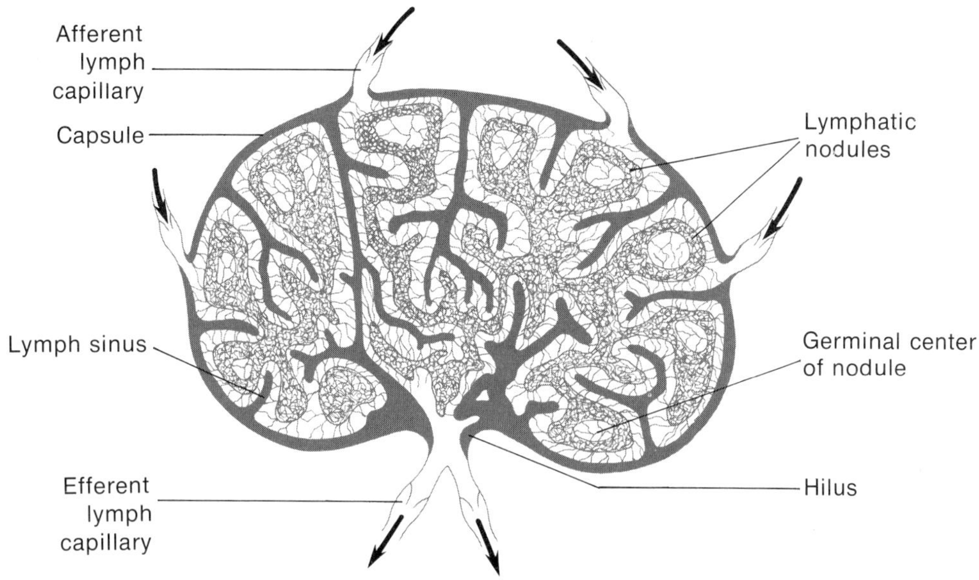

Figure 16–3 Anatomy of a lymph node. Lymphocytes develop in the germinal centers of the nodules.

most active during early life; after puberty it undergoes degenerative changes and is replaced by adipose tissue.

3. Spleen. This organ is situated in the left abdomen next to the stomach. It is designed both to filter blood and to produce lymphocytes that carry on phagocytosis. The spleen has many functions concerned with lymphatic and blood circulation. Prominent structures inside the spleen are round masses of lymphoid tissue, and these justify its classification as a lymphatic organ.

Lymphatic Circulation

There are both superficial and deep sets of lymphatic vessels. The surface lymphatics, immediately below the skin, often continue near the superficial veins (Fig. 16–4). The deep lymphatics usually are larger, and accompany the deep veins. All lymphatic vessels form networks, and, at certain points, carry lymph into the regional lymph nodes. There are groups of large nodes in the neck (cervical), under the arm (axillary), at the elbow (cubital), and in the groin (inguinal).

The lymphatics in the region of the small intestine (Peyer's patches) have the special function of absorbing digested fat. The inner wall of the small intestine is lined with finger-like projections, the *villi.* Each villus contains a central lymph capillary, or *lacteal,* which absorbs the end-products of fat digestion. Lymph carrying absorbed fat has a milky appearance, and is called *chyle.* The cisterna chyli is concerned with the collection of lymph from the lower limbs, the intestine, and the kidneys (Figs. 16–5 and 16–6).

Lymphatic vessels are named according to their location. Those on the thumb side of the forearm are called the radial lymphatic vessels; those on the medial part of the forearm are the ulnar lymphatic vessels. Nearly all the lymph from the upper extremity and from the breast is carried into the axillary lymph nodes, and eventually to one of the two large drainage ducts that empty into the bloodstream.

The *right lymphatic duct* is a short tube about one-half inch long which receives lymph that comes from the right side of the head and thorax and from the right upper extremity (Fig. 16–4, *A*). It empties into the right subclavian vein. The remaining three-fourths of the body is drained by a 16-inch long tube called the *left lymphatic* or *thoracic duct* (Figs. 16–4 and 16–5). It begins at the back of the abdomen below the attachment of the diaphragm, and its first section arises in the cisterna chyli. The duct then extends upward through the diaphragm and along the back wall of the thorax up into the root of the neck on the left side. Here it receives

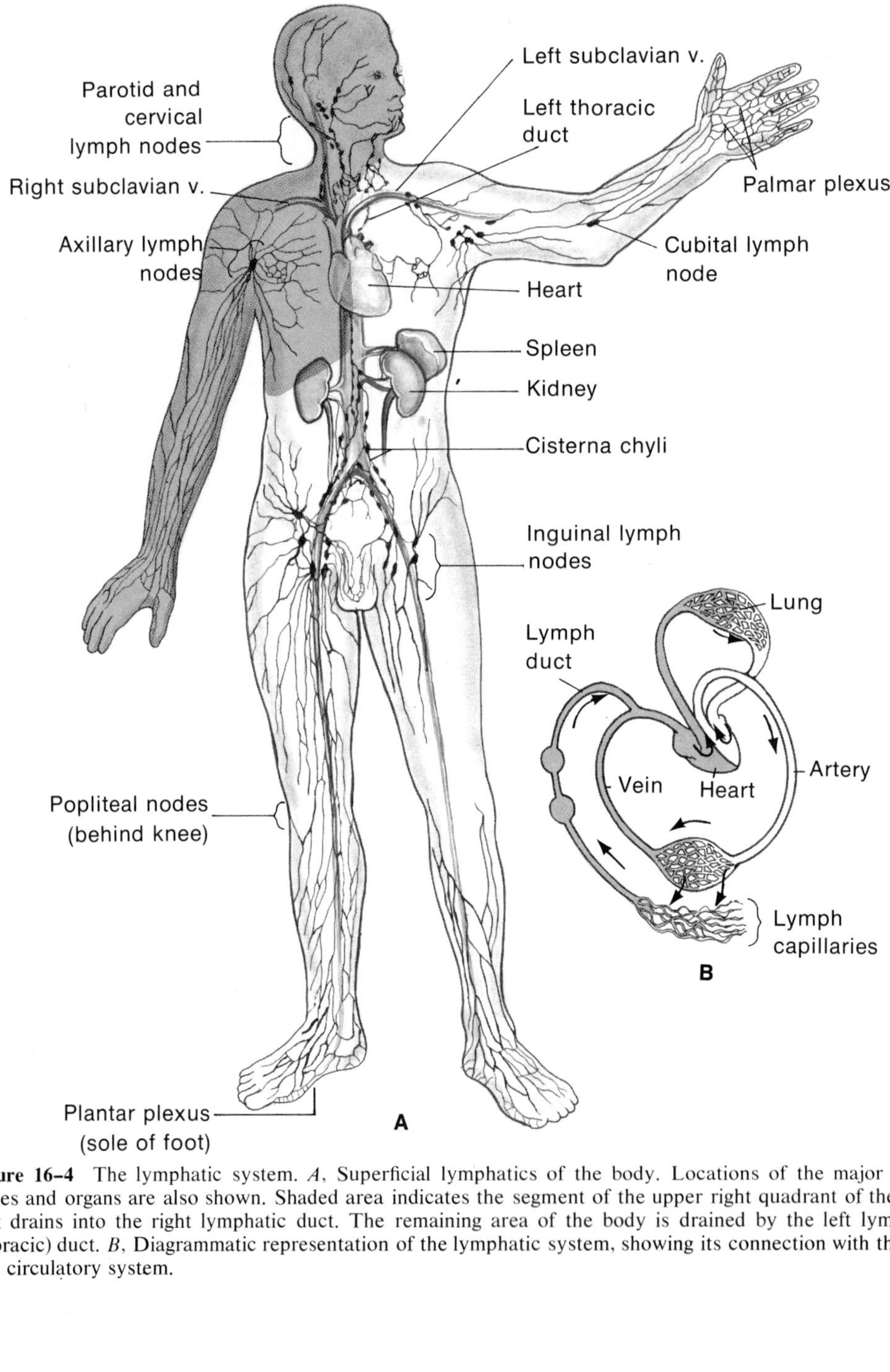

Figure 16–4 The lymphatic system. *A*, Superficial lymphatics of the body. Locations of the major lymph nodes and organs are also shown. Shaded area indicates the segment of the upper right quadrant of the body that drains into the right lymphatic duct. The remaining area of the body is drained by the left lymphatic (thoracic) duct. *B*, Diagrammatic representation of the lymphatic system, showing its connection with the general circulatory system.

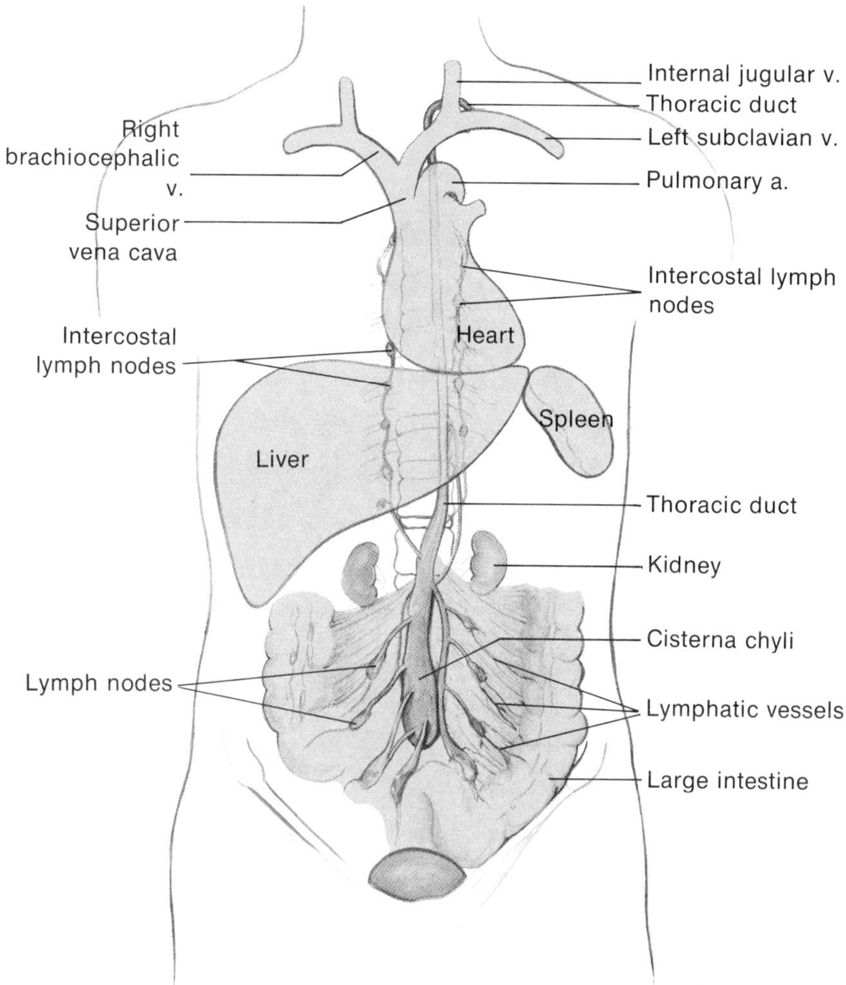

Figure 16–5 Lymphatic drainage to the cisterna chyli and thoracic duct.

the left jugular lymphatic vessels from the head and neck, the left subclavian vessels from the left upper extremity, and other lymphatic vessels from the thorax and its parts. It then opens into the left subclavian vein in the angle between the left subclavian and left internal jugular veins.

PHYSIOLOGY OF LYMPHOID TISSUE

Lymph Formation

Lymph is defined as the fluid that returns to the circulation from the tissue spaces by way of the lymphatics. Since lymph originates from plasma as the difference between filtration and absorption of fluid across the capillary wall, its composition is very similar to that of plasma. Indeed, except for variations in protein concentration and electrolytes, lymph and plasma are nearly identical. The average protein concentration of lymph is about 1.5 per cent, compared with 6 per cent in plasma. Since albumin is the smallest plasma protein, it permeates the capillary endothelium more readily than do globulin and other protein fractions. Consequently, more albumin accumulates in lymph.

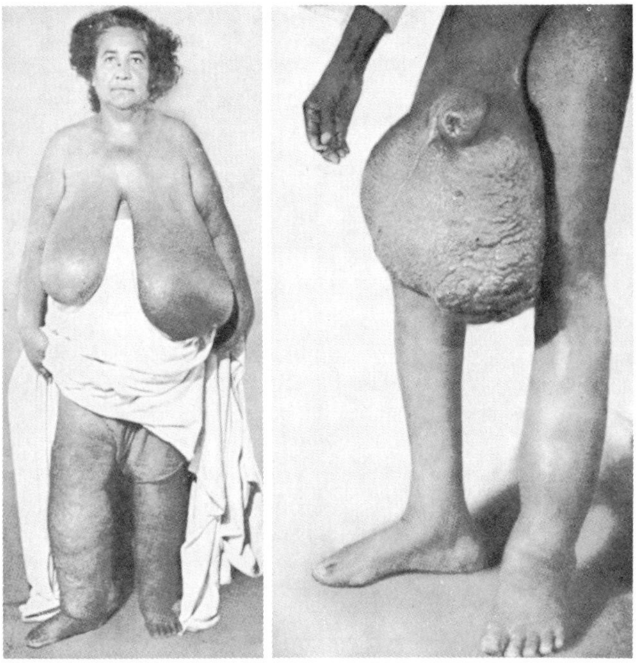

Figure 16-6 Elephantiasis. (From Hunter et al.: Tropical Medicine, 5th ed. Philadelphia: W. B. Saunders, 1975.)

Lymph is a clear fluid containing a low count of granular leukocytes and a varying number of lymphocytes. Its protein content is less than that of blood, as previously mentioned, and the total phosphorus and calcium content also is lower. Many enzymes concerned with digestion of the basic foodstuffs have been found in lymph, as has the enzyme concerned with clotting.

Lymph Flow

Lymph flows very slowly under normal circumstances, aided by muscular contractions and the pressure built up by filtration of fluid from the capillaries. Intravascular infusions of various solutions enhance the formation of lymph and speed lymph flow through increases in the volume of fluid in the circulatory system. Corresponding increases in arterial and venous blood pressure also occur. In addition, the volume of fluid that gains access to the capillaries increases. These mechanisms increase capillary pressure and stimulate filtration.

Function of the Lymphatics

Lymph and the lymphatic system serve many functions:

1. *Defense mechanism*—lymphocyte production phagocytizes pathogenic organisms.
2. *Immunity*—production of antibodies aids in combating disease and infection.
3. *Tissue drainage*—networks of vessels drain the spaces between the tissue cells and return the metabolites to the system, where they can be reused.
4. *Excretion*—the lymphatic system removes impurities such as carbon particles, waste products, and dead blood cells.
5. *Digestion*—intestinal lacteals absorb end-products of fat digestion and distribute them via the circulatory system.

CLINICAL CONSIDERATIONS

Lymphangitis

Inflammation of a lymphatic vessel or vessels is called lymphangitis. Red streaks can be seen extending along an extremity, usually beginning in the region of an infected or neglected injury. In such cases, the lymph nodes may not be able to arrest a serious infection, and blood poisoning (septicemia) may result. In most cases, the invading organism is a streptococcus.

Adenitis

Adenitis means inflammation of a gland, but the term is frequently used in reference to an enlarged or tender lymph node. Cervical adenitis occurs during measles, scarlet fever, septic sore throat, and the common cold. Chronic adenitis may be due to tuberculosis of the so-called "cold abscess."

Hodgkin's Disease

Hodgkin's disease is a chronic disorder characterized by enlargement of the lymph nodes. Enlarged nodes are found in the neck particularly, but often in the axillary and inguinal regions as well. The cause of the disease is not well understood, but research indicates that it may be due to some infection, such as tuberculosis, that is more common in young men. Hodgkin's disease is ultimately fatal, although radiotherapy has been used to check its progress temporarily.

Lymphosarcoma

Lymphosarcoma is a malignant tumor of lymphoid tissue that is likely to be rapidly fatal. Early diagnosis by a tissue examination (biopsy) and surgical removal of the neoplasm offer the only possible cure at this time. Radiotherapy has proved beneficial in the early stages of tumor growth.

If the lymph glands around the mammaries are involved, the corrective surgery is called a *radical mastectomy* (Fig. 16–7).

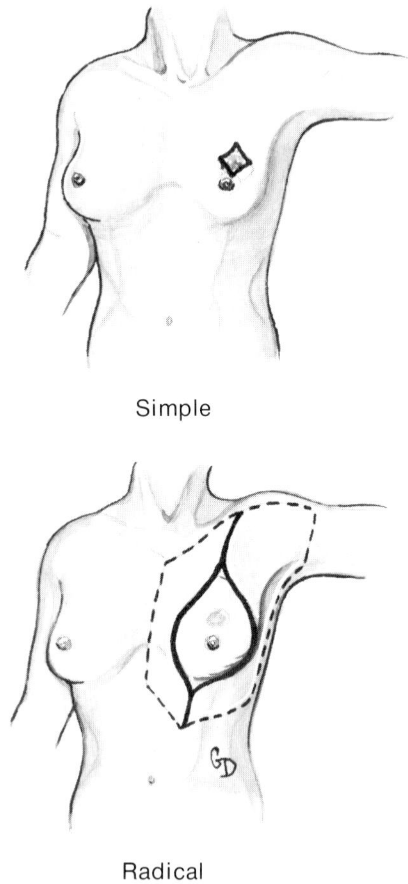

Figure 16–7 Simple and radical mastectomies.

This procedure requires the removal of all tissue around the breast, including the axillary lymph nodes, to prevent systemic metastasis.

Splenomegaly

Enlargement of the spleen may follow certain infectious diseases, such as scarlet fever, typhus, and syphilis. Splenic anemia also causes an enlargement of the spleen accompanied by abdominal hemorrhage from the stomach. In this disease and others like it, a splenectomy—total removal of the spleen—appears to constitute a cure.

Elephantiasis

Elephantiasis, an enlargement of the lower extremities, is due to the stoppage of lymphatic vessels by small worms called filariae. These tiny parasites, carried by insects, invade the tissues and grow in the lymph channels, thus obstructing lymph flow. Without proper back drainage of the lower extremities, the tissue fluids collect in the legs and scrotum, rendering a male incapacitated and impotent (Fig. 16–6). The disease is common in Asia, and no cure has been reported.

Edema

Edema is an excessive accumulation of fluids in the tissue spaces. It is of functional significance in that excess fluid between capillaries and cells hampers the exchange of nutrients and metabolites between cells and plasma (Fig. 16–8).

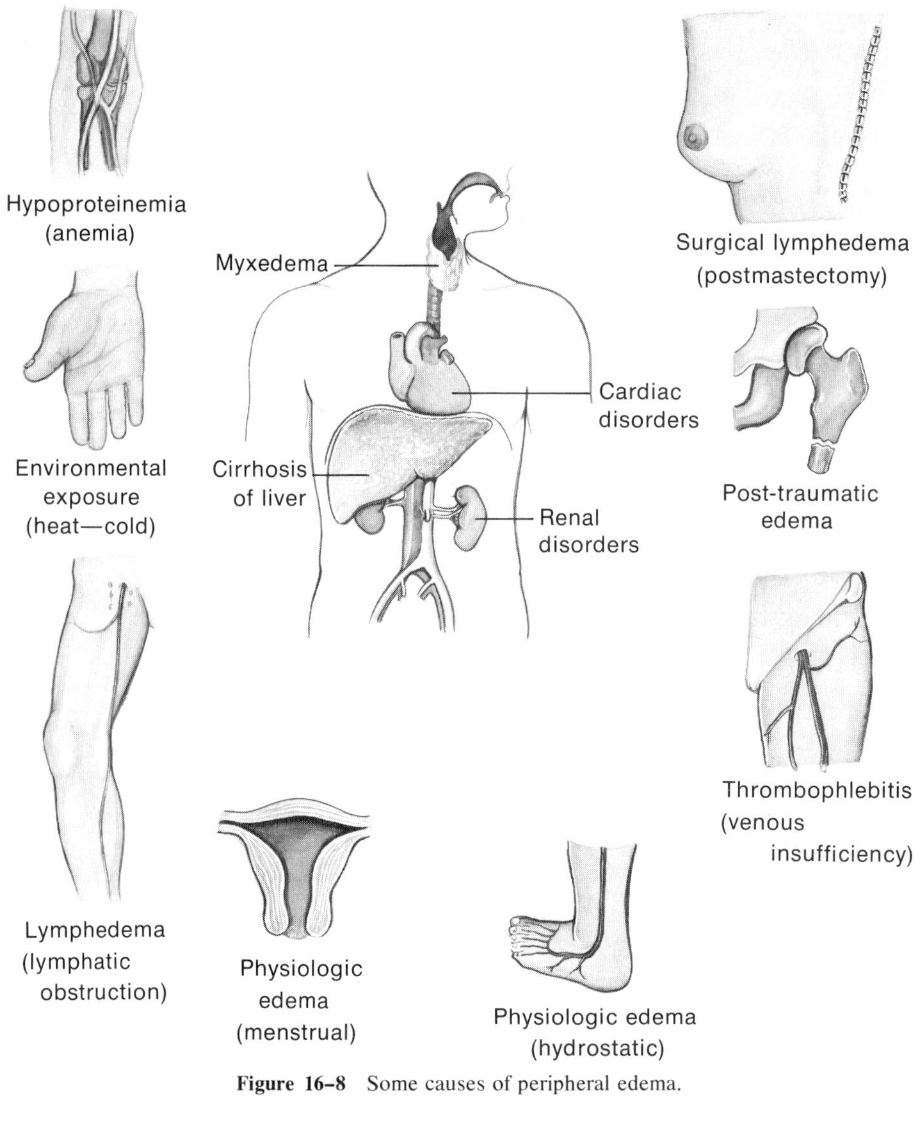

Figure 16–8 Some causes of peripheral edema.

In active tissues, the capillary pressure rises, often to the point where it exceeds the oncotic pressure throughout the length of the capillary. In addition, metabolites may temporarily accumulate in the interstitial fluid because they cannot be washed away as rapidly as they are formed. Both situations exert an increased tissue pressure that in turn increases the amount of fluid leaving the capillary and decreases the amount entering it, resulting in increased interstitial fluid—edema.

Interstitial fluid tends to accumulate in dependent parts owing to the effect of gravity. When a person is in an upright posture, the capillaries in the legs are protected from the high arterial pressure by the arterioles, but the venous pressure is transmitted to them through the venules. Skeletal muscle contractions keep the venous pressure low by pumping blood toward the heart when the individual moves about; but, if he stands still for long periods, fluid accumulates and edema develops. Ankles will swell when a person sits for prolonged periods, as when traveling.

Salt retention in abnormal quantities results in water retention. Both are distributed throughout the tissue spaces, increasing the interstitial fluid volume and resulting in tissue edema. Salt and water retention are major factors in the edema seen in heart failure, nephrosis (kidney disease), and cirrhosis (liver disease). In congestive heart failure there is usually an elevation in venous pressure, and the capillary pressure is consequently elevated. In cirrhosis of the liver, oncotic pressure is low because liver synthesis of proteins is depressed; in nephrosis, oncotic pressure is low because large amounts of protein are lost in the urine (proteinuria).

Another cause of edema is inadequate lymphatic drainage. Following the removal of the axillary networks of lymph vessels in a radical mastectomy, drainage from the arms is disrupted. In elephantiasis, filarial blockage of the inguinal lymph nodes creates an accumulation of tissue fluid in the lower extremities.

OUTLINE

I. Definition—a drainage system that returns extracellular fluid to the blood circulation
II. General Anatomy
  A. Lymphatic vessels
    1. Capillaries
      a. Single endothelial layer
      b. Closed terminal end
    2. Larger vessels
      a. Resemble veins
      b. Have valves
  B. Lymph nodes
    1. Oval-shaped
    2. Afferent vessels drain into node
    3. Efferent vessels leave only at hilus
    4. Spongelike sinuses
    5. Produce lymphocytes
  C. Lymphatic tissues
    1. Tonsils—protect oral region
      a. Palatine
      b. Pharyngeal
      c. Lingual
    2. Thymus
      a. Produces antibodies before puberty
      b. Manufactures lymphocytes
    3. Spleen
      a. Filters blood
      b. Produces lymphocytes
  D. Lymphatic circulation
    1. Superficial and deep vessel networks
    2. Groups of nodes at key locations
      a. Cervical, axillary, cubital, and inguinal
      b. Intestinal nodes (Peyer's patches)—involved in digestion
    3. Right lymphatic duct—drains upper right quadrant of body

4. Left lymphatic duct (thoracic duct)—drains remaining three-quarters of body

III. Physiology of Lymphoid Tissue
   A. Lymph formation
      1. Extracellular fluid accumulation
      2. Similar to blood plasma
         a. Decreased protein quantity
         b. Higher albumin ratio
      3. Low leukocyte content; no erythrocytes
      4. Many digestive enzymes
   B. Lymph flow
      1. Very slow under resting conditions
      2. Flow a result of muscular contractions
      3. Valves in large vessels prevent backflow
   C. Function of lymphatics
      1. Defense mechanism—phagocytosis
      2. Immunity—antibody production
      3. Tissue drainage—edema prevention
      4. Excretion—removal of waste products
      5. Digestion—absorption and circulation of end-products of fat digestion

IV. Clinical Considerations
   A. Lymphangitis—inflammation of lymphatic vessels
   B. Adenitis—enlarged lymph nodes or glands
   C. Hodgkin's disease—infected and enlarged lymph nodes
   D. Lymphosarcoma—malignant lymphatic tumor
   E. Splenomegaly—enlarged spleen following infectious disease
   F. Elephantiasis
      1. Parasitic worms (filariae) grow in lymph vessels
      2. Blockage of lymph vessels causes lower extremities to swell
   G. Edema—excess accumulation of interstitial fluid
      1. Increased accumulation of metabolites
      2. Gravity effect on dependent parts of body
      3. Salt retention
      4. Systemic failures
         a. Congestive heart failure
         b. Nephrosis—kidney
         c. Cirrhosis—liver
      5. Inadequate drainage
         a. Surgical removal of diseased lymph nodes and vessels
         b. Parasitic blockage of lymph network

STUDY QUESTIONS AND PROBLEMS

1. Explain why lymph capillaries, vessels, and nodes are important in everyday living.
2. What is the function of the thymus gland? What is its longevity?
3. What are antibodies? Where are they produced in the lymphatic system?
4. Why are tonsils and adenoids important in early childhood? Why aren't they removed the first time they become infected?
5. List the networks of nodes that are found at junctions of the appendages to the main body trunk. Why do you think they are primarily located in those regions?
6. What is lymph? How does it originate?

Chapter 17
The Respiratory System

Most animal cells require a continual supply of oxygen. They also require a means of getting rid of the surplus carbon dioxide that results from cellular activity. Any full consideration of respiration must include the processes involved in utilizing oxygen and producing and disposing of carbon dioxide. In the broadest sense, respiration refers to all the oxidative processes concerned with the release of energy from nutrients; in its narrowest sense, the term sometimes is used merely as a synonym for breathing. Oxidative processes within the cells are discussed in Chapter 18, and so the present discussion will cover the delivery of oxygen to the cells and the removal of carbon dioxide from them.

GENERAL ANATOMY

The Upper Respiratory Tract

Air drawn in through the nose passes over a complex series of surfaces formed by the nasal septum and the nasal conchae (Fig. 17–1). These surfaces not only warm and moisten the air, but also act to baffle its flow, causing dust particles to settle out on the mucous films that line the nasal passages. Their ciliated epithelium acts to move dust-laden mucus backward to the pharynx, where it is either swallowed or expectorated. Breathing through the mouth allows relatively cold and unfiltered air to reach the lungs.

Both the nasal and buccal cavities open into the pharynx, and the esophagus and larynx descend from the lower part of the pharynx. The trachea (windpipe), in turn, descends from the larynx. Mucous membranes of the superior conchae contain the endings of the olfactory nerve fibers. The nerve fibers for the muscles of the nose come from the facial nerve (Cr. VII), and the skin receives fibers from branches of the trigeminal nerve (Cr. V).

The partition separating the two nasal spaces from each other is called the *nasal septum* (Fig. 17–1). Since many of us have minor structural defects, it is not surprising that the nasal septum is rarely situated exactly at the midline. If it is markedly to one side, the anomaly is described as a *deviated septum.* Attacks of hay fever or colds, accompanied by swelling of the nasal mucosa, make breathing extremely difficult in the smaller of the two nasal cavities. Such occurrences may lead to surgical repair of the septum, a submucous resection (SMR), in which the cartilaginous septum is removed entirely, allowing the opposing mucous membranes to act as a septum and to be more flexible.

Nosebleeds (epistaxis) can result from an injury or blow to the nose. Growths and, occasionally, high blood pressure also can cause nosebleeds. Pressure applied to the nostril of the bleeding side and cold compresses over the nose usually alleviate the problem. If not, cotton packing dipped in epinephrine has proved helpful as an emergency measure.

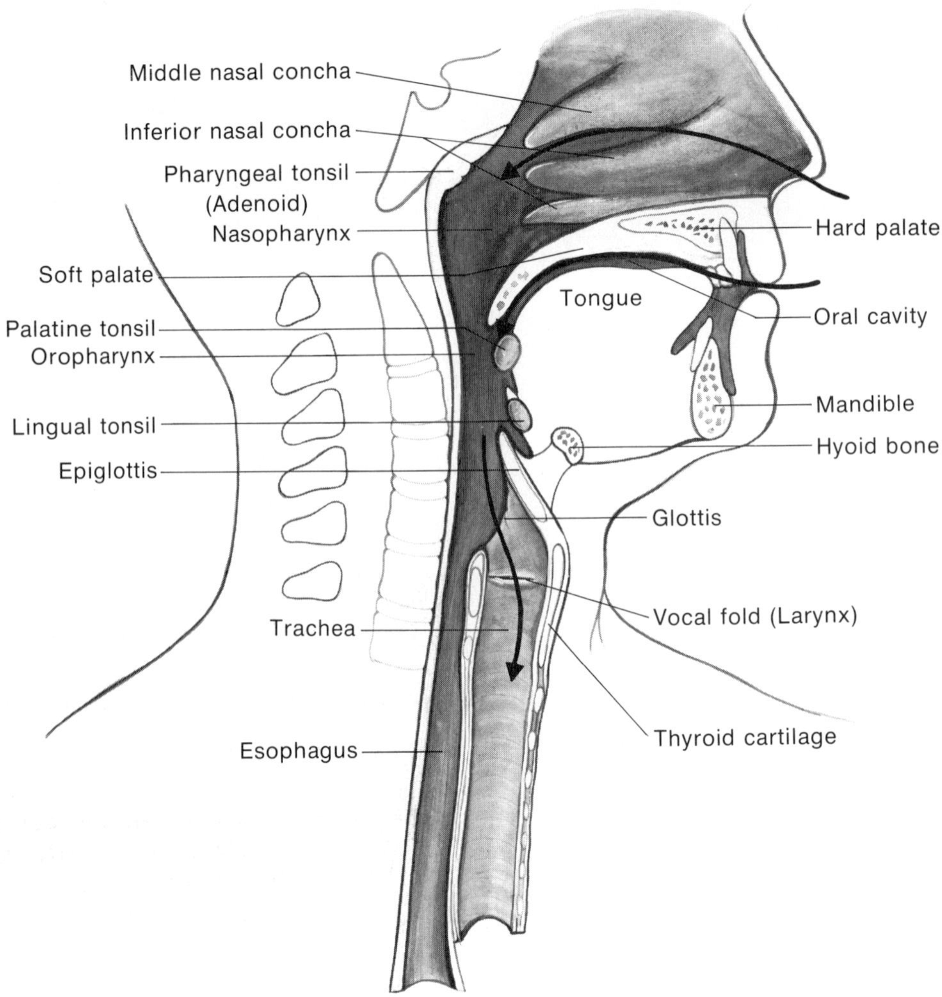

Figure 17–1 Sagittal section of the head and neck, showing the respiratory passage down to the bifurcation of the trachea.

The Respiratory Tract

In addition to the nose and nasopharynx already described, the parts of the respiratory tract are the larynx, trachea, bronchi, and lungs (Fig. 17–2).

The *larynx,* sometimes called the "voice box," lies just below the pharynx in the upper and front part of the neck (Figs. 17–2 and 17–3). It is shaped like a triangular box, with flat sides and a prominent ridge ("Adam's apple") in front, and is made up of many fibrocartilages united by ligaments and moved by muscles. It contains the vocal folds, and is protected from above by a movable cartilaginous lid called the *epiglottis,* which opens as one breathes and closes reflexly when one swallows, thus preventing food or drink from entering the respiratory tract (Fig. 17–3).

The opening between the vocal folds is V-shaped when the individual is resting; when he talks or sings, the folds are stretched or slackened, causing the aperture between them, the opening into the glottis, to become oval in shape (Fig. 17–3). A high-pitched note is formed by stretching the folds tightly; a low-pitched note causes the folds to be pulled apart, forming an oval. Vocal sounds are made

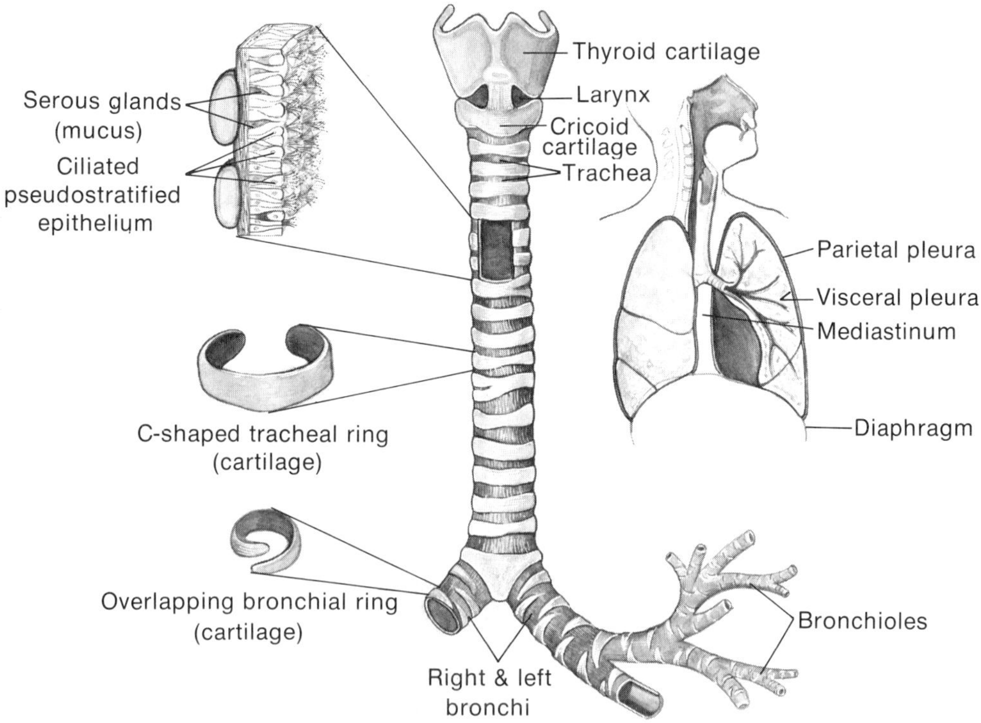

Figure 17–2 The human larynx, trachea, and bronchi, anterior aspect.

when air is forced over the folds, causing them to vibrate. An inflammation of the vocal cords, as in laryngitis, prevents free movement of the cords and causes temporary loss of voice.

A very useful instrument, the *laryngoscope,* permits the investigator clear visual access to the trachea. Insertion of a bladed tongue depressor allows the anesthesiologist an easy route for in-

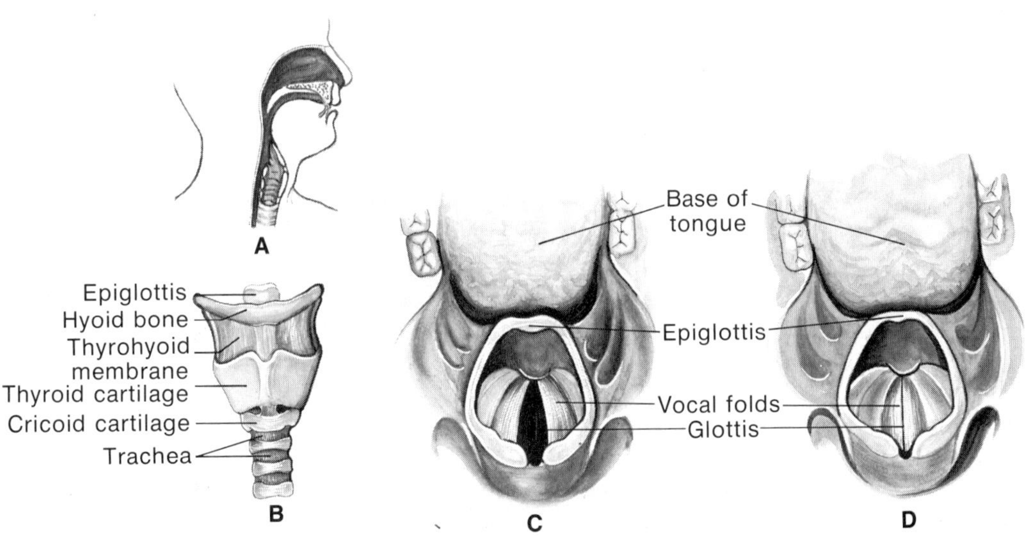

Figure 17–3 The larynx. *A,* In relation to the head and neck. *B,* Anterior aspect. *C,* Superior aspect with vocal folds open. *D,* Superior aspect with vocal folds closed.

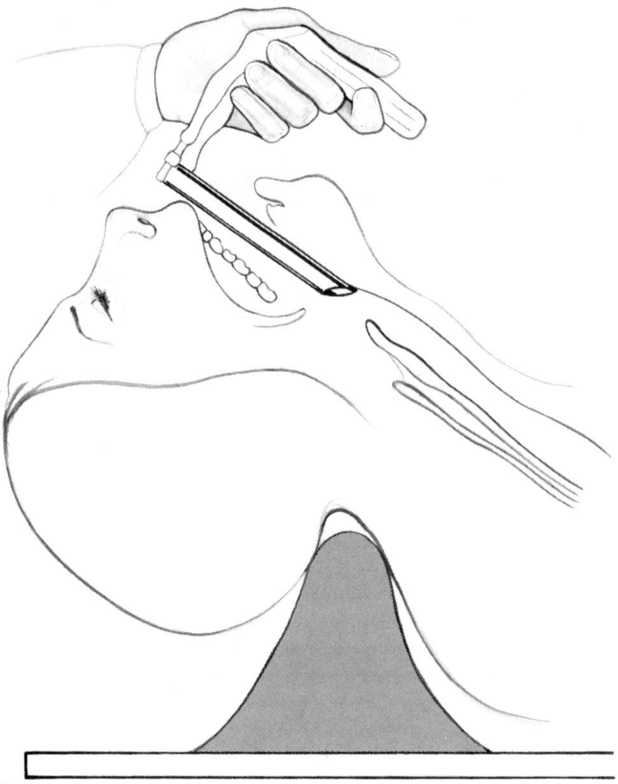

Figure 17–4 Laryngoscopy.

sertion of an endotracheal tube in general anesthesia (Fig. 17–4). The laryngoscope also is commonly used to locate and remove food from the larynx. Foreign particles will induce choking and gagging; and, if the cough reflex fails to dislodge them, emergency procedures are indicated.

Croup, a common respiratory problem of childhood, is characterized by sensations of suffocation, laborious breathing accompanied by spasms of the larynx (laryngospasm), a "barking" cough and a "crowing" sound during inspiration. There is much secretion of mucus in which infectious material is mixed. Muscle-relaxing drugs, such as curare, have been used in emergency situations to open the airway so that the patient can breathe more normally. A moist air or croup tent can be used in the home to alleviate the child's symptoms.

The *trachea,* a long, membranous tube supported by rings of hyaline cartilage and fibrous connective tissue, lies in front of the esophagus and extends from the larynx. Its C-shaped cartilaginous rings (Fig. 17–2) strengthen it and keep it open. Like the larynx, it is lined with mucous membrane and has ciliated epithelium on its inner surface. Particles of foreign matter in inhaled air become engulfed in mucus. Movements of the cilia continually sweep the dust-laden mucus upward into the pharynx.

Tracheotomy is a surgical procedure in which an opening is made into the trachea through which the patient may breathe (Fig. 17–5). It may be performed for the following reasons:

1. Inadequate air breathing space caused by a tumor, foreign body, or edema.

2. A need for removal of bronchial secretions.

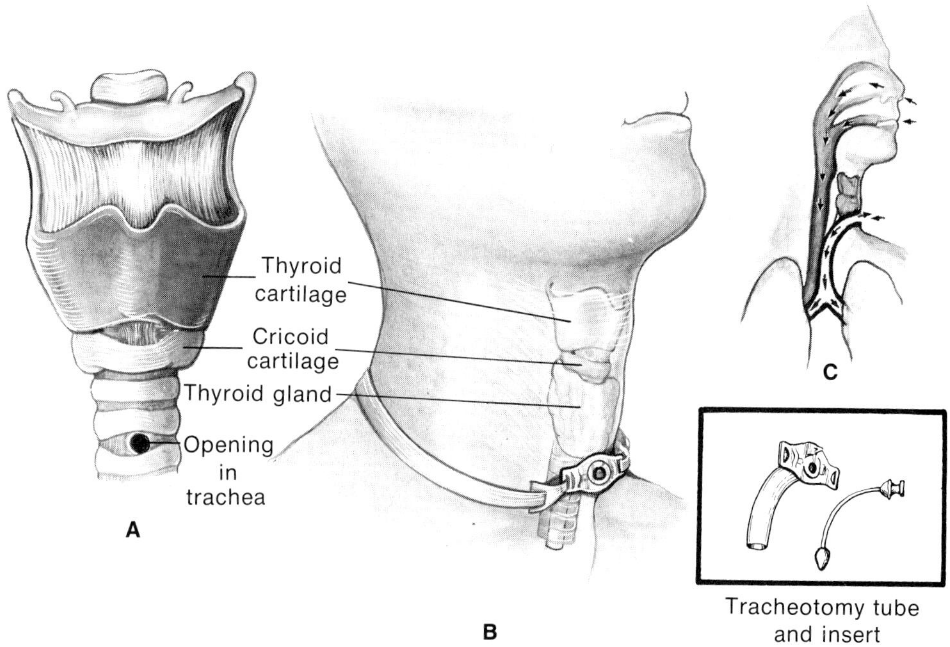

Figure 17–5 Tracheotomy. An opening is made in the trachea *(A)*, and a tracheotomy tube is inserted to allow the patient to breathe *(B)*. This tube allows free access to outer air *(C)*.

3. To afford greater air intake in cases of lung surgery or emphysema.

The tracheotomy tube may be removed and the opening closed surgically following a correction of the abnormality. If the opening is a permanent one, the term *tracheostomy* is used.

The trachea divides into two *bronchi* that enter the right and left lung, respectively, the right bronchus being shorter, wider, and more vertical than the left. Within the lungs the bronchi subdivide into smaller branches, the bronchial tubes, and finally into *bronchioles* (Figs. 17–2 and 17–6).

The two bronchi resemble the trachea in structure, except that the bronchi are supported by overlapping cartilaginous rings. As the bronchi subdivide into smaller tubes, the rings are replaced by fibrous tissue. Each bronchiole terminates in an elongated sac called the *atrium* (Fig. 17–6, *B* and *C*); these are lined with numerous pouches, the *alveoli* or air cells (Fig. 17–6). Alveolar walls are very thin and highly vascular. They are lined with squamous epithelium, and an exchange of gases occurs through the walls of the capillaries in the alveolus between the blood and the air.

In addition to the elaborate branching of the bronchial tubes, the lungs have an abundant circulatory network. The pulmonary artery divides into smaller and smaller branches until it finally resolves itself in capillary networks around the bronchioles and alveoli (Fig. 17–6). Pulmonary capillaries anastomose to form small venules as the blood leaves the alveoli. These then become the pulmonary veins, which carry freshly oxygenated blood back to the left atrium of the heart.

If a foreign body enters the smaller air passages of the lungs, it may obstruct the bronchi, causing a blockage of air in the portion of the lung they supply. This produces a partial collapse of the lung, causing symptoms of coughing and expectoration. In many patients it is possible to remove the foreign body following exami-

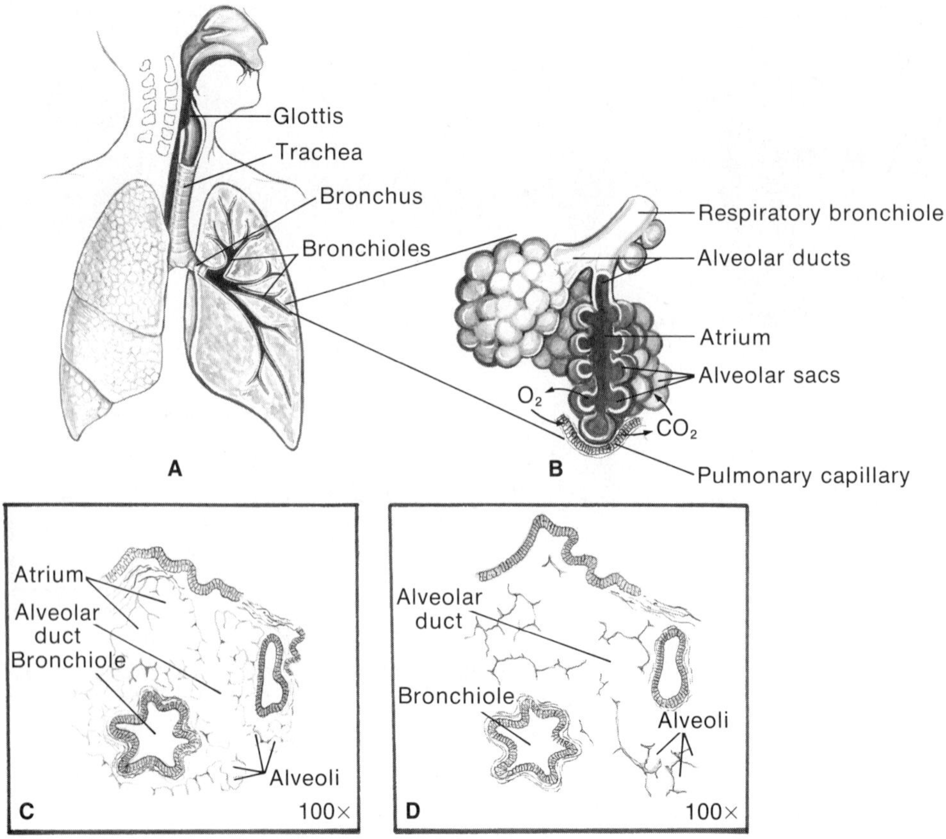

Figure 17–6 Internal structure of the lungs. *A*, Relationship of the lungs to the head and neck. *B*, External and internal appearance of a lung lobule, showing the atrium and alveolar sacs. *C*, Section of a lung, magnified 100 times. *D*, Section of a similar lung with emphysema. Note the decreased number of alveolar sacs, and consequent diminishing in the gas exchange area of lung tissue.

nation by *bronchoscopy*. With this method, a rigid, lighted tube is inserted through the mouth, pharynx, and trachea into the obstructed bronchus. After aspiration of the secretions, the foreign body may be removed with lung forceps (Fig. 17–7). Bronchoscopy is also used in the diagnosis and treatment of many intrathoracic diseases. Specimens from tumors of the air passages can be removed and examined (biopsy); secretions can be aspirated and bacterial studies made of them. In the study of some diseases of the lungs, such as bronchiectasis, liquids are injected to outline the air passages and an x-ray film taken (bronchogram). Esophagoscopy is the viewing of the interior of the esophagus through a lighted tube. Foreign bodies may lodge in the esophagus just as they do in the bronchi; and disease processes of the esophagus are similar to those of the bronchus.

Atelectasis is the collapse of an expanded lung. Primary atelectasis occurs when the lungs fail to expand completely at birth; secondary atelectasis occurs in adults following a bronchial obstruction, fluid in the chest cavity, or pneumothorax (air in the chest cavity).

Bronchiectasis is a dilatation of the bronchi. The disorder may be congenital, but it generally occurs as a result of infection and overdistention of one bronchus or a group of bronchi due to chronic bronchitis. This overdistention results from tearing of the elastic fibers in the bronchial wall. It is characterized by spasms of coughing and the production of large quantities of sputum containing pus.

Emphysema is a pathologic enlargement of the alveoli and air passages due to the rupture and tearing of the elastic alveolar walls, resulting in a lung that cannot collapse. The disease may be a consequence of bronchitis, pulmonary tuberculosis, whooping cough, or bronchial asthma. As it progresses, there is increased shortness of breath, and the thorax becomes enlarged and "barrel-shaped" as the patient attempts to inhale.

The thoracic cavity is divided from top to bottom by a double wall, the *mediastinum*. The heart, great vessels, esophagus, and many nerves lie within this area; the lungs are in *pleural cavities* on each side.

The two *lungs* vary in shape and size. The right lung is the larger, and has three lobes. It is wider than the left lung but a little shorter. The left lung is divided into two lobes and is somewhat narrower and longer than the right (Fig. 17–2).

A thin serous membrane covers each lung and continues over the thoracic wall, the diaphragm, and the mediastinum. This membrane is the *pleura* and has two components. The pulmonary (visceral) pleura covers the lung; the parietal pleura lines the thoracic cavity. These two layers are in constant contact, and a serous fluid is secreted by the membranes to lubricate the surfaces that move upon each other with each respiration. Inflammation of the pleural membranes is called *pleurisy*. This condition can be quite painful because the inflammation produces a sticky exudate that roughens the pleura of both the lung and the chest wall, causing an acute irritation. In pleurisy due to bacterial infection, pus will accumulate in the pleural cavity—a condition known as *empyema*. Removal of the fluid by syringe is often necessary. An abnormal accumulation of pleural fluid is *pleural effusion*. Aspiration of this fluid is done under closed conditions so as not to allow air into the chest cavity and collapse the lung. This process is called a *thoracocentesis*.

The substance of the lungs is porous and spongy because of the air in the sacs.

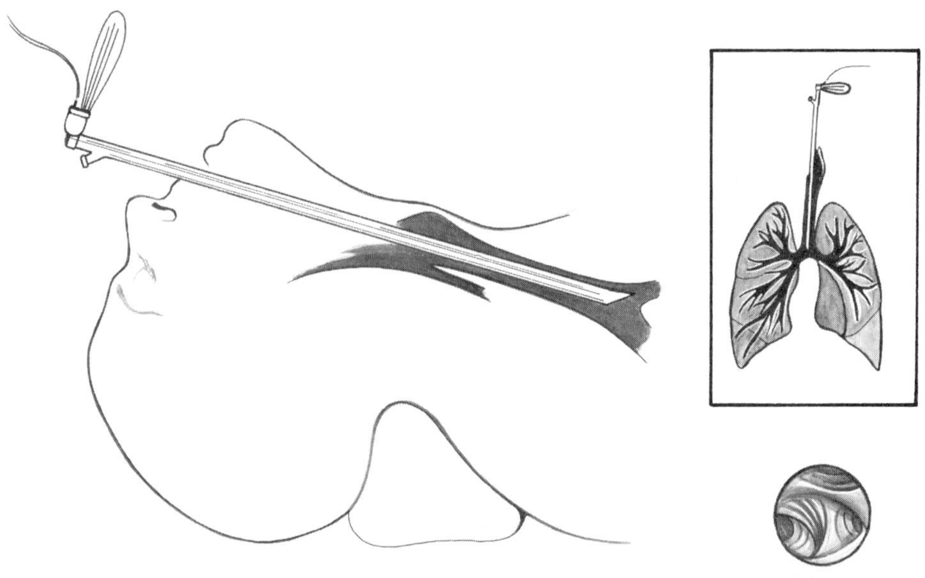

Figure 17–7 Introduction of the bronchoscope. *A*, The bronchoscope is a device used to explore the trachea and bronchi. *B*, View through the bronchoscope as it approaches the bifurcation of the trachea into left and right bronchi.

Each lobe of the lung is composed of many lobules, and into each lobule a bronchiole enters and terminates in an atrium. Surgical removal of a single diseased lobe is called a *lobectomy;* removal of a whole lung is a *pneumonectomy.*

MECHANISM OF BREATHING

The rhythmic changes in the capacity of the thorax are brought about by muscular action. Changes in lung volume with intake or expulsion of air follow passively.

Lung tissue is elastic but is unable to expand or contract of itself. The mechanism by which the lungs are filled is best illustrated by a model like that shown in Figure 17–8, *A*. The balloons are connected to the outside of the jar by the Y-shaped tube that passes through the stopper of the jar. The space between the outside of the balloons and the inside of the jar contains air, but is closed from the outside. The floor of the jar is made of rubber, and when it is pulled down, the same amount of air in the jar fills the greater space, thereby creating a negative pressure on the outside of the balloons. The inside of each balloon is still subject to the external atmospheric pressure, and, as this is now higher than before, air enters the balloons and inflates them until the pressure in the jar and the expanded elastic tension of the balloons are equal to the atmospheric pressure. When the rubber floor is allowed to return to its original position, the increased pressure in the jar squeezes air out of the balloons.

The filling and emptying of the lungs is very similar to but more complex than the model just described (Fig. 17–8, *B*). Two pressures are involved here. The intrapleural space is a potential space between the lungs and the chest wall. The pressure exerted there normally is less than atmospheric pressure to permit full expansion of the lungs. The intrapulmonary (intra-alveolar) pressure is that pressure in the alveolar sacs that rises and falls above or below atmospheric pressure according to the muscular action of breathing.

The rib cage, or thorax, is airtight and analogous to the jar; the diaphragm closes off the floor of the thorax as does the rubber in the model; and the bronchi and trachea connect the lungs with the outside atmosphere. At rest, the diaphragm is curved and bowed into the chest cavity; but when it is stimulated to contract, the muscle tightens, forming a relatively straight line and not a normal curve—thus increasing the air space in the thorax and decreasing the intrathoracic pressure (Fig. 17–8, *B*). The intercostal muscles are located between the ribs. Contrary to popular belief, the ribs are not rigidly fixed to the spinal column, and when the intercostal muscles are stimulated, the ribs pull together collectively, working in conjunction with the diaphragm to increase the thoracic volume.

Recognizing that the combined efforts of the diaphragm and intercostal muscles result in an increased thoracic volume and decreased intrathoracic pressure, and considering that the thorax is a closed chamber, the only way that the increased chest space can be filled is to draw air in from the outside, filling the lungs—this is called *inspiration*. Relaxation of the diaphragm and intercostal muscles returns them to their normal dome-shaped and extended positions, respectively, resulting in a decrease in thoracic volume and increased intrathoracic pressure. Air that had previously been "sucked" in is now "forced" out—*expiration*. A graphic interpretation of the mechanism of breathing illustrates the comparison of intrathoracic and intrapulmonary pressures relative to atmospheric pressure (Fig. 17–9).

Puncture of the chest wall, which permits air to enter the intrathoracic space between the pleura (pneumothorax), will increase the normal negative pressure, causing the lung to collapse. In cases of multiple gunshot wounds of the thorax, the patient can die of anoxia (insufficient oxy-

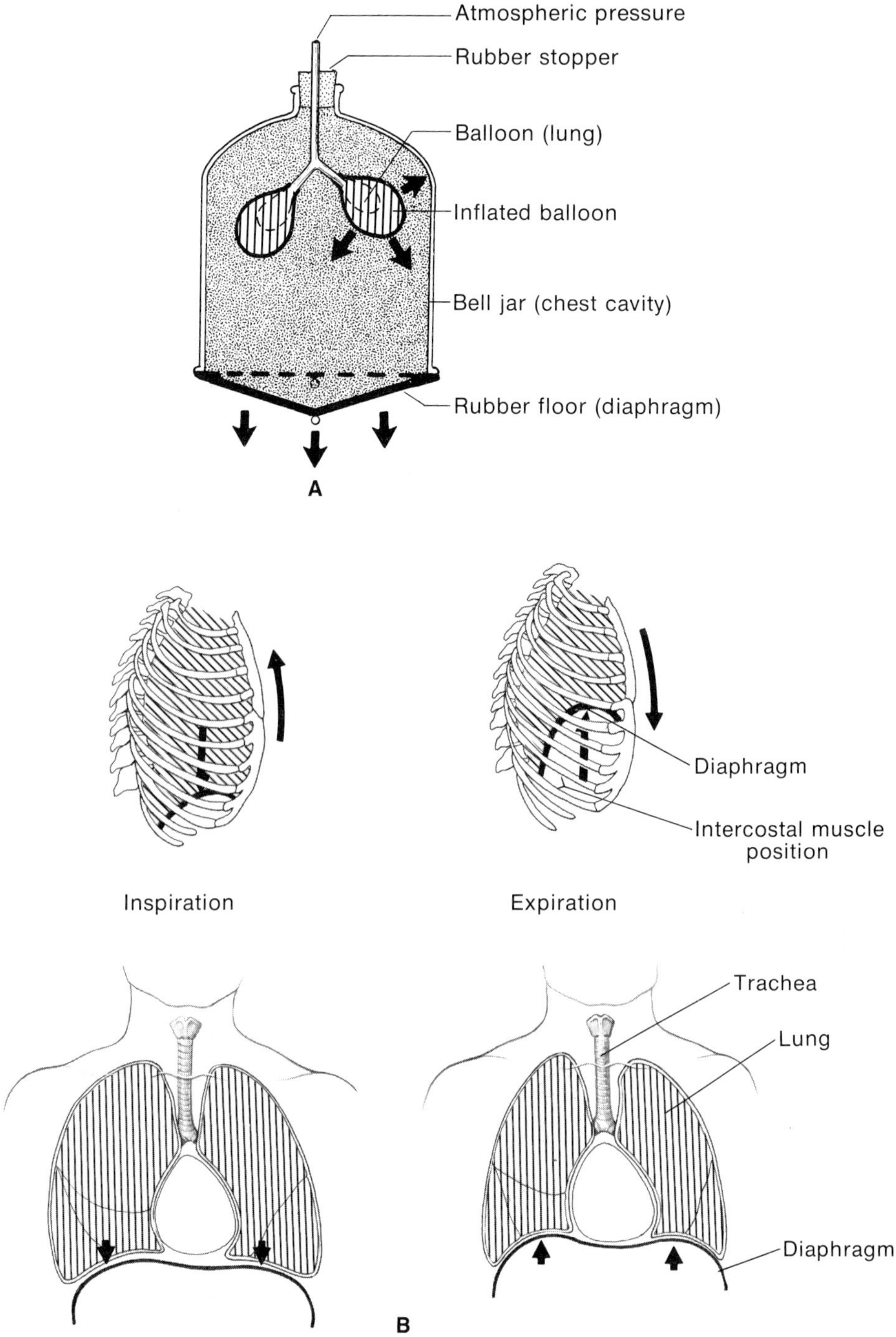

Figure 17–8 Mechanism of breathing. *A*, Bell jar apparatus used to simulate the action of the diaphragm. As the rubber floor (diaphragm) is pulled down, the balloons (lungs) are forced to inflate owing to the closed air cavity in the jar. *B*, Chest and diaphragm positions front and side views, in breathing. *Left*, the end of a normal inspiration, with chest expanded and diaphragm down; *right*, the end of a normal expiration, with chest cavity contracted and diaphragm relaxed.

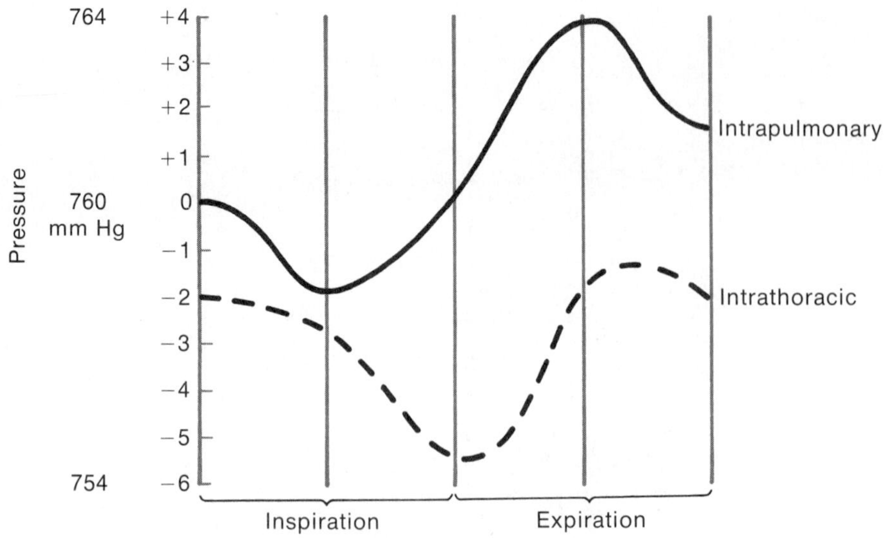

Figure 17–9 Graph showing changes in intrathoracic and intrapulmonary pressures during respiration.

gen) unless the holes are covered with a tight dressing. Artificial pneumothorax has proved beneficial in treating such lung diseases as tuberculosis by resting the infected lung.

LUNG VOLUMES

Much can be learned about lung function by measuring the volume of air taken in and exhaled at each breath. This is conveniently carried out by using a *spirometer* (Fig. 17–10, *A*), which consists of a drum inverted over a gas-filled container. A tube connects the air space in the bell with the subject's mouth. At inspiration, the removal of air causes the drum to sink; expiration causes the drum to rise. The movement can be observed on a scale and calibrated for volume of air moved. An added rotating drum (kymograph) gives a continuous record of volume changes within the drum.

An idealized record is shown in Figure 17–10, *B*. The small oscillations at the left are due to normal quiet breathing—the *tidal volume*. A maximal inspiration followed by a maximal expiration gives the next part of the record. The greatest volume that can be inspired is the *inspiratory capacity;* the greatest volume that can be expired is the *expiratory reserve*. The total of inspiratory capacity and expiratory reserve is the *vital capacity*.

On the whole, athletes tend to have larger than normal vital capacities. Small vital capacities are indicative of pulmonary disease. *Residual air* is that air still contained in the lungs after a maximal expiration. The sum of all components of lung volume is termed the *total lung capacity,* and is approximately 6 liters (6000 ml). The normal values for all lung components are given in Table 17–1.

By increasing the depth of breathing (i.e., tidal volume of air), the rate at which oxygen is brought in and carbon dioxide removed is increased. An increased frequency of breathing has a similar effect. Depth and frequency determine the *ventilation* of the lungs, or the amount of air moved in and out of the lungs per unit of time. When ventilation is increased, larger supplies of fresh air are made available more frequently to the respiratory epithelium. During exercise, tidal volume and respiratory rate both increase, thus increasing the minute volume. The amount of air exposed to the blood per minute

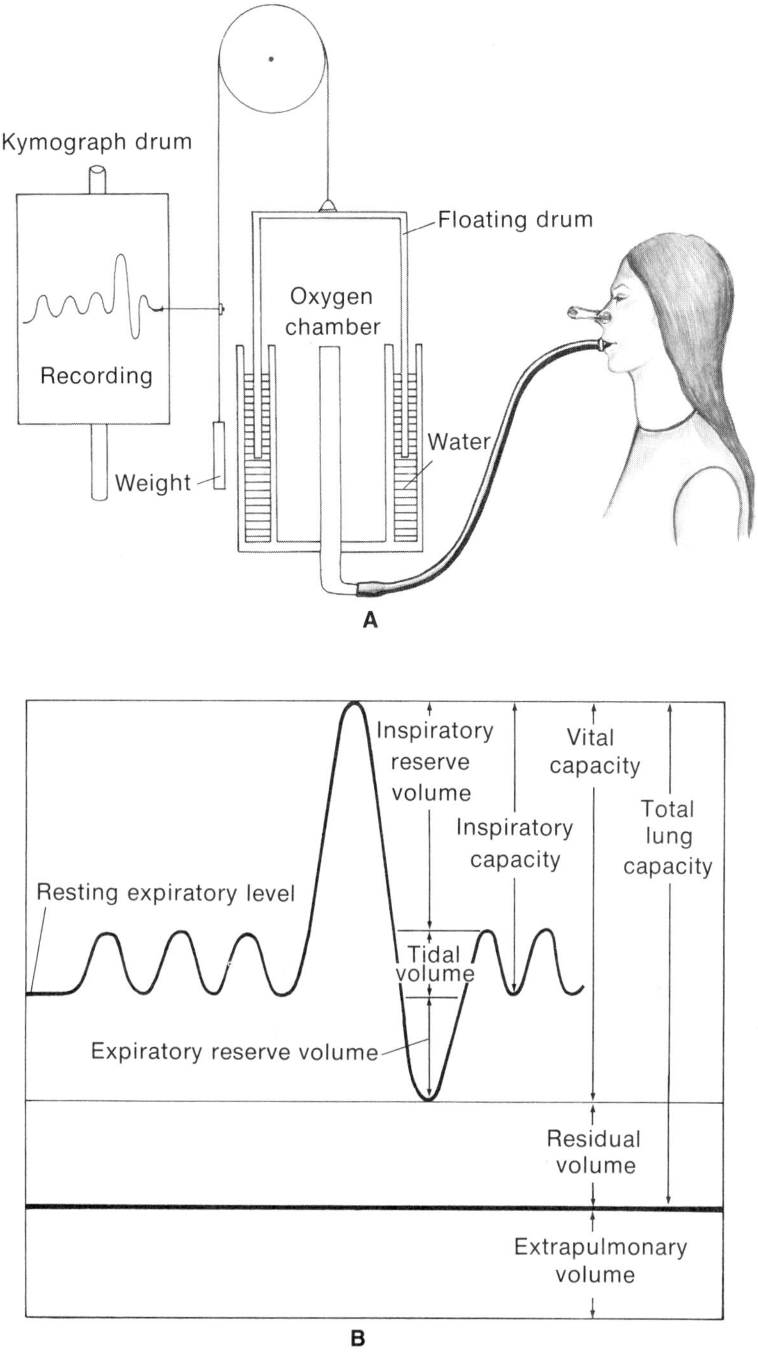

Figure 17–10 *A*, Determination of lung volumes by use of a spirometer. *B*, Spirometric graph showing respiratory capacities and volumes. Average values are given in Table 17–1.

therefore rises. Normal breathing *(eupnea)* is about 16 breaths per minute. At 0.5 liters per breath, the average ventilation is about 8 liters (8000 ml) per minute. The term *hyperpnea* means overbreathing due to abnormally rapid respiratory movements. *Apnea* is a temporary cessation of breathing. In some fevers the respiratory

Table 17–1 NORMAL VALUES FOR LUNG VOLUMES

Tidal volume	500 ml
Vital capacity	4800 ml
Inspiratory reserve volume	3100 ml
Inspiratory capacity	3600 ml
Expiratory reserve volume	1200 ml
Residual volume	1200 ml
Functional residual capacity	2400 ml
Total lung capacity	6000 ml

rate increases directly with the increase in body temperature. In the medical literature, the term apnea sometimes is used as a synonym for *asphyxia* or suffocation.

Usually the term *dyspnea* is reserved for difficult breathing, in which the expirations are active and forced. In *Cheyne-Stokes* breathing, respirations increase in force and frequency up to a point and then gradually decrease until they cease altogether. There is a short period of apnea; then the respirations begin again, and the cycle is repeated. Cheyne-Stokes respiration is associated with conditions that depress the respiratory center—especially brain, heart, and kidney diseases.

Hypoxia is an oxygen deficiency occurring when arterial oxygen saturation is low. *Anoxia* is a marked deficiency or absence of oxygen reaching the tissue cells. It may be due to (1) decreasing O_2-carrying capacity of hemoglobin; (2) interference with cellular respiration by chemicals in the blood such as cyanide; or (3) low oxygen tension in the lungs.

PHYSIOLOGY OF RESPIRATION

Neural Control of Breathing

The movement of air in and out of the lungs depends on the rhythmic contraction and relaxation of the diaphragm, the intercostal muscles, and other respiratory muscles. Because muscles contract in response to volleys of impulses over the nerves serving them, it is clear that there must be a control center. This control is automatic and unconscious.

Within the medulla of the brain are paired structures known as *respiratory centers* (Fig. 17–11). Each side bears both an inspiratory and an expiratory center. The inspiratory center sends out nerve impulses to the diaphragm and intercostal muscles via the phrenic and intercostal nerves, respectively, causing the muscles to contract. Contraction of these muscles, as we have seen, leads to inspiration. *Stretch receptors*—specialized nerve endings within the tissues (Fig. 17–11) are stretched as the lung tissue expands. Thus stimulated, they send impulses back to stimulate the expiratory center, which, in turn, inhibits the inspiratory center. When inhibited, the inspiratory center no longer sends out impulses, keeping the respiratory muscle contracted. Consequently, the lungs deflate passively and return to normal, forcing the air out—expiration. As the lungs empty, the stretch receptors are no longer stimulated, and the inhibitory effect of the expiratory center ceases.

Through the stretch receptors, the act of inspiration is automatically stopped: this is known as the *Hering-Breuer reflex.* In addition to this reflex, there is another automatic mechanism in the pons, the *pneumotaxic center* (Fig. 17–11). Some impulses from the respiratory centers excite the pneumotaxic center, and it in turn signals the expiratory centers in much the same way as do the stretch receptors in the lungs. The excited expiratory centers then inhibit the inspiratory centers.

Chemical Regulation of Breathing

During activity, the concentration of CO_2 diffusing from active muscles into the bloodstream increases greatly over normal levels and must be controlled. It is therefore reasonable to expect that this substance is a powerful stimulant to respiration.

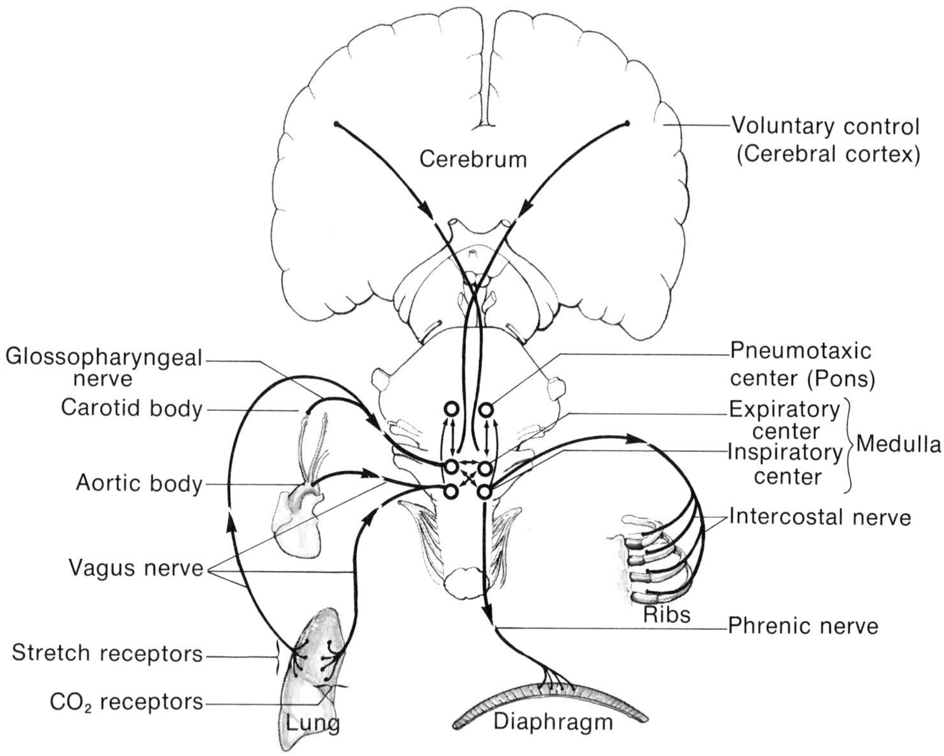

Figure 17–11 Chemical and nervous control of respiration. Arrows indicate direction of nerve impulses.

Carbon dioxide acts directly on the respiratory centers via the chemoreceptors in the carotid and aortic bodies (Fig. 17–11). A higher than normal concentration of CO_2 in this blood indicates that the CO_2 is not being removed from the lungs in sufficient quantities. The high levels of CO_2 in the blood bathing the respiratory centers stimulate discharge of impulses from these centers to the respiratory muscles, increasing both the depth of inspiration and the respiratory rate. This increased ventilation permits the blood to lose CO_2 more readily; consequently, the arterial CO_2 is returned to its resting level.

The CO_2 concentration in arterial blood can be reduced by breathing deeply and rapidly for a minute or two—a procedure called "voluntary hyperventilation." After such forced breathing, a moderate apnea results; that is, breathing is temporarily inhibited. When it resumes, it is irregular until the oxygen and carbon dioxide levels in the blood become stabilized again.

It is possible to hold your breath for minutes (seemingly) at a time by exercising voluntary control of the cerebral cortex (Fig. 17–11). Chemical stimuli, however, soon overpower all voluntary inhibition, and one is forced to take a breath.

RESPIRATORY FUNCTION OF THE BLOOD

External Respiration

External respiration refers to the exchange of gases in the lungs. There is a continuous flow of blood through the capillaries so that at least once or twice each minute all the blood in the body passes through the capillaries of the lungs. The air we breathe contains oxygen, nitrogen, and carbon dioxide. The significance of external respiration lies in the exchange of oxygen and carbon dioxide between the air and the blood. Inspired air contains oxygen that is taken up by the lungs, and expired air takes away carbon dioxide. The interchange takes place through the thin capillary walls in the alveoli (Figs. 17–6 and 17–12).

Table 17–2 lists the average composition of dry inspired and expired air with the person at rest. It is apparent that the amount of oxygen used is greater than the amount of carbon dioxide given off. This occurs because some of the oxygen used in the metabolic processes appears in the form of water (H_2O), not as carbon dioxide (CO_2).

The direction in which the gases move between lungs and tissues depends solely on the diffusion gradients established during CO_2 production and O_2 use by the cells. The gradient in the lungs is increased by the mixing action of external respiration; the gradient between the blood and tissue is increased by the circulation of the blood.

Gaseous Exchange

Cellular gas exchange takes place in all body tissues; cells take up oxygen from the blood and add carbon dioxide to it. When this occurs, blood becomes venous. As we have seen, this CO_2-rich blood then returns to the heart, entering via the vena cava. From the heart, venous blood is pumped through the short, paired pulmonary arteries to the lungs, where these arteries branch into extensive capillary networks spread over the alveoli. Pulmonary gas exchange takes place here: CO_2 leaves the blood and O_2 enters. Thus, blood becomes arterial. This O_2-rich blood now is brought back to the heart by the pulmonary veins, and this blood, in turn, is redistributed throughout the body via the aorta.

The direction in which the gases move is determined by the prevailing pressure gradients, or tension gradients, between blood vessel and organ cell. Specifically, atmospheric air in the alveoli contains only a little CO_2, but the venous blood that flows into the lungs from the body is virtually saturated with the gas. The pressure gradient of CO_2 is greater in the capillary; therefore, CO_2 diffuses into the alveoli (Figs. 17–12 and 17–13).

Predictably, the pressure pattern is the reverse for oxygen. Blood flowing into the lungs from the body is oxygen-poor. The air in the alveoli contains a maximal content of O_2; therefore, O_2 diffuses into the capillary from the alveolus to satisfy the increased alveolar pressure gradient.

Just as in the lungs, cellular gas exchange in the other body tissues is governed by tension gradients. Cells continuously use up oxygen in internal respiration, and the tension of this gas in cells is therefore low. The arterial tension is high, however (Table 17–2), and the O_2 diffuses

Table 17–2 AVERAGE COMPOSITION OF DRY INSPIRED AND EXPIRED AIR (PERSON AT REST)

	O_2		CO_2		N_2	
	Vol %	*(mm Hg)*	*Vol %*	*(mm Hg)*	*Vol %*	*(mm Hg)*
Inspired air	21	(159)	0.04	(0)	79.0	(601)
Expired air	16	(116)	4.0	(28)	79.2	(569)
Alveolar air	14	(100)	5.5	(40)	79.1	(573)
Arterial blood	19	(96)	50.0	(40)	79.1	(573)
Venous blood	12.5	(40)	56.0	(46)	79.1	(573)
Tissues	12–19	(30–40)	60–70	(50–60)	79.1	(573)

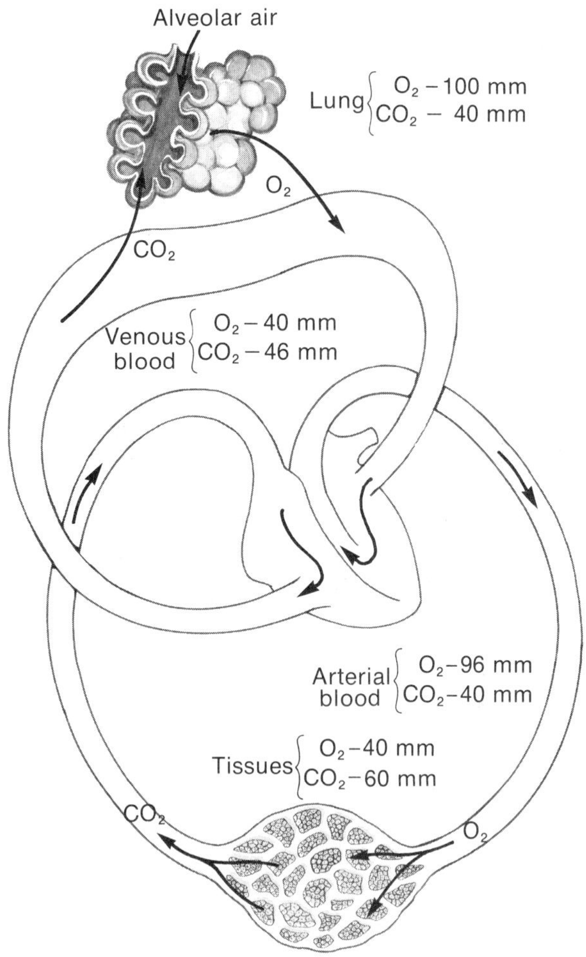

Figure 17–12 Movement of respiratory gases. Diagram indicates oxygen and carbon dioxide levels in the lungs and tissues as the gases move through the circulatory system.

from the blood into the tissue cells. At the same time, since respiratory CO_2 is produced as a waste product of cellular metabolism, the CO_2 tension within the tissue cells is high. Arterial blood has a low CO_2 tension, and the gas, therefore, diffuses from the tissue cells into the blood.

Small amounts of nitrogen are found in the blood, but, since nitrogen is almost an inert gas, it takes no part in metabolism. It is more commonly found in the form of amino acids that make up the important proteins of the body.

Gaseous Transport

Transport of respiratory gases requires a medium (H_2O), a number of inorganic ions, and red blood cells. The red blood cells derive their color from hemoglobin. As noted in Chapter 13, this complex pigment, symbolized as Hb, consists of two parts, heme and globin. Heme contains iron and is the functional fraction of hemoglobin. Globin is a protein that serves mainly as a carrier of heme.

Hemoglobin has the capacity to form a loosely chemical combination with oxygen:

$$Hb + O_2 \rightleftarrows HbO_2 \text{ (oxyhemoglobin)}$$

Oxygen is carried in blood largely in the form of HbO_2. A little oxygen dissolves in the water of blood. When HbO_2 reaches the tissues, the reaction shown above shifts from right to left. Cells are oxygen-poor compared with blood, and the O_2 breaks free from the Hb and is taken up by the tissues (Fig. 17–13). It should be noted here that hemoglobin may also transport carbon monoxide (CO). This union is achieved more easily than that with oxygen; accordingly, if CO is present in the atmosphere, Hb becomes HbCO rather than HbO_2. This means that little oxygen can be transported to the tissues, causing cellular poisoning and suffocation.

Cyanosis, a bluish color of the skin and mucous membranes, is caused by an insufficient amount of oxygen in the blood. It may result from asphyxiation or suffocation, but it can also be caused by other factors. Heart disease, for example, may inhibit circulation through the lungs, resulting in insufficient oxygenation.

Of the CO_2 released from the tissue cells, a small fraction dissolves physically in blood water; another small fraction combines with hemoglobin:

$$Hb + CO_2 \rightleftarrows HbCO_2 \text{ (carboxyhemoglobin)}$$

The $HbCO_2$, in turn, combines with an amine group (NH_2) in the hemoglobin to form carbonate ions (HCO_3^-):

$$CO_2 + H_2O \rightleftarrows \underset{\text{carbonic acid}}{H_2CO_3} \rightleftarrows H^+ + \underset{\text{bicarbonate ion}}{HCO_3^-}$$

Inasmuch as tissue cells constantly add CO_2 to the blood, this reaction proceeds from left to right in the tissues. Most of the CO_2 is therefore transported to the lungs in the form of HCO_3^-. In the lungs, the reaction reverses itself; CO_2 escapes into the alveoli, and more CO_2 is released for exhalation (Fig. 17–13).

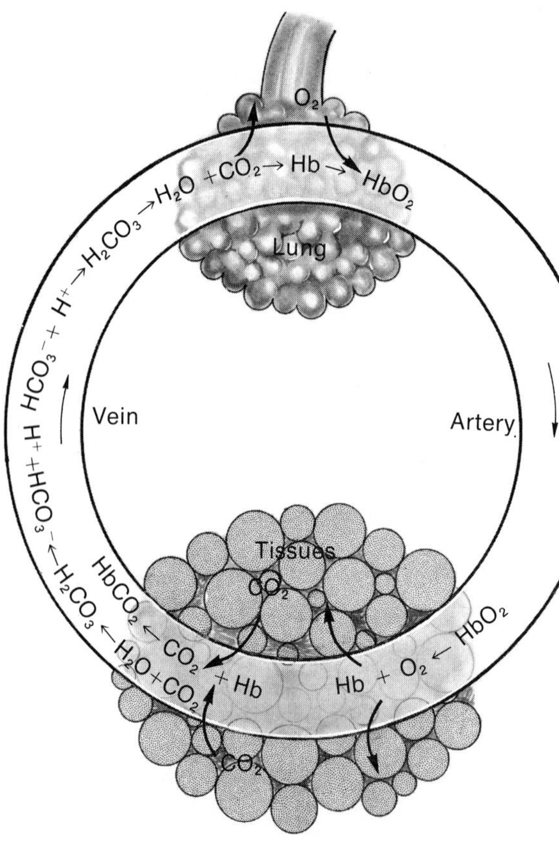

Figure 17–13 Carriage and transfer of oxygen and carbon dioxide.

Chloride Shift

Carbon dioxide enters the blood from the tissue cells by diffusion. Some of it remains in solution in the plasma, but most diffuses into the red blood cells, where the enzyme carbonic anhydrase catalyzes the formation of carbonic acid. This results in an increase in hydrogen ions and favors the decomposition of oxyhemoglobin and the escape of oxygen to the tissue cells. A further decrease of CO_2 results from the reaction of carbonic acid with the potassium salts of hemoglobin:

$$KHb + H_2CO_3 \rightarrow HHb + KHCO_3$$

During these reactions, the concentration of bicarbonate ions (HCO_3^-) within the red blood cells increases, creating an ionic imbalance between the red blood cells and the plasma. The ions, therefore, diffuse from the red blood cells into the plasma, thus building up the bicarbonate concentration in the plasma. In compensation for the exit of these ions, chloride ions (Cl^-) diffuse from the plasma into the red blood cells; this ionic adjustment is known as the *chloride shift* (Fig. 17–14).

When the blood reaches the lungs, oxygen is available and carbon dioxide can escape, and the sequence of reactions that took place at the tissue cells is reversed.

CLINICAL CONSIDERATIONS

Oxygen Therapy

A continuous supply of oxygen is essential for all cells throughout the body. Any shortage of oxygen interferes with the production of energy. The causes of oxygen deficiency are many and varied, but the conditions most frequently requiring oxygen therapy are the acute shortages resulting from lowered oxygen tension of blood plasma and a consequent decrease in the formation of oxyhemoglobin.

Prolonged general anesthesia, lung diseases, carbon monoxide poisoning, and a drug-depressed respiratory center are a few of the common problems necessitating oxygen therapy. The signs are not always clear. Cyanosis can be an indication, but it is not always accurate.

The purpose of oxygen therapy is to re-establish the oxygen tension of the lungs and to restore the oxyhemoglobin in the red blood cells to normal. This is done by providing an environment in which the oxygen content of the air is increased, such as in an oxygen tent, or by means of a mask or nasal catheter. Measures must be taken, however, not to produce *oxygen intoxication*. This can accompany an oversupply of oxygen in the blood, and results in a change in tissue cell metabolic rates. Brain cells are especially sensitive, and oxygen intoxication over a prolonged period may cause convulsions and/or brain cell destruction.

The Bends

Under conditions of high atmospheric pressure, the nitrogen of the blood diffuses slowly through membranes and is absorbed into tissues. Rapid decompression causes excessive nitrogen gas bubbles to form in the blood and tissues. These bubbles collect in the lung capillaries, causing severe cramps called *caisson disease* or "the bends." Deep sea divers must be careful to return to the surface slowly to allow absorbed nitrogen to return to solution.

Artificial Respiration

Artificial respiration is resorted to when an individual has either temporarily or permanently lost his capacity to perform the normal motions of respiration. Such emergencies include smoke asphyxiation, electric shock, drowning, and the like.

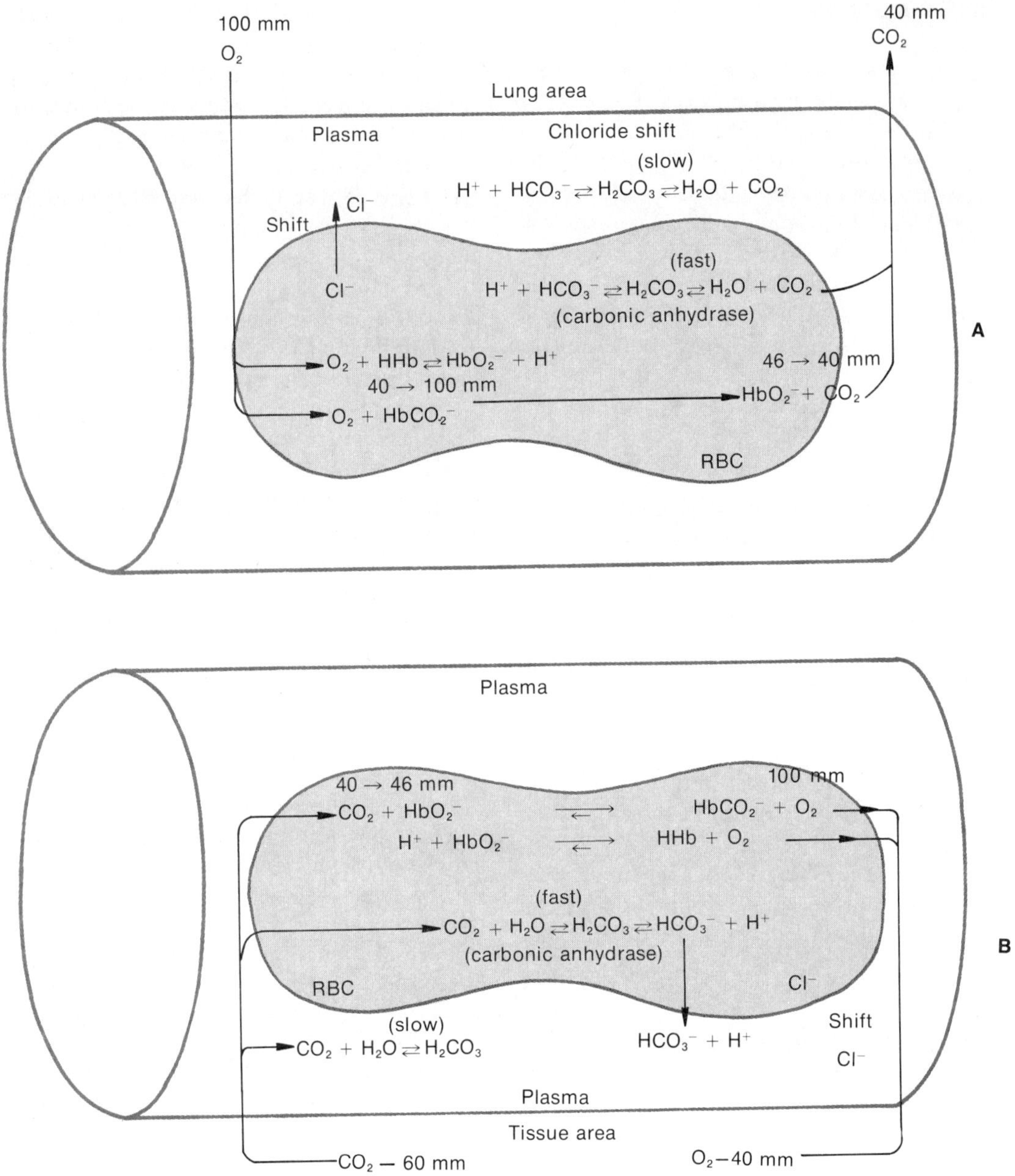

Figure 17–14 Diagram illustrating some of the chemical changes accompanying oxygen and carbon dioxide exchange at the lung *(A)* and tissue *(B)* areas.

Many people are acquainted with the *Schafer* (prone pressure) *method* of artificial respiration. The subject is placed on the ground in a prone position with both arms outstretched or with one arm flexed to cushion the forehead. The head is turned to one side and the tongue pulled forward. The operator kneels astride the patient, facing forward, and places his hands on the lowest ribs. Leaning forward, he lets his weight compress the subject's chest and abdomen, thus forcing air from the respiratory tract. He then raises his body slowly, at the same time relaxing the hand pressure. These rhythmic movements are repeated at 12 to 16 times per minute until normal breathing is restored.

The *Holger-Neilsen* (back pressure–arm lift) method of artificial respiration utilizes a positive push-pull technique. The subject is placed face down in the prone position. The elbows are bent, and one hand is placed on top of the other. The head is turned to one side, resting on the hands. The operator kneels at the head of the subject, placing his hands on the subject's back above the ribs. He rocks forward, and his weight exerts pressure on the subject's back, forcing air out of the lungs. The operator then rocks backward, grasping the subject's arms just above the elbows and exerting a pulling force on the chest cavity. This creates a sucking force into the lungs.

The *"mouth-to-mouth" technique* (Fig. 17–15) is now considered superior to any of the manual methods of artifical respiration. With the subject supine, the operator tilts the head back and pinches the nose shut. He methodically blows into the subject's mouth to inflate the lungs, watching the patient's chest rise. He then pauses to allow the lungs to deflate. Once again, the procedure is carried out at a normal rate, for an adult, of approximately 16 times per minute. With children, the rate should be slightly faster.

Patients with poliomyelitis are sometimes left with permanently paralyzed muscles, so that a *respirator* ("iron lung") is employed to provide a mechanical means of maintaining respiratory movements. This device is a metal chamber that envelops all of the patient except his head. The pressure within the respirator is changed rhythmically, much as the diaphragm changes the pressure within the thoracic cavity, allowing the patient to breathe normally with no effort.

Pneumonia

Pneumonia is an inflammation of the alveoli. It may be caused either by a diplococcoid bacterium known as the pneumococcus or by a virus. Any or all of the causative pathogens may be carried by a healthy person in the mucosa of the respiratory tract, and as long as his general health remains good, they will produce no ill effects. A weakened condition, however, affords the pathogens an opportunity to invade and damage the lung tissue. *Lobar pneumonia* involves one or more lobes or segments of a lung at one time, and is usually caused by a pneumococcus. *Bronchopneumonia* is generally caused by a virus or staphylococcus, and is widespread throughout the lung. This is the

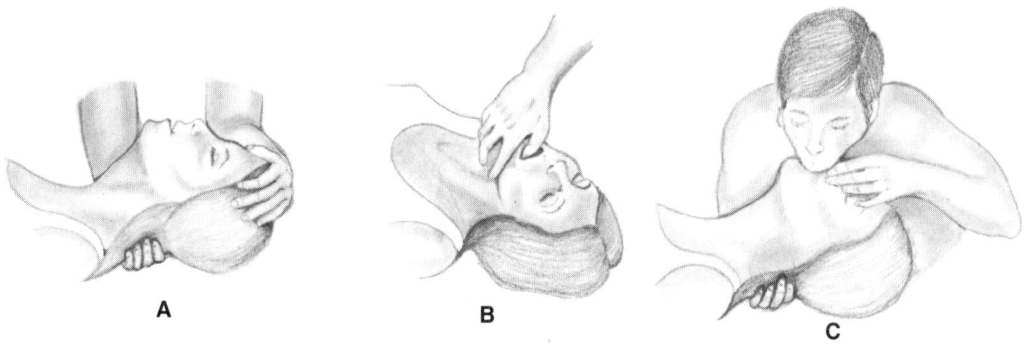

Figure 17–15 Mouth-to-mouth method of artificial respiration. *A*, With the subject on his back, the operator tilts the head backward. *B*, The operator next checks the oral cavity for any possible obstruction. *C*, He blows into the mouth, watching the chest rise. The subject's nose is then pinched shut, and the operator methodically blows into the subject's mouth to inflate the lungs and pauses to allow the lungs to deflate.

more common form of pneumonia, and it frequently occurs secondary to some infection that has lowered the subject's resistance.

A characteristic of most types of pneumonia is the formation of a fluid or exudate in the infected alveoli; this fluid consists chiefly of serum and pus cells. Sometimes so many air sacs become filled with fluid that the victim finds it difficult to breathe. Antibiotic drugs and oxygen therapy have greatly improved the outlook for pneumonia victims, although the disease can still be fatal in the very young or elderly patient.

Tuberculosis

Tuberculosis, an infectious disease caused by the tubercle bacillus (bacterium) *Mycobacterium tuberculosis,* continues to be a most serious and costly illness and a major killer among the communicable diseases. It occurs most frequently in the lung, but may develop in other parts of the body as well.

When the bacillus invades the tissue, the cells build up a protective capsule around the invading organism, effectively isolating it from the rest of the body. As with pneumonia, if the subject's resistance is high, the pathogen eventually will die. On the other hand, if the subject's resistance is lowered, the capsule (tubercle) containing the organism will destroy its enveloping tissue, thus causing a cavity where there once was living tissue. The bacilli spread to other parts of the body through the blood and/or lymphatics, and the destructive process is repeated.

The most important factor in the control of tuberculosis is the isolation of the infected patient from any susceptible persons. Drugs (chemotherapy) have proved quite successful, but absolute rest is often required as well. The most important aspect of cure is the destruction of the organisms, which gives the body an opportunity to repair damaged tissues.

Lung Cancer

The incidence of malignant lung tumors has increased markedly over the past 25 years. The rate is considerably higher in urban and industrial areas than in rural districts. Factors such as smoke particles, smog, exhaust fumes, and so forth have proved to be irritants to the lungs, thus increasing their susceptibility to disease.

Bronchogenic carcinoma, a common form of lung cancer, originates in a bronchus. The tumor may grow until a major bronchus is blocked, cutting off the air supply to that lung. The lung then collapses, and secondary infection results in pneumonia or the formation of a lung abscess. The only treatment that offers a possibility of cure is complete surgical removal of the lung (pneumonectomy). Early detection is essential if there is to be any hope of a cure. Routine chest x-ray examinations have saved many thousands of lives.

Acid-Base Balance

Respiratory "acidosis" (excess of carbon dioxide) may be defined as a primary increase in the carbon dioxide tension of the plasma. It may be encountered in certain pulmonary diseases, such as emphysema, bronchopneumonia, depression of the respiratory center, and convulsive apnea.

Respiratory "alkalosis" (deficit of carbon dioxide) is a decrease in the carbon dioxide tension of the plasma. This is most common in cases of hyperventilation and encephalitis, and can occur at high altitudes.

Pulmonary Embolism

A pulmonary embolism is an embolus or clot lodged in a pulmonary artery. It commonly produces *pulmonary infarction,* a hemorrhagic condition that causes tissue cell death owing to occlusion of the blood supply (Fig. 15-15).

As a consequence of the decreased pulmonary arterial capacity, the right side of the heart becomes overloaded and dilated. Poor filling at the left ventricle results in decreased cardiac output. Right-sided congestion, peripheral circulatory collapse, and even congestive heart failure may result. A large embolus can cause death within a few minutes or hours. A smaller embolus generally causes an infarction.

The most important part of the treatment is to prevent shock, hypoxia, and heart failure. Oxygen, digitalis, and anticoagulants are usually administered. In some patients, surgical removal of the embolus (embolectomy) has been successful.

OUTLINE

I. Introduction
 A. Respiration has two purposes
  1. Delivers oxygen to the tissues
  2. Removes carbon dioxide from the tissues
II. General Anatomy
 A. Upper respiratory tract
  1. Nose
   a. Ciliated epithelium
   b. Nasal septum
   c. Nasal conchae
   d. Mucous membranes containing olfactory nerve fibers
  2. Pharynx
   a. Membranous sac between the nasal and oral apertures and the esophagus
  3. Abnormalities
   a. Deviated septum
   b. Nosebleeds
 B. Respiratory tract
  1. Larynx—"voice box"
   a. Guards opening of trachea
   b. Contains vocal folds—speech
   c. Laryngoscopy
   d. Laryngospasm—croup
  2. Trachea
   a. C-shaped hyaline cartilage rings
   b. Connects larynx with bronchi
   c. Lined with ciliated mucous epithelium
   d. Tracheotomy-tracheostomy
  3. Bronchi
   a. Supported by overlapping cartilaginous rings
   b. Ciliated mucous epithelium
   c. Bronchoscopy
  4. Bronchioles
   a. Small bronchi
   b. No supporting structure
  5. Atria
   a. Elongated sacs at ends of bronchioles
   b. Lined with alveoli
    1) Thin-walled air cells
    2) Highly vascular
    3) Gas exchange occurs here
  6. Abnormalities
   a. Atelectasis
   b. Bronchiectasis
   c. Emphysema
  7. Mediastinum
   a. Divides thoracic cavity into two parts
   b. Right lung—larger, 3-lobed
   c. Left lung—smaller, 2-lobed
   d. Pleura—serous membrane
    1) Parietal pleura—lines thoracic cavity
    2) Pulmonary pleura—covers lungs
    3) Membranes secrete mucus to avoid friction between pleura
    3) Pleurisy — inflammation of pleural membranes
III. Mechanism of Breathing
 A. Anatomy
  1. Respiratory muscles
   a. Diaphragm—dome-shaped at rest

b. Intercostal—between the ribs
2. Nerve supply
a. Phrenic—controls diaphragm
b. Intercostal—controls intercostal muscles
3. Thoracic pressures
a. Intrathoracic pressure—in thoracic cavity
b. Intrapulmonary pressure—in lungs

B. Physiology
1. Stimulation of respiratory muscles
a. Contraction results in increased thoracic volume and decreased intrathoracic pressure
b. Decreased intrathoracic pressure causes air to be sucked in—inspiration
2. Relaxation of respiratory muscles
a. Thoracic cavity returns to normal shape and volume
b. Increased intrathoracic pressure forces air out—expiration
3. Pneumothorax—air in thoracic cavity.

IV. Lung Volumes
A. Spirometer and kymograph
1. Tidal volume
2. Vital capacity
3. Inspiratory capacity
4. Expiratory reserve
5. Residual air
6. Total lung capacity

B. Ventilation
1. Eupnea—normal
2. Hyperpnea—increased
3. Apnea—decreased or discontinued
4. Dyspnea—difficult
5. Cheyne-Stokes—hyperpnea–apnea
6. Hypoxia—deficiency of oxygen
7. Anoxia—absence of oxygen

V. Physiology of Respiration
A. Neural control
1. Brain—medulla
a. Inspiratory center—stimulates
b. Expiratory center—inhibits
c. Pmeumotaxic center—in pons
1) Safety mechanism
2. Sensory receptors
a. Stretch receptors - in lungs
b. Chemoreceptors—in alveoli and aortic and carotid bodies

B. Chemical regulation
1. Increased CO_2 in bloodstream stimulates inspiratory center to promote increased inspiration
2. Increased ventilation permits blood to lose CO_2 more readily

VI. Respiratory Function of the Blood
A. External respiration—exchange of O_2 and CO_2 between air and blood

B. Gaseous exchange
1. Exchange regulated by diffusion pressure gradients
2. In lungs
a. O_2 content higher in alveoli than in capillaries; therefore, O_2 diffuses into capillaries
b. CO_2 content higher in capillaries; therefore, CO_2 diffuses into alveoli
3. In tissues
a. Reverse of lungs—CO_2 diffuses into capillaries and O_2 diffuses into tissue cells.
4. Small quantities of nitrogen gas remain in inert form

C. Gaseous transport
1. Hemoglobin in erythrocytes acts as carrier
2. Oxygen transport
a. $O_2 + Hb \rightleftarrows HbO_2$ (oxyhemoglobin)

3. Carbon dioxide transport
   a. $CO_2 + H_2O \rightleftarrows H_2CO_3$ (dissolved CO_2)
   b. $Hb + CO_2 \rightleftarrows HbCO_2$ (carboxyhemoglobin)
   c. $HbCO_2 + NH_2 \rightarrow HbNH_2$ COOH (carbamino Hb)
   d. $CO_2 + H_2O \rightleftarrows H_2CO_3 \rightleftarrows H^+ + HCO_3^-$ (bicarbonate ion)
4. Chloride shift
   a. Carbonic anhydrase aids bicarbonate ion formation in red blood cells but not in plasma
   b. Bicarbonate ions diffuse from red blood cells into plasma
   c. Chloride ions diffuse from plasma into red blood cells in compensation
5. Cyanosis—lack of oxygen—increased carbon dioxide

VII. Clinical Considerations
   A. Oxygen therapy
      1. Utilized in lung diseases, carbon monoxide poisoning, etc.
      2. Restores oxygen tension of blood plasma
   B. The bends
      1. Nitrogen gas is found in solution in blood under normal atmospheric conditions
      2. Increased external pressures produce nitrogen gas bubbles
      3. Normal reduction of increased external pressure returns N_2 bubbles to solution
      4. Too rapid a decompression traps N_2 bubbles and they circulate in the body, causing pain and ultimate death
   C. Artificial respiration
      1. Support of normal breathing
      2. Schafer method—prone pressure
      3. Holger-Neilsen method—back pressure-arm lift
      4. Mouth-to-mouth method—superior to manual methods
      5. Iron lung—respirator chamber
   D. Respiratory abnormalities
      1. Pneumonia
      2. Tuberculosis
      3. Lung cancer
      4. Acid-base balance—acidosis vs. alkalosis
      5. Pulmonary embolism

STUDY QUESTIONS AND PROBLEMS

1. What is the purpose of respiration?
2. Describe the mechanism of breathing. What pressures are involved?
3. Trace the passage of air from the nose to the alveoli.
4. How do the vocal folds produce sound?
5. What is meant by inspiration and expiration?
6. What part do the diaphragm and intercostal muscles play in the mechanism of breathing?
7. Discuss the Hering-Breuer reflex.
8. Why do we need a pneumotaxic center?
9. Define chemoreceptor and pressoreceptor and describe their roles in respiration.
10. What are the effects of rapid decompression? Of hyperventilation?
11. What is meant by a diffusion pressure gradient?
12. What body organ is most sensitive to lack of oxygen?
13. Differentiate, structurally, between the trachea and the bronchi.
14. Mr. Jones is stabbed in the chest by a neighbor. What will happen to his respiratory mechanism?
15. What is the function of the pleura? What is pleurisy? Pneumothorax?
16. Mike had chest surgery and was sent back to his room with an underwater seal suction coming from his chest. Why?
17. If you saw a man choking on something, what would you do?
18. If a tracheotomy had to be done, as an emergency measure, how would you do it?
19. Discuss the reflex control of breathing. Can Mary sit in the corner and hold her breath until she dies? Why?
20. Differentiate among asphyxia, emphysema, atelectasis, and bronchiectasis.

Chapter 18

Digestion and the Alimentary Tract; Metabolism

It is apparent that the composition of the internal environment must depend ultimately on what is eaten. Nutritional substances required by the cells must be provided for them dissolved in interstitial fluid. A state of adequate nutrition exists when the internal environment contains all the materials the cells require in the correct proportions and concentrations.

CELLULAR REQUIREMENTS

The absolute necessity of water is self-evident. It is consumed as such or in beverages; it is also present in most foods.

In a resting state, the very existence of a cell as an island of highly organized material requires energy. Any form of work, such as muscular contraction, adds to this energy requirement. The most important source of energy for the cells is oxidized glucose (simple sugar); therefore, the diet must contain materials that will yield glucose. These materials are the carbohydrates.

Much of the cell's structure is protein. Because protein molecules are extremely large, they cannot readily diffuse out of the blood capillaries or across cell membranes. The internal environment therefore must contain units from which proteins can be constructed by the cell. These building blocks, the amino acids, are obtained by the digestion of dietary protein in the intestine, where they are readily absorbed and carried to the cells.

Biochemical catalysts—the enzymes—facilitate the many chemical reactions that proceed in living cells. Enzymes themselves are protein in nature, but many of them require the presence of other organic compounds—coenzymes—before they can perform their tasks. The essential dietary constituents from which many of the coenzymes are replenished are the vitamins. They are required only in minute quantities, unlike the proteins and carbohydrates.

In addition to organic coenzymes, most enzymes require a metal ion for their activity. For some enzymes, sodium, potassium, manganese, magnesium, or iron is required—all these metals are necessarily part of the cell's structure. Iodine is required as a constituent of the hormones produced by the thyroid gland (thyroxine).

The tissues of the body are rebuilt after the wear and tear of daily work from the nutritive material supplied to them in food. The sum total of all breaking down (catabolism) and building up (anabolism) of the tissues is called *metabolism*. If more energy is used than is provided by the intake of food, extra food must be taken to restore the balance. Otherwise, the body calls upon its reserves of energy, and body weight is lost.

THE FOODSTUFFS

Food may be animal or vegetable, but whatever its origin, it must contain certain constituents that the body can use as sources of energy and sustenance. These are:

1. Carbohydrates
2. Fats
3. Proteins
4. Vitamins
5. Mineral salts

Not all of these are found in every type of food, and the proportions in which they occur in different foods vary greatly. Thus it is necessary to ensure that the diet is adequate in both quantity and quality—in other words, "balanced." Water cannot be classified as a food, nor can oxygen in the air, but they must be included in this discussion because the body cannot utilize foodstuffs or manufacture energy without them.

Carbohydrates

Carbohydrates are compounds of carbon (C), hydrogen (H), and oxygen (O), the last two elements being present in the same proportion as in water—that is, two atoms of hydrogen to each atom of oxygen (H_2O). The simple sugars with six carbon atoms, such as glucose, are called *hexoses* (hex- = six; -ose = sugar).

Carbohydrates may be divided into three principal groups (Table 18–1):

1. Monosaccharides. These simple sugars consist of one hexose. *Glucose* ($C_6H_{12}O_6$), or grape sugar, is widely distributed in nature in the juices of fruits and in honey.

2. Disaccharides. These sugars are made up of two conjugated monosaccharides, like glucose:

$$\underset{\text{glucose}}{C_6H_{12}O_6} + \underset{\text{glucose}}{C_6H_{12}O_6} \rightleftarrows \underset{\text{disaccharide}}{C_{12}H_{22}O_{11}} + H_2O$$

Examples are cane or beet sugar *(sucrose)* and *lactose,* or milk sugar. Any disaccharide may be split into two monosaccharides if the missing water molecule is reintroduced under appropriate conditions. This kind of splitting, called *hydrolysis,* is accomplished in the digestive tract by the hydrolytic enzymes. Cane sugar, found in many fruits and vegetables, is the common household sugar used for sweetening purposes.

3. Polysaccharides. This category includes the starches and glycogen ($C_6H_{10}O_5$). Polysaccharides are formed by joining 10 or more monosaccharides into large molecules of endless complexity. One of the commonest, starch, is the form in which most plants store their reserves of food. Animals store their carbohydrate reserves as glycogen. Glycogen is found in liver and in other meats, but it does not

Table 18–1 CARBOHYDRATES

Formula Example	*Category*	*Compound Name*	*Where Found*
$C_6H_{12}O_6$	Monosaccharide	Glucose (dextrose) Fructose (levulose) Galactose	Grapes, fruits, honey Fruits Lactose, gums
$C_{12}H_{22}O_{11}$	Disaccharide	Sucrose (cane sugar) Lactose (milk sugar) Maltose (malt sugar)	Vegetables, fruits, juices Milk of mammals Cereals, malts
$(C_6H_{10}O_5)_n$	Polysaccharide	Starch Cellulose Glycogen Dextrin	Grains, tubers, roots Plant cell walls, grains Liver, muscles Starch hydrolysis

compare with plant starch as a dietary source of glucose.

Carbohydrates provide a ready source of energy and are an essential part of the diet. They are readily digested and absorbed, and can quickly undergo oxidation in the tissues, especially the muscles, to provide energy. Sugar also plays an important part in the metabolism of fats.

Fats

Fats, also called lipids, form a large class of organic compounds. Like carbohydrates, fats are made up of carbon, hydrogen, and oxygen, but the structure of the fat molecule is different. It is divided into two parts: one is a *glycerol unit;* the other is made up of three molecules of *fatty acid.* Collectively, the two portions represent a *triglyceride.* (*See below*)

Triglycerides may be simple, as in the example shown, where all three fatty acids are the same; or they may be mixed. Animal fat composed only of triglycerides of these types is called *neutral fat* to distinguish it from other lipids.

In addition to the simple and mixed triglycerides, there are many compound lipids. In these a typical fat molecule is combined with some other structure (Table 18–2). For example, phospholipids are triglycerides in which one of the fatty acid molecules is replaced by a complex phosphoric acid and nitrogen-containing group. Phospholipids are important constituents of cell membranes, and are believed to help control the selective permeability of the cells.

A further category of lipids—one rather different from the others—is the *sterols.* Many hormones are sterols, and therefore the chemistry of these compounds has been subject to intensive research. One of the more familiar sterols is *cholesterol.*

All fats and oils are insoluble in water, but will dissolve in organic solvents like ether, acetone, alcohol, and chloroform. The fat is then broken up physically into small fat globules, a process called *emulsification.* This is one of the processes that takes place, with the aid of bile, during the digestion and absorption mechanism in the small intestine.

Fats in the diet are derived from meat, fish, milk and milk products, eggs, nuts, and the germ of cereals. These are valuable sources of energy and fat-soluble vitamins. Fats are stored in the body in deposits in a layer under the skin and in the abdomen.

Fats and oils of animal and vegetable origin, which are used as foods, should be distinguished from mineral oils, such as petroleum jelly and paraffin, which have no nutritive value.

Proteins

Proteins are highly complex organic compounds made up of carbon, hydrogen,

---

$$\begin{array}{rcl} & \mathrm{H} & \\ & | & \\ \mathrm{C_{17}H_{35}COO}— & \mathrm{C} & —\mathrm{H} \\ & | & \\ \mathrm{C_{17}H_{35}COO}— & \mathrm{C} & —\mathrm{H} + 3\mathrm{H_2O} \rightleftarrows 3\ \underset{\substack{\text{stearic acid} \\ \text{(fatty acid)}}}{\mathrm{C_{17}H_{35}COOH}} + \underset{\text{glycerol}}{\mathrm{C_3H_8O_3}} \\ & | & \\ \mathrm{C_{17}H_{35}COO}— & \mathrm{C} & —\mathrm{H} \\ & | & \\ & \mathrm{H} & \\ & \text{triglyceride} & \end{array}$$

Table 18–2 FATS (LIPIDS)

CATEGORY	COMPOUND NAME	WHERE FOUND
Simple lipid	Palmitic acid Stearic acid Oleic acid Linoleic acid Linolenic acid	Edible fats and oils
Compound lipid	Phospholipid (lecithin) Glycolipid (cerebrosides) Sterol (cholesterol)	Egg yolk

oxygen, and nitrogen; in some cases, sulfur and phosphorus are also present.

Just as the polysaccharides are combinations of simpler units (monosaccharides), proteins are combinations of *amino acids*. In a protein, however, the 23 natural amino acids may occur in any number or combination; thus there is an almost infinite number of possible structures. Plant proteins are noticeably different from animal proteins, and there are many minor differences among the proteins of individuals within a species.

The amino acids normally found in proteins have structures as shown below.

In the natural amino acid, an amino group ($-NH_2$) is attached to the carbon on one side, and the carboxyl group ($-COOH$) is attached to the carbon on the other side. If we consider the general formula for amino acids, the variations among them are due to the nature of the R group.

The amino acids are linked together to form complex protein molecules through bonds between adjacent amino and carboxyl groups only. A molecule of water is removed in the formation of the peptide bond, as indicated below.

A compound of two amino acids is

amino group 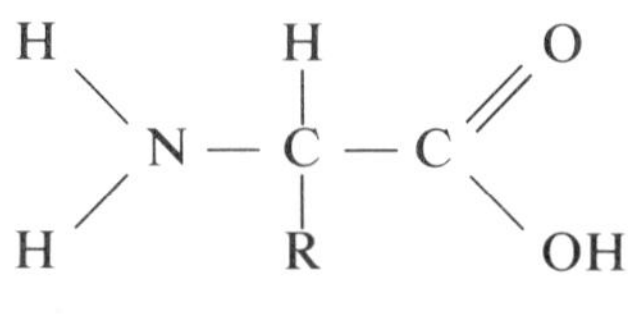 carboxyl group

an amino acid

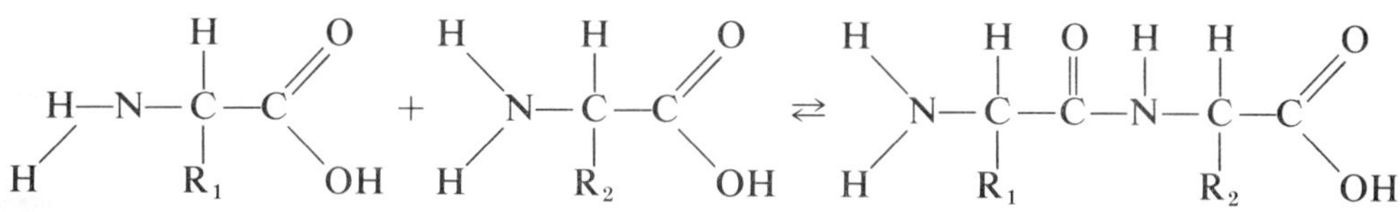

amino acid 1 + amino acid 2 → dipeptide + H_2O

known as a dipeptide; three as a tripeptide, and so forth. Proteins themselves are commonly called polypeptides; their molecular weights are all greater than 30,000. When they are broken down in digestion by the proteolytic enzymes of the stomach and intestine, they are ultimately hydrolyzed into polypeptides, and then into smaller amino acid molecules that are absorbed by the circulatory system and taken to the tissue cells. Within the tissues, some of the amino acids are reconstituted to form needed structural proteins; the remainder are used for fuel in much the same way as glucose.

Many amino acids can be manufactured by the body; some, however, cannot, and must be taken in with the food proteins—hence their name, *essential* amino acids. In their absence, young animals fail to grow and ultimately die. Ten out of the 23 amino acids are "essential" in this sense. They are found especially in proteins of animal origin, such as milk products, eggs, and fish. The essential amino acids that cannot be manufactured by the body are:

arginine	phenylalanine
histidine	threonine
isoleucine	leucine
methionine	lysine
valine	tryptophan

These are necessary not only for the promotion of growth in the young, but also for the repair of tissues destroyed by wear and tear throughout life.

Simple proteins are those containing only amino acids. In *conjugated* proteins, a simple protein is combined with another molecule or molecules; many important structural components of cells fall into this category. Nucleoproteins are combined with nucleic acids; glycoproteins contain carbohydrate units; and lipoproteins comprise lipid units. Hemoglobin consists of a simple protein, globin, combined with the iron-containing heme molecules (Table 18–3).

Vitamins

Vitamins, once known as "accessory food factors," are chemical substances found in fresh foods. They are of great importance for the maintenance of health, but are required by the body only in minute quantities. Vitamins act as organic catalysts in a manner somewhat similar to that of the digestive enzymes; that is, they assist in chemical reactions in the body without themselves being destroyed in the process. The vitamins constitute a "nutritional factor," which has been defined as a single substance performing a vital function in nutrition. In man, absence or insufficiency of certain vitamins in the diet will result in deficiency diseases of various kinds (Table 18–4).

Vitamins may be classified into two

Table 18–3 PROTEINS

CATEGORY	*COMPOUND NAME*	*WHERE FOUND*
Simple protein	Albumin (lactalbuminin)	Egg white Milk
	Globulin (legumin, tuberin)	Peas, lentils Potatoes
	Glutelin (glutenin)	Wheat
Conjugated protein	Nucleoprotein	Cell nuclei
	Glycoprotein	Mucin
	Phosphoprotein (casein, vitellin)	Milk, egg yolk

Table 18–4 PRINCIPAL VITAMINS AND THEIR FUNCTIONS

NAME	SOURCE	FUNCTION	EFFECTS OF DEFICIENCY
Fat-Soluble Vitamins			
Vitamin A (precursor: carotene)	Animal fats, milk, fish liver oils	Vision (rhodopsin)	Skin disorders, night blindness, bone and nerve disorders
Vitamin D	Fish liver oils, milk, eggs	Ca and P regulation	Rickets, osteomalacia
Vitamin E	Rice oils, wheat germ	Reproduction (in experimental animals)	Sterility
Vitamin K	Green vegetables	Blood clotting	Hemophilia
Water-Soluble Vitamins			
Thiamine (B_1)	Peas, yeast, beans, nuts	Metabolism of energy-rich carbohydrate molecules	Beriberi, neuritis
Riboflavin (B_2)	Liver, milk, meat, eggs	Oxidation	Skin lesions
Nicotinic acid (niacin)	Meat, fish	Cellular oxidation	Pellagra
Folic acid Pyridoxine (B_6) Vitamin B_{12}	Liver, cereals, green vegetables	Red cell maturation	Anemia
Vitamin C (ascorbic acid)	Citrus fruits, tomatoes	Capillary function	Scurvy

large groups—those soluble in fat and those soluble in water.

1. Fat-Soluble Vitamins. These require the emulsifying action of bile in the intestinal tract for their absorption.

Vitamin A and its precursor substance, *carotene,* provide a nutritional factor essential to growth and to the maintenance of epithelial tissues in particular. The vitamin itself is found in animal fats, in milk products, and in the oils of fish livers (e.g., cod liver oil). Carotene is found in green vegetables, carrots, and in some fruits; it is converted into vitamin A in the body.

Two forms of vitamin A are detectable—an acid and an aldehyde. Vitamin A acid is responsible for the maintenance of the epithelium; vitamin A aldehyde makes up part of the molecule of rhodopsin, the retinal pigment essential for night vision (see Chap. 12). Vitamin A is not destroyed by ordinary methods of cooking, and is therefore very stable. Lack of the vitamin in the diet leads to growth failure, night blindness, a thickening and roughening of the skin, and, in severe cases, to destruction of the cornea and blindness.

Vitamin D is the antirachitic factor. A lack of it causes *rickets,* a disease characterized by softening and deformation of bone. In adults, a vitamin D deficiency leads to a form of osteomalacia in which the bones become soft and fragile. This disease is due to a deficiency of calcium and phosphorus in the bones, for vitamin D assists in the absorption of these salts into the body from the contents of the small intestine.

Vitamin D is found in fish liver oil, in milk products, and in eggs. It is not destroyed by heat or by the canning process.

Vitamin E prevents sterility in experimental animals, but not necessarily in humans. This vitamin is found principally in vegetable oils obtained from the germ of wheat and rice.

Vitamin K influences the clotting of blood. It promotes the synthesis of prothrombin by the liver. Its absence or deficiency in the diet leads to a delayed clotting time and hemorrhage (see Chap. 13). The vitamin is found in a variety of foods, particularly such vegetables as cabbage, spinach, and tomatoes.

Diseases caused by deficiencies of

each vitamin are listed in Table 18–4. It is important to remember, however, that very large doses of vitamins A, D, and K are toxic. Hypervitaminosis A is characterized by gastrointestinal disturbances, loss of hair, patchy skin, and bone pain. Hypervitaminosis D is associated with weight loss and calcification of many soft tissues. Hypervitaminosis K also results in gastrointestinal disorders, and can cause anemia.

2. Water-Soluble Vitamins. The most significant problem associated with this group of vitamins is that they may be lost during the preparation of common foods. The water used to boil potatoes or to cook vegetables commonly is thrown out after the vegetables have been strained and served. Thus a significant quantity of the water-soluble vitamins is poured down the drain.

Vitamin B is not a single vitamin but a group of vitamins; therefore, the term "B complex" is used. The group includes vitamins B_1 and B_2, nicotinic acid (niacin), folic acid, vitamin B_{12}, and pyridoxine (vitamin B_6).

In its active form, vitamin B_1 (thiamine) is concerned with the metabolism of energy-rich carbohydrate molecules. Its deficiency is far-reaching, in that it causes a peripheral neuritis leading to paralysis and loss of appetite—a disease called *beriberi*. This disease is especially common in the Orient, where polished (white) rice is used extensively as the principal food. Beriberi can be prevented by using unpolished rice instead of polished rice, or by supplementing the diet with such other foods as peas, yeast, beans, ham, and nuts.

Vitamin B_2 (riboflavin) is present in eggs, liver, milk, and meat. It is a constituent of a series of enzymes, the flavoproteins, which serve vital roles in the oxidation of foods. A deficiency of riboflavin leads to lesions around the mouth, eyes, and nose.

Nicotinic acid (niacin) is a B vitamin widely distributed in several foodstuffs, including meat, fish, liver, yeast, and whole-meal flour. The vitamin is part of the structure of the coenzymes important in many different enzymic processes. Niacin deficiency leads to *pellagra,* a condition characterized by rough skin and lesions of the mouth and tongue, followed by a degeneration of the myelinated nerve fibers in the body.

Folic acid, vitamin B_{12}, and pyridoxine (vitamin B_6) collectively provide the *antianemic factor*. Folic acid is found in liver and in leafy vegetables. Vitamins B_6 and B_{12} are found in liver, cereals, and green vegetables. This group is essential in that it provides a metal, cobalt, necessary for the maturation of red blood cells. A deficiency causes pernicious anemia. Liver is a potent source of vitamin B_{12}, and patients of an earlier day were treated with crude liver extracts or "tonics." Today, vitamin B_{12} may be administered directly by injection.

Vitamin C (ascorbic acid) is found in fresh fruits and vegetables. Prolonged cooking and overheating destroys the vitamin, and deficiency causes bleeding from the subcutaneous tissues, especially in the gums and other mucous membranes—a condition called *scurvy*. In children, deficiency of ascorbic acid causes the teeth and bones to develop in an abnormal fashion. The chief lesion in scurvy is a reduced tissue strength leading to capillary fragility and hemorrhage.

It should be noted that, although one vitamin may have some bearing on the action of another, one vitamin cannot *replace* another. Each has its own function, and absence of any one leads to a sequence of signs and symptoms. A balanced diet allows the body to select what it needs at any given moment.

Minerals

The principal mineral salts are the chlorides, phosphates, and carbonates of sodium, potassium, calcium, and magnesium. Their widespread occurrence in all foods means that dietary deficiencies are noted only under bizarre conditions. It is estimated that about 5 per cent of the body weight is made up of mineral salts, which are found mainly in the bones and teeth as salts of calcium and phosphorus.

Calcium is the most abundant mineral in the body. Salts of calcium make up the hard parts of the skeleton and teeth; plasma calcium is important in the blood clotting mechanism; and myofibril calcium plays a major role in the contraction of skeletal muscle.

A fetus requires a large quantity of calcium for making bones, and the young child requires it for growth. Dietary deficiencies of calcium lead to decalcification of bone (rickets in youth, osteoporosis in the adult), and, in women, to possible dental troubles in the year or two following childbirth.

The best sources of calcium are milk, beans, wheat, and other cereals. Calcium absorption by the intestine, however, depends on the availability of vitamin D. Also, any defect in fat digestion causes calcium to be lost with the undigested fat in the feces.

Phosphorus is the body's second most abundant mineral. It is widely distributed in the soft tissue cells and occurs in relatively low concentrations in the extracellular fluid. Nucleic acids in the genetic materials of the cells contain phosphorus, as does the most important energy carrier, ATP.

Iron is present in red blood cells as hemoglobin, and is also a constituent of a whole series of respiratory enzymes responsible for controlled oxidation in the cells. The simplest symptom of anemia is a shortage of iron demonstrated by a low hemoglobin count. Eggs, meat, liver, and fruits are all good sources of iron.

Some mineral elements, such as copper, manganese, and cobalt, are known as "trace" elements because they are required in such minute amounts. They are, however, essential to health. Iodine is the major component of the thyroxine produced by the thyroid gland; copper, cobalt, and manganese act as catalysts in the chemical activity of the body; the gaseous element fluorine plays an important role in preventing tooth decay.

DIGESTIVE ORGANS AND PROCESSES

Before food can be absorbed into the body it must undergo a number of changes. Some of these are purely mechanical; others are of a highly complex chemical nature. Most foodstuffs are constructed of molecules too large to pass through the walls of the intestine, and the purpose of digestion is to convert these into smaller molecules.

During digestion, the chemical process of hydrolysis occurs. Complex molecules are split into simple molecules by the action of digestive enzymes, known collectively as the *hydrolases*. These enzymes are present in the secretions of the several digestive glands of the alimentary tract—the salivary glands, the gastric glands, the pancreas, and the intestinal glands.

The alimentary tract or canal (Fig. 18–1) extends from the mouth to the anus, and different names have been given to its various parts. The greater part of the tract has to do with the actual digestion of food, but it is also concerned with absorption of the nutrient part of food and with the excretion of waste products.

The Mouth

When food is taken into the mouth, it is cut and ground by the teeth, mixed with saliva, and finally molded into a semidigestible mass called a *bolus*. *Saliva* contains a large amount of water, the enzyme salivary amylase (ptyalin), and mucus, which helps to hold the food together and to lubricate it for passage down the esophagus. The secretions of the salivary glands (parotid,

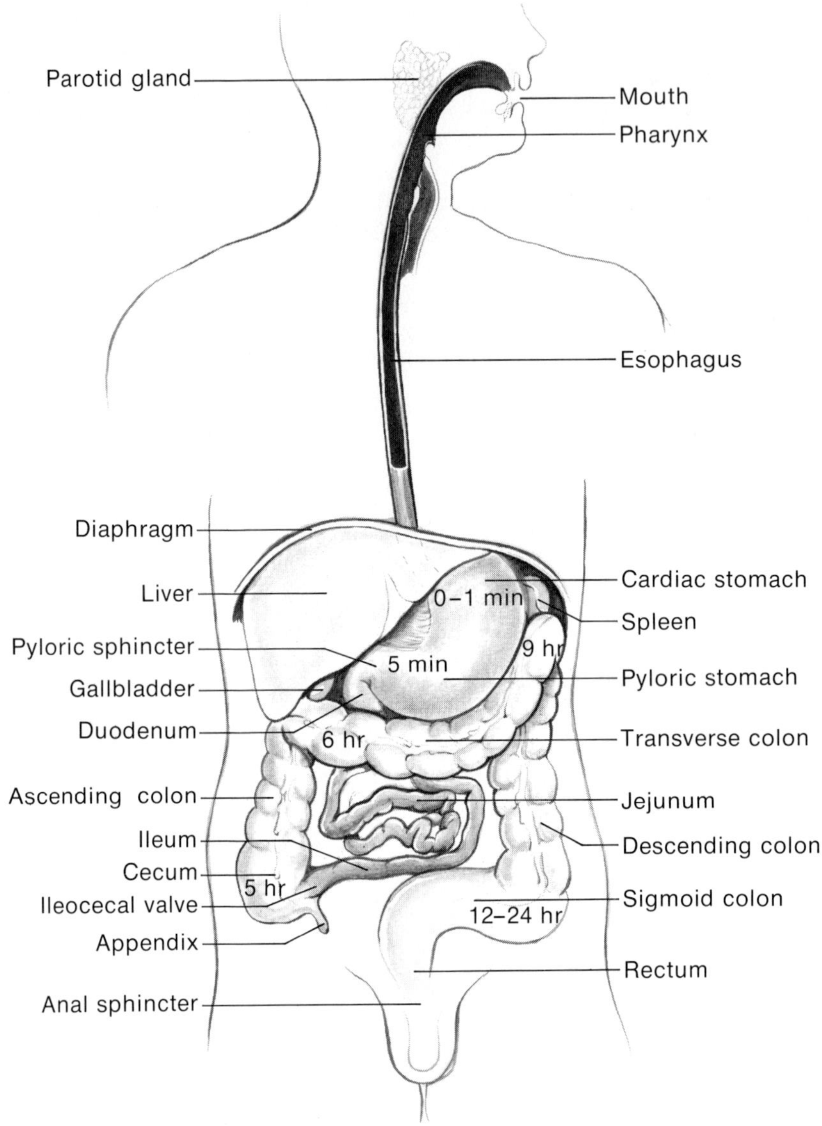

Figure 18–1 The human digestive tract. The times indicated along the tract represent how long it takes food to pass through each area during the process of digestion.

submaxillary, and sublingual) are basically neutral in reaction (pH 6.8 to 7.0). Salivary amylase hydrolyzes or splits starches into disaccharides—chiefly maltose and dextrin. This chemical process is facilitated by body heat and by chemically neutral conditions.

The mechanical breakdown of food (mastication) is carried out by the *teeth* (Fig. 18–2). There are several different types of human teeth, each adapted to a specific function. The knife-shaped incisors, four in the upper jaw (maxilla) and four in the lower (mandible), are used for biting. Next come the pointed canines, one on each side in each jaw, which are specialized for tearing food. Behind each canine, in adults, are two premolars and three molars. These teeth have flattened surfaces adapted for grinding, pounding, and crushing food. A child's first set of teeth *(deciduous dentition)* does not include all those just mentioned. The first teeth are lost during growth and are replaced as the permanent teeth erupt through the gums.

The muscular *tongue* manipulates food during chewing and forms the bolus in preparation for swallowing. It then pushes the bolus backward through the pharynx and into the esophagus (Fig. 18–1).

The Esophagus

The esophagus, sometimes called the "gullet" or "food tube," is the portion of the alimentary canal that connects the pharynx with the stomach. It is about 10 inches long, and passes down through the thorax until it reaches the diaphragm (Fig. 18–1), through which it enters the abdominal cavity. After continuing for about 2 inches, it ends at the cardiac (upper) portion of the stomach.

The esophagus has three coats—an outer fibrous coat, a middle muscular coat, and an inner mucosal lining of stratified squamous epithelium arranged in longitudinal folds.

Food moves through the esophagus, pushed along by the waves of muscular contraction called *peristalsis.* Circular muscles in the wall of the esophagus just behind the food bolus contract, squeezing the bolus forward. As it moves, the muscles it passes also contract; thus the region of contraction follows the bolus and constantly pushes it onward (Fig. 18–3).

The coordination of swallowing *(deglutition)* depends on nervous mecha-

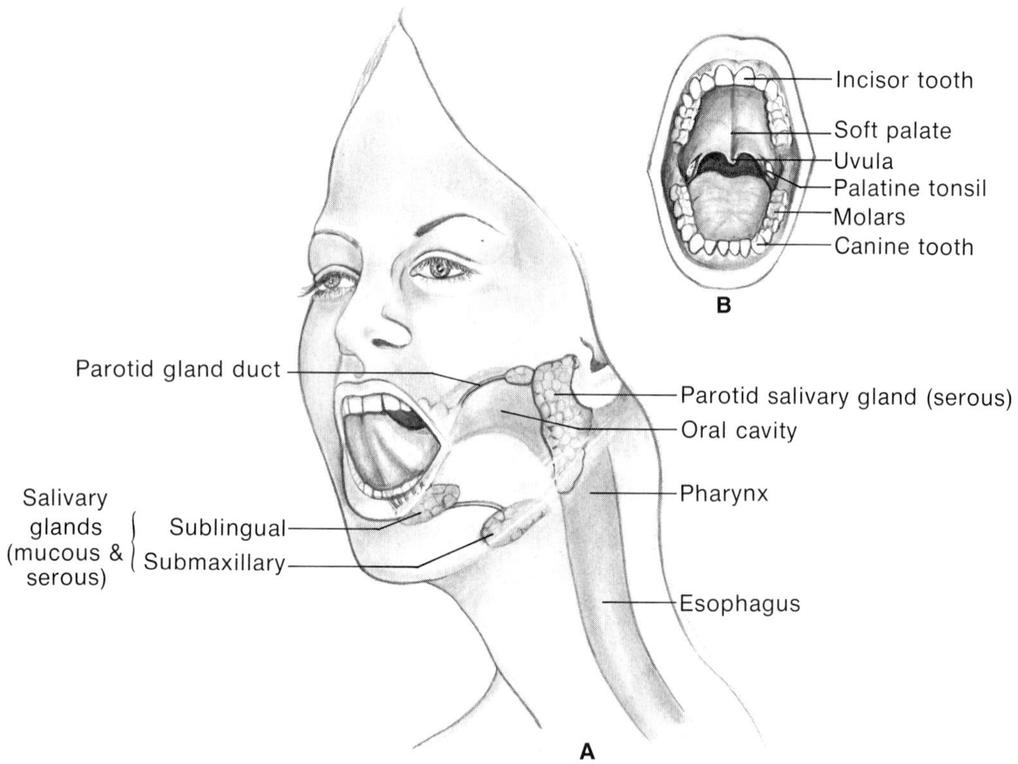

Figure 18–2 Mastication. *A*, Transparent view of the human mouth, showing the different salivary glands responsible for secreting digestive enzymes and mucin, which are found in saliva. *B*, The mouth, showing the teeth and tonsils.

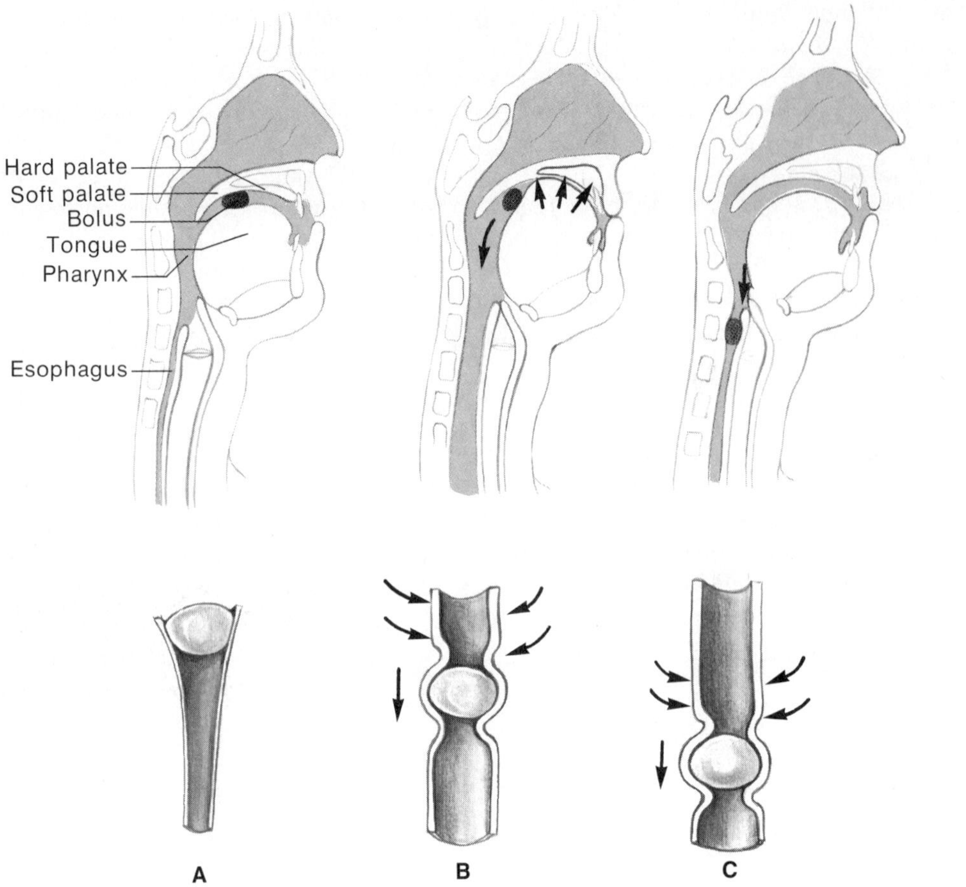

Figure 18–3 Swallowing. *A* and *B*, Voluntary stage: the food bolus stimulates nerve endings at the back of the tongue and pharynx, causing the tongue to draw upward and backward against the hard palate. The bolus is projected down past the soft palate into the upper part of the esophagus. *C*, Involuntary stage: the bolus stimulates nerve endings in the wall of the esophagus that set up peristaltic waves of contraction and push the bolus down the lower two-thirds of the esophagus into the stomach.

nisms. The first stage is initiated voluntarily, with the cerebral cortex playing the major role. The remainder of the swallowing response—and of all other movements of the alimentary tract except defecation—is involuntary. The swallowing center is located in the medulla, which is activated by stimulation of receptors in the mouth and pharynx. This center sends out impulses to the esophageal muscles to promote peristalsis.

A disorder called *esophageal varices* (Fig. 18–4) is caused by increased pressure (hypertension) in the veins of the esophagus. The varices are branches of the azygos vein in the lower esophagus and in the cardiac portion of the stomach. They drain into the superior vena cava. Increased pressure may result in hemorrhage. Violent physical exertion, ulceration of the vein, and trauma have also been known to cause varices.

Carcinoma of the esophagus occurs most often in men over 40. Its origin is obscure. The first and most common symptom is a difficulty in swallowing *(dysphagia)* solid food and an increasing difficulty in swallowing soft foods. This is caused by a growth narrowing the esophageal lumen. The pain is referred to the back, abdomen, neck, and face. Difficulty in swallowing gradually leads to loss of weight and strength. The lesion may spread to the blood vessels, causing massive hemorrhage and death.

The Abdomen

The abdominal cavity is the largest in the body. Abdominal muscles form its sides and front walls; the vertebrae form the back wall; the diaphragm and pelvis form the roof and floor, respectively. All the organs of digestion below the thoracic part of the esophagus are contained within the abdomen. To protect these organs and the abdominal wall itself, a serous membrane, the *peritoneum,* covers the surface of most of the abdominal organs, and lines the abdominal wall to form a parietal layer. The small intestine is connected to the posterior abdominal wall by folds of peritoneum, the *mesentery,* which is shaped like a fan and contains the blood vessels, nerves, and other structures that supply the small intestine (Fig. 18–8).

Another double-layered peritoneal structure, the *greater omentum,* hangs down from the greater curvature of the stomach and extends into the pelvic area of the abdominal cavity, where it is attached to the front of the transverse colon. It is well supplied with fat, and functions to insulate and protect the organs, much like a blanket. A smaller blanket, the *lesser omentum,* extends between the stomach and the liver.

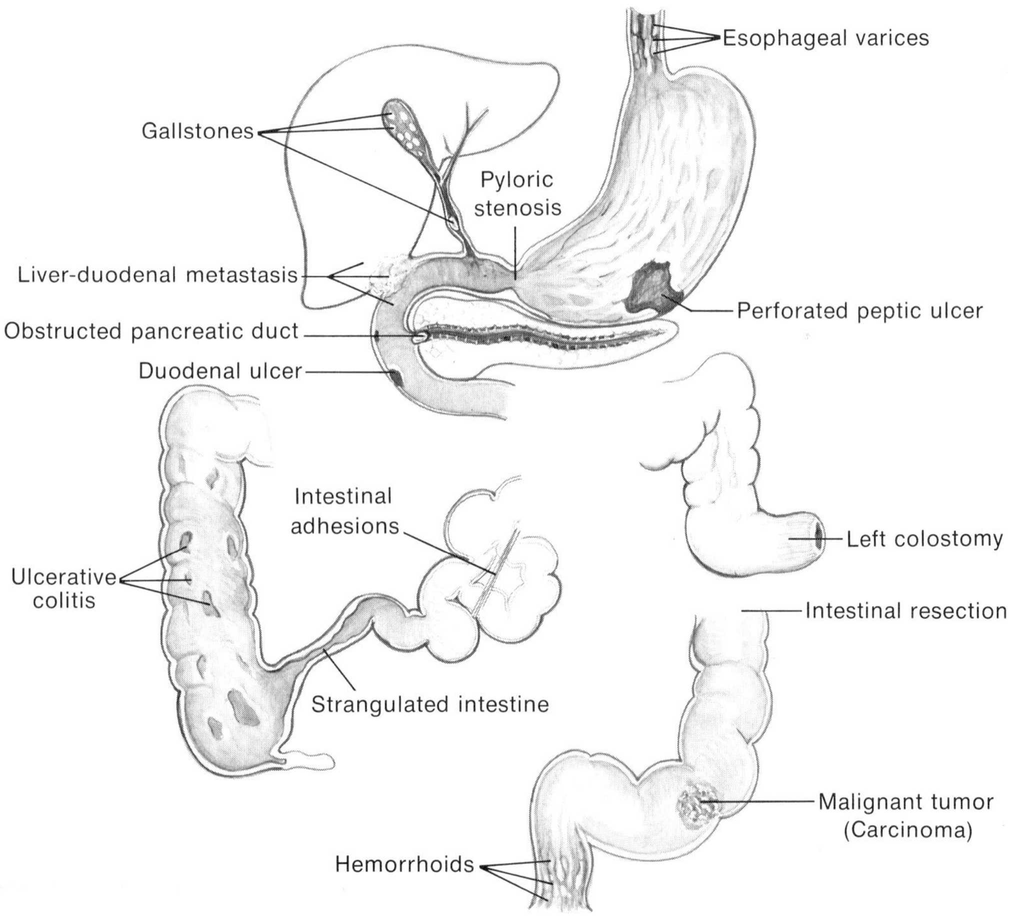

Figure 18–4 Pathologies of the alimentary tract.

Peritonitis, an inflammation of the peritoneum, is a life-threatening disease. Its most common cause is rupture of the appendix or another abdominal organ. A generalized peritonitis may be fatal because of the large quantity of bacterial toxins absorbed by the abdominal organs.

The Stomach

The stomach (Fig. 18–5) is an irregularly shaped, distended part of the alimentary canal. Although there is no well-defined muscular structure in the region where the esophagus joins the stomach, a *cardiac sphincter* is present at the entrance to the stomach. This region is normally contracted during rest, which prevents the escape of gastric juices into the esophagus. When a solid bolus passes down the esophagus by the wave of muscular contractions, the peristaltic movement causes relaxation of the sphincter and allows the bolus to enter the stomach. Once the bolus has passed through, the sphincter closes again.

The stomach lies slightly to the left side in the upper portion of the abdomen, just beneath the ribs and just below the diaphragm, which separates the abdominal and thoracic cavities. It serves two functions. First, it receives the bolus of swallowed food and passes this on to the intestine, where further digestion and absorption take place. Second, the stomach itself plays an important role in the digestion of proteins.

The upper end of the stomach—comprising the *cardiac* and *fundus* areas—is dome-shaped; the lower part, shaped like the letter "J," is called the *pylorus.* The

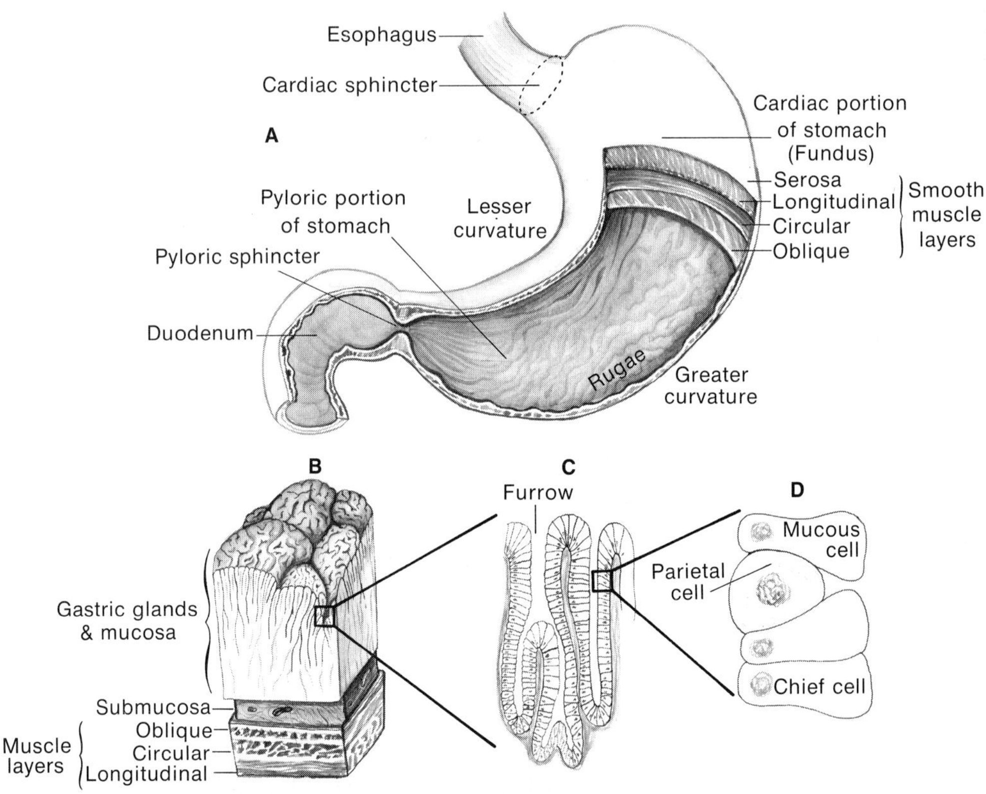

Figure 18–5 The stomach. *A,* External and internal anatomy of the stomach, showing the layers of muscle, the rugae, and the pyloric valve. *B,* Schematic representation of the gastric mucosa, showing all the layers of the stomach. *C,* Gastric glands from the greater curvature of the stomach. *D,* Detail of the gastric glands.

opening through which food passes from this part to the intestine is the *pyloric sphincter*. The middle or main part of the stomach is the *body*. The left border of the stomach forms a large curve called the greater curvature; the right border, which is much shorter, is called the lesser curvature.

The wall of the stomach is composed of three layers of smooth muscle (Fig. 18–5) lined by mucous membrane and covered by peritoneum. The fibers of the outermost layer of muscle run longitudinally and are continuous with the esophagus. This layer is very thin at both extremities of the stomach, but relatively thick along the curvatures. The middle layer is composed of circular fibers that encircle the stomach; the innermost layer of muscle fibers runs obliquely and does not form a complete coat, being confined largely to the fundus. The mucous membrane of the stomach is covered with nonciliated columnar epithelium, many cells of which secrete mucus for protection. At the lower end of the stomach the fibers of the circular muscle layer form the strong ring of the pyloric sphincter; at the upper end these fibers form the cardiac sphincter.

There are three kinds of cells in the mucosa of the stomach:

1. *Mucous* cells, located in the neck region of the stomach, secrete mucin, which protects the mucosa (inner lining) from the other highly acid gastric juices.

2. *Chief* cells (or peptic cells) are located along the tubule. These cells form glands for the secretion of pepsinogen.

Table 18–5 DIGESTIVE ENZYMES

LOCATION OF DIGESTIVE ACTIVITY	DIGESTIVE SECRETION	SOURCE	ENZYME	pH	ENZYME ACTION
Mouth	Saliva	Salivary glands	Amylase (ptyalin)	6.8–7.0	Starch → maltose
Stomach	Gastric juice	Stomach lining	Pepsin	1–4	Proteins → proteoses and peptones
					Nucleo proteins → peptones and nucleic acids
			Rennin	1–4	Casein → calcium paracaseinate
Small intestine	Bile	Liver	–	8–10	Emulsifies fats; no enzymatic activity
	Pancreatic juice	Pancreas	Trypsin	8–10	Peptones → polypeptides
			Pancreatic amylase (amylopsia)	8–10	Starch and glycogen → maltose
			Pancreatic lipase	8–10	Fats → fatty acids and glycerol
	Intestinal juice	Small intestine (crypts of Lieberkühn)	Enterokinase	8–10	Polypeptides → amino acids
			Sucrase	8–10	Sucrose → glucose and fructose
			Maltase	8–10	Maltose → glucose
			Lactase	8–10	Lactose → glucose and galactose
			Lipase	8–10	Fats → fatty acids and glycerol
			Nuclease	8–10	Nucleic acids → nucleotides

3. *Parietal* cells (oxyntic cells) located alongside the chief cells secrete hydrochloric acid (2 per cent HCl). These cells do not actually store and secrete the HCl; rather, they extract the component chemicals, H^+ and Cl^-, from the parietal cells and the blood cells, respectively. These then combine to form hydrochloric acid.

Hydrochloric acid secreted by the parietal cells combines with pepsinogen from the chief cells to form *pepsin*, the enzyme responsible for the chemical breakdown of proteins to proteoses and peptones (peptides).

Rennin is an enzyme that acts upon milk proteins, and it is thought to be produced by the chief cells in small children. It is not found in the adult. Rennin coagulates milk, producing a white mass called curd and a fluid material called whey. The action of rennin is to convert milk protein, casein, into a form suitable for digestion (Table 18–5).

Peristaltic contractions initiated in the esophagus are continued through the stomach. Within minutes after the bolus has entered the stomach, it begins to be macerated, mixed with the acid gastric juices, and reduced to a thin liquid mass called *chyme*. As the chyme is gradually propelled into the duodenum, the pressure of the fundus forces the food into the pyloric end.

The secretion of gastric juice is constant. It is a thin, colorless fluid with an acid pH ranging from 1 to 4. The quantity of acid secreted depends on the amount and type of food to be digested. It is important to note here the correlation between enzyme activity and the relative acidity of the digestive juices (Table 18–5). Salivary amylase, secreted in the mouth by the salivary glands, is basically neutral in reaction. The enzymes pepsin and rennin, secreted by the stomach's glandular cells, are effective in an acid medium. It is important to understand this now, as it is fundamental to the digestive process. Enzymes are specific in that they are effective only in their own pH range. Each enzyme works best in catalyzing a single chemical reaction.

Enzymes are frequently proteins. They act upon a substance called a *substrate*. Each enzyme catalyzes a chemical reaction most rapidly with a particular substrate, but also catalyzes reactions with other substances of the same general type. A means of identifying the enzyme is the suffix *-ase* at the end of the substrate's name; for example, the enzyme maltase breaks down maltose into simple sugars. Enzymes are affected by temperature as well as by pH. Their activity increases twofold with every 10° C rise in temperature. This rate of increase in activity continues until temperatures of 40 to 45° C are reached; then the peptide bonds rupture and activity ceases. It is therefore necessary to maintain the normal body temperature (37° C/98.6° F) to insure an efficient digestive process. Lowering the body temperature decreases the rate of enzyme activity and energy production; increasing the body temperature above 40° C (above 106° F) may be fatal if the condition is allowed to persist.

Hyperacidity of the stomach is usually a temporary condition resulting from overeating. Strong acids convert tissue protein to acid proteinase, which dissolves in concentrated acid. A common example of necrosis (tissue death) from acid is a *peptic ulcer*. Most peptic ulcers are located in the pyloric region. Ordinarily, peptic ulcers result from oversecretion of hydrochloric acid, but the condition may occur if insufficient protection is provided by the mucus secreted by the stomach cells for just that purpose. If the necrosis is very superficial, failing to extend completely through the mucosa, the lesion is termed an *erosion*. If the defect extends through and beyond the mucosa, it is an ulcer. Bleeding ulcers indicate that a vessel has ruptured in the defect, and blood is evident in the vomitus or feces. In *perforated ulcer* (Fig. 18–4), all the layers of tissue from the inside to the outside of the stomach wall are penetrated, causing extreme abdominal cramps and eventual collapse.

Surgical removal (gastrectomy) is indicated in perforated ulcer, whereas medicine and controlled diet may eventually cure general peptic ulcer. Regenerative activity is always present, and may at any time lead to healing of the ulcer.

A common congenital anomaly of the stomach is an obstructive narrowing of the pyloric valve, causing eventual closure of the opening *(pyloric stenosis;* Fig. 18–4) and preventing further passage of food. Infants who repeatedly vomit their feedings over a period of time may have this defect. If the milk does not have a place to go, it will come back up. Prolonged stenosis can result in constipation and rapid loss of weight. Dilation of the valve sphincter is indicated to relieve this problem.

The Small Intestine

The food leaves the stomach through the pyloric sphincter and enters the small intestine, where most digestion and absorption take place. The first section of the small intestine is the *duodenum* (Figs. 18–5 and 18–6). The duodenum leads into the

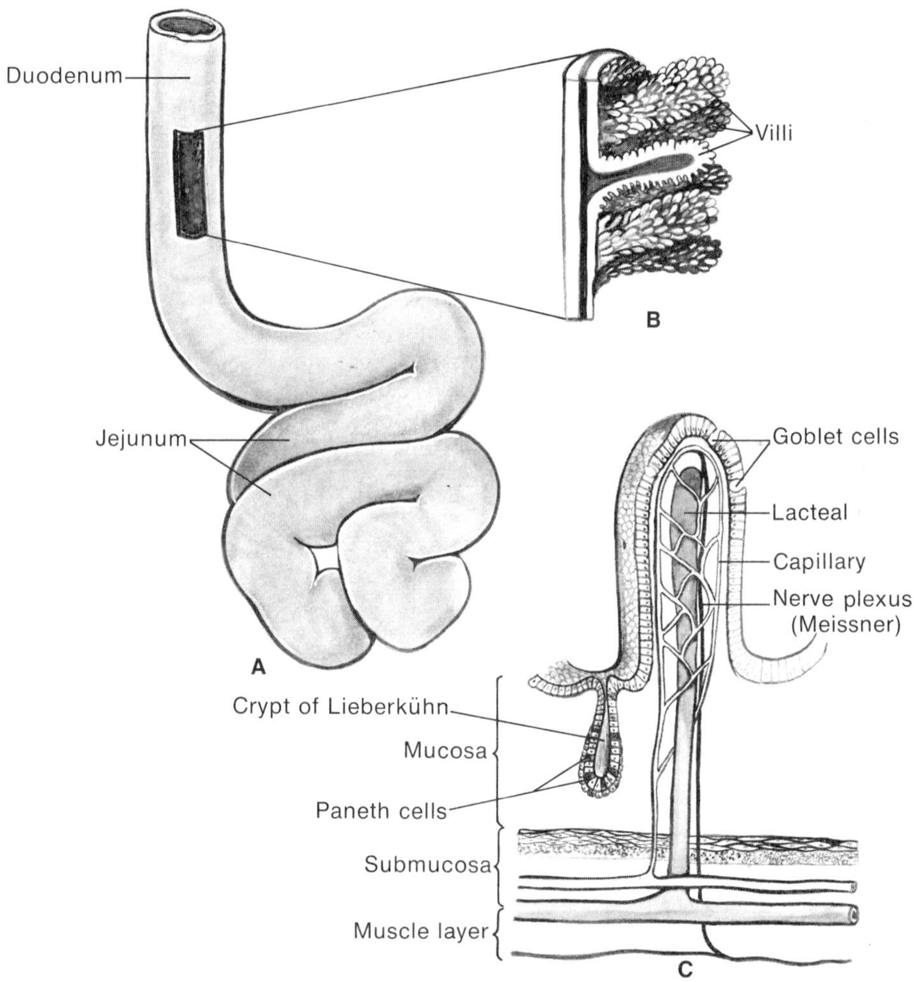

Figure 18–6 The small intestine. *A*, The long muscular duodenum and jejunum, two of the three parts of the small intestine. *B*, A section cut out of the wall of the duodenum, showing the many villi. *C*, Midsagittal section of a villus, showing the arrangement of its vascular elements and digestive cells.

middle portion, the *jejunum* (Fig. 18–6), and the small intestine finally terminates in the *ileum*. The entire small intestine of an adult male is about 23 feet long and one inch in diameter.

Partially digested food from the stomach passes into the duodenum, where the acidity of the food stimulates the intestine to secrete intestinal hormones. These, in turn, stimulate the pancreas to secrete pancreatic juices and the liver to secrete *bile*. (See page 389 and Figure 18–7 for more information about the liver.)

Bile salts are produced by the liver and stored in concentrated form in the gallbladder during periods when the intestine is devoid of food. The bile passes from the liver by way of the hepatic duct into the common bile duct and the gallbladder. Bile salts are derived from the metabolism of hemoglobin. The color of bile depends on the type and concentration of pigment that it contains. If the pigment is bilirubin, the bile is yellow; if the pigment is biliverdin, the color varies from green to black.

An important constituent of bile is cholesterol, which is also the chief constituent of *gallstones* (Fig. 18–4). Many people who show a tendency to form gallstones are fed bile salts. These help to maintain cholesterol in solution and to stimulate liver bile flow, thereby reducing the possibility that the cholesterol concentration will rise to a level sufficient to cause gallstone formation.

Acute *cholecystitis* is an inflammation of the gallbladder that may be caused by bacterial infection or by an increased retention of bile salts. The infection starts in the mucosa and extends into the wall of the gallbladder. Cholecystitis is generally responsible for the formation of gallstones in the gallbladder *(cholelithiasis)*. Pure cholesterol gallstones, described above, develop because of metabolic anomaly. Bilirubin gallstones develop as a result of hemolytic jaundice. The most common complications of cholelithiasis include obstruction (Fig. 18–4), perforation of the ileum or colon, and formation of a fistula (ulceration).

A *cholelithotomy* is an operative procedure whereby the cystic duct or common bile duct is opened in order to remove gallstones; the ducts may be dilated to allow passage of the stones into the duodenum for eventual expulsion in the feces. If the cystic duct is damaged, or if the stones present are of a sufficient size and number to indicate future problems, removal of the gallbladder *(cholecystectomy)* may be indicated.

Removal of the gallbladder diminishes markedly the quantity of bile that will enter the duodenum; therefore, a low fat or nonfat diet is indicated to prevent digestive discomfort.

The major function of bile salts is the digestion and absorption of fats. Bile salts act to emulsify the fats into minute fat particles, thus providing a larger surface area for the action of pancreatic lipase. Bile (pH 7 to 8) also aids in the conversion of the acid chyme to an alkaline medium (pH 8 to 10), which is the necessary pH range for intestinal enzyme activity. The intestinal juices secreted by the crypts of Lieberkühn and the succus entericus glands are always alkaline (pH 7 to 10) and assist the bile salts in maintaining a proper pH for intestinal digestion.

Since the small intestine is the place where absorption of the end-products of digestion occurs, special structural adaptations increase the absorptive surface area (Fig. 18–6). The surface of the inner mucous membrane is made up of tall columnar epithelium, and contains an abundance of goblet cells. The middle layer of the intestine, the submucosa, consists of loose connective tissue lying beneath the mucous membrane and uniting it to the outer muscular layer. This third layer contains blood vessels, lymphatics, and nerve networks (Auerbach's plexus) dispersed among layers of circular and longitudinal muscles. A fine serous coat *(serosa)* covers the entire intestine on the outside. This layer is made up of fibrous tissue and is formed by a visceral layer of peritoneum.

The innermost lining of the small intestine is not smooth, but has numerous finger-like projections, the *villi*, which greatly increase the absorptive surface. In addition, electron microscopy has shown that this surface is further increased by numerous microvilli (Fig. 18–6) to such an extent that the total surface area of an intestinal lining is estimated to be five times the surface area of the exterior of the body.

The muscularis mucosa continually contracts and relaxes rhythmically (peristalsis) during digestion. This causes corresponding shortening and lengthening movements of the villi that act to stir the intestinal fluids, thereby facilitating absorption of the digestive products.

Within each villus is a network of capillaries and a central lymph vessel or lacteal. Simple sugars, amino acids, vitamins, minerals, and water are absorbed by the capillaries. Fatty acids and glycerol are absorbed by the lacteal, and after absorption they again combine to form fat, the triglycerides found in the bloodstream.

The intestinal glands (Paneth cells in the crypts of Lieberkühn) (Fig. 18–6) secrete enzymes that assist the pancreatic enzymes to a marked extent:

1. *Enterokinase*, a protein-splitting enzyme, activates trypsin and completes the breakdown of peptides to amino acids.
2. *Maltase*, a carbohydrate-splitting enzyme, hydrolyzes maltose into glucose.
3. *Sucrase*, a carbohydrate-splitting enzyme, hydrolyzes sucrose into glucose and fructose.
4. *Lactase*, a carbohydrate-splitting enzyme, hydrolyzes lactose into glucose and galactose.
5. *Lipase*, a fat-splitting enzyme, splits emulsified fats into fatty acids and glycerol.
6. *Nuclease* acts on the nucleic acid component of available nucleoproteins.

Patches of lymphoid tissue, known as Peyer's patches or lymphatic nodules, are found in the mucosa of the small intestine.

The jejunum succeeds the duodenum. It is about 8 feet long and constitutes two-fifths of the coiled part of the intestine. The ileum, the remainder of the coiled part, joins the large intestine at the ileocolic valve (Fig. 18–8). Although the duodenum is the major portion of the small intestine concerned with digestion, the entire small intestine is lined with villi and acts to absorb end-products of digestion.

Meckel's diverticulum is an outgrowth or outpouching of the ileum, and has been referred to as a secondary appendix. *Duodenal ulcers* (Fig. 18–4) have the same causes as peptic ulcers. These ulcers are very painful, and are accompanied by vomiting and hemorrhage. Most often the ulcer is located about 2 inches below the pylorus along the duodenal wall, and may perforate, causing peritonitis.

Intestinal obstruction prevents the onward movement of the intestinal contents. The pathologic effects are fluid and electrolyte loss, strangulation (Fig. 18–4), vomiting, and probably acidosis. Malignant tumors of the small intestinal (adenocarcinoma) can cause obstruction and jaundice. An intestinal resection (Fig. 18–4)—the removal of the involved segment of the intestine—may be curative.

The Pancreas

The pancreas is the most active and versatile of the digestive organs (Figs. 18–7 and 18–9). In the absence of other digestive secretions, its enzymes alone are capable of almost completing the digestion of all foodstuffs. The pancreatic secretion into the duodenum is a complete digestive juice in that it contains all the enzymes needed to split foodstuffs:

1. Carbohydrate enzyme—*pancreatic amylase* (amylopsin) acts upon starch and glycogen, producing maltose.
2. Fat enzyme—*pancreatic lipase* (steapsin) is the main fat digestive enzyme in man. Each fat molecule is broken into a molecule of glycerol and three molecules of fatty acid.
3. Protein enzyme—*trypsin* is se-

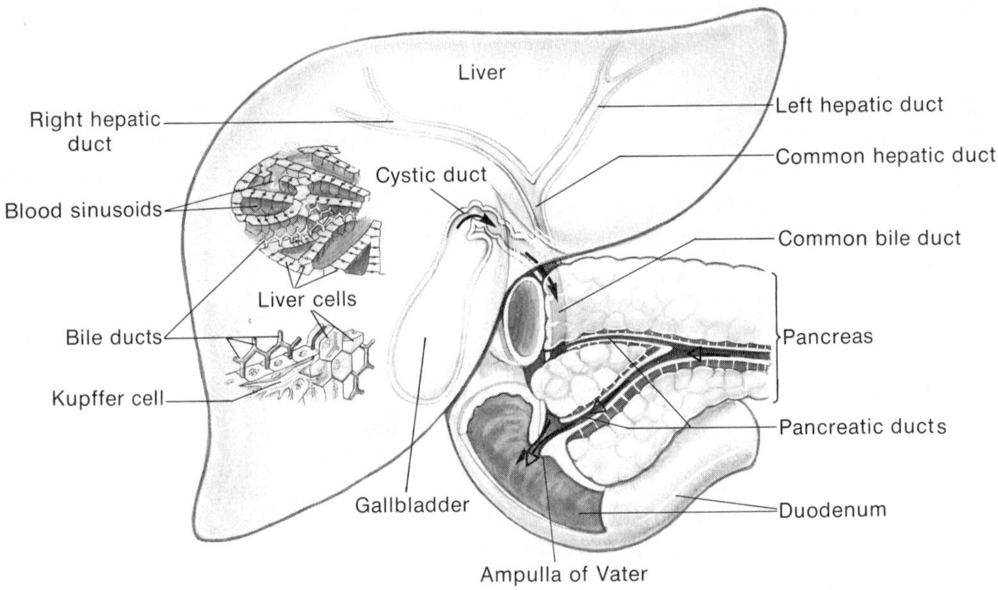

Figure 18–7 The liver and its interrelationship with the gallbladder, pancreas, and duodenum. A section has been removed from the liver and the area enlarged to show the arrangement of liver cells, bile ducts, Kupffer cells, and blood sinusoids to one another. Arrows indicate the direction of flow of bile from the gallbladder and liver and of digestive juices from the pancreas into the duodenum.

creted as an inactive precursor, trypsinogen, which is converted to active trypsin by *enterokinase,* an enzyme found in the intestinal juices. Trypsin digests proteins and peptides partially digested in the stomach. The end-products of trypsin digestion are polypeptides and amino acids.

A second proteolytic enzyme, *chymotrypsin,* is secreted by the pancreas as inactive chymotrypsinogen, which is converted to chymotrypsin by trypsin. Chymotrypsin acts in a manner similar to that of trypsin.

4. *Nuclease* is a nucleic acid-splitting enzyme that results in the production of mononucleotides.

The pancreas lies on the posterior wall of the abdomen; its expanded portion, or head, is positioned in the concavity formed by the duodenum. The pancreas is sometimes called the "salivary gland of the abdomen," in that it closely resembles (microscopically) the salivary gland. The secretions of the gland exit through a long tube, the pancreatic duct (Fig. 18–7). This duct opens into the duodenal mucosa, often at the ampulla of Vater, and its contents join the intestinal juices to further the digestive process.

Inflammation of the pancreas *(pancreatitis)* basically involves a lack of coordination between the secretion of pancreatic juice and its ability to flow through the pancreatic duct into the duodenum. Obstruction of flow, whatever its cause, results in a rupture of the cells within the gland owing to the increased pressure of the enzymes (Fig. 18–4). The enzymes amylase and lipase are measurable, and determination of their levels is clinically useful.

Interference with the secretion of the hormone insulin by the islands of Langerhans in the pancreas will result in *diabetes mellitus,* which is discussed in Chapter 10.

The Large Intestine

In man, the junction between the small intestine and the large intestine (colon or large bowel) is usually in the lower right portion of the abdominal cavity. A blind sac, the *cecum,* projects from the large intestine near the point of puncture, the *ileocecal (*ileocolic) *valve* (Fig. 18–8). It is here that a small, finger-like process, the *appendix,* projects from the cecum. As you know, the appendix frequently becomes infected, and sometimes must be removed surgically *(appendectomy).*

The cecum and appendix serve no purpose except as storage centers for microorganisms that aid digestion. Man cannot digest the complex cellulose found in plant cells without assistance; the stored microbes digest this cellulose for him, but most of the nutritional value of the material is lost. The duodenum, not the cecum, is the principal site of digestion.

The large intestine is about 5 to 6 feet long, and its sections are named according to the directions they travel and the shapes they assume. The cecum leads into the ascending colon, which is followed by the transverse colon, the descending colon, the sigmoid colon (shaped like the letter "S"), the rectum, and the anus (Fig. 18–8). The muscular coat of the large intestine contains three flat, longitudinal bands that extend from the vermiform appendix to the rectum and are known as the *taeniae coli* (Fig. 18–8).

The major function of the large intestine is the reabsorption of water into the body. If all the water in which enzymes are secreted into the digestive tract were lost in the waste material (feces), man would have a severe problem of dehydration.

A second function of the large intestine is the removal of the waste products of digestion (defecation). Occasionally, the colon becomes irritated, and peristalsis moves the material through it faster than

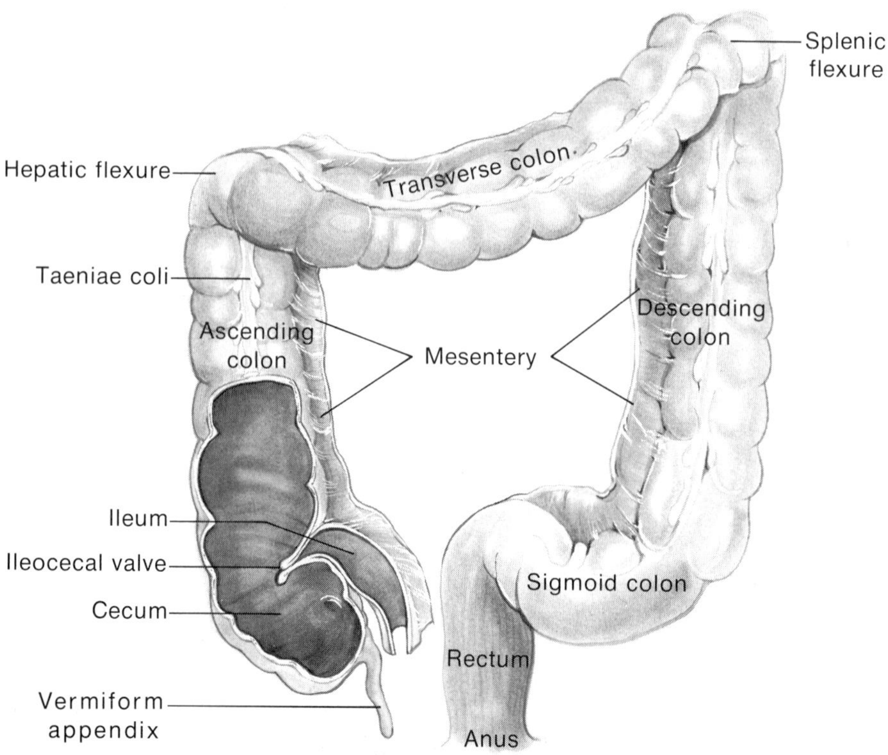

Figure 18–8 The large intestine. Each segment is named according to the direction it travels or according to its shape. Note the junction of the small intestine (ileum) with the cecum of the large intestine (bowel). It is here that the appendix is found.

normal. As a result, not enough water is reabsorbed, and a watery waste material (*diarrhea*) persists until the cause of the irritation is remedied. Conversely, if material moves too slowly, too much water is reabsorbed, and *constipation* results. Diarrhea may be the result of a bacillary dysentery or of many other causes. Constipation, on the other hand, results from an overstimulation of the intestinal muscle, narrowing the canal and preventing passage of fecal material. Nervous tension, excessive amounts of bulky foods in the diet, and use of laxatives increase the muscle tone of the intestines.

The large bowel is also the region where excess salts, such as those of calcium and iron, can be excreted from the body when their concentration in the blood is higher than normal.

The last portion of the large intestine, the *rectum*, stores the waste materials until defecation occurs. The feces are expelled through an opening called the *anus*.

The wall of the large intestine closely resembles that of the small intestine in structure, except in its mucosal layer. This is covered with columnar epithelium, but there are no villi; the glands are longer, contain larger numbers of goblet cells, but do not secrete digestive enzymes. Veins in the mucosa of the anal canal may become dilated (varicose), forming small swellings when a strain is exerted on the anal region. These swellings are called *internal hemorrhoids* (piles). Similar varicose swellings in the skin-lined lower part of the anal canal are called *external hemorrhoids* (Fig. 18–4). Medication, heat treatments, or surgery can be used to relieve the discomfort caused by this condition.

Inflammatory diseases of the large bowel *(colitis)* may be caused by tuberculosis, typhoid fever, or bacillary and amebic dysentery. *Ulcerative colitis* (Fig. 18–4) is a specific inflammatory disease of the colon that occurs most commonly in middle-aged people. It may be caused by allergy, infection, or cellular destruction (necrosis) of the colon.

Malignant tumors (adenocarcinoma) of the large bowel and rectum grow around the lumen, decreasing its size and causing an obstruction (Fig. 18–4). The colon above the obstruction becomes dilated, and a resection eventually is indicated. Massive tumor growth in the rectum and anal canal may necessitate surgical removal of the rectum, closure of the anus, and protrusion of the colon through the abdominal wall *(colostomy)* (Fig. 18–4). Defecation is then performed through the colostomy opening into a sterile bag strapped to the abdomen. Metastasis (spread of cancer cells to other organs) from adenocarcinoma of the large intestine commonly occurs in the regional lymph nodes and in the liver. Malignant growths of the abdomen spread via the circulation, resulting in liver involvement. The disease may spread as far as the lung and the brain.

Gastritis and enteritis are inflammations of the stomach and intestinal linings. Commonly both areas are involved, and the condition is called *gastroenteritis*.

THE HORMONAL CONTROL OF DIGESTION

Mechanical pressure of food against the stomach wall initiates endocrine (hormonal) control, the gastric phase of digestion. Food stretches the pylorus, and it secretes gastrin into the circulation. This has no particular effect until the secretion has returned to the parietal cells of the cardiac area of the stomach. Digestive juices and hydrochloric acid are secreted in response to the hormone (Fig. 18–9). The presence of HCl itself serves as a feedback mechanism to inhibit gastric secretion. Hydrochloric acid and the hormone enterogastrone act to inhibit the secretion of gastric juices. When fats enter the intestine, duodenal glands release enterogastrone into the circulation, and in this way partly digested food present in the intestine gradually slows gastric activity (Fig. 18–9).

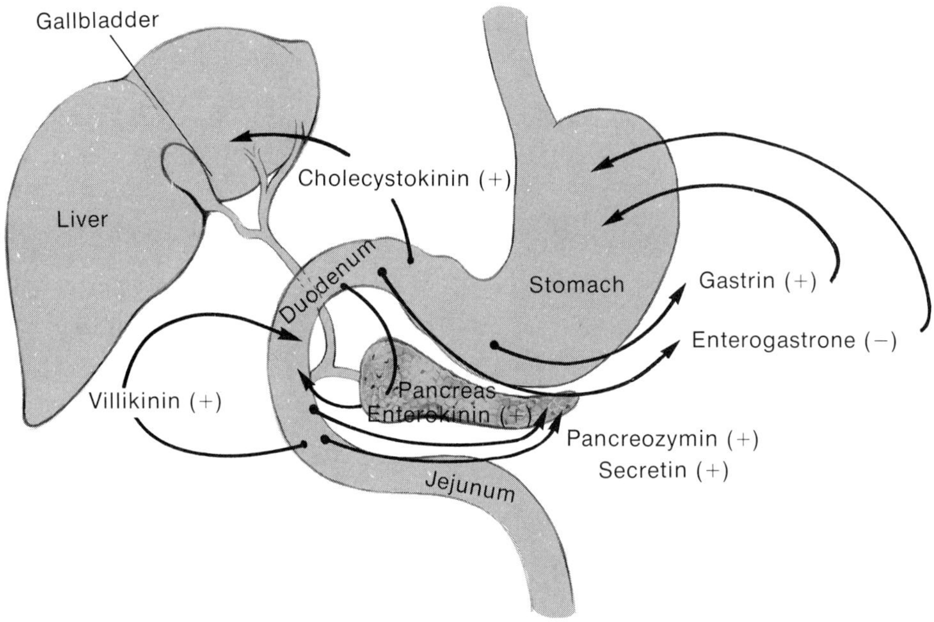

Figure 18–9 Hormonal control of digestion. The (+) sign indicates stimulation; the (−) sign indicates inhibition.

The presence of HCl and fat in the duodenum also triggers the release of a second intestinal hormone, cholecystokinin, which stimulates the gallbladder to release bile (Fig. 18–9). In this way, a bile sample is promptly added to the intestinal content in response to the fat which requires it for digestion.

Two very important hormones control the secretory action of the pancreas. The presence of acid chyme in the intestine stimulates the duodenal mucosa to secrete secretin; the products of carbohydrate digestion stimulate the duodenal mucosa to secrete pancreozymin. Both hormones stimulate the pancreas to secrete its digestive enzymes and alkaline salts (Fig. 18–9).

Two other digestive hormones are concerned with the action of the duodenum itself. Chyme in the upper intestine stimulates the release of enterokinin (duocrinin), which releases enzymes at several points in the small intestine. A second duodenal hormone, villikinin, stimulates the motility of the absorptive villi (Fig. 18–9).

NERVOUS CONTROL OF THE ALIMENTARY TRACT

Eating generates salivary reflexes of two types. An unconditioned (inborn) reflex is initiated by the presence of food in the mouth. Afferent messages from the taste receptors in the tongue activate the salivary control center in the medulla and the salivary glands secrete saliva.

A conditioned reflex involves the same pathway, but with the additional factors of sight, smell, or just the thought of food. Association pathways become established in the brain that evoke the salivary control center to activate the salivary glands until, eventually, the smell, sight, or sound of food preparation brings on salivation. This is called a conditioned (Pavlovian) reflex because it involves experience over the years.

The gastric glands, like the salivary glands, are regulated by nerve impulses. The most important nerve involving gastric (and also intestinal) glands is the vagus (parasympathetic) nerve. Impulses originate from the medulla and terminate in the

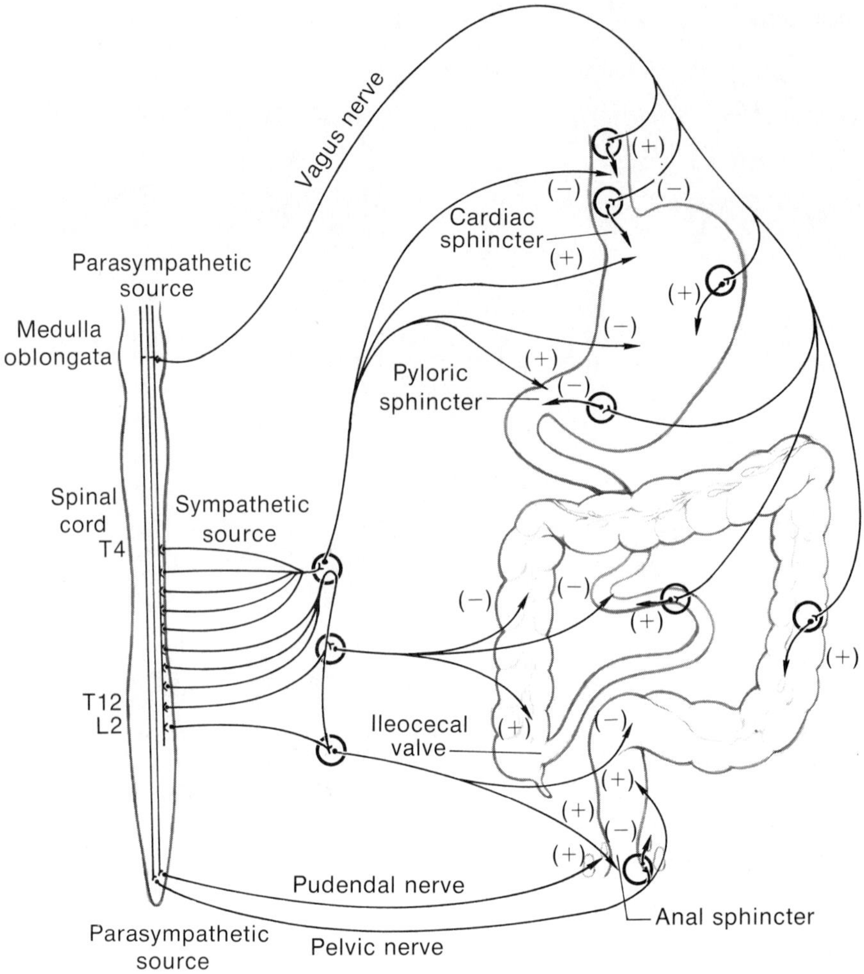

Figure 18–10 Nervous control of the alimentary tract. The parasympathetic nerves and smooth muscle sphincters of the stomach and intestine help to control the flow of digestive materials. The (+) sign indicates contraction, the (−) sign dilation, of the sphincters.

plexus beneath the gastrointestinal mucosa (Fig. 18–10). Impulses from this plexus stimulate the glands to secrete more juices. The control center in the medulla is stimulated by the presence of food in the mouth. Because of this reflex control, there are always gastric secretions awaiting the food that arrives in the stomach. If, on the other hand, secretion of hydrochloric acid and pepsin by the stomach is not followed by food, ulceration of the mucosal lining of the stomach may result.

The vagus nerve also controls movements of the stomach (Fig. 18–10). The enterogastric nervous reflex governs the peristaltic contractions passing through the stomach and into the duodenum. As chyme enters and distends the duodenum, the vagus nerve is inhibited, thereby temporarily inhibiting peristalsis and the movement of the bolus. This, in turn, depresses gastric action, slowing down the emptying of the stomach. As the duodenum empties, the vagus nerve is stimulated again, strengthening the peristaltic waves and increasing gastric motility.

The movements of the intestinal tract are carried out automatically by nerves of the autonomic nervous system. Sympathetic waves (Fig. 18–10), when active, release acetylcholine at the preganglionic fibers and norepinephrine at their endings in the muscle. This both inhibits, causing relaxation of the gut wall, and stimulates, causing contraction of the sphincters. Parasympathetic nerves release acetylcholine at their synapses and into the muscles, causing stimulation of the intestinal wall and inhibition of the sphincters.

Emptying (defecation) of the rectum is a complex reflex act. The passage of feces into the rectum distends its wall and initiates impulses that travel to the brain for a voluntary decision—to inhibit or permit reflex evacuation of the bowel. Outgoing nerve messages instruct the voluntary muscles of the abdominal wall and the diaphragm to raise intra-abdominal pressure, thus creating a "pushing" action. Simultaneously, messages travel to the internal and external anal sphincters by way of the pelvic and pudendal nerves, permitting the sphincters to relax and allow evacuation of the feces.

Like other "automatic" structures in the body, the intestinal wall is supplied by both sympathetic and parasympathetic nerves of the autonomic nervous system. These work together to achieve a balanced control of events.

THE LIVER

The liver, the largest of the glandular organs of the body, is situated in the upper right quadrant of the abdominal cavity. It lies to a large extent under the shelter of the lower ribs, its upper surface conforming to the lower surface of the diaphragm. The liver is divided into lobes—a large right and a smaller left lobe—by a fold of peritoneum called the falciform ligament (Fig. 18–11). Two smaller lobes, the quadrate and caudate lobes (unlabelled in Figure 18–11), are marked off on the posterior and inferior surfaces of the right lobe. The quadrate lobe acts to cushion and enfold the gallbladder.

Microscopically, the liver is composed

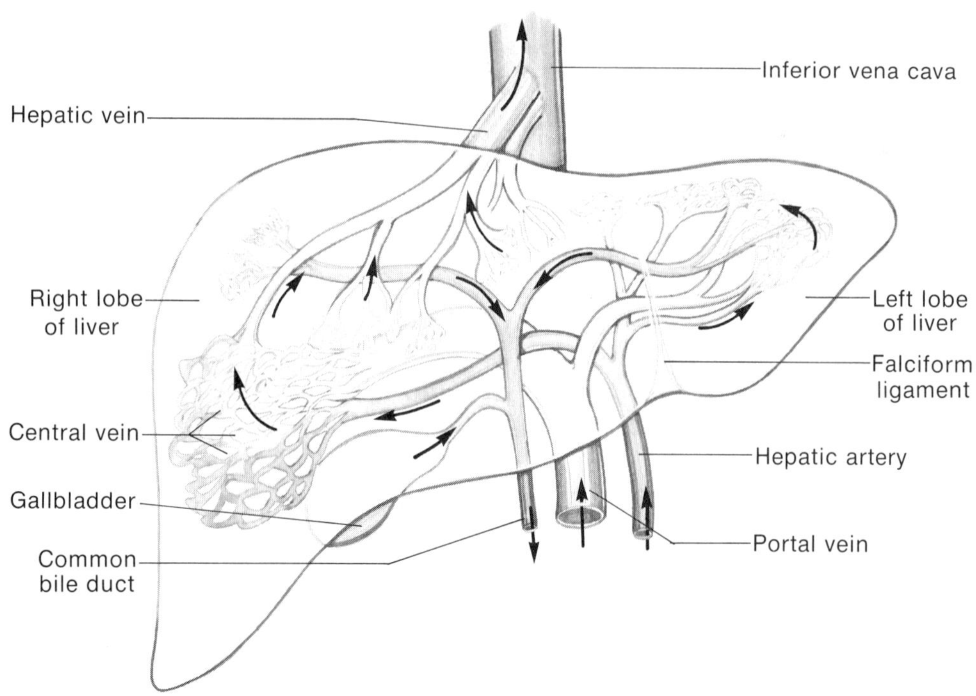

Figure 18–11 Vascular arrangement of the liver. Arrows indicate the direction of flow of blood in the arteries and veins, and of bile in the bile ducts.

of rows or cords of cubical cells that radiate from a central vein. On one side of each cord lies a blood vessel, considerably wider than a capillary, called a *sinusoid.* Blood from both the portal vein and the hepatic artery is brought into the sinusoid; it, in turn, empties into the central vein. On the other side of the single row of cells is a *bile capillary.* Bile pigment and other materials derived from phagocytosis are removed from the blood by the hepatic (Kupffer) cells, and deposited into the bile capillaries as bile. The blood received by the central veins of a number of neighboring liver units empties into larger veins, which unite to form the hepatic veins that in turn drain the blood into the inferior vena cava (Fig. 18–11).

Located in the sinusoids are the larger phagocytic *Kupffer cells.* These are macrophages capable of ingesting bacteria and other foreign materials from the blood. It is these cells that are commonly tested with a dye, bromsulphalein (BSP), to determine the extent of possible damage caused by liver disease. If the cells are damaged, they tend to absorb and retain larger than normal quantities of the dye; therefore, high retention of BSP can indicate liver cell damage.

The liver is a most remarkable organ and has numerous functions. The following are only the major ones:

1. The production of:
   a. Blood, plasma proteins
   b. Antibodies to fight off diseases
   c. Heparin, a substance that prevents blood clotting
   d. Bile pigments from red blood cells in the form of bilirubin and biliverdin
2. The storage of:
   a. Vitamins and minerals
   b. Glucose in the form of glycogen
3. The conversion and utilization of:
   a. Fats (desaturation)
   b. Carbohydrates (glycogenesis)
   c. Proteins (deamination)
4. The removal of:
   a. Old blood cells and toxins by phagocytosis (Kupffer cells)
   b. Waste products (urea and ammonia) from amino acids

Blood from the digestive tract has a special pathway all its own for return to the heart. Blood from the gastric and mesenteric arteries enters the capillaries of those organs, and digested food materials are absorbed into it. These capillaries then empty into the hepatic portal system described in Chapter 15. It will be recalled that a portal system is a blood vessel with capillaries at each end. One end of the hepatic portal vein collects from the capillaries of the stomach and intestine; the other end delivers blood to the capillaries of the liver, where food materials are utilized by the liver cells. Blood from the liver is then collected into the hepatic veins that drain into the inferior vena cava.

Inflammation of the liver is called *hepatitis.* Epidemic hepatitis is generally caused by a virus, and frequently occurs in densely populated areas, such as training camps or military institutions.

Cirrhosis of the liver is a chronic disease in which the active Kupffer cells are destroyed and replaced by scar tissue. Cirrhosis is commonly associated with repeated episodes of toxic injury or with continuous injury (as in alcoholism) to hepatic cells. Either may lead to progressive fibrous replacement of the damaged cells.

Jaundice, a yellowness of the skin and eyes, indicates the presence of bile pigment in the body tissues and a high concentration of bilirubin in the blood. Infectious hepatitis is usually accompanied by jaundice, owing to the increased content of bilirubin circulated by the infected liver.

MECHANISMS OF METABOLISM

A continual exchange of protoplasmic constituents with food materials occurs in the tissue fluids. The energy involved in maintaining this effort is called the *basal metabolism.* In children, the high rate of synthesis during growth requires considerably more energy than that needed by a fully developed adult. Thus, under comparable conditions, the basal metabolic rate in children is much higher.

The energy stored in food is released when it is burned (oxidized) by the cells. Sugar, for example, undergoes an oxidation by the cells to produce carbon dioxide, water, and energy:

$$C_6H_{12}O_6 + 6\ O_2 \rightarrow 6\ H_2O + 6\ CO_2 + \text{energy}$$

As a result of oxidation, the atoms become rearranged in a form less adapted to further change. This process, however, requires many steps regulated by enzymes distributed in orderly fashion in the cells. The chemical structures produced during the sequence of oxidation fulfill a variety of functions. They may replace worn-out parts in the existing protoplasm, provide charged particles in the cell membrane, or synthesize secretory products.

Calorimetry

If a person's heat production is to be measured directly, he must be placed in an insulated box, a *calorimeter,* which is surrounded by water to absorb the body's heat output. The increase in the temperature of the water provides a measure of the number of calories of heat produced. A *calorie* is the amount of heat required to raise the temperature of one gram of water one degree celsius. A kilocalorie (Kcal) is 1000 calories. Heat produced by the body is normally measured in kilocalories.

When sugar ($C_6H_{12}O_6$) is burned, it produces about 680 calories of heat for each molecule burned in the calorimeter; the same overall reaction occurs in the body. If you were to compare the energy potentials of different foodstuffs, you would in each case have to divide the number of kilocalories produced by the weight of the substance. For example, the gram-molecular weight of glucose is 180 gm. If you divide 680 calories by 180, the heat per gm equals approximately 4 calories per gm:

$$\text{Heat/gm} = 680\ \text{cal}/180 + \text{approx. } 4\ \text{cal/gm}$$

Fats have a heat per gm value of 9 calories per gm. Proteins have a heat per gm value of 4 calories per gm.

If the heat production of an individual is known, from either direct or indirect measurement, the overall rate of chemical action—the person's basal metabolic rate (BMR)—may be compared directly with the rate of another person with the same body dimensions. A set of values, or standards, can be derived and comparisons made to determine a patient's ability to utilize the energy in the food he has consumed. In persons who are gaining weight, oxidizable material is stored, not used, and the energy in food eaten exceeds the metabolic rate. Conversely, in persons losing weight, only part of the heat produced is obtained from the food eaten, and there is a net decrease in stored oxidizable material.

As mentioned in an earlier chapter, thyroxine is the main oxidative agent that regulates metabolic rate. Clinical tests utilizing isotopes (T_3 and T_4) and protein-bound iodine (PBI) studies have proved beneficial in diagnosing potential metabolic problems.

The term "metabolism" encompasses all changes that occur in the body involving foodstuffs. Thus, metabolism includes not only the breakdown of food (catabolism) but also its synthesis (anabolism). Energy derived from a catabolic reaction is said to be *exothermic*—that is, coming out of the reaction; an anabolic reaction in which energy is incorporated or stored is *endothermic.*

Carbohydrate Metabolism

The pancreas plays an important role in the regulation of carbohydrates. The endosecretion of insulin and glucagon from the islets of Langerhans provides a definite cellular control.

Following their absorption from the intestine, nutrient sugars are transported to the liver via the hepatic portal system. The liver cells can either convert most of the sugars into glycogen (glycogenesis) for storage, or, if levels of available glucose in the blood are lower than normal, the liver can break down the available glycogen back into glucose (glycogenolysis) to maintain a normal blood sugar level.

The greatest and most important body use of glucose is in oxidation. When glucose is oxidized by tissue cells, carbon dioxide and water are formed and energy is released.

Most tissues, apart from the liver, utilize glucose for their supply of energy. As soon as the glucose permeates the cell membrane, it is trapped in the cytoplasm. Phosphate bonds are attached to the glucose, forming a phosphate ester that cannot go back out through the cell membrane.

Insulin promotes the oxidation of glucose by the tissue cells. Most combustion and extraction of energy occur in the mitochondria of the cell, and it is here that the released energy is stored as ATP (Fig. 18–12). The mitochondrial membrane is permeable only to small molecules, however; thus a problem exists for nutrients larger in chemical structure than a three-carbon molecule. To facilitate entry of the larger-sized glucose (6C), lysosomes in the cytoplasm release phosphorylating enzymes that reduce the size of the 6-carbon glucose molecule to two 3-carbon molecules, with a net gain of two ATP molecules or 20,000 Kcal of energy. The products of this anaerobic process of glycolysis ("sugar breakdown") are two molecules of a 3-carbon pyruvic acid which, after an intermediate conversion to acetic acid, may enter the mitochondria, where each combines with a 4-carbon acid to form a 6-carbon citric acid.

The *Krebs citric acid cycle* is a series of reactions involving the breakdown of 6C citric acid to a 4C acid, with the release of carbon dioxide and energy in the form of hydrogen molecules (Fig. 18–12). This is the principal energy cycle in the body. The quantity of energy released by the cycle is enough to operate almost all of the other body cycles.

The next sequence of reactions, called the *electron transfer chain,* involves the transfer of electrons from the hydrogen bonds through a series of electron acceptors, each acceptor being at a progressively lower energy level. The enzymes are called cytochrome oxidases, or respirases, and effect the conversion of 12 hydrogen bonds into 36 molecules of ATP (Fig. 18–12). The 36 ATPs from the Krebs cycle, added to the two ATPs from glycolysis, produce a total of 38 ATPs formed by the complete oxidation of one molecule of glucose. This amounts to a considerable quantity of stored energy, because one molecule of ATP stores about 20,000 calories of free energy.

Glycogen is also formed and stored in the muscles. Such glycogen is used in muscles and not ordinarily returned to the blood.

Fat Metabolism

As we have seen, fat enters the system in the form of fatty acids, glycerol, phospholipids, and cholesterol. A small amount of fat is synthesized in the liver, but most is synthesized in adipose tissue. Because one gram of fat yields about 9 calories per gm as compared with 4 calories per gm of both protein and carbohydrate, the fat deposits in the body provide a concentrated source of energy. These fat concentrations furnish about 40 per cent of the energy used by a normal individual. The absorbed fats are acted upon by spe-

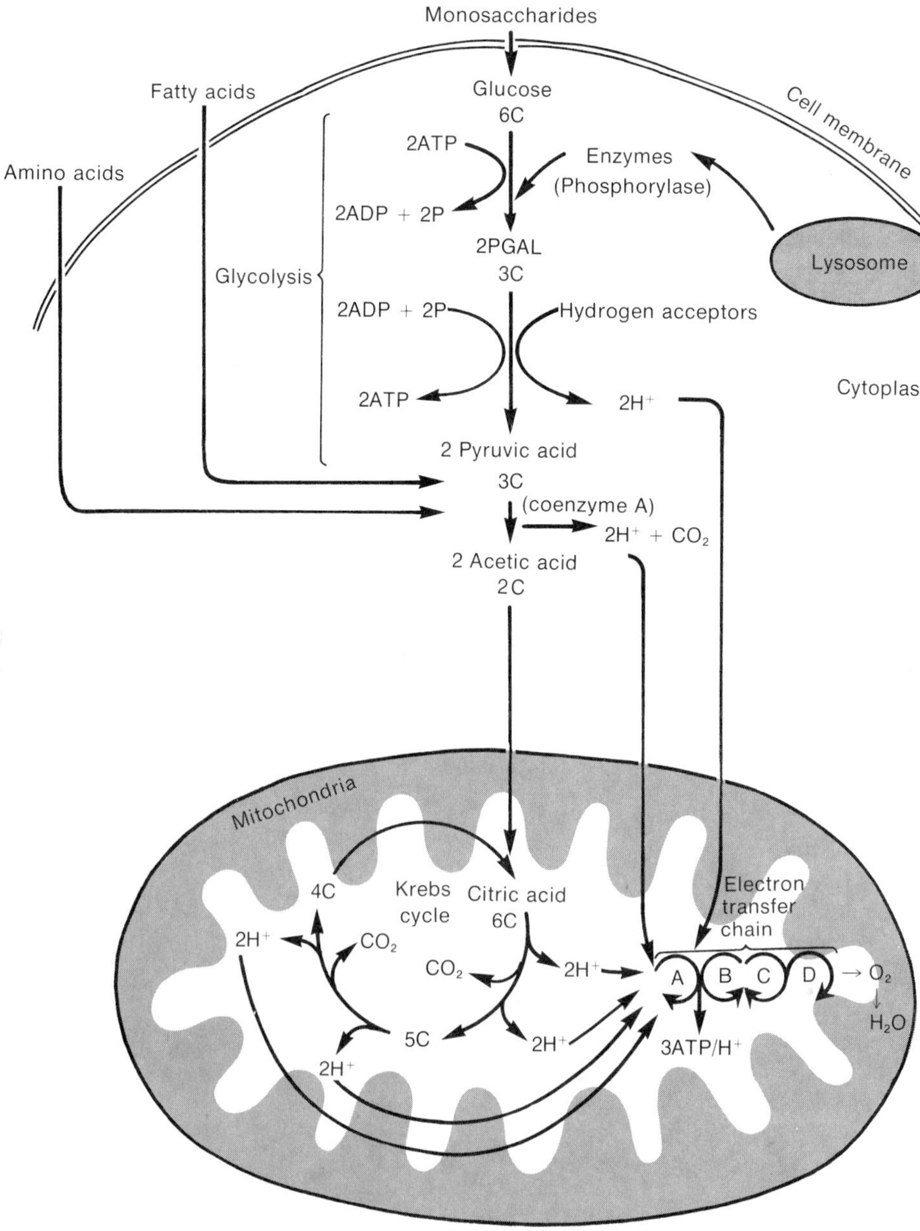

Figure 18–12 Metabolism of carbohydrates, proteins, and fats.

cial cells in the liver (lipolysis), and the resultant products are channeled into the Krebs citric acid cycle (Fig. 18–12). Synthesis of fats (lipogenesis) results from an excess of fatty acids and glycerol, which combine to form triglyceride. In addition to serving as a source of energy, stored fats act as cushions for some of the internal organs. Phospholipids are used in the formation of plasma membranes.

Fats are completely oxidized to carbon dioxide and water with a release of energy. This process occurs mainly in the liver by a mechanism that involves breakdown of long-chain fatty acids. The end-products of fatty acid oxidation are called *ketone* or *acetone* bodies. Under normal conditions, the quantity of ketones produced is excreted in the urine. When an excessive amount of fat is catabolized, ketone bodies are formed in massive quantities, and a condition known as *ketosis* results. This can occur in a person undergoing a complete fast, where there are insufficient carbohydrates and excess fats in the diet, and in severe diabetes. An excess accumulation of ketone bodies in the system requires neutralization with strong alkalis in order to preserve the pH of the blood and tissue fluids. An increased production of ketone bodies, as in diabetes, will result in a depletion of the alkali reserve, producing acidosis.

Cholesterol, as noted earlier, is a sterol found in fats derived from such animal foods as egg yolk and meats. Increased cholesterol levels in the blood (*hypercholesterolemia*) produce degenerative changes in the arterial walls (atherosclerosis; see Chap. 15).

Protein Metabolism

Amino acids absorbed from the intestinal tract are used by the metabolic mill for the synthesis of proteins, are converted to fatty acids and glycogen, or are oxidized as an energy source. All or a portion of the amino acid may be converted to glycogen or to ketone bodies. Those amino acids that yield pyruvic acid are channeled into the Krebs cycle (Fig. 18–12) as an energy source.

An important role played by amino acids is the donation of their amine group (NH_2), a process called *deamination,* and transfer of this group to other substances, a process called *transamination.* Deamination and transamination are important in that they are the first step in the metabolic breakdown of these compounds and in the synthesis of amino acids from other compounds.

The primary use of amino acids is in the formation and repair of cells. During growth and pregnancy, additional proteins are formed from the amino acids taken from the blood.

The major end-product of protein metabolism in man is *urea.* Other waste materials that are closely correlated are ammonia, creatinine, and uric acid.

Basal Metabolic Rate

The basal metabolic rate is the amount of energy expended by the body to maintain its vital functions when it is at rest. Although the needs of each individual vary, the average person requires about 3000 calories daily in his diet. Of this total, 1500 to 1800 calories are needed to maintain vital functions. The average person obtains about two-thirds of his energy from the oxidation of carbohydrates. From 45 to 50 per cent of the total calories of the diet should be furnished by carbohydrates, 35 to 40 per cent by fat, and 12 to 15 per cent by protein.

A calorie diet breakdown for a young man of average size who expends about 3000 calories per day could read:

1. Basal metabolism (70 calories per hour for 16 waking hours)	1120
2. Metabolism while alseep (64 calories per hour per 8 hours)	512
3. Allowance for week	1500
	3132

OUTLINE

- I. Cellular Requirements
  - A. Water
  - B. Energy
  - C. Protein
  - D. Enzymes
  - E. Minerals
- II. Foodstuffs
  - A. Carbohydrates
    1. Monosaccharides – glucose
    2. Disaccharides – sucrose, lactose
    3. Polysaccharides – glycogen

- B. Fats
  - 1. Triglycerides
    - a. Fatty acid molecules
    - b. Glycerol unit
  - 2. Compound lipids
  - 3. Sterols
- C. Proteins
  - 1. Amino acids
    - a. Manufactured in body
    - b. Essential—taken in with food
  - 2. Polypeptides
  - 3. Simple vs. conjugated proteins
- D. Vitamins
  - 1. Fat-soluble
    - a. A, D, E, and K
  - 2. Water-soluble
    - a. B complex and C
- E. Minerals
  - 1. Salts—chlorides, phosphates, and carbonates
  - 2. Calcium, phosphorus, and iron
  - 3. Trace elements—copper, manganese, and cobalt

III. Digestive Organs and Processes
- A. Mouth
  - 1. Salivary amylase—carbohydrates
    - a. pH 6.8–7.0
  - 2. Mastication—teeth
    - a. Incisors
    - b. Canines
    - c. Molars
- B. Esophagus
  - 1. Muscular tube
  - 2. Peristalsis
  - 3. Swallowing (deglutition)
  - 4. Esophageal varices and carcinoma
- C. Abdomen
  - 1. Peritoneum
  - 2. Mesentery
  - 3. Omentum
- D. Stomach
  - 1. Muscular layers
  - 2. Mucosa—mucous cells, chief cells, and parietal cells
  - 3. Pepsin, rennin, and HCl—proteins
  - 4. Gastric juice—pH 1–4
  - 5. Chyme
  - 6. Enzyme activity—specificity
  - 7. Peptic ulcer
  - 8. Pyloric stenosis
- E. Small intestine
  - 1. Duodenum, jejunum, and ileum
  - 2. Bile salts—fat emulsification
    - a. Gallbladder
      - (1) Bile storage
      - (2) Gallstones
      - (3) Cholecystitis
  - 3. Duodenal enzymes—assist pancreatic enzymes
    - a. Enterokinase—proteins
    - b. Maltase—carbohydrates
    - c. Sucrase—carbohydrates
    - d. Lactase—carbohydrates
    - e. Lipase—fats
    - f. Nuclease—nucleoproteins
  - 4. Villi
    - a. Mucosal projections
    - b. Contain capillaries and lacteals
  - 5. Duodenal ulcers and intestinal obstruction
- F. Pancreas
  - 1. Pancreatic enzymes
    - a. Pancreatic amylase—carbohydrates
    - b. Pancreatic lipase—fats
    - c. Trypsin—proteins
    - d. Nuclease—nucleoproteins
  - 2. Pancreatitis
- G. Large intestine
  - 1. Ascending, transverse, descending, sigmoid, and rectal colon
  - 2. Cecum and appendix
  - 3. Absorption of water
  - 4. Defecation of solid waste
  - 5. Diarrhea and constipation
  - 6. Hemorrhoids
  - 7. Colitis
  - 8. Tumors

IV. Hormonal Control of Digestion
- A. Stomach
  - 1. Gastrin
  - 2. Enterogastrone
- B. Pancreas

1. Secretin } Secreted in
2. Pancreozymin } duodenum

C. Gallbladder
 1. Cholecystokinin (secreted in duodenum)

D. Duodenum
 1. Enterokinin (duocrinin)
 2. Villikinin

V. Nervous Control of Alimentary Tract
 A. Salivary reflexes—activated by chewing
  1. Unconditioned reflex—medulla oblongata
  2. Conditioned reflex—activated by smell, sight, or thought of food
 B. Gastric glands—vagus nerve
 C. Peristaltic contractions—vagus nerve
 D. Intestinal movements
  1. Sympathetic nerves
  2. Defecation—complex reflex

VI. The Liver
 A. Largest gland in the body
  1. Sinusoids
  2. Hepatic portal system
  3. Phagocytic Kupffer cells
 B. Functions
  1. Produces plasma proteins, antibodies, heparin, and bile pigments
  2. Stores vitamins, minerals, and glycogen
  3. Converts fats, carbohydrates, and proteins to other metabolic forms
  4. Removes old blood cells and toxins by phagocytosis, and waste products from amino acids
 C. Hepatitis and cirrhosis

VII. Mechanisms of Metabolism
 A. Calorimetry
  1. Carbohydrate and protein oxidation give off 4 calories of heat per gm
  2. Fat oxidation releases 9 calories per gm
  3. Basal metabolic rate (BMR)
   a. T_3, T_4, and PBI
  4. Metabolism
   a. Catabolism—breakdown
   b. Anabolism—build-up
 B. Carbohydrate metabolism
  1. Insulin from pancreas
  2. End-products absorbed by villi and transported to liver
   a. Glycogenesis
   b. Glycogenolysis
  3. Glycolysis—anaerobic
  4. Krebs citric acid cycle—aerobic
  5. Electron transfer chain—stored energy
 C. Fat metabolism
  1. Fatty acids absorbed by villi
  2. Synthesized to form adipose tissue
  3. Found in form of cholesterol and triglycerides
  4. Lipolysis and lipogenesis supply 40% of body energy
  5. Ketone bodies—end-products of fatty acid oxidation
 D. Protein metabolism
  1. Amino acids absorbed by villi
  2. Used as building blocks of proteins
  3. Deamination—transamination
  4. Urea—major end-product of protein metabolism
 E. Basal metabolic rate
  1. Amount of energy used by body to maintain vital functions
  2. 3000 calorie normal adult diet
   a. 1500 to 1800 calories needed for vital functions

STUDY QUESTIONS AND PROBLEMS

1. Trace the path of an indigestible object from the mouth through all parts of the alimentary canal to the outside, describing what happens on the way.
2. Distinguish between deciduous and permanent teeth. Describe the relation between their shape and their function.

3. How does peristalsis begin? How is it maintained throughout the intestinal tract?
4. pH has an important role in digestion. Explain.
5. Where is energy located in a chemical compound? What are exothermic and endothermic reactions?
6. What is meant by enzyme specificity? Does pH have anything to do with it?
7. Differentiate between a peptic and a duodenal ulcer. What problems does a perforated ulcer present?
8. Trace the three major foodstuffs consumed in a normal balanced meal through their digestion, and correlate the specific enzymes involved.
9. What constitutes a balanced diet?
10. What are the end-products of digestion? Where are they absorbed and by which vessel?
11. How is the liver involved in metabolism?
12. What is the relation between the hepatic portal system and digestion?
13. What part does the pancreas play in digestion? What other role does the pancreas have in maintaining homeostasis?
14. What does insulin do in carbohydrate metabolism?
15. What is the source of bile pigments?
16. What is infectious hepatitis? What other conditions may produce jaundice?
17. What does the large intestine do? Explain the difference between diarrhea and constipation.
18. A person has a colostomy. Why must he be concerned with his electrolyte and water balance?
19. Why does hydrolysis play an important role in metabolism?
20. What is a calorie? Are the PBI and T_3 significant tests for a BMR?
21. Discuss ketogenesis and lipid metabolism. Why are cholesterol and triglyceride levels so closely watched?
22. List the major minerals and vitamins and indicate their importance.

Chapter 19
Excretion

The active life of every cell is necessarily accompanied by the production of waste materials, and it is imperative that these be removed quickly if the activities of the tissues are not to be impaired. In the chemistry of the body, some reactions produce starting materials for others, but some end-products may accumulate and disturb the sequence of essential reactions if they are not removed. These materials are the excretory substances, and they are produced in different amounts by all cells.

The process of excretion consists of the separation and removal of substances harmful to the body. Excretion is carried out by the skin, the lungs, the liver, the large intestine, and the kidneys (Fig. 19–1).

As we have seen, the lungs remove carbon dioxide from the blood; and other waste products of metabolism, such as ketone bodies (acetone), are removed by the respiratory system. The salivary glands excrete various salts, particularly sodium chloride; the sweat glands produce water, salts, amino acids, vitamins, and other materials that are carried from the body and deposited on the outer surface of the skin. Excretory functions of the liver include phagocytic action and the removal of bile pigments. The large intestine, following its principal function of water absorption, removes solid waste materials as feces.

THE URINARY SYSTEM

The two principal functions of the urinary system are the excretion of wastes and the regulation of the composition of blood.

Blood composition must not be allowed to vary beyond tolerable limits or the conditions in the tissues necessary for cellular life will be lost. Regulation of the composition of the blood involves not only the removal of harmful waste products, but also the conservation of water and metabolites in the body.

The urinary system (Fig. 19–2) consists of two large kidneys that secrete urine; the ducts leading from them, the ureters; a large urinary reservoir, the bladder; and the tube from it to the surface of the body, the urethra.

The Kidneys

The kidneys are deeply embedded in the abdominal cavity in the lumbar region (Fig. 19–1). Both are protected at the back by the last two pairs of ribs. They are surrounded by a layer of fat, the *perirenal fat*; in front they are covered by peritoneum, the *capsule*.

Each kidney is a dark red, bean-shaped organ about four and one-half inches long, two and one-half inches broad, and a little over one inch thick. An adult kidney weighs about five ounces. At the middle of the kidney's inner concave border, there is a deep depression called the *hilus* (Fig. 19–3). It is here that the renal artery enters and the renal vein leaves the kidney to take blood to the inferior vena cava. The ureter also leaves at

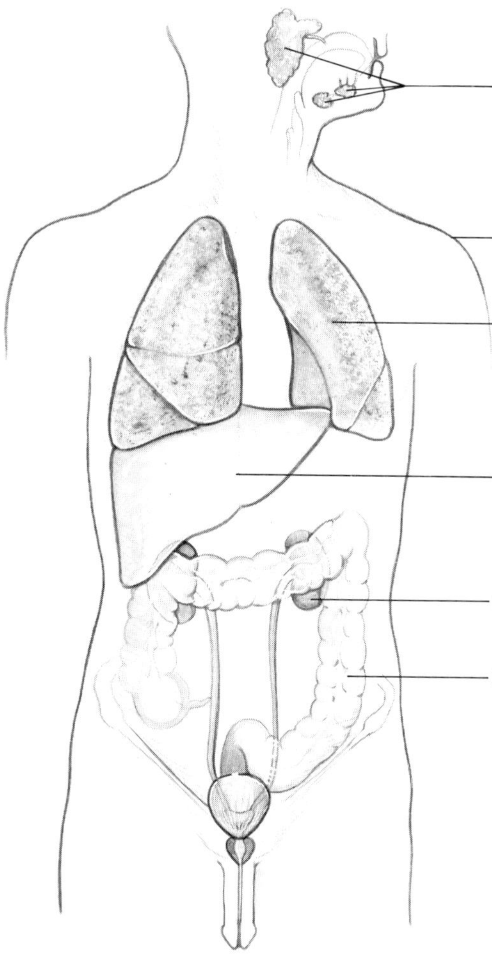

Figure 19–1 Component organs of the excretory system.

the hilus to connect the kidney with the bladder. Nerves, lymphatic vessels, and blood vessels control kidney function. Under resting conditions, about one-fourth of the blood that leaves the heart passes through the kidneys for regulation. The urine formed in the kidney is passed down the ureter to the bladder by peristaltic contractions of the muscular ureter.

In vertical section, the outer *cortex* of the kidney is darker in color than the inner *medulla.* The inner surface of the medulla is folded into projections, the *pyramids,* which empty their contents into the collecting space called the *renal pelvis* (Fig. 19–3). Numerous collecting tubules bring the urine from its sites of formation in the cortex to the pyramids, where it is discharged into the pelvis. The ureter actually is no more than an extension of the funnel-shaped pelvis.

The renal tubules or nephrons are the functional units of the kidney, and in man each kidney has more than one million of these. Each nephron tubule is about 12 mm long; its thin walls are made up of epithelial cells (Fig. 19–4). It is open at one end, where it connects with a collecting tubule; at the closed end, the wall forms a cup-shaped concavity known as *Bowman's capsule.* The capsule drains into the proximal convoluted tubule, which is followed by a straighter portion, *Henle's loop.* Henle's loop enters the medulla of

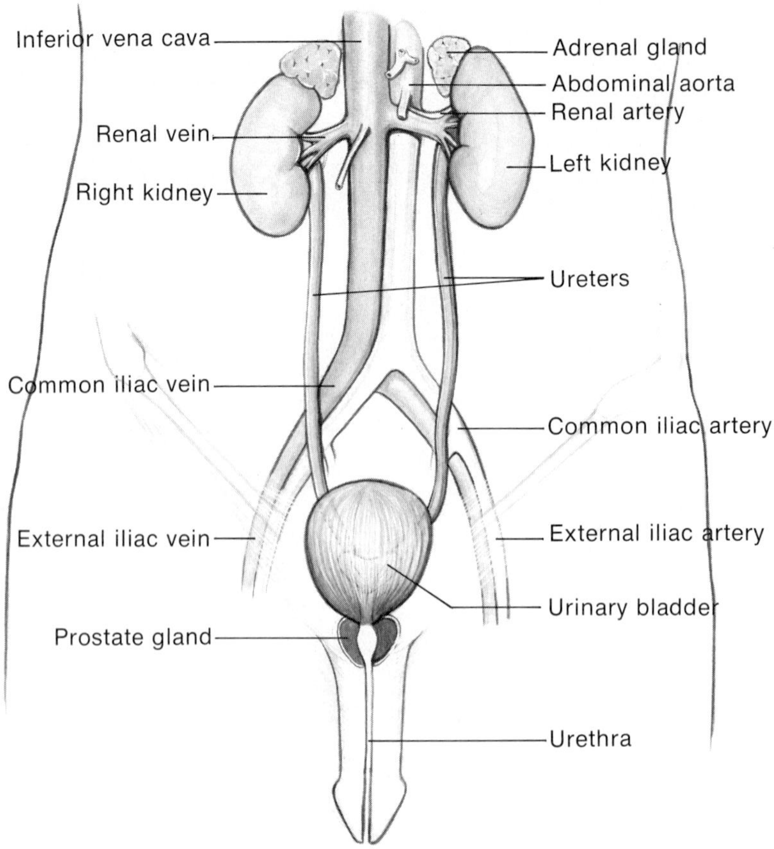

Figure 19–2 The urinary system with blood vessels.

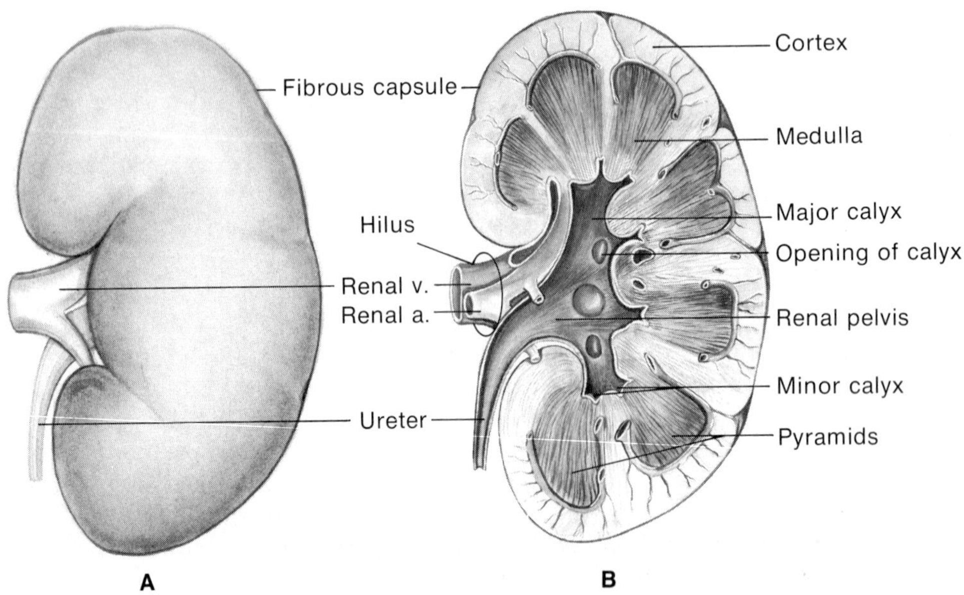

Figure 19–3 External *(A)* and internal *(B)* anatomy of the left kidney.

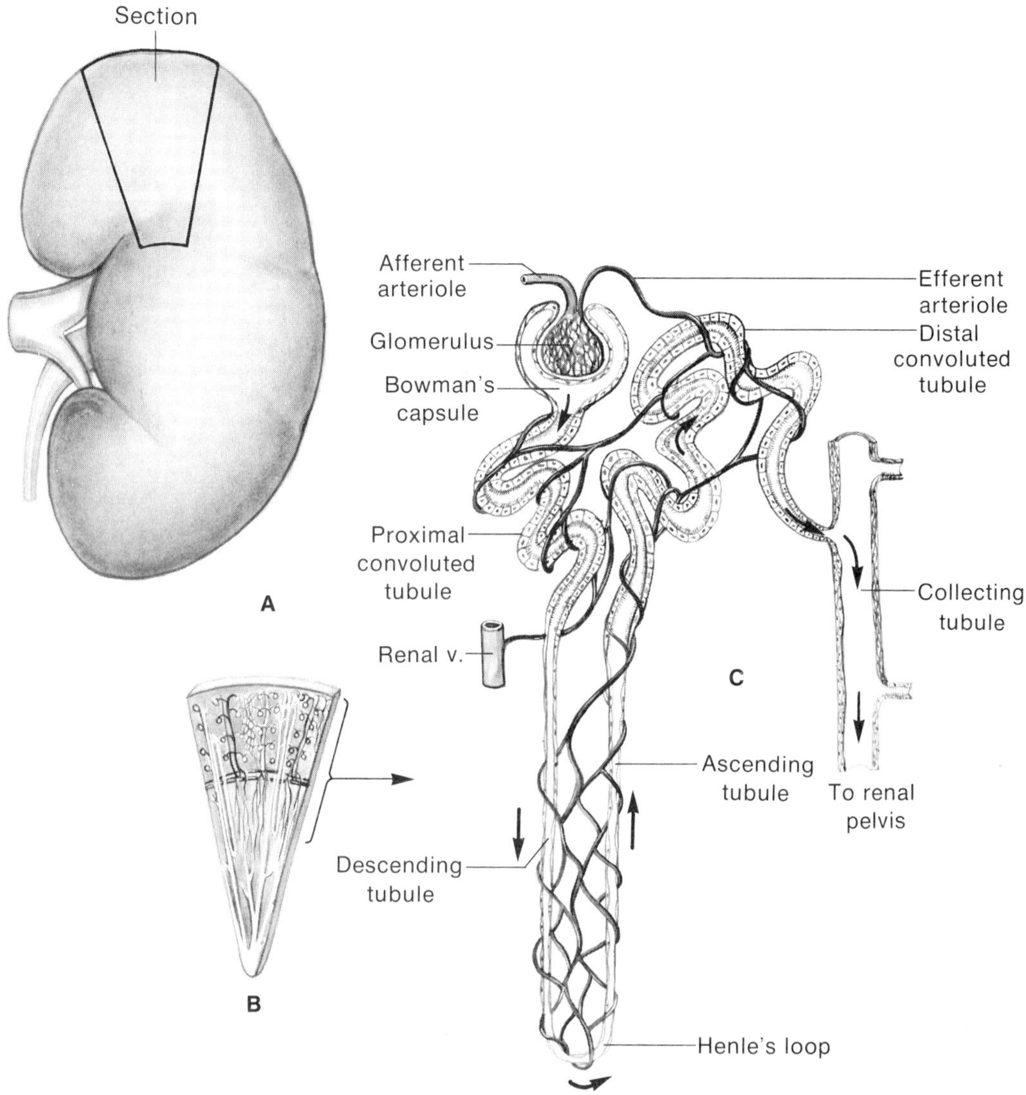

Figure 19–4 The kidney. *A*, External diagram of the left kidney. *B* and *C*, A section of the kidney has been enlarged to show the details of a nephron unit. Arrows indicate the direction of flow of urine through the nephron unit.

the kidney and continues as the distal convoluted tubule, which joins other similar tubules to form the collecting tubes.

Blood supply to the nephrons begins at the renal artery, which comes directly from the abdominal aorta. The renal artery subdivides within the kidney, and a small vessel (afferent arteriole) enters Bowman's capsule, where it forms a tuft of capillaries, the *glomerulus,* which entirely fills the concavity of the capsule (Fig. 19–4). The blood leaves the glomerulus via the efferent arteriole, which again subdivides into a network of capillaries that surround the proximal and distal tubules and eventually unite as veins which become the renal vein. The renal vein returns the now "cleansed" blood to the general circulation.

By movement of substances between the nephron and the capillaries of the tubules, the composition of the blood filtrate moving along in the tubules is changed, and it becomes a more concentrated solution. From the nephrons, the fluid moves to collecting tubules and into the ureter leading to the urinary bladder, where the urine is stored.

Urine Formation

It is customary to discuss the function of the nephron in terms of the formation of urine.

1. Filtration. The first process in the formation of urine is *glomerular filtration.* The walls of the glomerulus are permeable to water and to other small molecules, but they are not permeable to formed blood elements (cells) and proteins. Since the blood entering the glomerulus is under pressure, some of the nonprotein constitutents pass through into Bowman's capsule. The glomerular filtrate has a composition similar to plasma, since blood cells are not filtered. It contains the waste products which it is the kidney's job to remove from the plasma.

Plasma, which has been selectively modified, emerges from the kidney as urine. Approximately 600 ml of plasma flows through each kidney per minute, but only about 125 ml of glomerular filtrate is formed, owing to a glomerular pressure estimated to be 69 mm Hg. Glomerular filtration provides a large volume of essentially protein-free filtrate of plasma, which is largely resorbed in its passage through the nephron. The forces involved in glomerular filtration have been estimated in the following equation:

$$\underset{25\text{ mm Hg}}{P_f} = \underset{60\text{ mm Hg}}{P_g} - (\underset{30\text{ mm Hg}}{P_{op}} + \underset{5\text{ mm Hg}}{P_c})$$

where P_f is the filtration pressure, P_g is the glomerular hydrostatic pressure, P_{op} is the colloid osmotic pressure, and P_c is the capsular hydrostatic pressure (Fig. 19–5).

The most obvious factor likely to affect the glomerular filtration rate is the number of functioning glomeruli, which may be reduced by disease. The blood pressure is the driving force for filtration; therefore, with high blood pressure, the fraction of plasma filtered at the glomerulus is increased. The kidneys maintain a constant filtration rate over a wide range of systemic blood pressures. By appropriate vasoconstriction and vasodilation of the smooth muscles in the afferent and efferent arterioles, blood flow through the kidney may be adjusted to maintain a constant filtration rate, despite a fluctuating blood pressure.

The glomerular filtration rate in an average-sized man is, as mentioned earlier, 125 ml per minute. This is 7.5 quarts per hour, or 180 quarts per day, whereas the normal urine volume is about one quart per day. Thus, 99 per cent or more of the filtrate is resorbed. At this rate of 125 ml per minute, the kidneys filter in one day an amount of fluid equal to four times the total body water, and 69 times the plasma volume.

Some of the factors that affect the glomerular filtration rate are:

a. Changes in glomerular hydrostatic pressure due to changes in blood pressure or constriction of the afferent or efferent arterioles.

b. Changes in Bowman's capsule hydrostatic pressure due to ureteral obstruction or edema of the renal capsule.

c. Changes in colloid osmotic pressure of plasma proteins due to dehydration or hypoproteinemia.

d. Increased permeability of glomerular filtration due to disease.

e. A decrease in the total area of the glomerular capillary bed due to destruction of glomeruli from disease or from a partial nephrectomy.

2. Resorption. An equally important physiologic process carried on by the nephron unit is resorption. The filtrate does contain such metabolic wastes as urea and creatinine, but most of the substances found in it (water, salts, glucose, amino acids, etc.) are extremely important to the body. If the water, for example,

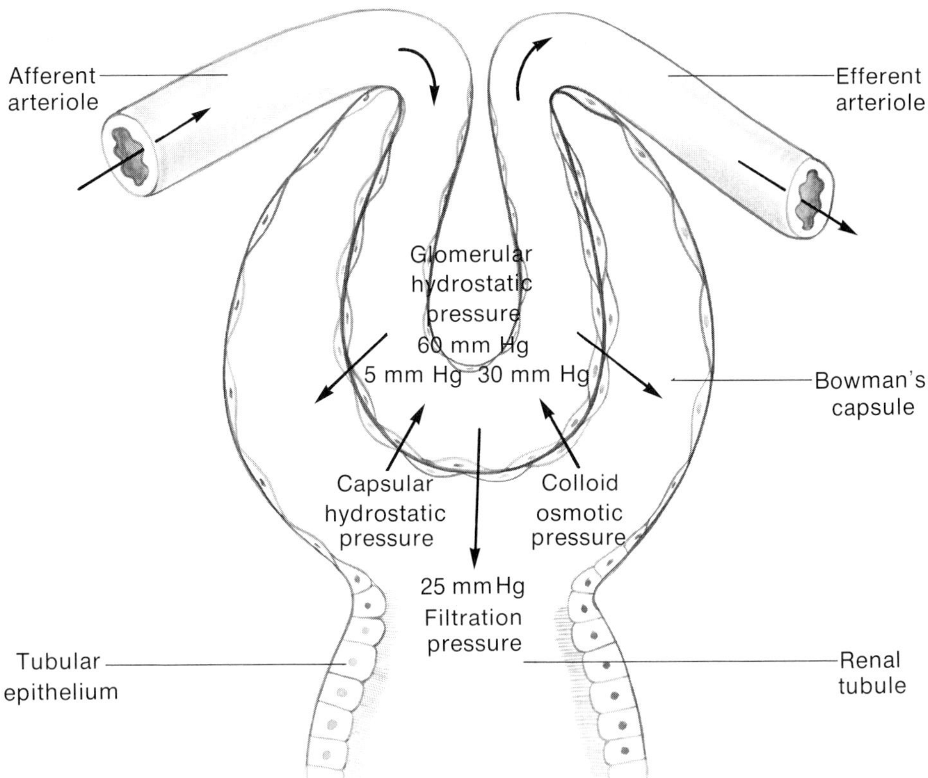

Figure 19–5 Urine formation: filtration. The glomerular membrane acts as a simple filter that allows passage of dissolved materials only because of a greater pressure on one side of the membrane than on the other.

were not resorbed, the 180 quarts that normally are resorbed would have to be replaced by drinking. The movement of water and ions—especially sodium—across different parts of the nephron tubules makes it possible for man to form urine that is more concentrated than his blood and tissue fluids, thus conserving water.

Blood is filtered in Bowman's capsule. Blood proteins and blood cells do not pass into the kidney tubule, but everything else does. The filtrate entering the proximal convoluted tubule is isotonic (same osmotic pressure) with the blood. As the filtrate grows down toward Henle's loop, it becomes hypertonic because Na^+ diffuses passively into the filtrate from the surrounding tissue fluids (Fig. 19–6) where the Na^+ concentration is high. When the filtrate passes through the ascending loop, Na^+ is actively pumped from the filtrate back into the surrounding tissue fluids. The filtrate becomes hypotonic and the surrounding fluids develop a high Na^+ concentration. The Na^+ pumped back into the surrounding fluids can then diffuse into the filtrate flowing through the descending loop. In this way, Na^+ is constantly recycled, and the loop of Henle is continuously surrounded by a salty fluid. As hypotonic filtrate flows out through the collecting tubule, water diffuses from the filtrate into the surrounding fluids, which have a high Na^+ concentration. The final filtrate, urine, is hypertonic.

The net effect of passage of the filtrate through the loop of Henle is the removal from it of much sodium but very little water. Therefore, the urine in the distal tubule becomes less concentrated than the initial filtrate. But, now, the urine flows into the collecting tubule, which runs through the medulla from the cortex, or

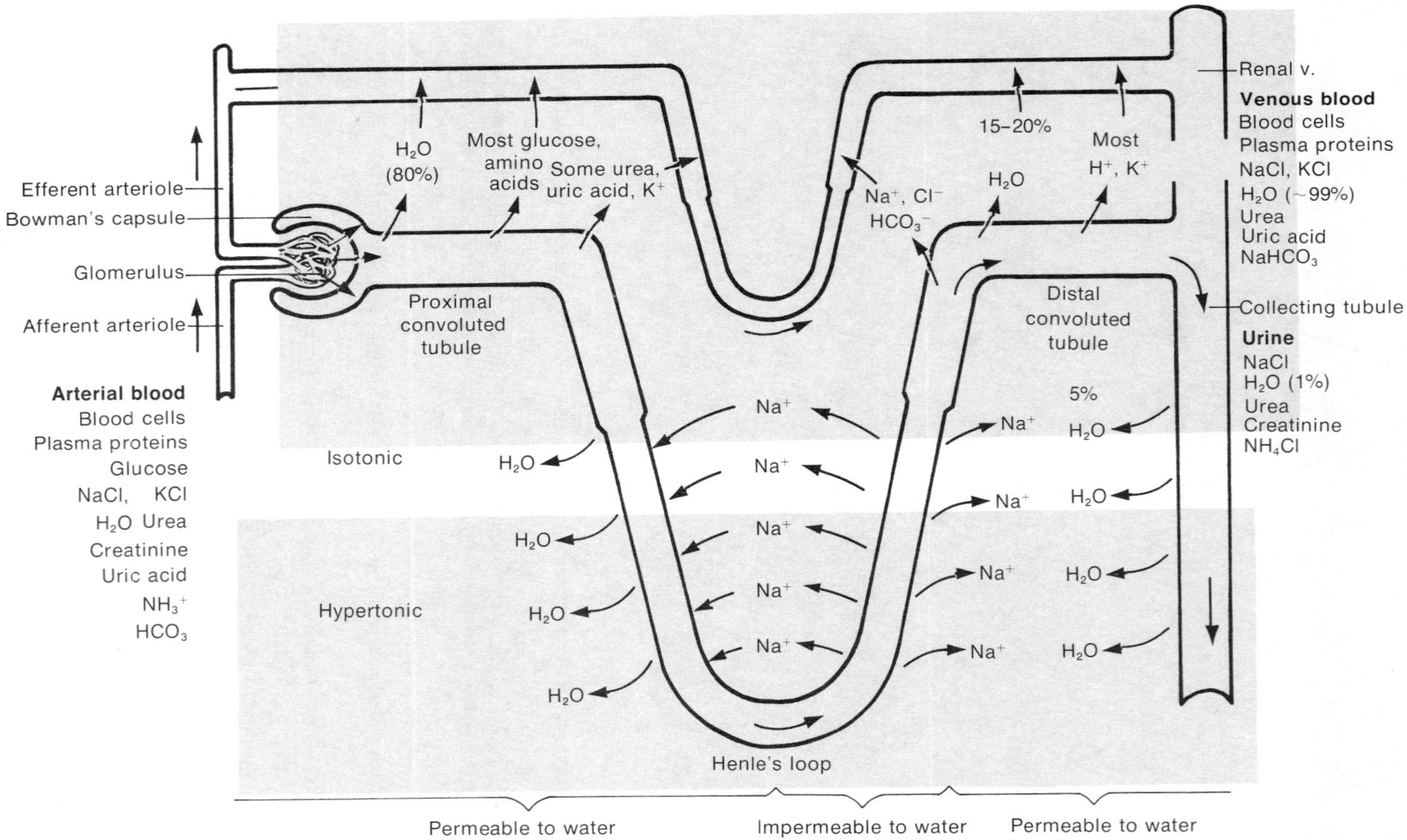

Figure 19–6 Urine formation: resorption. As the filtrate passes along the tubule, its concentrated and essential elements are reabsorbed by the capillary network. The entire mechanism depends on the recycling of Na^+, which causes the fluids in the tissues surrounding the renal tubules to be very salty and permeable to water.

through a region of increasing sodium concentration in the tissue fluid. Since the walls of the tubule are permeable to water, water moves by passive osmosis from the diluted urine in the tubule to the tissue fluid. This "final" urine is much more concentrated than the initial glomerular filtrate, and is hyperosmotic to the blood plasma.

Glucose is typical of substances removed from the urine by active transport. Almost all of it is filtered, but essentially all of it is resorbed in the first part of the proximal tubule (Fig. 19–6). The amount resorbed is proportional to the amount filtered. The renal threshold for glucose is the plasma level at which the glucose first appears in the urine—this level is about 100 mg per cent in venous plasma. As the concentration is raised further, the resorption mechanism becomes saturated. Although the details of the renal glucose transport mechanism are still under investigation, a widely accepted theory has it that glucose involves the expenditure of energy (ATP) in the cells that make up the lumen of the proximal tubule. In patients with insufficient levels of insulin from the pancreas, an elevated plasma glucose level exists. If this level surpasses the renal threshold, glucose can be detected in the urine (glycosuria). A patient with diabetes mellitus may stabilize himself at home by performing daily urine tests to determine the level, or lack of it, of glucose in the early morning urine.

Sodium has a marked effect on the secretion of potassium (K^+) and chloride (Cl^-). Potassium generally is completely resorbed by the distal tubule (Fig. 19–6). An important consideration, however, is the fact that K^+ is exchanged for Na^+; thus the degree of exchange is determined by the amount of Na^+ in the distal tubule. Secretion of K^+ decreases if there is little Na^+ available, and it increases if there are large quantities in the distal tubule. Chloride is the main anion because it accompanies the Na^+ through the tubular epithelium. Cl^- is secreted along with the active transport of Na^+; therefore, its secretion also depends on the amount of Na^+ available.

The two principal waste products of the excretory mechanism are *urea* and *creatinine*. Urea, the highly nitrogenous end-product of protein metabolism, is produced in the liver and is readily filtered by the glomerulus. This is not surprising, since urea diffuses rapidly through all membranes of the body. It diffuses into the descending Henle's loop and out of the distal tubules and collecting ducts. However, the urea concentration in the tubule is still much higher than in the blood vessel; therefore, some urea is resorbed into the vessel and returned to the blood circulation. The urea clearance test is a measure of the amount of urea excreted by the kidney. It has proved to be a better index of kidney function than the blood urea nitrogen (BUN) test, which not only measures the quantity of urea excreted but also indicates the production of urea by the liver.

In summary, the renal resorption of huge amounts of water during each minute occurs by simple osmosis. The osmotic forces responsible are a direct result of the active resorption of sodium. This withdrawal of water from the tubules concentrates any solute left behind. Any of these concentrated solutes that cannot easily penetrate the tubular walls are found in high concentration in the urine. Because urea is permeable to most membranes, it will diffuse from the tubular fluid back into the blood to a certain extent.

Regulatory Mechanisms

Hormonal Control. A hormone from the posterior pituitary gland, called *antidiuretic hormone* (ADH), plays a central role in the regulation of body water. It acts directly on the kidney tubules, causing an increase in water absorption. Osmoreceptors in the hypothalamus control ADH release (Fig. 19–7) in response to changes

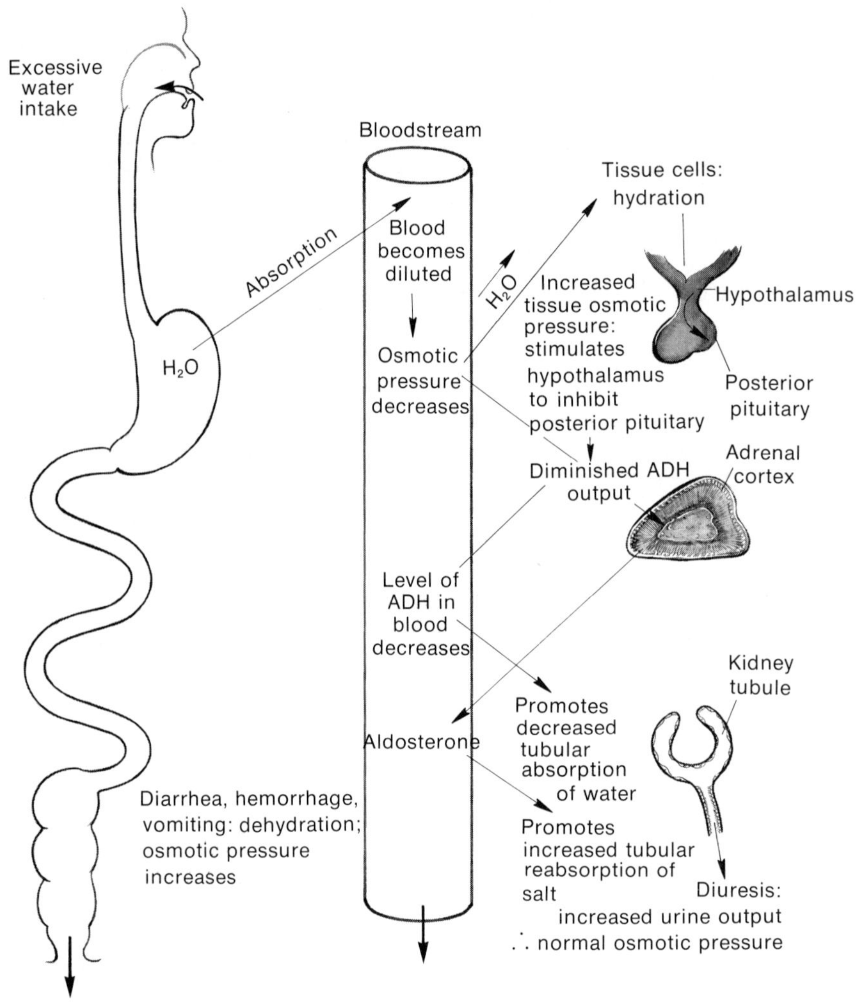

Figure 19–7 Body water regulatory mechanism. As the quantity of water and/or electrolytes in the body varies, the kidney adjusts the water level through excretion, thereby restoring the body fluids to their normal composition and volume.

in the osmotic pressure of the blood circulating through it. ADH acts by increasing water permeability and hence promotes the removal of water from the urine.

Several of the adrenal steroids affect the electrolyte balance, but *aldosterone* is the most active and appears to be the most important mineralocorticoid secreted by the adrenal cortex. Its action is on the distal convoluted tubule, where it slows down the Na^+/K^+ exchange by promoting the absorption of sodium and favoring the secretion of potassium (Fig. 19–7).

The effects of this ADH control mechanism are seen in water diuresis (urine output). After drinking a large volume of water, the individual's urine output increases rapidly, returning to normal only after the excess fluid has been excreted. In this case, the extra urine output (*polyuria*) results from reduced active resorption of water in the distal tubules.

Lower concentrations of ADH in the blood cause diminished water resorption and increased urine volume. This means that more water is removed from the plasma passing through the kidneys, a condition called *diabetes insipidus*. Conversely, if water loss exceeds intake, as in diarrhea or vomiting (Fig. 19–7), the extracellular fluid becomes more concentrated and the osmotic pressure increases. This increase in osmotic pressure and the resultant hypertonicity have a reverse effect on the osmoreceptors, and ADH secretion is increased, restoring the normal body fluid volume.

Electrolyte Regulation. Water is the vehicle for resorption of many of the body's solutes in dissolved form. It is the active transport of sodium, however, that triggers the resorption processes. Sodium is one of the most abundant solutes in plasma. It contains an electropositive charge, and when it is pumped back into the blood by the cells lining the convoluted tubules, an electropositive situation is created in the blood (Fig. 19–6). This, in turn, attracts the negatively charged ions that are available—chlorine and bicarbonate. Approximately 80 per cent of NaCl and 20 per cent of $NaHCO_3$ are resorbed by means of the active transport mechanism of sodium. It is the salinity of the blood that regulates ADH secretion; the resorption of water influences other solutes in the tubules in that their concentration is increased when water is lost to the tissues.

Acid-Base Balance. In Chapter 1, an acid was referred to as a chemical that can dissociate or yield H^+ ions in solution. Large amounts of acids are produced in cells during metabolism. The maintenance of a constant pH, therefore, depends on renal excretion of appropriate amounts of hydrogen ion derived from the diet. Together with the respiratory system, the kidneys are responsible for maintaining this acid-base equilibrium.

The acids are first neutralized by buffering agents (carbonates and phosphates) in the bloodstream. They cannot be excreted in this form, however, or the body would lose its stock of alkalis. The kidney conserves its alkali reserves by the tubular secretion of hydrogen ions (Fig. 19–6) in exchange for Na^+. The H^+, in turn, combines with Cl^- and NH_3 to form ammonium chloride, which is then excreted in the urine:

$$\begin{aligned} NaCl &\longrightarrow Na^+ + Cl^- \text{ (in tubule)} \\ Na^+ &\longleftrightarrow H^+ \text{ (exchange)} \\ NH_3 &\longrightarrow \text{tubular secretion} \\ NH_3 + H^+ + Cl^- &\longrightarrow NH_4Cl \text{ (ammonium salts)} \end{aligned}$$

The sodium, in turn, combines readily with HCO_3^- in the blood to form sodium bicarbonate ($NaHCO_3$), which acts as a buffer in the blood.

As we have seen, acid-base (pH) derangements can occur in either direction—resulting in acidosis or alkalosis. Respiratory acidosis results from a reduction in area of the lung surface for gas exchange, as in emphysema or pneumonia. Carbon dioxide accumulates in the alveoli, resulting in an increased carbonic acid concentration in the body fluids and eventual lowering of the pH. Respiratory alkalosis, on the other hand, results from *hyper*ventilation. The body secretes more carbon dioxide than normal, thus lowering the carbonic acid content of the blood.

Metabolic acidosis is a common result of diarrhea in children. Intestinal secretions are alkaline, and their continued loss creates an excess of acid in the body. Most metabolic acidosis results from an accumulation of keto-acids and a lowering of the pH, as in diabetic acidosis.

Metabolic alkalosis can be caused by excessive vomiting. Loss of gastric acid or consumption of large amounts of sodium bicarbonates may also lead to alkalosis. In either acidosis or alkalosis, the electrolyte balance between the cells and the extracellular fluids shifts, allowing such important metals as calcium and potassium to redistribute themselves. This results in changes in excitability, which, if not corrected, can lead to neuromuscular disorders, paralysis, and eventual death.

The Ureters

The urine in the pelvis of each kidney passes to the urinary bladder through the ureter (Fig. 19–8). The paired ureters are narrow, muscular tubes of small diameter and about 10 inches long. They descend from each of the two kidneys to the posterior abdominal wall, and then into the lower third of the bladder. Like the kidneys, the ureters lie in connective tissue behind the peritoneum. Their walls contain three layers: (1) an inner mucous epithelium; (2) a middle layer of smooth muscle; and (3) an outer fibrous coat of connective tissue. The muscular coat propels the urine by its rhythmic contractions; therefore, urine enters the bladder in small spurts, not in a steady stream.

The Urinary Bladder

The urinary bladder, a large muscular bag that lies in the pelvis behind the pubic bones, has three openings: two posterior and lateral for the ureters, and one anterior for the urethra (Fig. 19–8). The interior of the bladder is lined with highly elastic transitional epithelium. When the organ is full, the lining is smooth and stretched; when it is empty, the lining is a series of folds.

In the middle layer, a series of smooth muscle coats distends as urine collects, and contracts to expel urine through the urethra. Urine is produced almost continuously, and accumulates in the bladder until the increased pressure stimulates the organ's nervous receptors. When the muscles of the bladder contract, urine leaves by way of a single tube, the urethra—a process called *micturition* or *urination.*

Inflammation of the bladder, called *cystitis,* is often secondary to infection in some other part of the urinary tract or

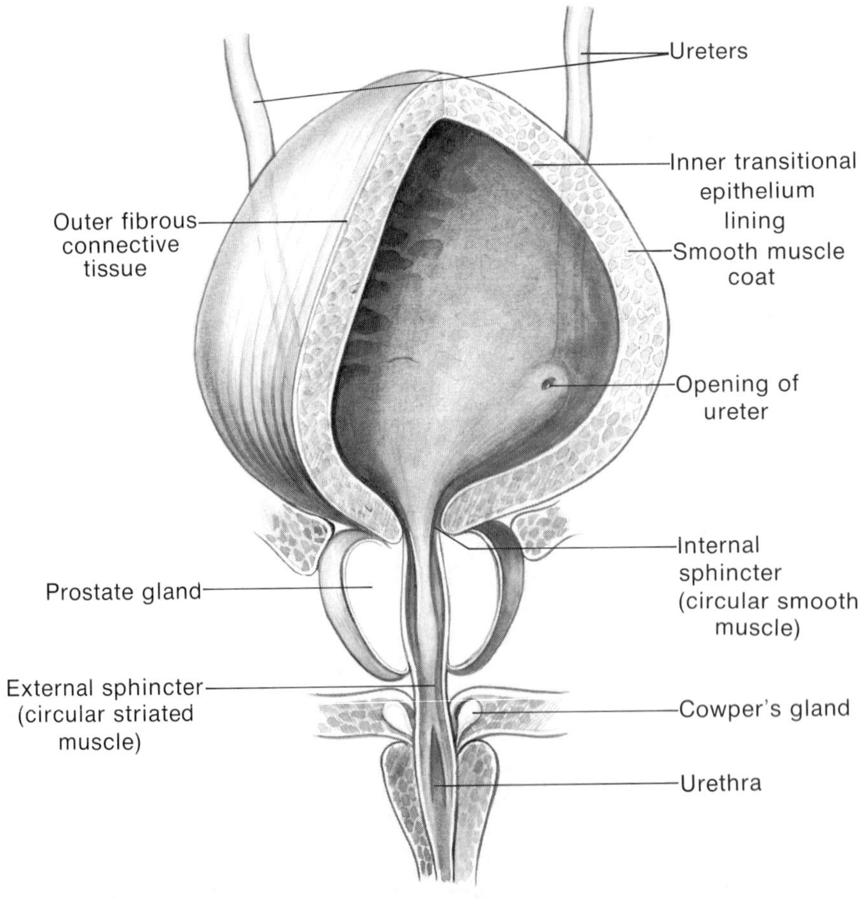

Figure 19–8 The urinary bladder, ureters, and urethra of an adult male.

elsewhere in the body. Acute cystitis is commonly caused by gonorrhea (a type of venereal disease) or by trauma to the bladder. Chronic cystitis may result from an enlarged prostate gland, pregnancy, or a stricture of the urethra that produces back pressure. Examination of the inner mucous membrane of the bladder is performed by means of a long tubular instrument, a *cystoscope*, which has a light at one end and an eyepiece at the other. Bladder tumors (neoplasms) may necessitate open-bladder surgery (*cystostomy*) or removal of the bladder (*cystectomy*).

The Urethra

The urethra is a membranous tube that passes from the anterior part of the urinary bladder to the outside of the body. In the female (Fig. 20–4) it is only one and one-half inches long; in the male it is about eight inches long (Fig. 20–2) and serves to conduct both urine and sperm to the outside. Two sphincters are located in the urethra (Fig. 19–8):

1. The *internal sphincter* is formed of circular smooth muscle, and guards the exit from the bladder. It is controlled by the autonomic nervous system.
2. The *external sphincter* is circular striated muscle, and is under the voluntary control of the central nervous system.

Congenital anomalies may involve the external opening of the urethra. The opening may be too small, or the urethra may be narrowed to such an extent that back pressure of urine may cause serious damage to the bladder if the defect is not corrected. In adults, a *stricture* or closure of the urethra also can prevent normal micturition. The stricture may be caused by muscle spasms or by *urethritis*, an inflammation of the mucous membrane. This condition is more common in the male, and dilation of the inner tract may be required to permit urination.

If urine is retained in the bladder for a prolonged period, a back pressure is established, causing the kidneys to malfunction. Insertion of a rubber tube, a *catheter*, will temporarily relieve the problem. Persistent *dysuria* (difficult urination) may require continued catheterization until the cause of the stricture is found and treated.

MICTURITION

The movement of urine collected in the pelvis of the kidney down the ureters to the bladder is the result of both the force of gravity and the peristaltic contractions in the muscle layers of the ureter. By stretching the bladder, a considerable amount of fluid—about 300 ml—can be accommodated without a significant rise in pressure. As the volume increases, the tension rises until, finally, stretch receptors are excited. These initiate an impulse that creates the sensation of distention and the desire to urinate. Sensory signals are conducted to the sacral segments of the spinal cord through the pelvic nerves and then back again to the bladder through the parasympathetic fibers in these same nerves (Fig. 19–9).

As the sensory impulses increase in number and frequency, the motor impulses to the bladder, via the pelvic nerves, cause a reflex contraction of the bladder and a relaxation of the internal sphincter. The external sphincter is controlled by the pudendal nerve (motor). The bladder wall contains many stretch receptors, the afferent nerves of which enter the spinal cord and eventually synapse with these parasympathetic and motor neurons, stimulating the former to contract and inhibiting the latter, causing a relaxation.

The sequence of events just described shows how micturition occurs in infants. In the adult, higher brain centers control and inhibit the reflex mechanism so that micturition can be delayed or induced voluntarily. It is by learning to control these pathways that a child achieves a capacity to inhibit his spinal reflexes.

The loss of voluntary control over

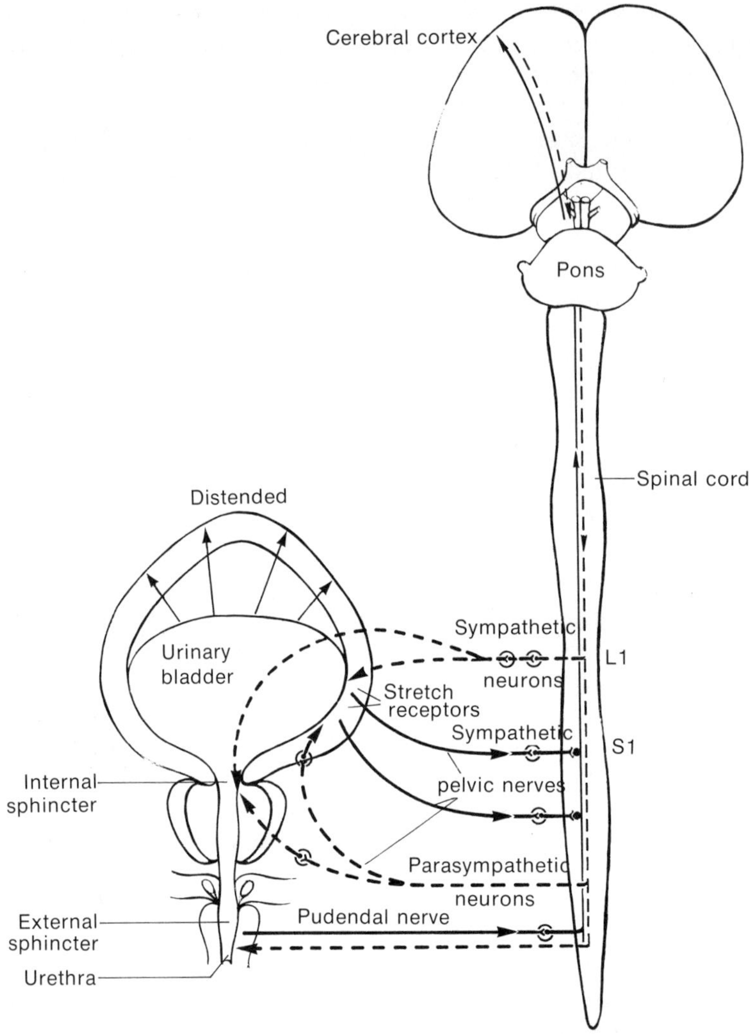

Figure 19–9 Micturition. Distention of the urinary bladder from accumulation of urine stimulates the stretch receptors in the bladder wall. The afferent fibers of the pelvic nerve conduct this impulse to the brain, which institutes a reflex mechanism. The brain controls the storage and expulsion of urine in the bladder by smooth muscle contraction or relaxation of the internal and external sphincters.

micturition is called *incontinence*. Incontinence is normal in young children, since the infant voids whenever the bladder is sufficiently distended to arouse a reflex stimulus. The term *nocturnal enuresis* is used to describe the involuntary discharge of urine during sleep.

The normal output of urine per day ranges from 1000 to 2000 ml. Any suppression of urine flow is referred to as *oliguria,* as in urine retention in the kidney; any increase in urine flow is called *polyuria,* as in diabetes insipidus. Complete suppression of urination is *anuria.* Inflammatory conditions that prevent normal blood flow to the glomerular filtering beds are a common cause of anuria. Any stricture or blockage of the ureters will also result in anuria.

URINE

Macroscopic Examination

The urine formed by the processes described earlier in this chapter is a hyper-

Table 19–1 COMPOSITION OF URINE

Constituent	*GRAMS/PER DAY*
Water	1500.0
Organic substances	
Urea	30.0
Creatinine	1.2
Uric acid	0.6
Ammonia	0.8
Inorganic substances	
Sodium	5.0
Chloride	6.0
Calcium	0.2
Potassium	2.2
Phosphates	1.5
Sulfhates	1.6
Magnesium	0.2

tonic solution of inorganic salts, nitrogenous waste products, and a few gases (Table 19–1). A good index of the degree of concentration of the dissolved materials is the specific gravity. The normal range, using a hydrometer, is between 1.015 and 1.025 (with reference to water as 1.000). An adult excretes an average of 1500 ml of urine per day. Higher and lower specific gravities may occur, depending on fluid intake and the quantity of dissolved substances in the urine. An elevated specific gravity generally indicates urine retention and an increased concentration of materials. Conversely, a low specific gravity is usually due to polyuria. When liquids are consumed at a greater than normal rate, urine forms more rapidly, and its specific gravity may approach that of water, descending to about 1.005. In severe diabetes mellitus, however, the high sugar content of the urine will keep its specific gravity around 1.040, even though urine output is very high.

The color of urine is generally pale to dark yellow (straw yellow), and the fluid is very clear. Bile pigments cause the color, which can vary from a brown-yellow to olive-green, depending on the pigment concentration. Red hues, due to hemoglobin in the red blood cells, indicate that a degree of hemolysis has occurred. Urine containing fat droplets or pus may have a milky appearance.

The odor of urine is clinically insignificant. Normal urine is slightly aromatic. A smell of ammonia is usually due to bacterial action after voiding, and can be very unpleasant. A sweet-smelling urine indicates a highly concentrated ketone level, especially acetone. Diabetic ketosis is a common cause.

Urine normally varies in pH from 5 to 8, and usually is slightly acidic. Reportedly, Western man, who lives on a high protein (animal) diet, has a urine pH of 5 to 6. Eastern man, the vegetarian, has a slightly alkaline (7 to 8) urine. Urinary acidosis is indicative of such problems as diabetic ketosis, starvation, and diarrhea; it also occurs in emphysema, in which carbon dioxide retention is high. Alkalosis, on the other hand, results from most of the urinary infections and from high blood oxygen, as in hyperventilation.

One of the more important clinical tests of urine is that for protein. This is derived from plasma proteins, and usually comes from the kidney. Normally, there should be no protein in the urine. Gross bleeding from the lower urinary tract can add large quantities of protein to the urine *(proteinuria)*. Glomerular filtration problems, renal disease (nephritis), and inadequate resorption of extracellular protein by the lymphatic system all produce traces of albumin in the urine *(albuminuria)*.

Normally, simple sugar (glucose) is completely resorbed by the convoluted tubules. In diabetes mellitus, blood sugar is not broken down in the body cells because of a deficiency of insulin, and therefore is excreted in the urine *(glycosuria)*. The "Clinitest" is a simple, rapid, and accurate technique that can be used to detect sugar in the urine. This test may be done at home, under the supervision of a physician, with quite reliable results. It allows a diabetic to regulate his diet and medication.

The presence of ketone bodies in the urine *(ketonuria)* is especially significant in diabetes. If the test for glucose is positive, a follow-up should include determination of the presence or absence of acetone. Uncontrolled diarrhea, starvation, and an altered carbohydrate metabolism that uses protein and fat as the energy source in place of carbohydrates can cause this condition. It also may occur in children after prolonged vomiting.

Blood *(hematuria)* and pus *(pyuria)* appear in the urine during kidney or bladder infections. Their presence is of sufficient importance to warrant further testing.

Nitrogenous waste products, which include urea, uric acid, and creatinine, are classified as *organic* compounds (Table 19–1) because they are formed by the breakdown of cells in the body and from food proteins. In other words, they originated from living organisms. *Inorganic* substances include the chlorides, phosphates, sulfates, and so forth that make up the body's salt and mineral content. These do not originate in living organisms.

Chorionic gonadotropin, a hormone produced by the placenta during pregnancy, can be detected in the urine four to six weeks after conception.

Microscopic Examination

Examination of urine sediment is undoubtedly the most valuable part of a routine urinalysis. A fresh urine specimen will show red blood cells and *casts*. Casts are cylindric structures or molds formed within the renal tubules. A hyaline cast contains protein and has the same clinical significance as proteinuria—renal damage. Other casts contain red blood cells, pus, or epithelium; each is indicative of specific renal problems.

Under normal conditions urine is free from bacteria. When bacteria are found, infections of the bladder (cystitis) or urethra (urethritis) may be present.

KIDNEY FUNCTION TESTS

The *PSP (phenolsulfonphthalein)* test is primarily a test of tubular secretory function. Normal kidneys excrete at least 25 per cent of this dye in 15 minutes, and a total of at least 65 per cent of the dye in two hours. Since PSP excretion is related to renal blood flow as well as to tubular function, low values occur in certain heart diseases. Obstructive lesions of the lower tract as well as delayed or incomplete emptying of the bladder will make it impossible to interpret the test in terms of renal function.

Clearance tests may be applied to any substance present in the blood and excreted in the urine. These tests make it possible to gauge the progress of renal damage. Creatinine and urea are filtered through the glomerulus, and, to some degree, are subject to tubular resorption; therefore, both creatinine and urea clearance tests are used as guides to the glomerular filtration rate. They are said to be sensitive guides to kidney function. In early kidney disease, a fall in the clearance rate may occur before the serum creatinine level or the blood urea nitrogen (BUN) rises out of the normal range. It must be remembered, however, that although these two clearance tests can signify anomalies earlier than other tests do, they still yield positive results only after the disease is advanced. They cannot diagnose kidney disease but merely determine its progress.

Urea itself is the nontoxic chief end-product of protein metabolism. Elevated urea concentrations in the blood found by the BUN test signify inadequate kidney function.

Creatinine and uric acid are nonprotein nitrogenous compounds found in low concentrations in the blood. In kidney dysfunction and uremia, creatinine is retained in the blood, and plasma creatinine levels may increase markedly—an indication of early renal disease.

CLINICAL CONSIDERATIONS

Uremia is a clinical condition resulting from severe loss of renal excretory

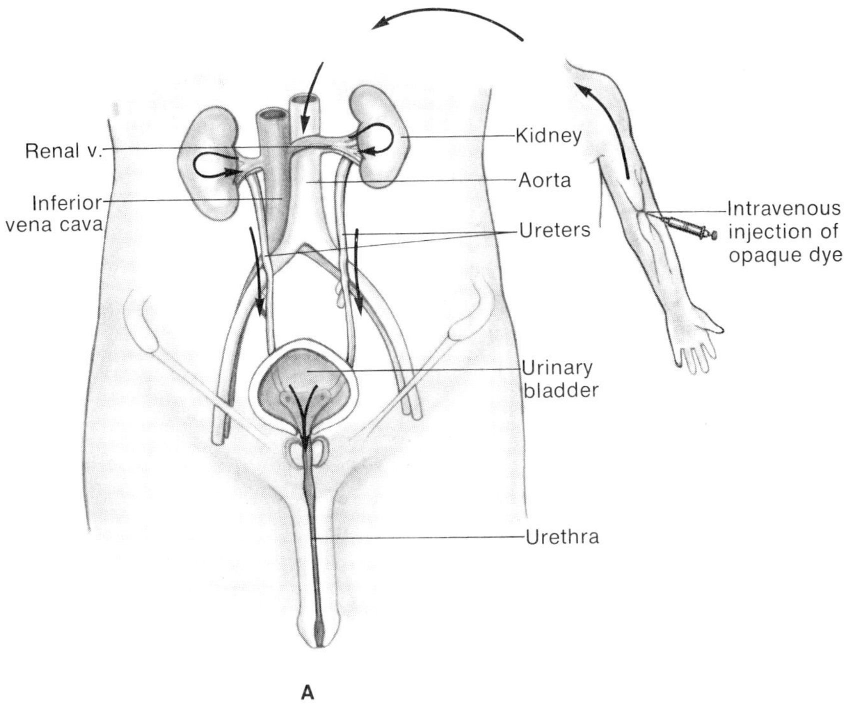

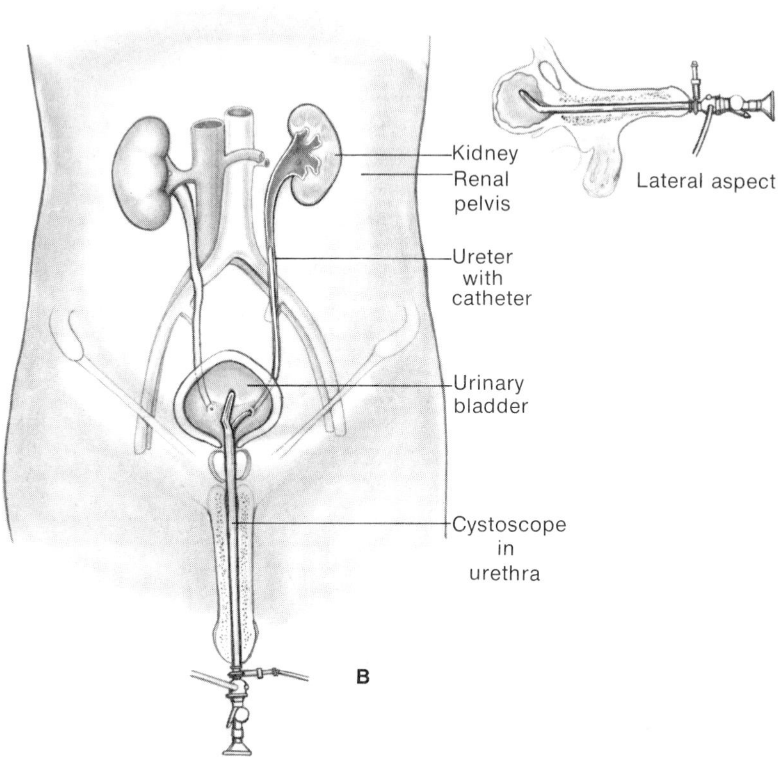

Figure 19–10 Pyelogram. *A*, An opaque dye is injected intravenously to determine the function, filling, and emptying of the upper urinary tract. *B*, Retrograde pyelogram is a direct inspection of the lower urinary tract using a cystoscope. A catheter is inserted to check the ureters and renal pelvis for any pathologic condition or obstruction.

function. This is a toxemia of the blood produced by the retention of urinary constituents. Although nitrogenous substances, such as urea, are a good index to the severity of renal dysfunction, there is little evidence to justify the assumption that urea, creatinine, or uric acid contribute to clinical symptoms. The most common causes of uremia are the acute and chronic renal diseases. Symptoms of visual disorders, nausea, vomiting, headaches, and anuria may lead to convulsions and coma.

Renal Calculus

Urinary calculi (stones) may be formed in the renal pelvis or in the bladder. A stone is basically an organic material covered by layers of salts. The most common stones are urate or calcium oxalates; these are extremely hard and may be spiney. Urinary calculi may be formed as a result of infection, high salt concentration, vitamin A deficiency, or parathyroid tumors. The calculus generally forms in the pelvis of the kidney, and either remains there or passes down the ureters to the bladder. If the stone blocks the opening of the ureter, hydronephrosis (distention of the kidney) will result. If the stone lodges in the ureter and causes an obstruction, pyelonephritis will result. A calculus lodged either in the kidney or in the ureter creates pressure that irritates the sensitive mucosal lining, producing ulceration. If stones are lodged in the pelvis of the kidney or in the ureter, they may be removed by surgery (nephrolithotomy or pyelolithotomy).

Stones that originate in the bladder generally are of phosphate origin; those that enter the bladder from the kidney usually are formed of uric acid or calcium oxalate. Infections, ulceration, and tissue breakdown typically accompany bladder stones. The most characteristic symptom is hematuria resulting from irritation of the mucosal lining of the bladder. *Renal colic* is the term used to describe the excruciating pain caused by the passage of one or more stones through the ureter. If the stones are not able to pass out through the urethra, they may have to be removed surgically.

In a *pyelogram,* an x-ray of the kidney and ureter (Fig. 19–10), an opaque substance is injected intravenously or perfused into the ureters through a cystoscope. The pyelogram will show either a clear, open pathway from the kidney to the

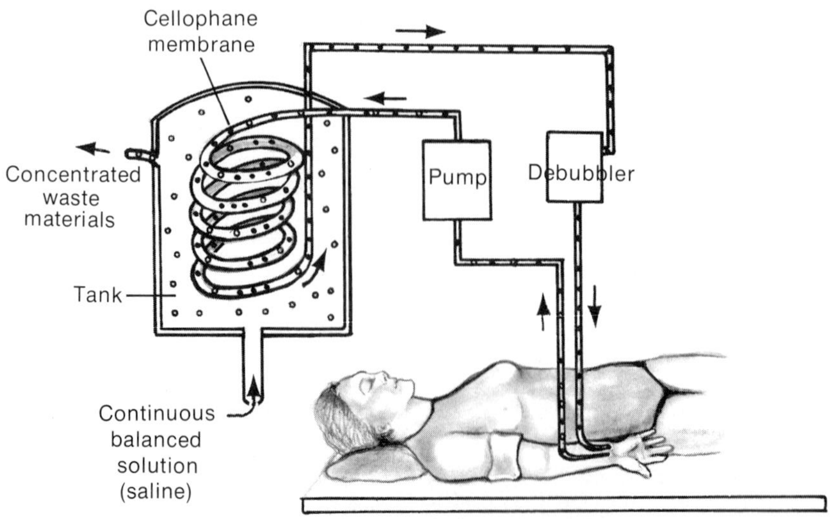

Figure 19–11 Hemodialysis. Schematic diagram of the artificial kidney.

bladder or an obstruction caused by calculi or stricture.

Hemodialysis

The artificial kidney is a continuous-flow dialysis system (Fig. 19–11) of large surface area. A stream of blood from an artery spreads out in a film over a thin cellophane membrane in contact with a balanced salt solution. Exchange of all diffusible components of the two fluids occurs across the artificial semipermeable membrane. Blood is returned to the body through a venous catheter, and the dialyzing medium is rapidly renewed from a large reservoir.

Any substance diffusible in higher concentration in the blood also diffuses into the dialysis fluid; therefore, the artificial kidney extracts toxic waste products from the blood. The reverse procedure also works: diffusible substances, for example, bicarbonate ions in acidosis, can be added to the blood.

The rate of dialysis depends on the instrument used, the surface area of the dialyzing membrane, and the concentration of waste materials in the blood. A patient may be on the dialysis machine for six to eight hours at a time or for longer periods.

The artificial kidney is not a substitute for the patient's own kidney, but it does afford a temporary therapeutic device in the continuing effort to save the lives of patients whose kidneys are diseased or destroyed.

OUTLINE

I. Excretory Organs of the Body
  A. Skin—salts
  B. Lungs—carbon dioxide
  C. Liver—bile pigments
  D. Large intestine—solid waste materials (feces)
  E. Kidneys and urinary tract—urine
II. Urinary System
  A. Functions
    1. Excretion of wastes
    2. Regulation of composition of blood
  B. Kidneys
    1. Abdominal—lumbar region
    2. Bean-shaped
    3. Anatomy
      a. Cortex
      b. Medulla
      c. Pyramids
      d. Hilus
      e. Renal pelvis
    4. Nephron units—functional units
  C. Urine formation
    1. Filtration
      a. Walls of glomerulus are semipermeable
      b. Glomerular filtration force
      c. Factors affecting glomerular filtration rate
    2. Resorption
      a. Water
      b. Sodium control
      c. Urea and creatinine—principal waste products
      d. "Final" urine more concentrated than initial glomerular filtrate
  D. Regulatory mechanism
    1. Hormonal control
      a. Posterior pituitary—antidiuretic hormone (ADH)
      b. Adrenal cortex—aldosterone
    2. Electrolyte regulation
      a. Sodium transport
      b. Electropositive charge in blood
      c. Chloride and bicarbonate (negative) attraction
    3. Acid-base balance
      a. Buffers
        1) Carbonates
        2) Phosphates
      b. Acid-base derangements
        1) Respiratory acidosis
        2) Respiratory alkalosis
        3) Metabolic acidosis
        4) Metabolic alkalosis
  E. Ureters—Rhythmic muscle contractions propel urine

F. Urinary bladder
   1. Muscular reservoir
   2. Middle layer of smooth muscle contracts to expel urine
   3. Openings for ureters and urethra
   4. Cystitis, cystotomy, and cystectomy

G. Urethra
   1. Size difference in male and female
   2. Internal and external sphincters
   3. Urethritis
   4. Strictures

III. Micturition
   A. Gravity
   B. Neurochemical control
      1. Sensory receptors—stretch
      2. Motor impulses—pelvic nerves
      3. Reflex mechanism in infants
      4. Incontinence
   C. Oliguria, polyuria, and anuria

IV. Urine
   A. Macroscopic examination
      1. Specific gravity
      2. Color
      3. Clarity
      4. Odor
      5. pH
      6. Protein
      7. Glucose
      8. Ketone bodies
      9. Blood or pus
      10. Gonadotropic hormones
   B. Microscopic examination
      1. Blood cells
      2. Bacteria
      3. Casts
      4. Pus

V. Kidney Function Tests
   A. PSP
      1. 15 minute and 2 hour tests
      2. Dye retention level
   B. Creatinine and urea clearance tests
      1. Compared with serum creatinine and blood urea nitrogen levels
      2. Determine progress of kidney disease but cannot diagnose it

VI. Clinical Considerations
   A. Uremia
      1. Blood retention of urinary constituents
      2. May be present in acute or chronic kidney disease
   B. Renal calculus
      1. Urate or calcium oxalate—formed in kidney
      2. Phosphate oxalate—formed in bladder
   C. Pyelogram—x-ray of kidney and ureters
   D. Hemodialysis
      1. Continuous-flow dialysis system
      2. Semipermeable membrane
      3. Concentration gradient principle
      4. Removes wastes and supplies metabolites to the blood

STUDY QUESTIONS AND PROBLEMS

1. In what way might secretion and excretion be the same, and how are they different?
2. What do we mean when we say that the kidneys are not in the abdominal cavity?
3. Differentiate between the cortex and medulla of the kidney.
4. What are calculi, and where do they come from?
5. What is meant by hemodialysis?
6. What is renal colic?
7. What is the significance of peristaltic ureteral contractions?
8. How do the female and male urethras differ in structure and function?
9. What is the role of the kidney in homeostasis?
10. Describe the anatomy of a nephron unit.
11. What substances are filtered by the glomerulus?
12. Is all resorption passive? Describe the physiology of resorption.
13. What is nocturnal enuresis?
14. Describe the hormonal control of the kidney. What is diabetes insipidus?
15. How does sodium affect the acid-base balance of the body?
16. List the abnormal contents of urine and correlate them with specific diseases.
17. What is urea? How is it formed. Is all of it excreted from the blood into urine?
18. Describe the neurochemical control of micturition.
19. What does the specific gravity of urine have to do with excretion?
20. What is incontinence?

Chapter 20

Reproduction, Growth, and Development

The simplest form of reproduction is the splitting of a cell in half to yield two daughter cells, each identical to the original parent cell. The cells of every animal species contain a definite and characteristic number of *chromosomes*. A roundworm, the Ascaris, has only two chromosomes; a moth has 224. In the human—both male and female—there are 46 chromosomes in each cell. It is essential to remember, however, that these are *double* sets of chromosomes: 22 pairs called *autosomes* and two *sex chromosomes*. In the female the sex chromosomes are identical and are designated XX. The male sex chromosomes are not identical; they are designated XY.

The *somatic* cells of the human body are those primarily concerned with growth and differentiation. These cells form the main structural tissues of the body—the bones, muscles, nerves, and the like. They are said to be *diploid* because each somatic cell contains double sets of chromosomes (23 sets, or a total of 46). Mitosis, as we saw in Chapter 2, produces new cells with exactly the same number of chromosomes as the parent cell. Hence, whenever a somatic cell divides, the new cells produced have 46 chromosomes—two of each type. Reproduction in this sense, then, means the growth or replacement of tissues.

A sexual type of reproduction also occurs, whereby two cells join (fertilization), allowing for the continuation of the species. This form of reproduction is unlike mitosis because union of two cells results in a new organism not identical to its parents.

We know that in sexual development two cells (the egg and the sperm) unite to produce the first cell (*zygote*) of the new individual. We also know that the chromosome number remains constant within a species. Thus the new zygote, although formed from two separate cells, must contain a total of no more than 46 chromosomes. It follows, therefore, that prior to fertilization a different kind of cell division must take place. This second kind of division would be restricted to the *germinal* (sex) cells, since these are the only cells concerned with fertilization. It must reduce the chromosomal number by half (*haploid* cells) so that, when the egg and sperm unite in fertilization, the normal diploid number of chromosomes is restored. This second special process of cell division is called *meiosis*.

MEIOSIS

The sexual, haploid forms of the species produced by meiosis are known as *gametes*. Meiosis itself is also known as *gametogenesis* (Fig. 20–1). More specifically, the production of male sperm is *spermatogenesis*, and *oogenesis* is the production of the female egg. Spermatogenesis begins at puberty and continues until

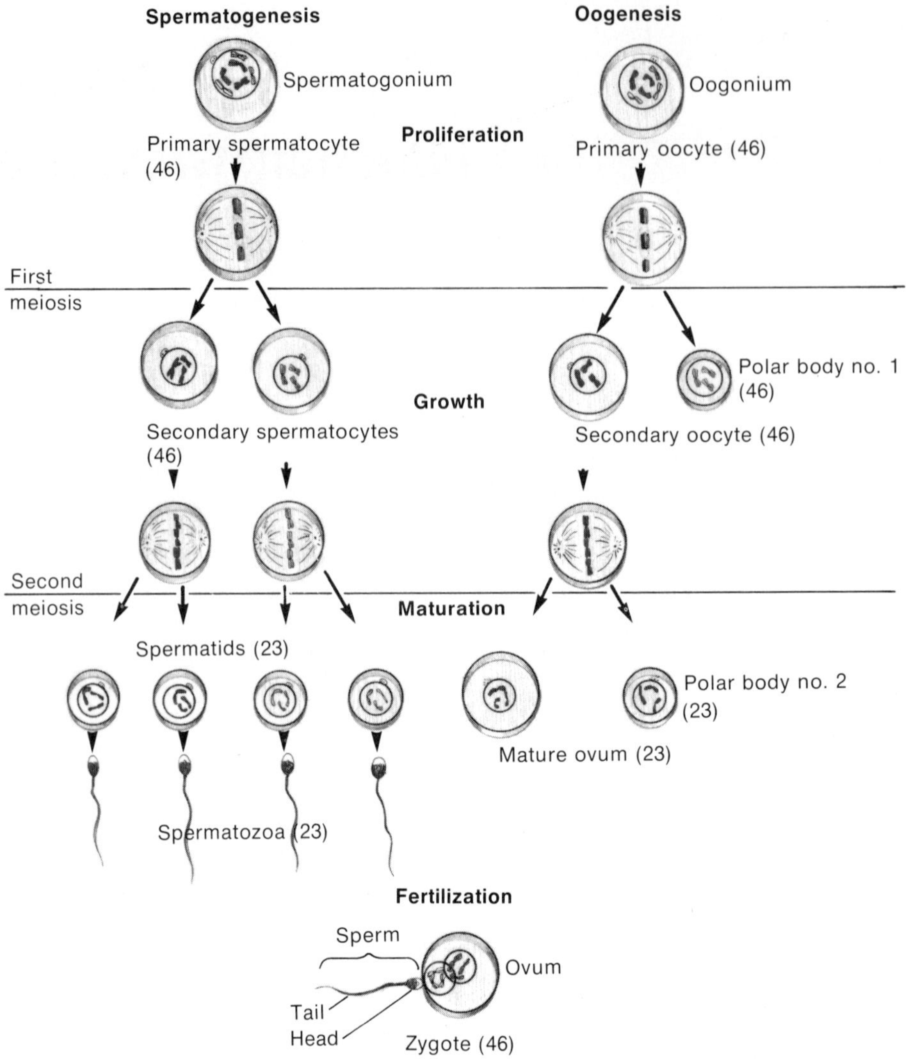

Figure 20–1 Meiosis (gametogenesis). *Left,* The various stages of spermatogenesis that give rise to four viable, mature sperm, each having 23 chromosomes. *Right,* Oogenesis, the production of a single, viable ovum (egg) and the two polar bodies from each oogonium. The union of sperm and egg (fertilization) produces a zygote. The numbers in parentheses indicate the number of chromosomes.

senility; oogenesis, on the other hand, is a cyclic process, starting at puberty and terminating at *menopause,* which occurs between the ages of 35 and 50 in most American women.

First Meiotic Division

The two meiotic divisions (Fig. 20–1) occur in a series of stages similar to those of mitosis (see p. 23—interphase, prophase, metaphase, anaphase, and telophase. The chromosomes in prophase are long and thin, and each pair of like chromosomes moves together and unites longitudinally—a process known as *synapsis.* The synaptic pairs become shorter and thicker, and it is at this point that it becomes evident that they are doubled. At metaphase, the paired chromosomes (tetrads) arrange themselves so that like pairs line up at the equatorial plate. It is this maneuver that distinguishes meiosis from mitosis. In mitosis like chromosomes line up and separate, resulting in one identical chromosome going to each respective daughter cell. In the metaphase of meiosis, like chromosome pairs line up and separate, but the result, in this stage, is a reduction in the number of chromosomes in each cell. The rest of meiosis I is mechanical, the product being two daughter cells that together have the same number of chromosomes as the parent.

Second Meiotic Division

The second meiotic division follows a mechanical sequence identical to that of the first. Maturation and division of the two intermediate-stage daughter cells result in four haploid cells or gametes. The chromosomes do not synapse because no like chromosomes are available (Fig. 20–1). Each double-stranded chromosome moves into the metaphase stage, and the new single-stranded chromosomes separate. The second meiotic division is essentially a mitotic division.

In summary, the first meiotic division produces two haploid cells containing double-stranded chromosomes. Each of these cells divides in the second meiotic division, producing a total of four new haploid cells containing single-stranded chromosomes.

It is evident now that the chromosomes are not all functionally equal. Chromosomes in adult human cells do occur in pairs with similar genetic function, but the pairs constitute 23 distinct and unique sets of genetic information. Thus, we see the necessity for a second remarkable aspect of meiosis. Not only does meiosis halve the chromosome number, it does it in such a way that each mature germ cell receives one member of each pair of chromosomes. The zygote, therefore, receives a member of each pair of chromosomes from each parent. Sexual reproduction is advantageous in that it provides for genetic variability—a mixture that can produce countless variations in the offspring.

Spermatogenesis and Oogenesis in Mammals

At the time of reproduction, meiosis produces gametes, which then unite in fertilization to produce a zygote. The zygote then grows mitotically to produce the new diploid individual. The sperm and eggs are, therefore, the only haploid stage in the life cycle of the individual.

In males, sperm cells (spermatozoa) are produced by the germinal epithelium of the testes (Fig. 20–3). When one of the epithelial cells undergoes meiosis (Fig. 20–1), the four haploid cells that result differentiate into swimming sperm cells. This is spermatogenesis.

In the female, eggs are produced in the follicles of the ovaries (Fig. 20–6) by a comparable process called oogenesis. There is one major difference, however. When a cell in the ovary undergoes meiosis, the haploid cells are unequal in size. The first meiotic division produces one relatively large cell (primary oocyte) and a tiny dead one, the first polar body (Fig. 20–1). The second meiotic division of the living larger cell produces a secondary oocyte (the egg) and a smaller dead secondary polar body. Thus, when an ovarian cell undergoes meiosis, only one viable mature egg or ovum is produced. The polar bodies are not functional, and ultimately are absorbed into the metabolic system as food.

REPRODUCTION IN THE HUMAN MALE

Sperm Production

The organs involved in the production and liberation of human sperm are illustrated in Figure 20–2. Sperm are produced in the *testes,* which are contained in the *scrotum* outside the abdominal cavity.

The coiled *seminiferous tubules* make up the mass of each testis (Fig. 20–3). The epithelial lining of their walls contains con-

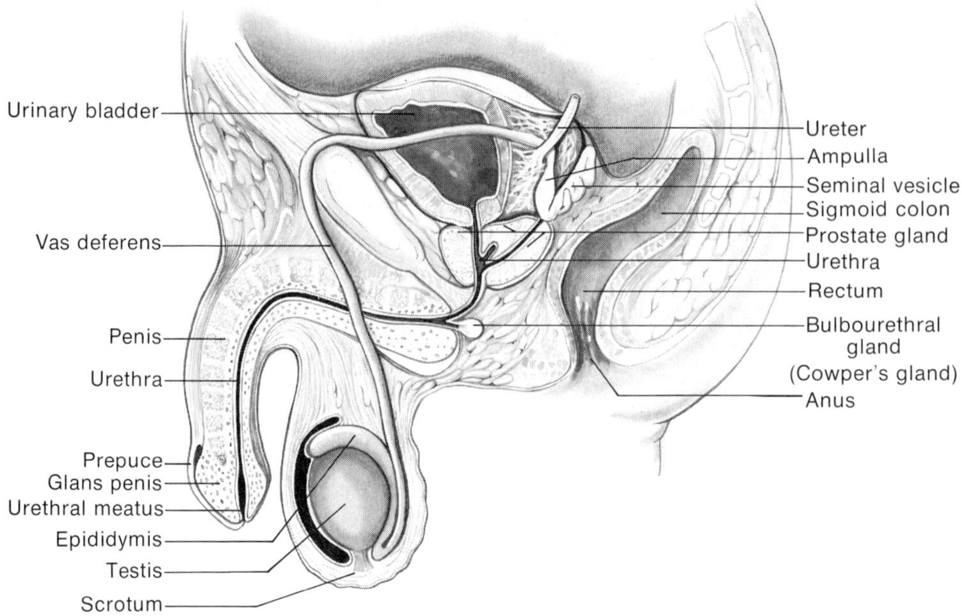

Figure 20–2 Midsagittal section of the male reproductive system.

tinuously proliferating cells called *spermatogonia.* These develop to form primary spermatocytes, which undergo meiotic division to yield two secondary spermatocytes each (Fig. 20–3). These divide again, mitotically, to yield spermatids and motile sperm. The newly formed sperm pass along the *vas efferens* into the *epididymis,*

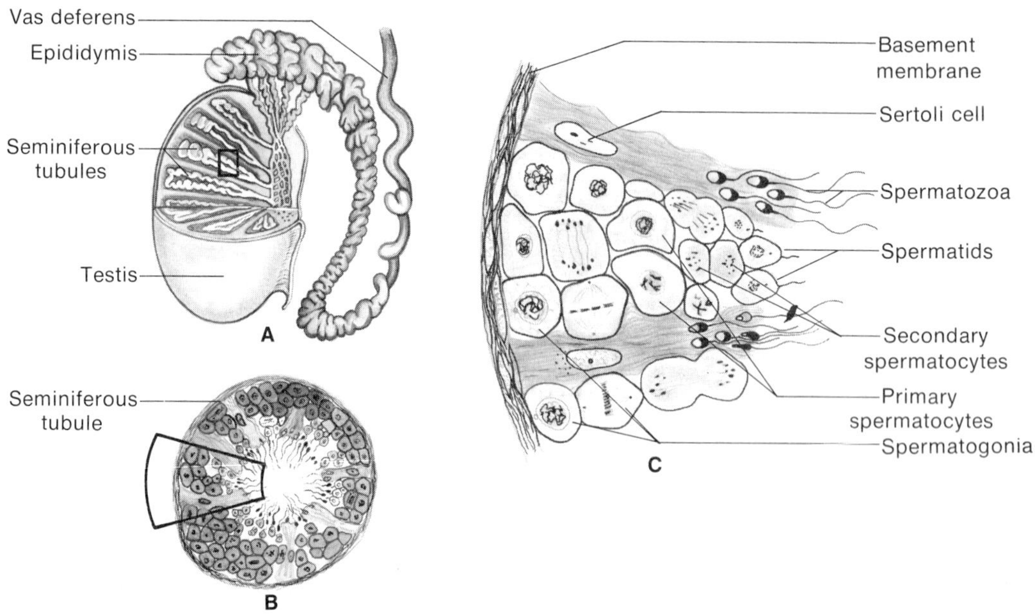

Figure 20–3 The testes. *A,* Diagram showing the duct system of the human testes. *B,* A cross-section of a seminiferous tubule of the testes. *C,* Enlarged diagram of a seminiferous tubule, showing the development of spermatogonia.

where they remain for a short time. If they are not ejaculated, they soon degenerate and are absorbed in these tubules. Some sperm are stored in the *ampulla,* an enlargement of the excretory duct of the testis *(vas deferens).* Adjacent to the ampulla are two *seminal vesicles* that secrete a nutritive fluid for the sperm. From the ampulla, the duct passes through the prostate gland into the urethra, a tube that passes out of the abdominal cavity down the middle of the penis (Fig. 20–2).

Interspersed between the seminiferous tubules of the testes are a number of interstitial cells, called *Leydig* cells, which produce and secrete the male hormone, *testosterone.* Testosterone passes directly into the bloodstream to control the process of spermatogenesis and the development and activity of the secondary sex organs. The hormone also promotes skeletal and muscular growth during adolescence and is necessary for the development of the secondary sex characteristics—deepening of the voice, growth of pubic hair, and so forth.

The third cell of importance in the testis is the *Sertoli* cell, which is located in the seminiferous tubule between the spermatogonia (Fig. 20–3). It is primarily responsible for supplying the developing germinal cells with any necessary nutritional support. Sertoli cells are commonly called "nurse cells."

The *prostate* (Fig. 20–2) is fairly large—about 4 to 5 cm in diameter—and almost completely surrounds the urethra. In prostatitis, the gland may become so large that it constricts the urethra and prevents urination. This problem is common in old age, and usually results in the surgical removal of the prostate gland (prostatectomy). Just below the prostate gland is the *Cowper's gland.* The two glands collectively secrete material that mixes with the sperm and provides it with nourishment, a proper alkaline medium (semen), and support for the sperm's swimming motions.

The *penis* (Fig. 20–2) is a tubular organ plentifully supplied with large venous sinuses that can fill with blood, causing the organ to become erect under the stimulation of sexual excitement.

Hormonal Function of the Testes

The interstitial cells (Leydig) of the testes are stimulated by the ICSH (interstitial cell-stimulating hormone) of the anterior pituitary to produce *testosterone,* a male hormone (androgen). Little or no testosterone is secreted before puberty, which normally occurs between the ages of 12 and 16. There is evidence to indicate that ICSH is stored in the pituitary until this age, but the hormone is not released until then. A second gonadotropic hormone, FSH (follicle-stimulating hormone), is also secreted by the anterior pituitary. FSH assists ICSH in stimulating spermatogenesis in the testes.

If the testes are absent because of castration, congenital defects, or hyposecretion of gonadotropins at the time of puberty, there is no reproductive potential, and the secondary sex characteristics do not develop. Such an individual is called a *eunuch.* Loss of the testes after puberty produces sterility, but there is only some recession of the secondary sex characteristics.

It is well known that spermatogenesis occurs best at temperatures slightly lower than that of the normal body, and the scrotum is suspended outside the body where the temperature is a few degrees lower than body temperature. In the embryo, the testes develop within the body cavity, but normally they descend before birth down through the inguinal ring to their location in the scrotum. Sometimes one or both testes do not descend, a condition called undescended testicle or *cryptorchidism.* An undescended testis will develop its hormones, but at puberty it will not produce sperm, presumably because of the high temperature within the abdomen. In some patients, hormonal treatment may stimulate spermatogenesis after the testis has been moved surgically to its normal position in the scrotum. At least 300 million sperm are contained in the average ejaculation, and therefore the loss of one testis does not render a male sterile. Sperm counts of 60 million per specimen are considered to be at the lower border of the fertile range. It is remarkable to note that only one sperm fertilizes the ovum, whereas the other millions degenerate in a short time—reportedly within 72 hours after ejaculation.

Sperm

A human sperm is microscopic in size (0.06 mm). It is a specialized, motile germ cell that carries the male chromatin material and consists of an oval head, short neck, and body. The tail is long and capable of violent swimming actions (Fig. 20–1). This vigorous activity is facilitated by closely packed rows of mitochondria found in the neck.

Disorders of the Male Reproductive System; Male Sterilization

Undescended testes or cryptorchidism has already been discussed. The most common infection of the male urethra is *gonorrhea.* The infection, which displays itself by a purulent discharge from the urethra, may spread along the mucous membrane into the prostate gland. If the infection is not treated, scar tissue may form in the reproductive organs, causing sterility. Bacilli and staphylococci also may invade the organs, producing an inflammation of the testes, *orchitis.* In metastatic orchitis, an infection carried from the bloodstream may be due to mumps, to typhoid fever, and, occasionally, to smallpox. Traumatic orchitis is an inflammation caused by injury or surgical intervention.

Phimosis is a congenital condition in which the prepuce (foreskin) is too tight around the head of the penis and cannot be drawn back. This situation normally may be remedied by removing all or part of the foreskin, a process called *circumcision.*

Vasectomy is a surgical procedure resulting in sterilization of a male. A portion of the vas deferens (Fig. 20–2) on each side is removed to prevent sperm from reaching the urethra. The sperm are absorbed into the system, and the individual retains his ability to produce hormones as well as to perform the sex act.

Many categories of *malignant tumors* of the testes have been reported. Benign tumors are very rare. The malignant tumor characteristically metastasizes to the lymph nodes in the inguinal region. From here a gradual involvement of all lymphatic areas occurs, with eventual extension to the lungs and brain. The disease ultimately results in death.

A *hydrocele* is a collection of fluid in the epididymis. Acute cases are usually due to a spread of gonorrhea or to tubercular infections. Chronic hydrocele can be due to a low-grade infection. Chemotherapy and surgical repair of the water sac are the recommended treatments. *Hematocele* is the name given to a hemorrhage into a hydrocele following trauma.

Hermaphroditism is a blending of the male and female sex organs. In true hermaphroditism, both sex organs are found in the same person. There are no known cases in which this condition has resulted in fertility, either as a man or as a woman. In false hermaphroditism, the genital glands of one sex are evident, but the secondary sex characteristics of the opposite sex predominate.

REPRODUCTION IN THE HUMAN FEMALE

The Female Reproductive Organs

The female reproductive organs are illustrated in Figure 20–4, *A* and *B*. There are two *ovaries,* and it is within these that

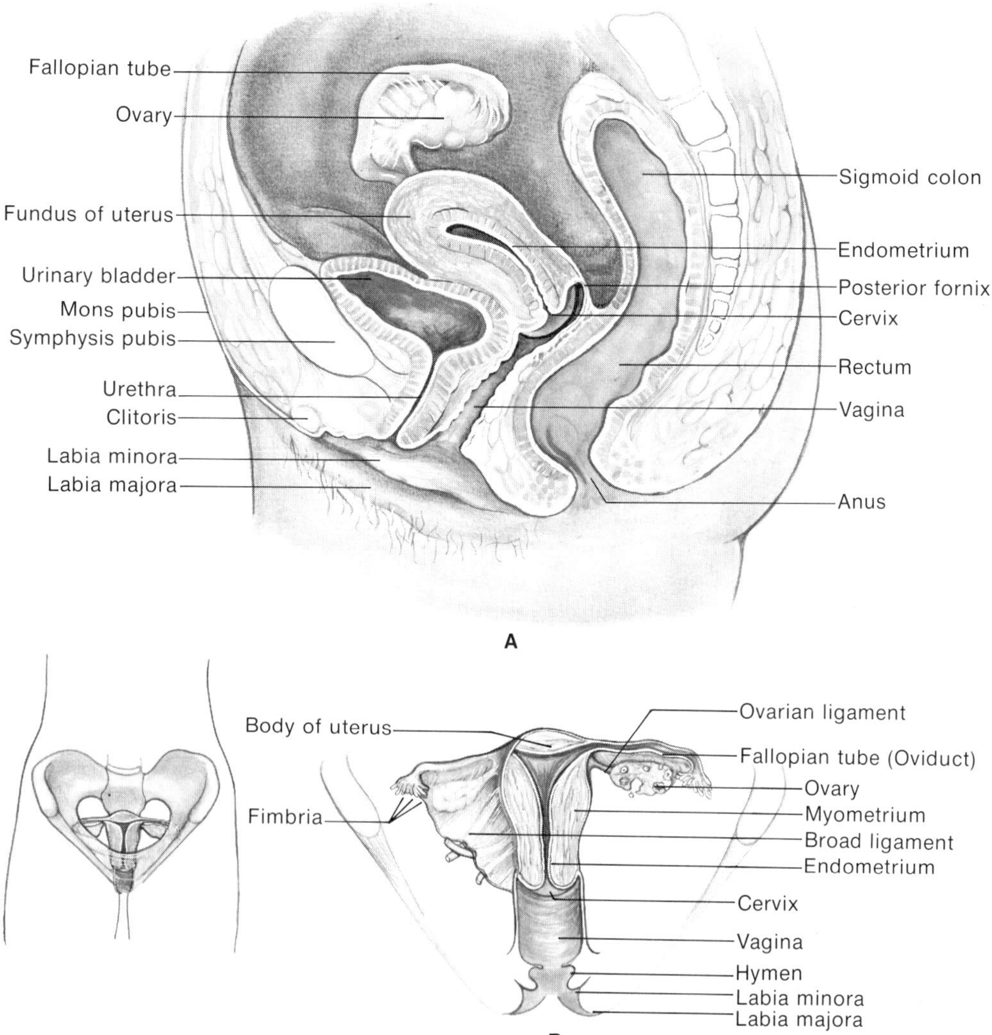

Figure 20-4 Female reproductive system. *A*, Midsaggital section of the reproductive organs of the human female. *B*, Anterior aspect of the female reproductive organs, with the left tube, ovary, and entire ureters cut away to show their internal anatomy. At the left is the position of the female reproductive organs in relation to the pelvis.

the female sex gametes, the ova, are formed by oogenesis. The ovaries are the counterparts of the testes in the male. They are small bodies located in the pelvic part of the abdomen, and are attached to the uterus by two layers of peritoneum called the *broad ligament* (Fig. 20-4, *B*). All the eggs to be produced—reportedly 750,000 of them—during the lifetime of a woman are already present in the ovaries of the female child. At the beginning of puberty the eggs begin to mature, one at a time. Approximately every 28 days an egg passes down the oviduct (fallopian tube) toward the uterus. The egg lands on the endometrium of the uterus and, if not fertilized, is expelled, along with the endometrial lining, through the cervix, down the vaginal canal, and to the outside. This process is called *menstruation,* and, as mentioned, occurs about once every month. This regularity is controlled by the pituitary gland.

The *fallopian tubes* are small and muscular, and are lined with cilia (Fig. 20–4, *A* and *B*). They are nearly 5 inches long, but are not directly connected to the ovaries; their distal, open end is fragmented into finger-like projections called *fimbria* (Fig. 20–4, *B*). Once inside the tube, the ova are propelled along toward the uterus by the combined motions of the cilia and peristalsis.

The pear-shaped *uterus* is located between the urinary bladder and the rectum (Fig. 20–4, *A*). It is approximately 3 inches long, 2 inches wide, and 1 inch deep. Its upper, wider end is called the *fundus;* the lower, narrow end that opens into the vagina is the *cervix*. The interior lining of the uterus is a specialized epithelium known as the *endometrium,* and it is this layer that is concerned with either menstruation or the development of the placenta. The endometrium is supported by a thick muscular layer, the *myometrium,* which is under the control of the autonomic nervous system (Fig. 20–4, *B*).

The *vagina,* a muscular tube about 3 inches long, connects the uterus with the outside (Fig. 20–4, *A* and *B*). Its lining is a folded type of mucous membrane that allows the vagina to enlarge greatly at childbirth. At or near the opening of the vaginal canal to the outside there is a folded membrane of connective tissue called the *hymen*. An intact hymen prevents insertion of the penis during sexual intercourse (coitus), and sometimes must be removed surgically (hymenectomy).

At either side of the vaginal opening are two glands, the *Bartholin glands,* which secrete a mucus-like substance that serves to lubricate the canal. Occasionally, these glands may become infected, and surgical incision and drainage may be necessary.

The region of the female external genitalia is called the *vulva* (Fig. 20–5). At each side of the vestibule there are two pairs of fleshy lips, the *labia majora* and *labia minora*. The *clitoris,* located near the

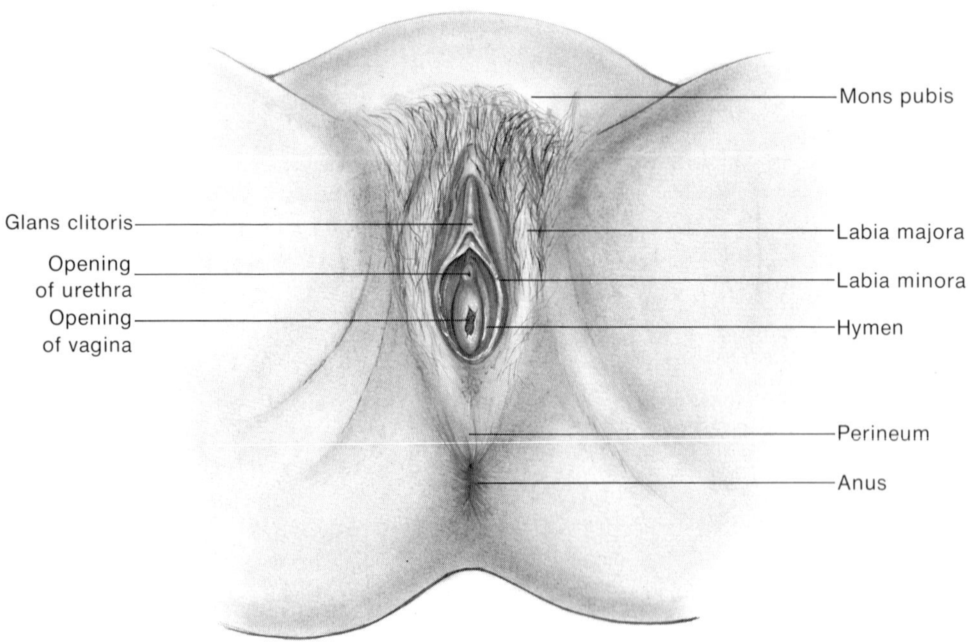

Figure 20–5 Perineal view of the female genitalia.

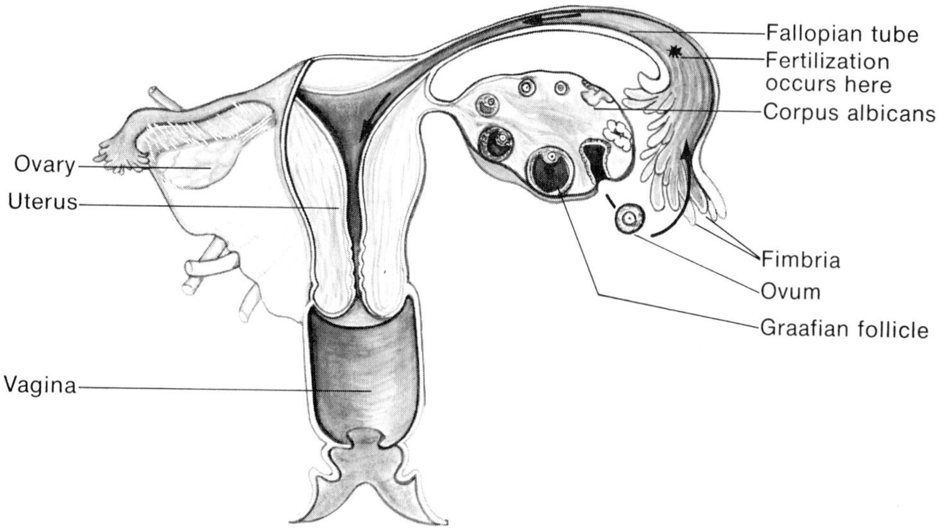

Figure 20–6 Schematic drawing of the female reproductive organs, showing path of the egg from the ovary into the uterus.

anterior junction of the labia minora, is composed of erectile tissue, as is the penis in the male. The region between the external genitalia and the anus in both sexes is the *perineum.*

Production of the Ovum

As we have seen, the germinal epithelium of the ovaries yields many thousands of ova, but only one or two develop each month. The designated ovum becomes surrounded by a mass of cells that form a follicle, or capsule, which acts as the "home" for the ovum during its maturation (Fig. 20–6). At ovulation, the follicle ruptures and expels the ovum. During the two-week period spent in the follicle, the ovum has developed by a process of meiosis and has released two polar bodies that died during the process. The ovum now enters the oviduct and is propelled down the tube to the uterus, where expulsion or implantation occurs.

After expulsion of the ovum, the cells of the ruptured follicle do not regenerate but rather proliferate greatly, developing into the *corpus luteum,* a solid mass of yellow tissue. This new structure acts as an endocrine gland and secretes progesterone. If fertilization does not occur within 12 to 14 days, the corpus luteum dies and becomes scar tissue, a *corpus albicans* (Fig. 20–6).

Hormonal Function of the Female Reproductive System

The Menstrual Cycle

Menstruation, ovulation, and their associated monthly body changes begin in the 11- to 13-year-old girl, and are controlled primarily by the anterior pituitary gland. The cycle can best be described if it is considered in its two phases—preovulatory and premenstrual:

1. Preovulatory Phase. In the preovulatory phase, the anterior pituitary secretes *follicle-stimulating hormone* (FSH), which acts to initiate the development in the ovary of a follicle *(graafian follicle)* containing an ovum. As the ovum matures, the cells of the follicle secrete *estrogen,* which, in turn, is largely responsible for the development of the secondary sex characteristics in the maturing female. Estrogen also supports oogenesis in the ovary and initiates the growth of the endometrial wall in the uterus, preparing it for the implantation of the fertilized egg. Additionally, this hormone, like testosterone in the male, effectively directs tissue growth—the characteristically wide pelvis and breasts of the female are good examples of its influence. A high blood level of estrogen prevents further secretion of FSH from the pituitary, and therefore acts as a control mechanism. The rising estrogen level also causes the pituitary to begin to release *luteinizing hormone* (LH). It is, in fact, the ratio of FSH to LH in the blood that triggers the release of the egg (ovulation).

The first half of the menstrual cycle just outlined takes from 12 to 14 days. By the end of this stage, the ovum is mature, haploid, and ready for fertilization. The pressure of fluid inside the follicle rises and the follicle bursts, ejecting the egg into the finger-like projections (fimbria) of the nearby oviduct.

The first phase is now complete, but before we go any further, we should correlate what has happened throughout the female reproductive system (Fig. 20–7):

a. The anterior pituitary has secreted FSH and, later on, LH to effect the development of the graafian follicle.

b. The follicle in the ovary has secreted estrogen to promote oogenesis and to initiate the development of the uterine endometrium.

c. The uterus "expects" an embryo to be implanted; therefore, the endometrium becomes very fleshy and engorged with blood vessels that will afford a proper means of exchange between the embryo and the mother.

2. Premenstrual Phase. At this point, subsequent events in the cycle are determined by whether or not the egg has been fertilized. First we shall consider the normal menstrual cycle that occurs when the ovum is not fertilized; later, fertilization and pregnancy can be integrated with these events.

Following ovulation, LH transforms the pre-existing follicle cells into a mass of yellow tissue. This mass, the corpus luteum, becomes a new endocrine gland in

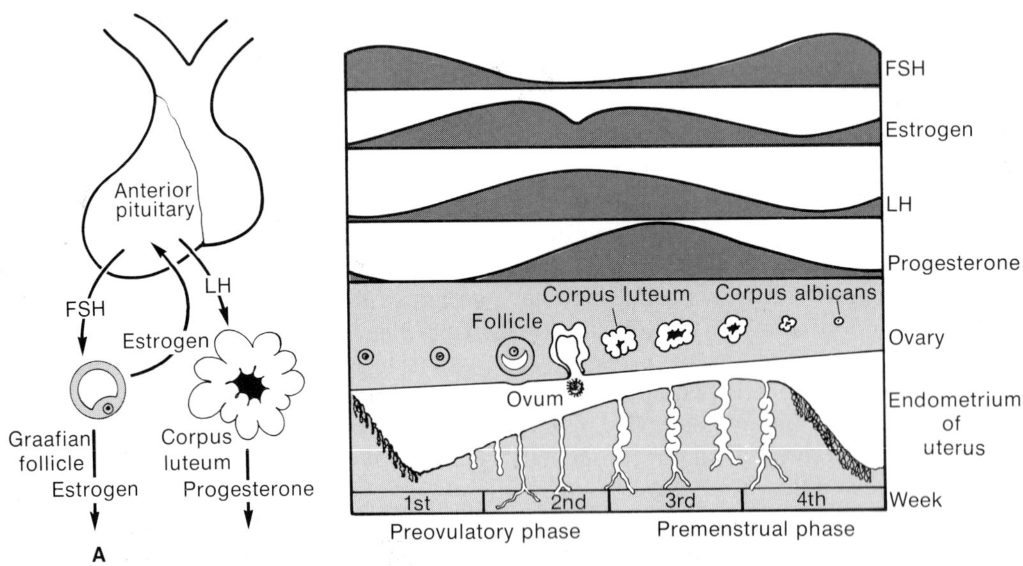

Figure 20–7 Relationships among the anterior pituitary, ovarian, and endometrial cycles. *A*, Menstrual cycle of a nonpregnant woman. The hormone levels in the blood are graphically illustrated to show their effect in the ovary and on the endometrium. Note the correlation of FSH and estrogen in the preovulatory phase, and of LH and progesterone in the premenstrual phase.

Illustration continued on opposite page

that it secretes progesterone. Progesterone's primary role is to continue the development of the endometrium in the uterus; its secondary role is to aid the other female gonadotropic hormones in maintaining the secondary sex characteristics.

Maintenance of the corpus luteum depends on a continuing supply of hormones from the pituitary. The supply of pituitary hormones is not adequate to keep the new glandular tissue alive for more than 12 to 14 days. As we have seen, in the absence of adequate support from the pituitary, the corpus luteum degenerates and becomes scar tissue, a corpus albicans. A resultant drop in progesterone secretion initiates a sloughing of the uterine endometrium, and menstruation ensues. A low progesterone level in the blood, in turn, triggers the pituitary to secrete FSH again, and the next month's cycle begins.

The entire cycle is summarized in the following diagram:

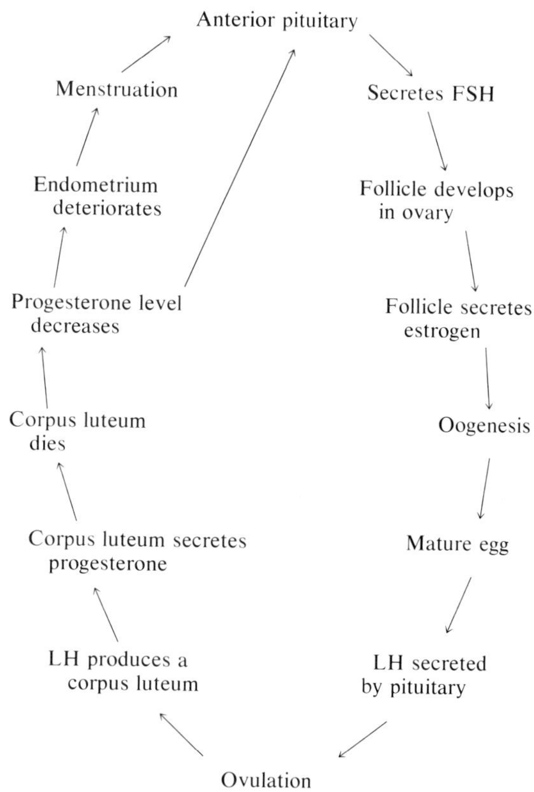

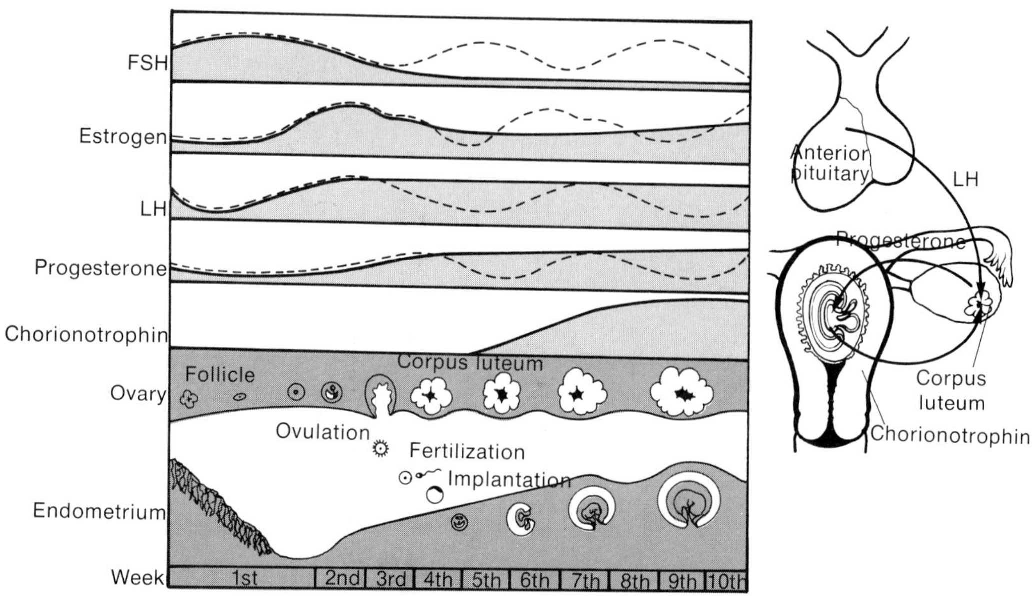

B

Figure 20–7 *Continued B*, Cycle during pregnancy. The addition of chorionotropin from the embryo eliminates the premenstrual phase by keeping the corpus luteum alive and continually secreting progesterone to the endometrium. Abbreviations: LH, luteinizing hormone; FSH, follicle-stimulating hormone.

When an egg is fertilized, the cells resulting from the growth of the embryonic membranes secrete chorionotropic hormone (chorionotropin) which, like the pituitary hormones, maintains the existence of the corpus luteum and its progesterone secretion. The continued secretion of progesterone keeps the endometrium intact and prevents the pituitary from secreting further FSH. This development will be discussed in detail later in this chapter.

The regularity of the menstrual cycle can be interrupted in two ways—by pregnancy, as already mentioned, or by menopause, when the uterine cycles begin to change. During menopause FSH is still produced, but ovulation ceases. The woman in menopause must adjust both physiologically and psychologically to the withdrawal of estrogen. It is not rare, by any means, for a doctor to prescribe diminishing doses of estrogen, over a period of time, to assist a woman experiencing a troublesome menopause.

The Mammary Glands and Lactation

The mammary glands—the breasts—are accessory glands of the female

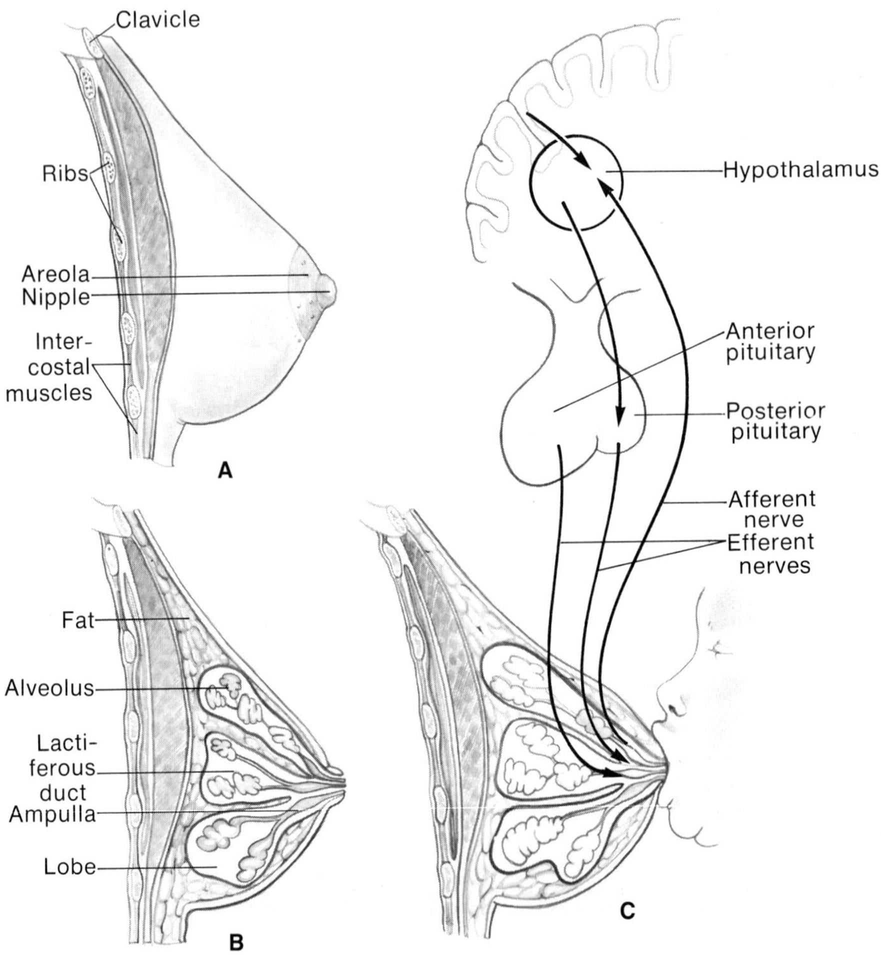

Figure 20-8 Mammary glands. *A* and *B*, Lateral aspect and sagittal section of the mammary glands, showing the external and internal anatomy of the milk (lactiferous) glands and duct system. *C*, Lactation—The child's suckling gives rise to afferent nerve impulses which, in turn, stimulate the mammary glands to secrete milk.

reproductive system. Their function is to provide nourishment for the newborn baby. The secretion of milk by the mammary glands at this period is known as *lactation.* Each mammary gland is divided into a number of lobes composed of glandular tissue and fat. The secretions of the lobes are conveyed through *lactiferous ducts*, all of which converge on the nipple (Fig. 20–8, *A* and *B*).

In childhood, the breasts are small and undeveloped; the rudimentary ducts lie surrounded by fibrous tissues. When a girl reaches puberty, estrogen from the developing ovarian follicle stimulates the ducts to grow and branch. By maturity, when the ovarian hormonal cycle is well established and ova are being produced regularly, progesterone from the corpus luteum stimulates the specialized acinar glands to develop.

In pregnancy, progesterone and estrogen stimulate further growth of the acinar glands and their ducts. The zone around the nipple, the areola, and the nipple itself become darkly pigmented. After childbirth, the fall in blood levels of estrogen and progesterone triggers the release of prolactin (lactogenic hormone) from the anterior pituitary. Prolactin stimulates the acinar glands to secrete the milk produced by blood flowing through the glands.

Lactation starts three to four days after childbirth, and is maintained by prolactin. Afferent nerve fibers convey impulses established by the child's suckling (Fig. 20–8, *C*). These impulses give rise to efferent nerve impulses to the posterior pituitary gland for the release of oxytocin, which is carried by the blood to the acinar glands to maintain secretion of milk. If continued lactation is not in the mother's best interest, it can be suppressed by injections of estrogen.

Physiologically, the advantages of breast feeding are clear-cut in the first few days after childbirth. The first milk secreted is called *colostrum.* Its nutritional content is a little different from that of ordinary milk, but, more importantly, it also contains antibodies from the mother that can be absorbed by the infant to provide a short-term passive immunity against disease.

Cancer of the breast is one of the most common forms of malignant disease in women. Any lump should be reported to a physician at once. Commonly, some or all of the lump is removed and examined (biopsy) while the patient is in the operating room. If it is found to be malignant, removal of the affected breast and its associated axillary lymph nodes and pectoral muscles has been the usual treatment, a procedure called *radical mastectomy* (Fig. 16–7, p. 339). The operation often is followed by x-ray or cobalt therapy in order to ensure the destruction of any stray cancer cells that may remain after surgery.

A *simple mastectomy*—removal of the affected breast but not of the lymph nodes or pectoral muscles—is preferred by some surgeons.

Disorders of the Female Reproductive System; Female Sterilization

Leukorrhea is a nonbloody vaginal discharge that is whitish in color. It is most often a symptom of infection or congestion of the cervix or vagina, and is caused by a single-celled animal, *Trichomonas vulgaris.* The organism can be detected by a vaginal smear. The condition is potentially dangerous in that any cervical disorder can erupt into more severe problems, such as cancer.

Absence of menstrual flow is called *amenorrhea,* and is symptomatic of a hormone deficiency. *Dysmenorrhea,* on the other hand, is painful or difficult menstruation. Common causes are pelvic congestion, endocrine disturbances, faulty position of the uterus, or any form of obstruction of the reproductive tract. In some cases, the first pregnancy has proved to be the best remedy.

Uterine tumors, such as fibroids, are common. It is reported that over 50 per cent of women over 50 years of age or older have one or more uterine growths. Problems arise when these growths become malignant or when they interfere with pregnancy. Fibroids may become large enough to cause grave disorders. For this reason and others already mentioned, it may be necessary to remove the entire uterus *(hysterectomy)* or, at times, the entire reproductive system (salpingo-oophorectomy with hysterectomy).

Abnormal vaginal bleeding and irregular discharge are among the first symptoms of uterine cancer. Routine cancer detection studies of the upper vagina, such as the *Papanicolaou stain slide test* (Pap smear), have saved many lives.

Salpingitis is an inflammation of a tube, and is commonly caused by gonorrhea or the tubercle bacillus. The infection can produce sterility because it obstructs the tubes, preventing passage of the egg. *Oophoritis* is an infection of an ovary. This condition or an ovarian tumor may necessitate surgical removal of both the ovary and the tube.

Causes of sterility in women correspond directly with those in men. Inadequate quantities of hormones (hyposecretion), abnormal development, trauma, or intentional ligation of the tubes all result in a degree of sterility. In most cases of difficulty in achieving pregnancy, the husband should be examined first, since procedures for detecting sterility in the male are far simpler than those used in women.

In *endometriosis,* endometrium-like masses are formed in a variety of sites within the female pelvis and abdominal cavity. Such lesion masses occur only during the active reproductive period of a woman's life. Most women who experience endometriosis are sterile. The lesions tend to hemorrhage periodically, forming scar tissue wherever they may be located.

COPULATION, FERTILIZATION, AND EARLY DEVELOPMENT

Copulation

Ejaculation is the expulsion of semen from the male urethra during sexual intercourse (copulation) or during sleep (nocturnal emission). Ejaculation is impossible without erection of the penis, which is controlled by autonomic impulses from the thalamus and cerebral cortex. Psychological or physical stimulation results in the dilation of the arteries and blood sinuses of the penis so that they become engorged with blood and an erection occurs. The friction of the copulatory act stimulates the nerve endings of the glans penis, and these impulses activate a reflex mechanism that terminates in ejaculation itself. During this act, the vas deferens, seminal vesicles, and prostate gland all contract, propelling the semen and sperm into the vagina. This process of ejaculation and the feeling of intense pleasure accompanying it are termed an *orgasm.* The sperm cells, being extremely motile, swim up the vagina toward the uterus and eventually enter the tubes. It is said that the sperm can swim from the vagina to the upper end of the tubes in less than one hour.

Fertilization

The microscopic sperm cells are very fragile and susceptible to damage and destruction before they encounter the egg. To preserve the sperm for at least 48 to 76 hours, the seminal fluid contains energy sources, pH buffers, and other chemicals. Of the 300 million sperm in each ejaculation, several thousand are necessary for fertilization. These high numbers may sound ridiculous, but it must be remembered that millions are destroyed during the act of copulation itself, other millions are destroyed by the normal acidity of the vagina, and the remaining number of sperm is halved because both tubes receive sperm and only one contains an egg.

When the minute ovum ruptures from the ovarian follicle, it is surrounded by numerous cells, the *corona radiata* or protective zone. Sperm arriving near the egg secrete an enzyme, *hyaluronidase,* necessary to break down an organic acid barrier, *hyaluronic acid,* secreted by the cells of the corona radiata. Individual sperm contain the enzyme, but it takes thousands of sperm to secrete enough enzyme so that one sperm can successfully enter the egg and accomplish fertilization. Once this sperm has entered the egg, its tail is absorbed and its nucleus joins the nucleus of the egg to complete the transfer of genetic information.

When the process of fertilization is complete, the fertilized egg or zygote is restored to the diploid (46 chromosomes) number. The zygote, therefore, contains a new arrangement of genes on the chromosomes—an equal contribution from each parent. If the ovum was fertilized by a sperm bearing a Y chromosome, the result will by XY, a male; if the sperm contained an X chromosome, the offspring will be XX, or female (Fig. 20–1).

The site of fertilization in the tubes is important. Normal embryonic growth and development will occur if fertilization occurs one-third of the way down the tube from the ovary (Fig. 20–9). The embryo needs the following few days of development to produce the chorionic sac which, in turn, secretes chorionotropic hormone. This hormone, you will remember, is the principal hormone needed to maintain the corpus luteum for a three-month period. The corpus luteum is then able to secrete enough progesterone to maintain the endometrium awaiting the new embryo. Failure of the embryo to develop the chorionic sac in sufficient time will result in normal menstruation and loss of the embryo.

Sometimes the zygote becomes lodged in the tube, and partial development occurs there. Sometimes sperm swim the length of the tube and fertilize an egg ejected from the ovary and dropped into the pelvic cavity. In either case, the pregnancy is said to be *ectopic* ("out of place"). In the first instance, surgical removal of the tube and ovary is necessary. Degeneration of the zygote occurs in the latter case.

More than one egg may be fertilized if more than one is present. This is typical in such animals as dogs and cats, in which ovulation releases many eggs. When a human female releases eggs from both ovaries during the same cycle, the conception of twins is not abnormal. The twins in

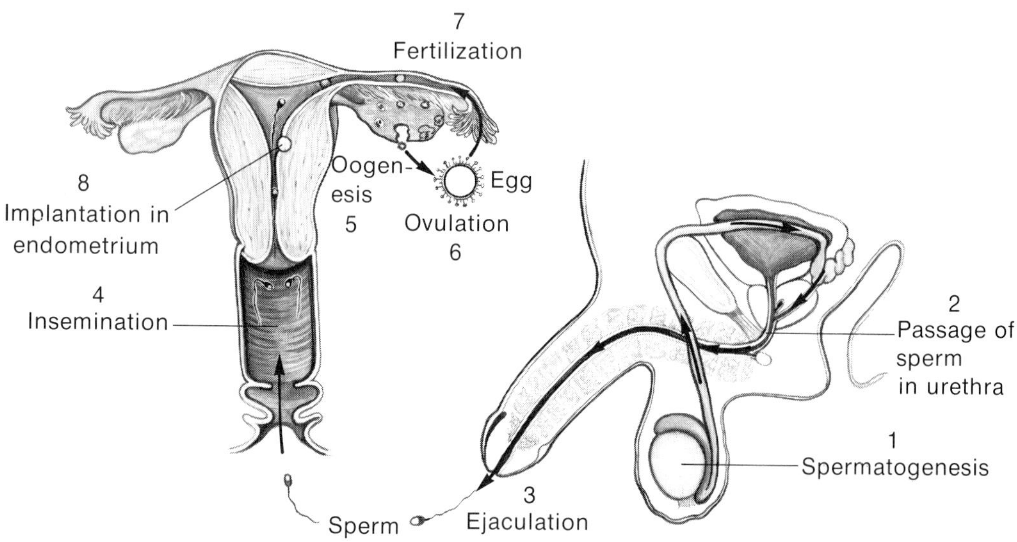

Figure 20–9 Conception. Schematic diagram showing the events leading up to fertilization of the egg and eventual implantation in the uterus. The arrows indicate the direction of flow of the swimming sperm; the numbers indicate the sequence of events.

this case would be *fraternal,* meaning that each child grows separately from its own placenta. *Identical twins,* triplets, and so forth occur when the zygote divides and more than one embryo develops from the same original egg.

Family planning today sometimes involves the use of hormones (FSH-derivative) to produce many eggs during a cycle—the so-called "fertility drugs"—for couples having difficulty in conceiving a child. The intent is to narrow the odds for the woman hoping for fertilization. Multiple births have occurred following use of these drugs.

Early Development

The zygote passes through a series of distinct developmental stages. *Cleavage* is the first step after fertilization. This involves a process of cell division (mitosis) without any significant growth. The one-celled zygote divides and passes through stages described as two-cell, four-cell, eight-cell, and so forth (Fig. 20–10). Since the cells do not grow very much between divisions, cleavage makes them smaller and smaller. At the end of the cleavage stage, the zygote consists of a ball of many small cells, each cell bearing a complete set of genes from the original zygote.

By the time cleavage has produced several hundred cells, the zygote has become a ball-shaped structure consisting of a single layer of cells surrounding a fluid-filled cavity—a *blastula* (Fig. 20–10). The blastula (blastosphere) is commonly referred to as the "job-assignment" stage. The time element is now becoming important. Blastula formation took approximately five days, and the zygote is now located just inside the horns of the uterus, ready for implantation. The corpus luteum, meanwhile, is about half way through its 10- to 12-day lifespan in the ovary, and the quantity of pituitary hormones has already begun to diminish.

In the next stage, *gastrulation,* the cells of the zygote begin to grow and differentiate into cells clearly different from one another. At this point there is an inner layer of cells, the *endoderm,* and an out-

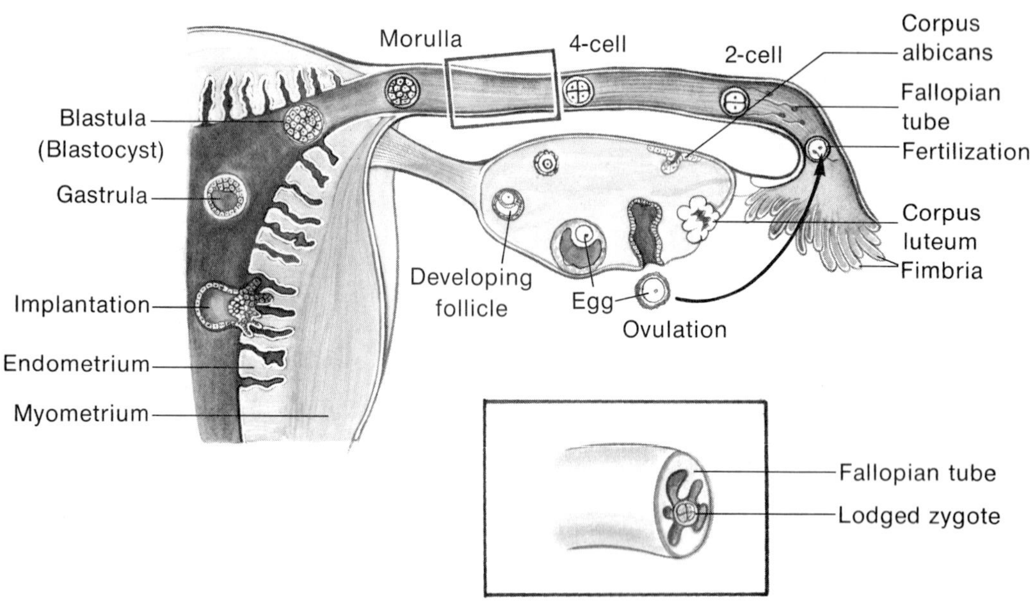

Figure 20–10 Implantation of the fertilized ovum. Approximate stages of the fertilized egg during development and implantation in the endometrium. The inset shows a section of the fallopian tube, with partial development and eventual surgical removal of an ectopic pregnancy.

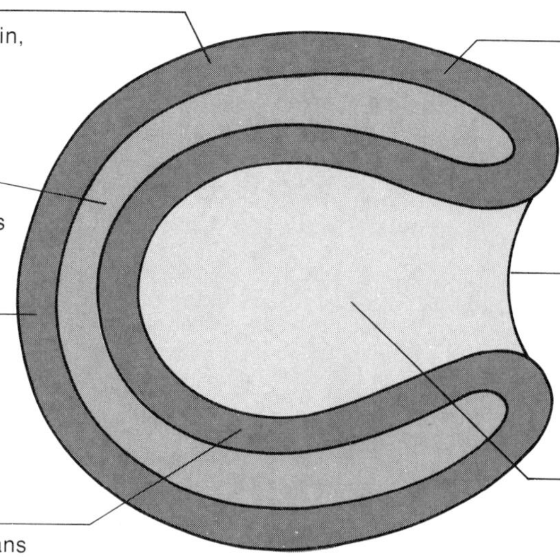

Figure 20–11 Early differentiation of the various tissue layers as seen in the gastrula stage of the human embryo.

side layer, the *ectoderm*. Gastrulation represents the tissue formation of the new embryo, which, seven days following fertilization, becomes implanted in the endometrium (Fig. 20–10).

During gastrulation, one side of the hollow wall of cells becomes indented, and this indentation continues until a cup-shaped structure is formed. The resulting cavity now becomes the inner layer, and the previous inner layer becomes a new middle layer of cells—the *mesoderm*. As the embryo continues to differentiate and develop, the mesoderm will produce bones, muscles, the heart and circulatory system, and the excretory and reproductive systems (Fig. 20–11). The endoderm differentiates into the intestinal tract, stomach, pancreas, liver, and lungs. The ectoderm produces cells that will become the skin, brain, and nervous system.

FROM IMPLANTATION TO BIRTH

Once the developing mass of cells has reached the uterus, it generally takes four to five more days before it becomes firmly implanted in the uterine lining. Implantation begins when the embryo digests some of the endometrium. The embryo, in its gastrula form, sinks into the soft lining, and later becomes firmly attached. When implantation is complete, the cells of the embryo and the endometrium form a structure called the *placenta* (Fig. 20–12) that enables the mother and embryo to exchange materials throughout gestation. The lining of the uterus thickens, and finger-like projections, the *villi,* develop. These function in the exchange of products of metabolism between the mother and the embryo. Loops of blood capillaries are connected to the embryo through the umbilical cord. These capillary loops are bathed in tiny pools (sinuses) of blood supplied by the mother. It is important to remember that the maternal and fetal circulations are separated by thin membranes and do not mix. The exchange of nutrients and waste products occurs by diffusion through these membranes.

Fetal Membranes

The developing embryo forms two principal extraembryonic membranes, the *amnion* and the *chorion* (Fig. 20–13).

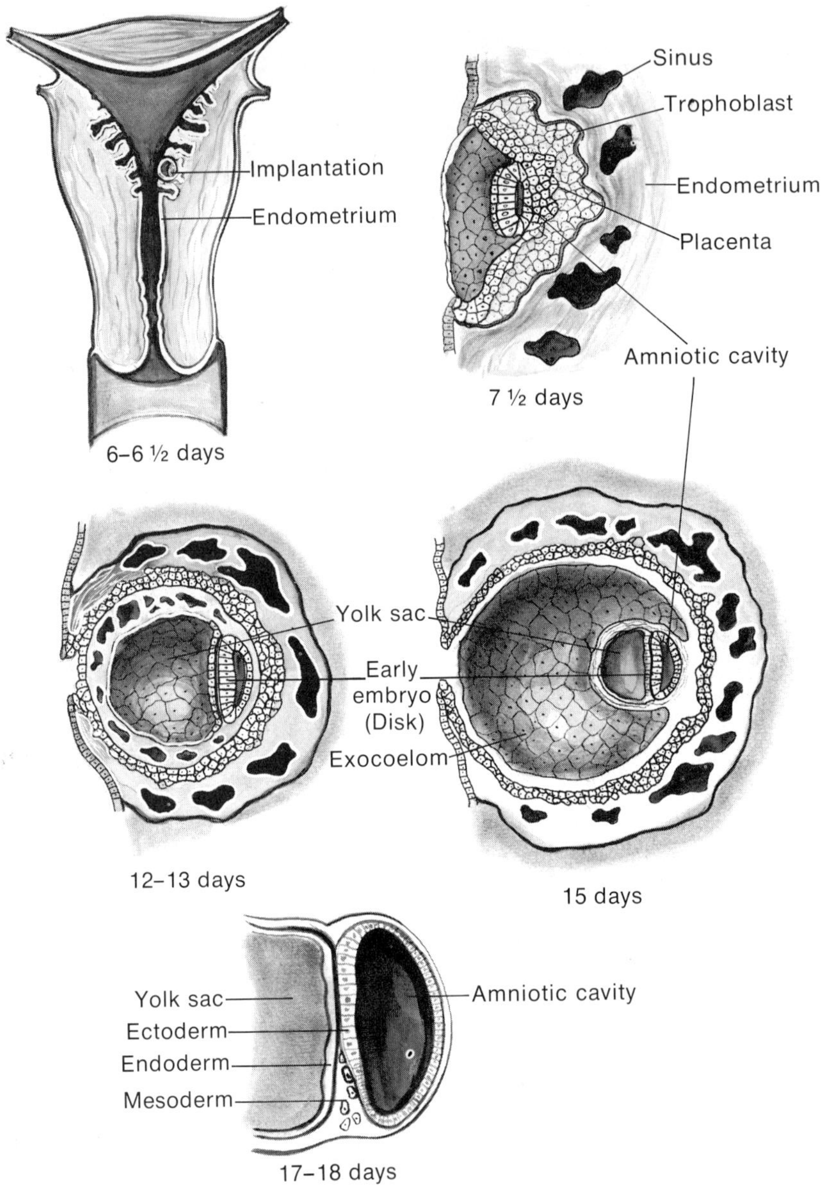

Figure 20–12 Implantation. Schematic representation of the development of an embryo up to three weeks. The three primary embryonic germ layers are indicated.

The amnion is a thin protective layer that directly covers the embryo and is filled with amniotic fluid. The amniotic sac is very small at first, but expands during the embryo's development until it eventually fills the entire extraembryonic cavity.

The chorion is a thicker membrane that invades the endometrium with villi and forms the embryonic portion of the placenta. It is this membrane that produces the all-important chorionotropic hormone that will maintain the corpus luteum for three months. The placenta requires two months to develop sufficiently to produce adequate levels of progesterone; therefore, the first two months of endometrial stimulation must be supplied by the corpus luteum. During the third month of its existence, the corpus luteum works in conjunction with the placenta to maintain the progesterone level. As an outer membrane, the chorion is also responsible for the interchange of nutrients and waste products between maternal and fetal circulations.

The human ovum does not contain the large quantity of yolk present in the eggs of many lower animals. A *yolk sac* develops (Fig. 20–13), but it is evident only during early embryonic existence. It eventually becomes incorporated into the umbilical cord, where it degenerates. A second rudimentary sac, the *allantois* (Fig. 20–13), also degenerates as the embryo grows, and reportedly functions as an excretory organ during its short existence.

The Course of Human Development

The cells inside the blastosphere (Fig. 20–10) form a plate, the *embryonic disk*

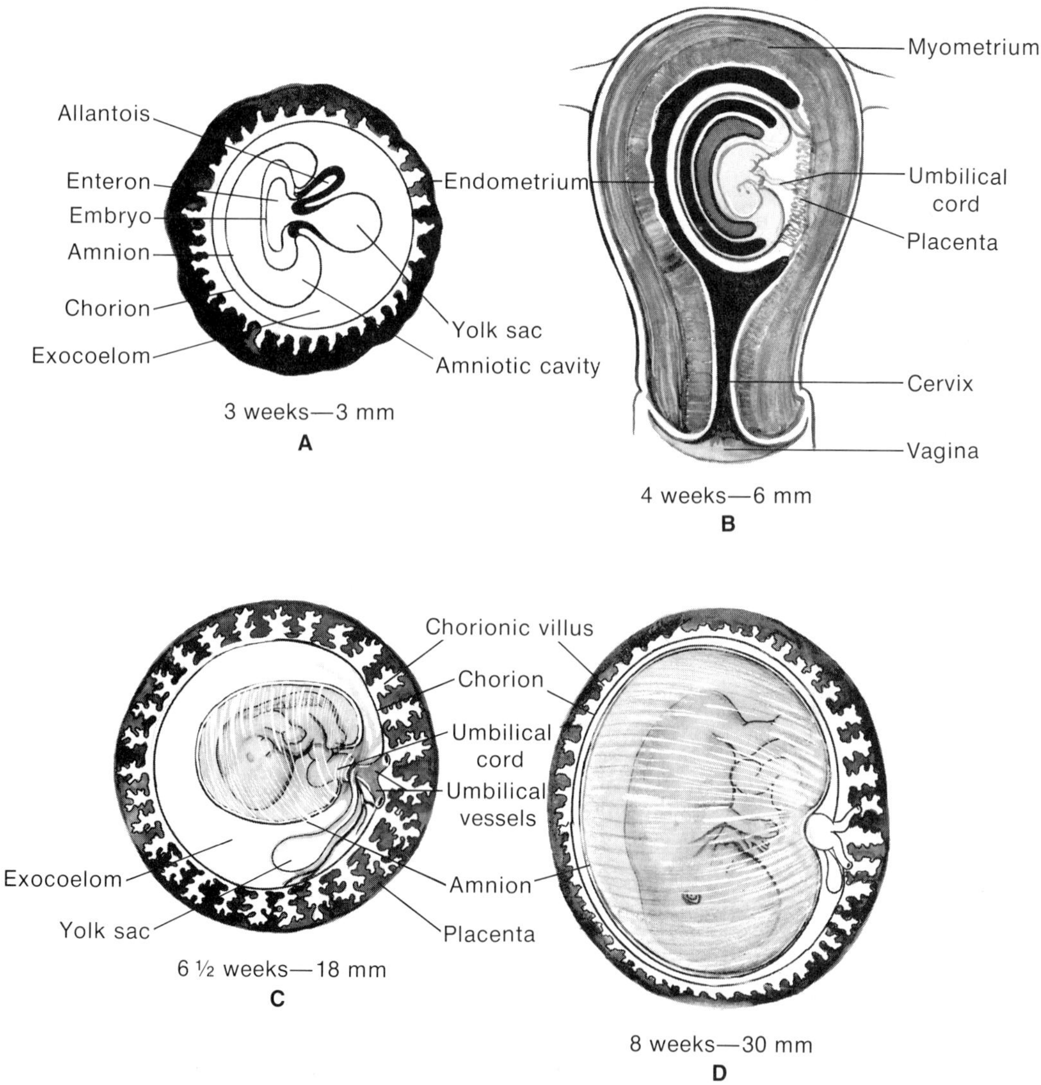

Figure 20–13 Fetal membranes and early human development. *A*, Schematic representation of a three-week-old embryo enclosed within its fetal membranes. *B*, Coronal section of a pregnant uterus. *C* and *D*, Development of the human embryo through its sixth to eighth weeks of growth.

(Fig. 20–12). It is this disk that eventually undergoes gastrulation between eight and 20 days after fertilization, becoming an embryo. By the time a human embryo is one month old, it is about 1/4 inch long, and the developing features of the brain, eyes, heart, stomach, and liver can be distinguished. Its heart beats regularly and its weight is many thousand times what it was at the moment of conception—but still less than that of an envelope.

A convenient way to follow the development of the embryo is to indicate its course of development chronologically, rather than by structure. The following sequence begins one week after fertilization and terminates at birth:

1 week—Beginning of implantation.

2 weeks—Implantation complete; placenta begins to form; fetal membranes functional.

3 weeks—Embryo is 1/8 inch long; growth centers (somites) appear; mesodermal tissue develops into major filtering organs (liver, spleen, lymph nodes); heart begins development; a two-lobed pancreas develops from the intestine; an ear pit starts development internally; the embryo demonstrates a forward and tail end.

4 weeks—Embryo is 1/4 inch long; heart has fully developed into a four-chambered pump; amniotic and chorionic sacs are clearly visible; eye cup forms externally.

5 weeks—Embryo is 2/5 inch long; placenta is 1 inch in diameter; forebrain and midbrain differentiate into their respective areas; arm and leg buds appear, the arm bud developing faster; bronchial tubes begin development; alimentary tract differentiates into stomach and intestine, with one loop of the enormously long intestine pushing out into the umbilical cord; head seems to elongate; eye cup appears on the neck.

6 weeks—Embryo is 2/3 inch long; a dark retina and eyelids appear; ear centers now visible as the eye cup begins to assume its normal position; esophagus elongates, differentiating into pharynx and pharyngeal pouches; embryo has a cartilage form for a skeleton; external genital features appear but not enough to determine sex; a large heart and liver are prominent.

7 weeks—Embryo is 1 inch long; fingers and toes are evident, but no joints present; eyes are now in a forward position; a diaphragm separates the pleural and peritoneal cavities; the two-lobed pancreas fuses to form one; external ear structures appear; a shimmering spinal cord can be seen bordering the embryo's back.

8 weeks—Embryo is 1 1/5 inches long; an important period—the embryo is now called a *fetus* because the last functional system, the bony skeleton, has begun to form from cartilage; muscles become evident in the contracting arms and legs; the heart is very active at this point; the fetus is now a recognizable human being that will stir the amniotic fluid with its weak kicking.

9 weeks—Fetus is 1 1/2 inches long; a prominent liver makes up about 10 per cent of the body weight; the eyes will now close and remain closed until delivery; skull formation is evident.

10 weeks—Fetus is 2 inches long; bone formation has markedly increased (almost double in two weeks); the extruding loop of the intestine is now pulled back in from the umbilical cord; the male and female sex organs are formed but are not evident for another five to six weeks.

11 weeks—Fetus is 2 1/2 inches long; now weighs approximately 3/4 ounce (weight of a letter), yet has been functional for quite a few weeks.

12 weeks—Fetus is 3 inches long; nail beds develop on fingers and toes.

13 to 14 weeks—Fetus is 4 inches long; ears have now moved into proper position with final formation of head shape.

16 weeks (4 months)—Fetus is 6 inches long; feet and legs start to become chubby; skin hair begins to develop; fetus is now becoming larger and quite active—the mother should feel the first signs of life at this point; respiratory bronchioles are developing.

17 weeks—Fetus is 8 inches long; solid trunk formation as final systems are fully developed.

18 weeks—Fetus is 9 inches long; sex organs are evident at this point and can be identified; the first hair becomes evident.

20 weeks (5 months)—Fetus is 11 inches long; fingernails are fully developed (in fact, a fetus has been known to scratch its way through the amniotic sac); muscles are very active; umbilical cord measures about 20 inches.

6 months—Fetus is 13 inches long; hair is evident over the body; oil (sebum) is secreted to protect the skin and hair from drying; fetus is quite active (it sleeps and wakes just as if it had been born); remaining functional structure of the respiratory tract is developed—alveolar sacs are present, and the lungs are somewhat functional despite their airless quarters.

7 to 9 months—Fetus is 14 to 20 inches long at delivery; this is the time of growth and putting on weight—all the systems needed for life are developed and ready. If the fetus is delivered after the seventh month, it has a chance to live; before the seventh month, survival is very improbable.

Birth

The birth of a baby is preceded by considerable hormonal changes in the mother—particularly by secretion of a de-

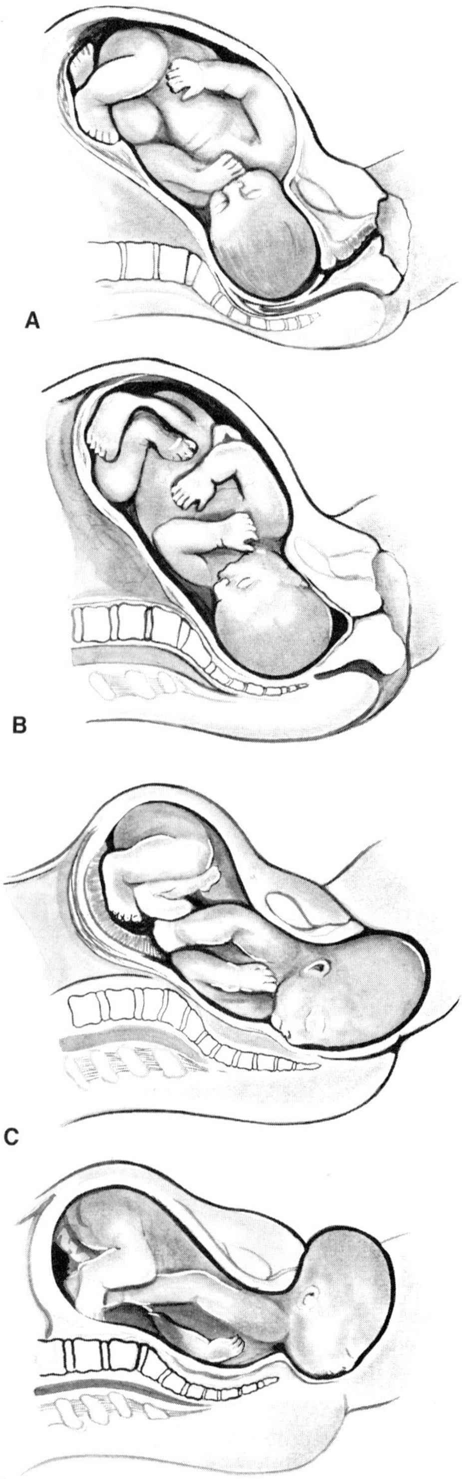

A

B

C

D

Figure 20–14 Birth.

A, At childbirth (parturition), the cervix dilates, and the fetal membranes rupture.

B, Rhythmic uterine muscle contractions move the fetus through the dilated cervix into the birth canal.

C, The head now rotates so that it faces the pubic symphysis.

D, The final phases of birth are well underway.

creased proportion of progesterone relative to that of estrogen. These changes act to increase the contractility and irritability of the uterus, thus facilitating delivery (Fig. 20–14).

Approximately 280 days after fertilization, the uterine hormone levels suddenly drop, and the birth process starts. The baby cannot pass out of the uterus into the vagina unless the cervix is first dilated. Contractions of the uterus force the baby's head against the cervix, providing the main factor in cervical dilation. During this process the amniotic sac is also forced into the cervix, contributing to its dilation. This stage usually requires several hours, and the strong uterine contractions that accompany the initial stages of childbirth give rise to the pains commonly called "labor pains."

At the end of this first stage the amniotic sac usually ruptures, allowing the fetus, head first as a rule, to pass through the cervical canal and the vagina to the outside (Fig. 20–14). During the third and last stage, the umbilical cord is cut, freeing the fetus from the mother. Finally, the placenta (afterbirth) is expelled.

The newborn baby is no longer protected by its mother's womb. It must now breathe, eat, and grow, and rely on its own homeostatic mechanisms to maintain its own life.

Further sources of discomfort during and after childbirth are the rips or tears that can occur in the vagina and other parts of the birth canal. The vagina will stretch many times its size to allow passage of the fetus. However, if the baby's head is larger in circumference than the vagina's distention, the vaginal wall may tear. The area most likely to tear is in the lower, more delicate region leading to the anus. When tears occur in the anal sphincter, contamination and complications may arise. To prevent this, the obstetrician will make a small cut in the vaginal wall, angling out and down away from the anus to allow expansion in that direction, away from the anus. This procedure is called an *episiotomy* (Fig. 20–15).

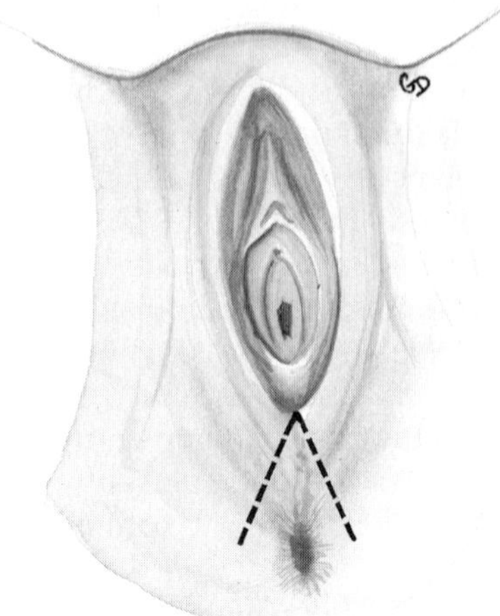

Figure 20–15 Episiotomy.

Abnormalities of Birth

Placenta praevia is an abnormal condition in which the placenta is attached to or near the internal opening of the cervix. Normally it is attached to the upper area of the uterus, and a low attachment can cause an early separation from the uterine wall, resulting in hemorrhage as well as danger to the continued development of the fetus. Surgical removal (cesarian section) of the fetus through an incision made in the abdominal wall and into the uterus is generally the indicated treatment in a case of placenta praevia.

Instances of fetal death have prompted a new terminology to distinguish this phenomenon from abortion or miscarriage. As accepted now, an "early fetal death" occurs when the fetus or embryo is less than 20 weeks old; an "intermediate fetal death" when the fetus is more than 20 weeks but less than 28 weeks old; and a "late fetal death" when the fetus is more than 28 weeks old. A *premature* infant is a baby who weighs less than 5½ pounds or whose gestation has lasted less than 37 weeks instead of the normal 40 weeks.

An *abortion* is the loss of the embryo or fetus during the first 20 weeks, regardless of cause. A *spontaneous abortion* is an expulsion of the embryo due to natural causes, such as infection or disorders of the reproductive tract. Abortions induced by artificial means result in thousands of fetal deaths each year.

A *miscarriage* is generally considered to be a spontaneous abortion occurring between the 20th and 28th weeks of pregnancy.

CLINICAL TESTS INVOLVING THE REPRODUCTIVE SYSTEM

Acid Phosphatase

Cancer of the prostate gland is most often diagnosed when the patient seeks medical advice because of urethral obstruction, as mentioned earlier. The obstruction involves the prostate gland, urethra, and bone in the area, and results in pain as well as dysuria. Phosphatase is an enzyme produced by the prostate, and elevated levels of acid phosphatase in the blood serum suggest cancer of the prostate. Since the acid phosphatase is not absorbed by the blood until the cancer has extended into other tissues, this is not a good test for early diagnosis. Prostatectomy is the most common corrective measure.

A-Z (Aschheim-Zondek) Test

The A-Z test is based on the knowledge that chorionotropic hormone is secreted by the growing embryo after the chorionic sac has been developed. Sufficient levels of the hormone must be present for the test to be meaningful; therefore, it should be taken only after five to six weeks of pregnancy. If the test is given in the first few weeks, false negative results may prove emotionally traumatic. The test is done on an early morning urine specimen because it is more concentrated. If the hormone is present in the urine it will produce a reaction either on a slide test (immediate) or after the urine is injected into an animal (24-hour reaction on the ovaries of the animal).

The A-Z test can also be used to diagnose a disorder of the male reproductive system. Unfortunately, in this instance a positive test indicates a terminal cancer of the testes or prostate gland.

OUTLINE

I. Introduction
   A. Cell division types
      1. Binary fission
      2. Mitosis
   B. Types of tissue cells
      1. Somatic – body
      2. Germinal – sex
II. Meiosis
   A. Gametogenesis
      1. Spermatogenesis – production of male sperm
      2. Oogenesis – production of female egg
   B. First meiotic division – Stages similar to those of mitosis
      1. Long homologous chromosomes unite longitudinally – synapsis
      2. Chromosome pairs line up at midplate – metaphase
      3. Resulting daughter cells together have same number of chromosomes as the parent
   C. Second meiotic division – Basically a mitotic sequence ending in four haploid cells or gametes
      1. Each daughter cell of the first division divides into two daughter cells, each containing single stranded chromosomes
      2. Four haploid daughter cells result for every diploid parent
   D. Spermatogenesis and oogenesis
      1. Four viable sperm are produced from every germinal epithelial cell in the testes
      2. A single viable egg and two polar bodies are produced from every germinal cell in the ovaries

III. Reproduction in the Human Male
   A. Gross anatomy
      1. Testes
      2. Vas deferens
      3. Seminal vesicles
      4. Prostate gland
      5. Cowper's gland
      6. Penis
      7. Epididymis
   B. Microscopic structure
      1. Seminiferous tubules
      2. Spermatogonia in germinal epithelium—undergo spermatogenesis
      3. Associated cells
         a. Leydig—interstitial cells between the seminiferous tubules; produce and secrete testosterone
         b. Sertoli—dispersed between germinal spermatogonia to nourish them
   C. Hormonal function
      1. Anterior pituitary
         a. Secretes ICSH, which stimulates Leydig cells in testes
      2. Testes—Leydig cells, in turn, secrete testosterone to promote spermatogenesis
      3. Abnormalities
         a. Hyposecretion before puberty produces sterility and lack of male development (eunuchism)
         b. Hyposecretion after puberty results in sterility and some regression of virility
         c. Congenitally undescended testes result in sterility due to increased temperature
      4. Sperm
         a. Microscopic—0.06 mm (60 μ)
         b. Specialized for motility—tail
         c. Contains many mitochondria in its body and neck to facilitate vigorous activity
   D. Disorders of male reproductive system; male sterilization
      1. Cryptorchidism—undescended testes
      2. Gonorrhea—symptom is urethritis
      3. Phimosis—remedied by circumcision
      4. Vasectomy—ligation of vas deferens
      5. Orchitis—inflammation of testes
      6. Hydrocele—fluid in epididymis

IV. Reproduction in the Human Female
   A. Gross anatomy
      1. Ovaries—egg production
      2. Oviduct—fallopian tubes
      3. Uterus—endometrial lining
      4. Vagina—Bartholin glands
      5. External genitalia—Vulva, labia majora and minora, clitoris
   B. Production of ovum
      1. Many thousands of ova are present at birth, but only one or two develop each month after puberty
      2. FSH from anterior pituitary stimulates formation of follicle
      3. Follicle secretes estrogen to promote genesis of egg from ovaries' germinal epithelium
      4. LH secreted from anterior pituitary assists in expulsion of haploid egg from ovary—ovulation
   C. Hormonal function—menstruation
      1. Preovulatory phase
         a. FSH stimulates follicle development
         b. Follicle secretes estrogen to promote oogenesis
         c. LH, from anterior pituitary, and remaining FSH promote ovulation
      2. Premenstrual phase
         a. LH promotes formation of corpus luteum from pre-existing follicle

b. Corpus luteum secretes progesterone
c. Corpus luteum dies after 12 to 14 days and menstruation occurs.

3. Accessory effects
   a. Estrogen promotes development of endometrium in uterus
   b. Progesterone maintains endometrial development
   c. Degeneration of corpus luteum and decrease in progesterone level promote endometrial repression and ultimate menstruation
4. Hormonal effects
   a. Hyposecretion before puberty results in sterility and lack of development
   b. Hyposecretion after puberty results in sterility and regression of femininity
   c. Menopause at 40 to 50 years is a gradual regression of hormonal production, especially of estrogen, necessitating physiologic and psychological adjustments

D. Mammary glands and lactation
1. Gross anatomy
   a. Many lobes composed of fat and glandular tissue
   b. Lactiferous ducts—convey milk to nipple
2. Estrogen production at puberty stimulates ducts to grow and branch
3. Progesterone stimulates specialized acinar glands to produce milk
4. During pregnancy, both hormones promote acinar gland growth
5. After childbirth, fall in estrogen and progesterone levels stimulates anterior pituitary to secrete prolactin
6. Prolactin promotes milk production—lactation
7. Child's suckling action stimulates posterior pituitary to secrete oxytocin, which maintains lactation
8. Breast abnormalities
   a. Growths—malignant—radical or simple mastectomy
   b. Growths—benign—simple excision

E. Disorders of femal reproductive system; female sterilization
1. Leukorrhea—vaginal discharge
2. Amenorrhea—absence of menstrual flow
3. Dysmenorrhea—painful menstruation
4. Fibroids—tumors
5. Uterine cancer
6. Salpingitis—oophoritis
7. Tubal ligation
8. Endometriosis—endometrium-like masses formed in abdominal cavity

V. Copulation, Fertilization, and Early Development

A. Copulation
1. Erection
2. Ejaculation
3. Autonomic control

B. Fertilization
1. Sperm
   a. Swim in seminal fluid; nourished by semen
   b. Live expectancy only 48 to 72 hours
   c. Numbers in millions
   d. Each sperm head secretes enzyme hyaluronidase
   e. Thousands of sperm needed to supply enough enzyme to penetrate egg barrier
2. Egg
   a. Corona radiata, a protective shield of cells
   b. Shield secretes hyaluronic acid
3. Fertilization occurs when one sperm penetrates the shield and its nucleus joins with the

egg nucleus to form a zygote
4. Location of fertilization
    a. Should be one-third down oviduct for normal development
    b. Abnormal fertilizations are ectopic
5. Multi-egg production may result in multi-births
    a. Fraternal twins—more than one egg
    b. Identical twins—same egg divides in two

C. Early development (first 5 to 7 days)
1. Cleavage—division of zygote into 2, 4, 8-cell, etc.
2. Blastula—hollow ball: Single-layered zygote
3. Gastrula—cup-shaped form
    a. One side of hollow ball becomes indented
    b. Endoderm, mesoderm, and ectoderm—three cell layers

VI. Implantation to Birth

A. Implantation
1. 4 to 5 day period following blastula formation
2. Gastrula digests endometrium
3. Placenta formation
    a. Finger-like villi
    b. Exchange of nutrients and wastes
    c. Blood sinus pools

B. Fetal membranes
1. Amnion—covers and protects fetus
2. Chorion
    a. lines endometrial cavity and forms embryonic portion of placenta
    b. Secretes chorionotropic hormone, which maintains corpus luteum for 3 months
3. Yolk sac—rudimentary food source
4. Allantois—rudimentary excretory organ

C. Course of human development
1. First month
    a. Embryo develops systems essential for life
    b. Embryo grows to about 1/4 inch long and rapidly multiplies its weight
2. Second month
    a. Nervous system differentiates
    b. Arm and leg buds appear; fingers and toes develop
    c. Embryonic skeleton formed of cartilage
    d. Alimentary tract differentiates into stomach and intestine
    e. Diaphragm separates pleural and peritoneal cavities
    f. Eye and ear structures first appear
3. Third month
    a. Embryo now a fetus—a living human being
    b. Bony skeleton develops
    c. Fetus now over 2 inches long and weighs as much as a letter
    d. Liver is prominent
    e. Eyes close
4. Fourth month
    a. Rapid growth to 4 to 6 inches
    b. Skin, nails, and facial characteristics develop
    c. Fetus very active—first signs of life
5. Fifth month
    a. Fetus over 8 inches long
    b. Fetus very active—putting on weight
6. Sixth month
    a. Fetus 13 inches long
    b. Hair and sebum evident
    c. Fetus sleeps and wakens
7. Seventh to ninth months
    a. Survival possible if birth occurs during this period

b. Fetus 14 to 20 inches long and up to 7½ lbs.
c. Time of growth and development of systems begun before seventh month
8. Birth (parturition)
a. Hormone level decreases
b. Fetal head pushes into cervix, causing dilation
c. Uterus contracts—childbirth
d. Afterbirth delivered
e. Episiotomy performed, if needed

D. Abnormalities of birth
1. Placenta praevia
2. Fetal death: early, intermediate, or late
3. Prematurity
4. Abortion
5. Miscarriage

VII. Clinical Tests Involving Reproductive System
A. Blood test—acid phosphatase—prostatic cancer
B. Urine test A-Z
1. In women—positive result indicates pregnancy
2. In men—positive result indicates terminal cancer

STUDY QUESTIONS AND PROBLEMS

1. Describe gametogenesis of the sperm and the egg. How do the two processes differ?
2. Where does fertilization normally occur? Why is it important that it occur at this location?
3. What is an ectopic pregnancy?
4. Describe the menstrual cycle.
5. What part do the anterior pituitary hormones play in the reproductive cycle?
6. Discuss the effect of estrogen on the ovum, endometrium, secondary sex characteristics, mammary gland development, and lactation.
7. What is cleavage?
8. What stage must be initiated for the corpus luteum to remain functional?
9. What is chorionotropic hormone? Where does it come from?
10. Describe implantation and the eventual blood supply connection between the fetus and the mother.
11. Explain the function of the fetal membranes.
12. What are meant by primary and secondary sex characteristics?
13. What is the importance of the male prostate gland to the life expectancy of sperm?
14. If diseased, how does the prostate gland interfere with genitourinary tract?
15. How do Leydig cell secretions affect spermatogenesis?
16. What is ejaculation? Describe its reflex action and autonomic control mechanism.
17. Discuss ovulation. How does the Pill affect it?
18. List the three embryonic germ layers and their derivatives.
19. Describe, briefly, the periods of pregnancy, on a developmental basis.
20. What is lactation? How does the baby's suckling affect it? How would you "dry out" the mother if she were not going to breast-feed?

Glossary

A

Abdomen The part of the body extending from the diaphragm to the pelvis.

Abortion The termination of a pregnancy before the fetus can fully develop.

Abrasion A rubbing or scraping off, as of the skin.

Absorption The process of assimilating substances into or across tissues.

Acapnia A condition of diminished carbon dioxide in the blood.

Acetone bodies The substances acetoacetic acid, beta-hydroxybutyric acid, and acetone, increased levels of which are found in blood and urine in certain diabetic conditions and other disorders. Also called "ketone bodies."

Acetonuria Excretion of acetone in the urine.

Acidosis A pathologic condition in which acid is present in the blood and body fluids in proportionally greater amounts than base.

Acne An inflammation of the sebaceous glands, ordinarily called "pimples."

Acromegaly A condition caused by excessive secretion of pituitary growth hormone, marked by enlargement of the skeletal tissues of hands, feet, and face.

ACTH Adrenocorticotropic hormone. A hormone produced in the pituitary gland, used to treat rheumatoid arthritis and certain allergies.

Acuity Sharpness or clearness, especially of vision.

Acute Showing severe symptoms and a rapid development; sharp.

Adaptation Adjustment to an environment or a stimulus, especially the ability of the eye to adjust on exposure to different intensities of light.

Addison's disease A usually fatal disease caused by hypofunction of the adrenal glands, characterized by a bronze-like skin, progressive anemia, digestive disturbances, and low blood pressure.

Adenitis Inflammation of a gland.

Adenocarcinoma A malignant adenoma.

Adenoma An epithelial tumor of glandular origin and structure, usually benign.

Adenopathy Enlargement of a gland.

Adenosarcoma A malignant tumor in connective tissue with glandular elements.

Adhesion Abnormal union of two surfaces as a result of inflammation.

Adolescence The period between puberty and adult life.

Adrenalin See epinephrine.

Aerobic Growing in the presence of oxygen.

Afterbirth The placenta with umbilical cord and membranes, expelled from the uterus after the birth of a child.

Agranulocytosis A symptom complex characterized by reduction in the number of granular leukocytes and by lesions of the throat and other mucous membranes, gastrointestinal tract, and skin.

AHF Antihemophilic factor. A substance normally present in blood plasma that promotes clotting.

Albumin A simple, water-soluble protein, coagulated by heat, and found in most animal and many vegetable tissues.

Albuminuria The presence of albumin in the urine.

Aldehyde One of a class of substances derived from primary alcohols by oxidation and containing the CHO group.

Alimentary Pertaining to nutritive substances or to the organs of digestion.

Alkali reserve The total amount of alkaline salts (especially sodium bicarbonate) in the plasma and corpuscles, used in maintaining normal blood alkalinity.

Alkalosis A pathologic condition in which the alkali (mostly bicarbonate) concentration in blood and body fluids is above normal.

Allergy A hypersensitive state acquired through exposure to a particular antigen.

Ameboid movement Movement of cells, such as leukocytes, by the extension of cell cytoplasm as pseudopodia ("false feet"), in the manner of the single-celled organism "ameba."

Amenorrhea Absence of or abnormal cessation of menstruation.

Amniocentesis Surgical perforation of the uterus during pregnancy, either to withdraw some of the amniotic fluid or to inject medication.

Amorphous Without definite shape or visible differentiation in structure; not crystalline.

Ampulla A sac-like dilatation of a tubular structure such as a canal.

Anabolism The constructive phase of metabolism, during which simple substances are converted by living cells into more complex compounds, such as hormones.

Anaerobic Chemical reaction or growth in the absence of oxygen.

Analgesia Absence of sensitivity to pain.

Anaphylaxis An exaggerated allergic reaction to foreign protein or other substances, following which susceptibility to the substance is greatly increased.

Anastomosis A communication between blood vessels without an intervening capillary network; a surgical connection of blood vessels or other body parts such as the intestines.

Anemia Lack of blood; insufficient red blood cells or erythrocytes.

Anesthetic A drug that produces loss of feeling or sensation.

Aneurysm A sac formed by the dilatation of the walls of a blood vessel or the heart.

Angina Spasmodic choking or suffocative pain, as in angina pectoris.

Angioma A tumor consisting of blood vessels.

Ankylosis Fusion or consolidation of the bones of a joint, with resultant immobility.

Anorexia Lack or loss of appetite.

Anoxia A diminished supply of oxygen to the body tissues.

Antagonist A muscle that balances the effects of another muscle.

Antibiotic Pertaining to destruction of life; a chemotherapeutic substance used to destroy infectious microorganisms.

Antibody Any substance produced in the blood in reaction to the presence of an antigen.

Antidote A remedy to counteract a poison.

Antigen Any substance which, when introduced into the blood or the tissues, incites the formation of antibodies or reacts with them.

Antiseptic Any substance that will inhibit the growth of microorganisms without destroying them; preventing decay or putrefaction.

Anuresis Cessation of urination; retention of urine in the bladder.

Apex The pointed portion or tip of a conical structure, as in the apex of the lung or heart; the top of any organ or part.

Aphasia A loss of the faculty of language in any of its forms.
Apnea Cessation of respiration, especially temporary.
Apoplexy A stroke; a sudden impairment of brain function due to cerebral hemorrhage or occlusive brain lesions.
Appendage An attached part, such as the upper and lower extremities.
Aqueduct A passage or channel in a body structure, especially for the conduction of a liquid.
Aqueous Watery; containing water.
Areola Any minute space in a tissue.
Arteriosclerosis A group of diseases characterized by thickening and loss of elasticity of arterial walls.
Arthritis Inflammation of a joint.
Articulate To join together to permit motion between parts; to form a joint.
Aschheim-Zondek test A test for the determination of pregnancy, consisting of subcutaneous injection of urine from women into immature female mice.
Ascites An accumulation of serous fluid within the abdominal cavity.
Asepsis Absence of bacteria; freedom from infection.
Asphyxia Loss of consciousness due to inadequate oxygen supply; suffocation.
Aspiration The removal of fluids or gases by suction; accidental inhalation of an object into the trachea.
Assimilation The absorption of food; constructive metabolism.
Asthenia Weakness; loss or lack of strength.
Asthma A condition due to spasmodic contraction of the bronchi, characterized by attacks of difficulty in breathing with coughing and a feeling of suffocation.
Ataxia Absence of the power of muscular coordination.
Atelectasis Insufficient expansion of the lungs in the newborn; collapse of the lung in the adult.
Atresia Congenital absence or closure of a normal body opening or tubular organ.
Atrophy A wasting or decrease in the size of a part of the body.
Auricle The flap of the ear, the portion of the external ear not contained within the head; a small pouch forming the upper portion of each atrium of the heart.
Avitaminosis Any disease due to a lack of vitamins.
Axial Pertaining to the central portion or axis of the body or a body part.
Axilla Armpit.

B

Benign Not malignant; favorable for recovery.
Beriberi A disease due to deficiency of thiamine (vitimin B_1), found chiefly in East Asian cultures where rice is the staple food.
Bilateral Pertaining to both sides of the body or a body part.
Biopsy The removal and examination of tissue from a living body, to establish precise diagnosis.
Biosynthesis The formation or building up of a chemical compound in living tissues.
Bolus A rounded mass of food or a pharmaceutical preparation, ready to swallow or pass through the digestive system.
Botulism A type of food poisoning caused by eating improperly canned or preserved foods.
Breech extraction Delivery of a child buttocks or feet first.
Bronchiectasis Chronic dilatation of the bronchi.

Bronchopneumonia Inflammation of the bronchi and lungs.

Buccal Pertaining to or directed toward the cheeks.

Buffer Any substance which tends to resist a change in pH (acidity or alkalinity) when acid or base is added.

Bursa A fluid-filled sac or cavity that facilitates movement in joints and other areas, located in tissue in which friction would otherwise develop.

C

Cachexia A profound and marked state of general ill health and malnutrition.

Caffeine A stimulant, found especially in coffee and tea.

Calcification Hardening of the tissues by a deposition of calcium salts.

Calculus An abnormal concrete mass formed in any portion of the body, such as a gallstone.

Calorie A unit of heat; the amount of heat required to raise 1 gm of water one degree centigrade.

Canaliculus An extremely narrow tubular passage or channel.

Cancellous A lattice-like reticular or spongy structure, as bony tissue.

Cancer A malignant tumor or new growth.

Carbaminohemoglobin The chemical compound formed by the combination of carbon dioxide with hemoglobin, found in the blood.

Carcinoma A malignant new growth made up of epithelial cells, tending to invade surrounding tissue.

Caries Decay of a tooth or bone.

Casein The principal protein found in milk.

Castration Removal or destruction of testes or ovaries.

Catabolism The "tearing-down" phase of metabolism, during which complex substances are converted into simpler compounds within the cells.

Catalyst A substance that affects the speed of a chemical reaction but does not become a part of the final product.

Cataract A loss of transparency of the crystalline lens of the eye or of its capsule.

Catharsis A cleansing or purging.

Catheter A rubber, metal, or glass tube for insertion into a cavity through a narrow passage, allowing withdrawal of fluids from (or introduction of fluids into) the body.

Caudal Pertaining to a position or direction away from the head; closer to the cauda (tail).

Celiac Pertaining to the abdomen.

Centimeter One-hundredth part of a meter.

Centriole An organelle within the cell center, that migrates to the pole of a cell during mitosis.

Cephalic Pertaining to the head, or to the head end of the body.

Cerumen Earwax.

Cervix The neck; the constricted part of an organ, especially the lower cylindrical portion of the uterus.

Cesarean section Surgical removal of the fetus by means of an incision through the abdominal wall and the uterus.

Chalazion A small mass on the eyelid, resulting from inflammation of a meibomian gland.

Chemoreceptor A nervous-system receptor that is stimulated by chemical substances.

Chemotherapy Treatment of disease by chemical agents.

Chiasma An X-shaped crossing, such as the crossing of the optic nerve fibers from the medial halves of the retinae.

Cholesterol A substance found in all animal fats and oils, which comprises a large part of most gallstones and occurs in cysts, carcinomatous tissue, and arteriosclerotic conditions.

Chondroma A tumor consisting of cartilage cells.

Chondrosarcoma A malignant chondroma.

Chromosome A body of chromatin in the cell nucleus, containing a linear thread of DNA, which transmits genetic information.

Chyle The milky fluid taken up by the lacteals from the food in the intestine after digestion, consisting of lymph and emulsified fat.

Chyme The semifluid mass of partly digested food passed from the stomach into the duodenum.

Ciliary Relating to any hairlike structure, such as the ciliary muscle of the eye.

Cirrhosis A disease marked by pathologic changes in the connective tissues of an organ, particularly the liver.

Coagulation The process of clot formation.

Coenzyme A nonprotein substance that activates an enzyme.

Collagen The chief structural protein of skin and connective tissue.

Collateral Accessory to, a branch of, or running parallel to.

Colloid A gelatinous substance which, when dissolved in a liquid, will not readily diffuse through a membrane and which does not affect the freezing or boiling point of the liquid.

Colostrum The thin, yellow, milky fluid secreted by the mammary glands following childbirth, before lactation begins.

Coma A state of profound unconsciousness.

Commissure A site of union of corresponding parts, such as the structure formed by groups of nerve fibers connecting opposite portions of the central nervous system.

Concave Curved inward.

Congenital Existing at or from birth.

Congestion Excessive collection of fluid, especially blood, in tissues.

Constriction Compression of a tubular part.

Contraception The prevention of pregnancy.

Contraction A shortening, especially of a muscle.

Convex Curved outward.

Convoluted Twisting, turning, or coiled.

Convulsion A violent involuntary contraction of muscles.

Corium The true skin, below the epidermis; the dermis.

Corpus The body as a whole; the main portion of a body part or organ.

Cortex The outer layer of an organ.

Cortisone A hormone from the adrenal cortex, which when converted to cortisol is active in carbohydrate metabolism and other processes, and used as an anti-inflammatory agent.

Costal Pertaining to a rib.

Crenation The formation of abnormal notching in the edge of a red blood cell, caused by its shrinkage after suspension in a solution.

Crepitus A rustling or crackling sound. *Bony crepitus* is the sound heard when fragments of a fractured bone rub against each other.

Cretinism A chronic condition due to congenital lack of thyroid secretion, marked by arrested physical development, mental deficiency, and other defects.

Croup A condition resulting from acute obstruction of the larynx, especially by inflammation, marked by difficulty of breathing and spasms of barking cough.

Cryosurgery Destruction of tissue by the application of extreme cold.
Curare A highly toxic extract from certain plants that paralyzes muscle and that acts on the motor endplates.
Curette A scoop or scraper for removing diseased tissue.
Cutaneous Pertaining to the skin.
Cyanosis A dark, bluish appearance of the skin, lips, mucous membranes, and nails, due to inadequate oxygenation of the blood.

D

Deamination A chemical reaction in which the amino group is split off the amino acid.
Debridement The removal, from a wound or other lesion, of all foreign matter and dead or contaminated tissue.
Decubitus ulcer A pressure sore.
Defecation Expulsion of fecal matter from the bowel.
Degeneration Deterioration with chemical change in the condition of the body tissues.
Dehydration A pathologic condition resulting from excessive loss of water from the body cells and tissues.
Delirium A temporary mental disturbance marked by hallucinations, illusions, incoherence, and restlessness, resulting from high fever, drug overdose, shock, alcoholism, and other toxic states.
Dentin That portion of a tooth beneath the enamel.
Desiccant An agent that promotes dryness.
Diabetes A disease marked by increased secretion of urine.
Diagnosis The process of identifying a disease by its signs and symptoms.
Dialysis The separation of crystalloids from colloids in solution by using a selectively permeable membrane.
Diapedesis The passage of blood or of leukocytes through the unruptured walls of the blood vessels.
Diaphoresis Perspiration, especially excessive perspiration.
Diastole The rhythmic period of relaxation and dilatation of the heart.
Diathermy The therapeutic production of heat in the tissues by means of high-frequency currents.
Diffusion The passage of a liquid or gas from a region of greater concentration of its molecules to a region of lesser concentration of its molecules.
Digitalis A drug used as a cardiac stimulant.
Diploid Having two sets of chromosomes.
Diplopia Double vision.
Disinfection The destruction of disease-producing germs.
Diuresis Increased secretion of urine.
Diurnal Daily.
Diverticulum A pouch or sac branching out from a main tubular structure or organ cavity.
DNA Deoxyribonucleic acid. A ribonucleic acid found in all cells, that is the carrier of genetic information.
Dysmenorrhea Painful menstruation.
Dyspnea Difficult or labored breathing.
Dystrophy A degenerative disease of the body tissues, caused by defective or faulty nutrition.

E

Eclampsia An attack of convulsions during pregnancy or delivery.
Ectopic Located away from the normal position.
Eczema A superficial skin disease marked by redness, itching, a watery discharge and the later formation of scales.
Edema The presence of abnormally large amounts of fluid in the tissues of the body, producing swelling.
Effusion The escape of fluid from the blood vessels or the lymphatics into the tissues.
Electrocardiogram EKG or ECG. A graphic record of the electric current produced by the excitation of the heart muscle.
Electroencephalogram EEG. A graphic record of the electrical activity of the brain ("brain waves").
Electrolyte A substance which dissociates (ionizes) in solution and is then able to conduct an electric current.
Embolism An obstruction in a blood vessel by air, fat, a blood clot, or other foreign body.
Embryo Term applied to a developing organism up to the age of two months; after that, the term fetus is used.
Emesis Vomiting.
Emphysema Dilatation of the pulmonary air vesicles, usually through atrophy of the septa between the alveoli, producing labored breathing and increased susceptibility to infection.
Empyema Accumulation of pus in a body cavity, especially the pleural cavity.
Encephalon The brain.
Endemic Prevalent and recurring in a particular locality. Applied to a disease that appears regularly among certain groups.
Endocrine Secreting internally; pertaining to any of the ductless or endocrine glands.
Energy The capacity for doing work.
Enucleation Removal of an organ, tumor, or other body from its cover or sac, especially removal of the eyeball.
Enuresis Involuntary passage of urine.
Enzyme Any of numerous proteins manufactured by living cells, which produce or accelerate certain chemical changes in organic substances by catalytic action.
Epilation The removal of hair by the roots.
Epilepsy A nervous disorder characterized by recurring attacks of convulsions, loss of consciousness, psychic and sensory disturbances, and other malfunctions; *grand mal*: a severe form of epilepsy marked by sudden loss of consciousness followed by generalized convulsions; *petit mal*: a mild form of epilepsy.
Epinephrine An extract from the adrenal gland that stimulates the sympathetic nervous system, secreted in response to hypoglycemia, and used as a cardiac stimulant.
Epistaxis Nosebleed.
Erectile Capable of being made rigid and elevated when filled with blood, as vascular tissue in the penis, nasal cavities, and elsewhere.
Erythema Redness of the skin due to dilatation of the capillaries.
Etiology The study of the cause of disease.
Evagination Protrusion of a part or organ.
Exfoliation The process of scaling.
Exocrine Secreting externally; pertaining to ducted or exocrine glands.

Exophthalmos Abnormal protrusion or prominence of the eyeball.
Expectorant An agent promoting the ejection of sputum.
Extravasation A discharge or escape, as of blood, from a vessel into the tissues.
Extrinsic Originating outside of the part in which found.
Exudate A substance thrown out or discharged.

F

Fascia A sheet of connective fibrous tissue.
Fasciculus A small bundle of nerve, muscle, or tendon fibers.
Febrile Pertaining to fever.
Fetus The name given to a developing organism after the second month of pregnancy.
Fibrosis The formation of fibrous tissue; fibrous degeneration.
Filtration The passage of a liquid through a filtering device.
Fimbria A fringelike part or structure.
Fissure Any cleft or groove, normal or otherwise.
Fistula An abnormal passage or opening, often leading from the body surface to an internal organ.
Flaccid Flabby, soft.
Flatus Gas or air in the stomach or the intestine; commonly used to denote passage of gas by rectum.
Foramen A natural opening or hole in a body part.
Fossa A depression or pit.
Fovea A small depression or pit in the surface of a structure or organ.
Fracture A break.
Fundus The bottom of a sac or hollow organ.
Fusion, spinal The fusion of two or more vertebrae.

G

Ganglion A knotlike mass, especially a group of nerve cells located outside the central nervous system, or a similar group within the brain, along the course of a sensory nerve.
Gangrene Death and decay of tissue, usually in a limb, due to failure of blood supply, injury or disease.
Gene The biologic unit of heredity, self-reproducing, located in the chromosome, and composed mostly of DNA.
Geriatrics The study of old age and treatment of its diseases.
Gestation The period of development of the young, from conception; pregnancy.
Glaucoma A disease characterized by abnormally high pressure within the eye, resulting in blindness.
Glioma A malignant growth of the retina.
Glossitis Inflammation of the tongue.
Glycosuria The presence of an abnormal amount of sugar in urine.
Goiter Enlargement of the thyroid gland, causing a swelling in the neck.
Gonad A gland serving the reproductive system; an ovary or a testis.
Gram The basic mass (weight) unit of the metric system, about .035 ounce.
Groin The region between the abdomen and the thigh.
Gustation The sense of taste.

H

Hematemesis The vomiting of blood.
Hematoma A localized collection of blood, usually clotted, in an organ, space, or tissue, due to a break in a blood vessel wall.
Hematuria The presence of blood in the urine.
Hemiplegia Paralysis of one side of the body.
Hemoglobinemia The presence of excessive hemoglobin in blood plasma.
Hemoglobinuria The presence of hemoglobin in the urine.
Hemophilia A hereditary tendency to hemorrhage due to deficient clotting of the blood.
Hemoptysis The expectoration of blood.
Hemorrhage Bleeding; the escape of blood from the vessels.
Hemorrhoid A varicose dilatation of a rectal vein, causing painful swelling at the anus.
Hemostasis The arrest of flow of blood through any part of the body.
Hemothorax An accumulation of blood in the chest cavity.
Hepatitis Inflammation of the liver.
Heredity The transmission of qualities from parent to offspring.
Hermaphroditism The condition of having both male and female sex organs.
Hernia The protrusion of a part of an organ or tissue through an abnormal opening.
Herpes An inflammation of the skin characterized by the formation of small blisters.
Heterosexual Pertaining to the opposite sex.
Hiatus Any gap, fissure, or opening.
Histology The study of the minute structure, composition, and function of body tissues.
Hives A skin disorder, often a sign of allergy, marked by reddish wheals with an itching or stinging sensation; urticaria.
Homeostasis The state of or tendency to stability and equilibrium in the organism.
Homologous Corresponding in structure, position, and origin.
Hormone A chemical substance produced in one organ, which regulates the function of another organ.
Hyaluronidase An enzyme that promotes the breakdown of cells and tissues, which promotes greater fluid absorption during subcutaneous injections.
Hydrolysis Chemical breakdown by the addition of water.
Hydronephrosis An accumulation of urine in the kidney due to an obstruction.
Hydrostatic Pertaining to a liquid in a state of equilibrium.
Hydrothorax The presence of fluid in the chest cavity.
Hypercalcemia Excess of calcium in the blood.
Hyperemia An excess of blood in a part; engorgement.
Hyperglycemia An abnormal amount of sugar in the blood.
Hyperopia Farsightedness.
Hyperpyrexia Very high fever.
Hypertrophy The abnormal enlargement or excessive growth of an organ or part, due to an increase in size of its cells.
Hypnotic A drug that causes sleep.
Hypochondrium The region at each side of the abdomen below the lowest rib.
Hypodermic Applied or administered underneath the skin (as an injection).
Hypoglycemia Deficiency of sugar in the blood.
Hypophysis The pituitary gland.

I

Icterus Jaundice.

Immune Being highly resistant to a particular disease.

Impulse A wave of excitation passing over a nerve, muscle, or other body tissue.

Infantilism A condition marked by retarded mental development, underdeveloped sexual organs, and at times stunted growth.

Infarct Local deterioration of tissue due to an obstruction in the supplying blood vessel.

Inflammation A localized protective response produced by injury or destruction of tissues, marked by pain, heat, redness, swelling, and sometimes loss of function.

Influenza An infectious, sometimes epidemic viral disease of the respiratory tract; the grippe.

Infundibulum A funnel-shaped structure or passage.

Inguinal Relating to the region between the abdomen and the thigh.

Innervation The supply of nerves to the different parts of the body; the impulses sent through these nerves.

Insufficiency, cardiac Inadequate function of the heart muscle.

Insulin A hormone produced in the pancreas, which regulates carbohydrate metabolism.

Intussusception The prolapse of one segment of the intestine into another adjoining segment.

Invagination The infolding of one part within another, specifically the process in which a region infolds to form a double-layered cup.

Involution The return of an enlarged organ to normal size, as of the uterus after pregnancy.

Ion An atom or group of atoms bearing a positive or negative electrical charge.

Ipsilateral Situated on or pertaining to the same side.

Isotonic Two solutions having the same concentration of solute and solvent and, therefore, identical osmotic qualities; a muscle contraction that involves no change in tonus.

Isotope A chemical element which has the same atomic number as another but a different atomic mass.

J

Jaundice A syndrome characterized by yellow discoloration of the skin and eyes, caused by any of several disorders in which normal processing of bile is interrupted.

K

Ketosis A condition marked by excessive production of ketone bodies.

Kilogram 1,000 grams; about 2.2 pounds.

Kinesthetic Pertaining to muscle sense or to the perception of movement, weight, and position of body parts.

Kyphosis Abnormal outward curvature of the spine; hunchback.

L

Lactation The production of milk by the mammary glands.

Lacuna A small hollow, depression, or pit, especially within or between other body structures.

Lamella A thin layer of plate, as of bone.

Lamina A thin, flat layer, especially in a portion of tissue consisting of layers of cells; a flat plate, e.g., the laminae of vertebrae.
Lavage Irrigation or washing out of an organ, such as the stomach or bowel.
Lesion Any pathologic or traumatic destruction of tissue or loss of function of a part.
Leukemia A progressive, malignant disease of the blood-forming organs, marked by increase in the number of leukocytes (leukocytosis) in blood and bone marrow.
Lipoma A benign tumor composed of fat cells.
Loin The part of the back between the thorax and pelvis; the lumbar region.
Lordosis Abnormal curvature of the spinal column toward the front of the body; sway-back.
Lumbago Pain in the lumbar region.
Lumen The space in the interior of a tubular structure such as an artery or the intestine.
Lymphosarcoma A malignant tumor of the glands or tissue of the lymphatic system.

M

Macula A spot or thickening.
Malaise A feeling of bodily discomfort or uneasiness.
Malignant Tending to become progressively worse and to result in death.
Manometer An instrument for measuring the pressure of gases or liquids, such as the blood.
Mastication The process of chewing food.
Meatus A passage or channel.
Meiosis Cell division occurring at maturation of the sex cells, during which each new nucleus receives half the normal number of chromosomes; reduction division.
Meninges The three membranes that envelop the brain and the spinal cord.
Menopause Cessation of menstruation; change of life.
Menorrhagia Excessive menstrual flow.
Menorrhalgia Painful menstruation.
Menses The monthly flow of blood from the genital tract of women.
Metabolism The complex of physical and chemical processes by which living organisms produce the necessary energy to maintain life.
Metastasis The spreading of disease from one portion of the body to another.
Meter The basic unit of length in the metric system; the equivalent of 39.371 inches.
Microgram One one-millionth of a gram, or 1/1,000 of a milligram.
Micron One one-millionth of a meter, or 1/1,000 of a millimeter: 1/25,400 of an inch.
Micturition Urination; the act of voiding urine.
Millimeter 1/1,000 of a meter; about 1/25 of an inch.
Miotic An agent that causes contraction of the pupil.
Mitosis Cell division during which each new nucleus receives the same number of cells characteristic of the species.
Morphine The chief narcotic principle of opium; used as an analgesic.
Mucus The viscid watery secretion of the mucous glands; it consists of water, mucin, epithelial cells, leukocytes, and inorganic salts.
Murmur An abnormal sound, usually in the thoracic region, indicating a pathologic condition of the heart or of the lungs.
Myasthenia Muscular disability.
Mydriatic An agent that causes dilation of the pupil.
Myopathy Any disease of the muscle.

Myopia Nearsightedness.
Myringotomy Surgical incision of the tympanic membrane or eardrum.
Myxedema A condition marked by dry, waxy swelling, with abnormal deposits of mucin in the skin associated with deficient activity of the thyroid gland.

N

Narcosis A reversible condition of stupor or unconsciousness, especially that produced by some narcotic drug.
Nausea An unpleasant sensation in the stomach and abdominal region, often culminating in vomiting.
Necrosis Local death of tissue.
Neoplasm Any new and abnormal growth; a tumor.
Nephrosis Degenerative disease of the kidneys.
Neuralgia Pain which extends along one or more nerves.
Neurosis An emotional disorder due to unresolved conflicts, characterized chiefly by anxiety.
Noxious Harmful; poisonous.
Nystagmus Involuntary rapid movement of the eyeball; may be horizontal, rotary, or vertical.

O

Occlusion The closing or obstruction of an opening or passage.
Oogenesis The process of formation of ova or egg cells.
Orifice The opening or outlet of any cavity in the body.
Osseous Bony or bonelike.
Ossification The process of bone formation; the conversion of fibrous tissue or cartilage into bony material.
Osteoarthritis Degenerative joint disease, occurring chiefly in older persons, marked by degeneration of joint cartilage, hypertrophy of bone, and other changes.
Osteochondritis Inflammation of bone and cartilage.
Osteomalacia A condition marked by softening of the bone, resulting from vitamin D and calcium deficiencies of various causes.
Osteomyelitis Inflammation of bone or of marrow and bone.
Osteoporosis A condition in which a bone has become abnormally porous because of demineralization.
Ostium An opening into a tubular organ or between two distinct cavities of the body.
Otosclerosis The formation of spongy bone in the middle ear, resulting in progressively increasing deafness.
Oxidation The combining of oxygen with the products of digestion, in the tissues.

P

Palpitation Unduly rapid pulsation of the heart.
Papilloma A benign tumor of the skin.
Paracentesis Surgical puncture of a cavity to draw off fluid.
Paralysis Loss of function in a muscle, nerve, or other body part.
Paresthesia A perverted or abnormal sensation, such as burning or prickling.
Parietal Pertaining to the walls of a cavity.

Parturition The act of giving birth to young.
Pathogenesis The development of a disease.
Pathologic Indicative of or caused by a disease.
Pathology The science that deals with the essential nature of a disease, including resulting structural and functional changes.
Pectoral Pertaining to the breast or the chest.
Pellagra A syndrome due to deficiency of niacin, affecting the skin and nervous system.
Percussion A diagnostic method that consists of striking a part of the body with short, sharp blows and determining the condition of the underlying parts by the sound obtained.
Perimeter An instrument for determining the extent of the field of vision on a curved surface.
Perineum The portion of the body extending from the pubic arch to the coccyx and between the ischial tuberosities, the pelvic floor and associated structures.
Periphery The surface of a body or part.
Perspiration The secretion of sweat.
Petechia A small spot under the skin due to subcutaneous bleeding.
pH The symbol commonly used to express hydrogen ion concentration.
Phagocyte A cell which ingests other cells or foreign particles.
Phagocytosis The engulfing of other cells, microorganisms, or foreign particles by phagocytes.
Phlebitis Inflammation of a vein.
Phlebothrombosis Presence of a clot in a vein without inflammation of its walls.
Physiotherapy Treatment of disease and injury by mechanical means such as manipulation and massage and the application of heat, water, electricity, or light.
Plasmolysis Shrinking of a cell due to loss of water by osmotic action.
Pleurisy Inflammation of the pleura.
Plexus A network or tangle of interweaving nerves, veins, or lymphatic vessels.
Pneumonia Inflammation of the lungs.
Pneumothorax An accumulation of air or gas in the pleural cavity.
Poliomyelitis An acute viral disease which causes inflammation of the gray substance of the spine, especially of the anterior horns, which may lead to paralysis.
Postmortem Occurring after death.
Process A prominence or projection, as of bone.
Prognosis A prediction of the probable outcome of an illness.
Prolapse The falling down or sinking of an organ or other part, such as the uterus.
Proliferation The reproduction or multiplication of similar forms, especially cells.
Prophylactic Preventing or protecting from disease.
Protoplasm The essential substance of the cell.
Pruritus Itching.
Psoriasis A hereditary skin disease marked by the appearance of red, scaly patches on the scalp, elbows, knees, and shins.
Psychogenic Originating in the mind; psychic.
Psychosis A severe disease or disorder of the mind, characterized by derangement of the personality and loss of contact with reality.
Psychosomatic Pertaining to the mind–body relationship; having bodily symptoms of a psychic, emotional, or mental origin.
Pterygoid Shaped like a wing.
Puberty The age at which the reproductive organs become functionally operative.
Purpura A condition marked by the formation of purple spots due to hemorrhage under the skin.

Pus A product of inflammation, consisting of a liquid containing leukocytes and the debris of dead cells.

Putrefaction Decay or decomposition due to microorganisms.

Pyemia A diseased state due to the presence of pus-forming microorganisms, with the formation of multiple abscesses.

R

Rabies An infectious, usually fatal disease of the central nervous system, affecting certain animals such as bats and dogs, communicable to man by the bite of an infected animal.

Receptor A sensory nerve ending which responds to stimuli.

Reduction The process of setting a fracture or treating a dislocation.

Reflection The throwing back of a ray of light from a surface that it does not penetrate.

Reflex A predictable, involuntary response to a stimulus.

Refraction The bending of a ray of light as it passes from one medium into another of different density.

Regurgitation The backward flowing of blood through incompletely closed heart valves; the return of small amounts of food from the stomach.

Remission A temporary improvement, or lessening of symptoms of a disease.

Reticular Netlike, resembling a net.

Retroversion The tipping of an entire organ backward.

Rh factor Any of several antigens present on the surface of red blood cells, named for the rhesus monkey, in which it was first discovered. An individual who has this antigen is Rh positive; one who does not is Rh negative.

Rheumatism An inflammation or degeneration of connective tissue structures.

Rickets A softening of the bones, due to lack of calcium caused by vitamin D deficiency.

Rubella German measles; a mild viral infection, which when contracted by a woman in early pregnancy can cause deformities in the unborn child.

S

Sarcoma A usually malignant tumor that arises from connective or nonepithelial tissue such as bone and muscle.

Schizophrenia A severe mental disorder characterized by withdrawal from reality with variable behavior and intellectual disturbances.

Sciatic Pertaining to or located near the lower hip bone or ischium, such as the sciatic nerve.

Sclerosis The formation of hard patches of connective tissue in the nervous system or blood vessels. *Multiple sclerosis* is sclerosis throughout the central nervous system, particularly the spinal cord.

Scurvy A pathologic condition due to vitamin C deficiency, marked by weakness, anemia, spongy gums, and other symptoms.

Sebum The secretion from sebaceous (oil) glands.

Sedative An agent that has a quieting effect.

Seizure A sudden attack, as of a disease or of convulsions.

Semen The ejaculatory fluid secreted by the male reproductive organs, consisting of sperm cells and secretions of the prostate and bulbourethral glands and seminal vesicles.

Septicemia A pathologic condition due to bacteria or their poisons in the blood; blood poisoning.

Shingles An acute inflammatory disease caused by the virus of chickenpox, characterized by a series of small blisters on the skin, along the root of a nerve.
Sinus A channel for the passage of blood; a hollow cavity in a bone or other tissue.
Sinusoid A blood space in certain organs, as the spleen and the liver.
Somatic Pertaining to the nonvisceral portions of the body.
Spasm An involuntary, convulsive, muscular contraction.
Spermatogenesis The process of formation and development of spermatozoa.
Sphincter A circular band of muscle fibers that serves, when in a state of normal contraction, to close a body orifice.
Spondylitis Inflammation of a vertebra.
Sputum Matter expectorated from the respiratory tract.
Squamous Scaly or platelike.
Stasis A stoppage or lessening of the flow of blood or other body fluid.
Stenosis Narrowing of an opening, duct, or canal.
Sterility Infertility; the inability to produce offspring.
Stimulus Any chemical or physical agent or influence in the internal or external environment, that produces a response in a receptor or other tissue.
Strabismus A visual defect in which one eye cannot focus with the other on an object because of imbalance of the eye muscles; squint.
Stricture A circumscribed narrowing of a tubular structure such as a canal or duct.
Stroma The supporting tissue or matrix of an organ, as distinguished from its functional portion.
Stye A localized, inflammatory, staphylococcal infection of the sebaceous glands of the eyelid.
Subcutaneous Beneath the skin.
Suppository An easily melted, medicated preparation, usually cone-shaped, for insertion into the vagina or rectum.
Susceptibility The state of being readily acted upon or affected by disease.
Synapse The anatomical relation between nerve fibers.
Syncytium A multinucleate mass of protoplasm produced by the merging of cells.
Syndrome A set of symptoms occurring together.
Synovial Pertaining to or secreting *synovia,* the viscid fluid of a joint or similar cavity.
Syringomelia The existence of abnormal cavities filled with fluid, within the spinal cord.
Systole The contraction or period of contraction of the heart muscle.

T

Tabes A progressive wasting of the body or a part of it.
Tactile Pertaining to the sense of touch.
Tetanus An acute infectious disease characterized by rigidity and spasmodic contraction of the voluntary muscles; lockjaw.
Tetany A condition caused by abnormal calcium metabolism, marked by intermittent tonic muscular contractions of the extremities.
Therapy The science and art of healing.
Thrombophlebitis Inflammation of a vein associated with the formation of a thrombus or intravascular clot.
Thrombosis The formation of a clot within a vessel.
Thrombus A blood clot in an artery or vein.
Tinnitus A ringing, buzzing, or singing sound in the ears.
Titer The amount of a substance required to produce a reaction with a certain amount of another substance.

Tonicity The state of tissue tone or tension.
Torticollis A contracted state of the cervical muscles, resulting in a contortion of the neck.
Tourniquet A device for the compression of a blood vessel, by application around an extremity, to prevent blood flow to or from the part.
Toxemia A general poisoning of the blood due to the presence of toxins following a local infection.
Toxic Pertaining to a poison, or harmful to the body as a poison.
Trabecula A supporting or anchoring strand of connective tissue, such as one extending from a capsule to the enclosed organ.
Traction A drawing or pulling.
Transfusion The transfer of blood from the veins of one person to those of another.
Trauma A wound or injury.
Tumor A swelling or enlargement; a new growth of tissue in which the multiplication of cells is uncontrolled and progressive.

U

Ulcer An inflammatory lesion on the surface of an organ or tissue, produced by the sloughing of inflammatory necrotic tissue.
Ultrasonic Pertaining to acoustic frequencies or waves above the range of the human ear; these waves are sometimes used in physiotherapy.
Unilateral Located on only one side of the body.
Umbilicus The navel.
Uremia A pathologic condition caused by the retention of excessive amounts of toxic substances in the blood.
Urticaria Hives; a vascular skin reaction marked by elevated, itching, white patches.

V

Varicocele Enlargement of the veins of the spermatic cord.
Varicose veins Unnaturally swollen and tortuous veins.
Vascular Pertaining to blood vessels.
Vasomotor Regulating the contraction and dilatation of the blood vessels.
Vertigo A form of dizziness; an illusion of movement, as if the external world were revolving around the patient, which may result from disease of the inner ear or disturbances in central nervous system pathways.
Vesicle A small bladder or sac containing liquid.
Vestibular Pertaining to a vestibule, or a space or cavity at the entrance to a canal, such as the inner ear, larynx, or vagina.
Virulence The degree of ability to produce infection.
Viscera The internal organs of the body.
Volvulus An intestinal obstruction due to knotting or twisting of the bowel.

W

Wheal A raised, whitish or reddish part of the skin, often with severe itching, resulting from skin reaction to pressure, irritation, or allergy.

Prefixes and Suffixes

PREFIXES

a- lack of
ab- away from
ad- to, toward
aden- gland
ambi- both
andr- man
ante- before
anti- against
arthr- joint
auto- self

bi- two; life
bil- bile
blast- a growing thing in its early stages.
blephar- eyelid
brachi- arm
brachy- short
brady- slow
bronch- windpipe
bucc- cheek

carcin- cancer
cardi- heart
cephal- head
chol- bile
chondr- cartilage
chromo- color
circum- around
cleid- clavicle
colp- hollow, vagina
contra- against, opposed
cost- rib
crani- skull
cut- skin
cyst- bladder
cyt- cell

dacry- tear
de- down from
derm(at)- skin
di- two, double
dis- apart, away from
dors- back
dys- bad (pain), difficulty

ect- outside
en- in, on
encephal- brain
end- inside
enter- intestine
epi- upon, in addition
erythr- red
ex- out of
extra- outside of, beyond

febr- fever

galact- milk
gastr- stomach
gloss- tongue
glyc- sweet (sugar)
gyn(ec)- woman

hem(at)- blood
hemi- half
hepat- liver
heter- other, unlike
hist- tissue, web
hom- same, common
hydr- water
hyper- above, extreme
hypo- under, below
hyster- uterus

idi- separate, distinct
in- in, on; not
infra- beneath
inter- between, among
intra- within, inside
intro- into, within

kerat- horn
kine- move

lact- milk
later- side
leuk- white
levo- to the left
lith- stone

macr- large, long
mal- bad, abnormal
mast- breast
mega(l)- great, large
melan- black
mes- middle
meta- beyond, accompanying
micr- small
mon- one, single
multi- many, much
my- muscle

ne- new, young
necr- corpse
nephr- kidney
neur- nerve

odont- tooth
olig- little, few
oo- egg
oophoro- ovary
ophthalm- eye
orth- straight, normal
oss- bone
ost(e)- bone
ot- ear

pan- all, every
para- beside, beyond
path- disease
ped- child
per- through
peri- around
phleb- vein
pod- foot
polio- gray matter (of nervous system)
poly- many
post- after, in time or place
pre- before, in time or place
pro- in front of, before
proct- anus
pseud- false
psych- mind
py- pus
pyel- pelvis, basin
pyr- fire

re- back, again
ren- kidney
retro- backwards

salping- tube
semi- half
somat- body
sten- narrow
sub- under, below
super- above, extreme
supra- above, over
syn-, sym- with, together

tachy- fast, swift
teno-, tento- tendon
tens- stretch
therm- heat
tox- poison
trache- windpipe, trachea
trans- across, through, beyond

ultra- beyond, excess
un- one

vas- vessel
ventro-, ventri- belly; to the front

SUFFIXES

-algia pain
-ase enzyme
-blast embryonic, young

-cele tumor, hernia
-cide kill (cause death)
-coccus seed, pill
-cele, -coele swelling; chamber, ventricle

-ectomy excision, removal of
-emia blood
-esthesia sensation, feel

-ferent bear, carry
-fuge flee, drive away

-genic be produced, origin
-gram tracing, record
-graph write, record

-itis inflammation

-logy science or study of
-lysis lose, set free

-malacia softening
-meter measure

-oma tumor
-osis process (disease), increase
-(o)stomy creation of an opening
-(o)tomy cutting into

-pathy disease of
-penia lack of
-pexy fixation
-phage consume, eat
-phobia fear
-plasty mold, shape
-plegia strike (paralysis)
-pnea air or breathing
-poiesis make, produce

-rrhagia break, burst
-rrhaphy suture
-rrhea flow, discharge

-scopy examine, inspect

-taxis order
-trophy nourish

-uria urine

Index

Note: In this index, page numbers in *italics* refer to illustrations; page numbers followed by (t) refer to tables.